T0271213

Statistical Principles for the Design of Experiments

This book is about the statistical principles behind the design of effective experiments and focuses on the practical needs of applied statisticians and experimenters engaged in design, implementation and analysis. Emphasising the logical principles of statistical design, rather than mathematical calculation, the authors demonstrate how all available information can be used to extract the clearest answers to many questions. The principles are illustrated with a wide range of examples drawn from real experiments in medicine, industry, agriculture and many experimental disciplines. Numerous exercises are given to help the reader practise techniques and to appreciate the difference that good design can make to an experimental research project.

Based on Roger Mead's excellent *Design of Experiments*, this new edition is thoroughly revised and updated to include modern methods relevant to applications in industry, engineering and modern biology. It also contains seven new chapters on contemporary topics, including restricted randomisation and fractional replication.

R. MEAD is Emeritus Professor of Applied Statistics at the University of Reading.

S. G. GILMOUR is Professor of Statistics in the Southampton Statistical Sciences Research Institute at the University of Southampton.

A. MEAD is Senior Teaching Fellow in the School of Life Sciences at the University of Warwick.

This series of high-quality upper-division textbooks and expository monographs covers all aspects of stochastic applicable mathematics. The topics range from pure and applied statistics to probability theory, operations research, optimisation and mathematical programming. The books contain clear presentations of new developments in the field and also of the state of the art in classical methods. While emphasising rigorous treatment of theoretical methods, the books also contain applications and discussions of new techniques made possible by advances in computational practice.

A complete list of books in the series can be found at www.cambridge.org/statistics. Recent titles include the following:

Statistical Principles for the Design of Experiments

R. Mead
University of Reading

S. G. Gilmour
University of Southampton

A. Mead
University of Warwick

CAMBRIDGE
UNIVERSITY PRESS

Shaftesbury Road, Cambridge CB2 8EA, United Kingdom

One Liberty Plaza, 20th Floor, New York, NY 10006, USA

477 Williamstown Road, Port Melbourne, VIC 3207, Australia

314–321, 3rd Floor, Plot 3, Splendor Forum, Jasola District Centre, New Delhi – 110025, India

103 Penang Road, #05–06/07, Visioncrest Commercial, Singapore 238467

Cambridge University Press is part of Cambridge University Press & Assessment,
a department of the University of Cambridge.

We share the University's mission to contribute to society through the pursuit of
education, learning and research at the highest international levels of excellence.

www.cambridge.org
Information on this title: www.cambridge.org/9780521862141

First published 2012

A catalogue record for this publication is available from the British Library

Library of Congress Cataloging-in-Publication data
Mead, R. (Roger)
Statistical principles for the design of experiments : applications to real experiments / R. Mead, University of
Reading, S.G. Gilmour, University of Southampton, A. Mead, University of Warwick.
pages cm. – (Cambridge series in statistical and probabilistic mathematics)
Includes bibliographical references and index.
ISBN 978-0-521-86214-1 (hardback)
1. Experimental design.   I. Gilmour, S. G.   II. Mead, A. (Andrew)   III. Title.
QA279.M38825   2012
001.4´34 – dc23   2012023181

ISBN   978-0-521-86214-1   Hardback

Contents

Part II First subject

Part III Second subject

Preface

Our aim in this book is to explain and illustrate the fundamental statistical concepts required for designing efficient experiments to answer real questions. This book has evolved from a previous book written by the first author. That book was based on 25 years of experience of designing experiments for research scientists and of teaching the concepts of statistical design both to statisticians and to experimenters. The present book is based on approximately a combined 100 years of experience of designing experiments for research scientists, and of teaching the concepts of statistical design both to statisticians and to experimenters.

The development of statistical philosophy about the design of experiments has always been dominated by mathematical theory. In contrast the influence of the availability of vastly improved computing facilities on teaching, textbooks and, most crucially, practical experimentation has been relatively small. The existence of statistical programs capable of analysing the results from any designed experiment does not imply any changes in the main statistical concepts of design. However, developments from these concepts have often been restricted by the earlier need to develop mathematical theory for design in such a way that the results from the designs could be analysed without recourse to computers. The fundamental concepts continually require reexamination and reinterpretation outside the limits implied by classical mathematical theory so that the full range of design possibilities may be considered. The result of the revolution in computing facilities is that the design of experiments should become a much wider and more exciting subject. We hope that this book will display that breadth and excitement.

The original development of the earlier book was particularly motivated by teaching postgraduate students specialising in statistics. However, the intention of this book is to reach a much wider audience. Understanding the fundamental concepts of design is essential for everybody involved in programmes of research and experimentation. In addition to this general need for an understanding of the philosophy of designing experiments there are particular aspects of design, such as the definition of experimental units (Chapter 5) or levels of replication (Chapter 6), which are relevant to virtually all research disciplines in which experimentation is required. Because of the concentration on basic concepts and their implications the book could be used for courses for final-year undergraduates provided such courses allow sufficient time for the concepts to be thoroughly discussed.

Parts of the book could also be used as text support for various tertiary education courses or short courses. Thus a course on linear models with an emphasis on data from designed studies could use Chapters 4, 9 and 18, with some examples taken from other chapters. An introductory course on the design of experiments could use Chapters 1, 2 and 3, followed by some small initial parts of Chapters 7, 8, 13 and 16. A course for final-year undergraduates could include Chapters 5 and 6, followed by large parts of Chapters 7, 8, 10, 12, 13, 15 and 16. It would even be possible to construct a course on advanced design by including the later parts of Chapters 7 and 8, with Chapters 11, 13, 17, and some parts of 19 and 20, before presenting Chapter 0.

This book concentrates on the ideas of design rather than those of analysis, on the statistical concepts rather than the mathematical theory and on designing practically useful experiments rather than on developing classes of possible design structures. Obviously it is also necessary to consider how the data from designed experiments will be analysed and the philosophy and methods of analysis are discussed in the introductory first part of the book. There is a further chapter, 9, concerned with the analysis of data from experiments with multiple levels of information, once the ideas of designs with information at multiple levels have been considered in Chapters 7 and 8. Of course, examples of analysis punctuate many later chapters of the book. However, in all the development of design ideas, it is assumed that the analysis of data from designed experiments is not difficult when good designs and modern computing facilities are used. Consequently ideas of analysis are introduced only when they illuminate or motivate the design concepts.

The formal language of statistics is mathematical. Thus it is not possible to discuss the design of experiments without some mathematically complex formulation of models and ideas. Some of the mathematical language used in the book requires a sound mathematical background beyond school level. However, in all parts of the book it is the statistical concepts which are important, and the structure of the book hopefully allows the less-mathematical reader to bypass the more complex mathematical details. Throughout the book the development of concepts relies on many examples. We hope that readers will consider the detailed arguments of these examples. By trying to solve the problems which underlie the examples before reading through the explanation of the solutions, we believe that readers will start to develop the intuitive understanding of design concepts which is essential to good design. For the mathematically sophisticated reader the mathematical details provide additional support for the statistical concepts.

Most importantly, the book is intended to show how practical problems of designing real experiments should be solved. To stimulate this practical emphasis real examples of design problems are described at the beginning of most chapters of the book. The final chapter of the book attempts an overall view of the problem-solving aspects of design.

The areas of application used in examples in this book inevitably reflect the personal experience of the three authors. The earlier book had a strong bias towards agricultural experimentation, but we believe that that bias is much reduced in this new book and that the examples and the approach to design is much broader than previously. Whether or not that belief is justified, all the examples are intended to illustrate particular forms of problem that will be relevant in many fields of application. We hope that statisticians and research scientists in a wide range of experimental disciplines will be able to interpret and adapt the concepts discussed in the book to their own requirements through the use of analogy when the examples discussed are not directly relevant to their discipline.

For those readers familiar with the earlier book it may be helpful to explain where this new book is clearly different. Compared to the previous book, there are several new chapters, 5 on 'Experimental units', 9 on 'Multilevel analysis', 11 on 'Restricted randomisation', 14 on 'Fractional replicates', 19 on 'Multiple experiments and new variation' and 20 on 'Sequential aspects of experiments'. Some chapters in the earlier book have disappeared, some of their material being included in a shortened form in the new book; these include 'Covariance' and 'Computer analysis programs' as sections in Chapter 4, and 'Mathematical theory for confounding and fractional replication', some parts of which appear in the new Chapters 14 and 15. The chapter on model assumptions and general models has been omitted, being not strictly relevant to a book on designing experiments.

The book is divided into an overture and two main subjects with a final coda to bring together all the previous material. Chapters 1–4 constitute the overture, providing a general introduction and the basic theory necessary for analysis of experimental data. Chapters 1, 2 and 3 should be familiar to readers who have taken an elementary course in the design of experiments. Alternatively, for those readers without any previous training in this topic, these three chapters provide an introductory presentation of the two most important ideas, blocking and factorial structure. Chapter 4 is the mathematically heavy chapter, providing the necessary theory for general linear models and the analysis of data from designed experiments, with an initial explanation of the important results at a rather simpler level. Chapter 4 also explores the universal use of computer programs and packages for the analysis of data from designed experiments and reflects on the implications of this for designing experiments.

The first main subject is unit variation and control. In Chapter 5 we examine the concept of an experimental unit, and the diverse forms of experimental units. The fundamental concepts of replication, and blocking (with either one or two systems of control) are developed in depth in Chapters 6, 7 and 8. The ideas of randomisation and restricted randomisation are explored thoroughly in Chapters 10 and 11. Our aim throughout these chapters is to distinguish the purposes and practical relevance of each concept and to eliminate the confusion about these concepts which seems to be common in the minds of many of those needing to design effective experiments. The need for analysis methods and computer programs to cope with data from designed experiments with variation occurring at more than one level is explored in Chapter 9 of this part.

The second main subject is treatment questions and structure. Chapter 12 presents an overview of the need for statisticians to be involved in all stages of discussions about the choice of treatments and the interpretation of results. The classical ideas of factorial structure and multiple and single replicates are presented in Chapter 13. In Chapter 14 we explore the uses of fractions of full factorial structures. Chapter 15 examines the combination of factorial structures with the need for units to be grouped into relatively small blocks. The choice of experimental treatments for the investigation of the responses to quantitative factors is discussed in Chapters 16 (mainly for single factor response functions) and 17 (for multifactorial response surfaces). In Chapter 18 we investigate a variety of design approaches in which different levels of experimental units are deliberately used to investigate different sets of treatment factors.

Finally in the coda, Chapter 19 explores the use of sets of experiments, often in multiple locations or at different times, and also the deliberate introduction of additional variation in experimental programmes. Chapter 20 explores various sequential aspects of experiments and experimental programmes, in particular focussing on the concept that individual experiments do not exist in isolation from past and future experimentation. Finally Chapter 0 seeks to draw the concepts of the two main subjects together to provide guidance on designing effective experiments to satisfy particular practical requirements. A book on the practical design of experiments should start with the approach of Chapter 0 but this chapter requires knowledge from the previous chapters before it can be read and understood. Hence the number and position of this chapter.

We owe a considerable debt to many consultees and collaborators, both for the stimulus to consider why the problems they presented should be covered by a book on designing experiments and also for the many examples they have provided. There are too many for us to thank them individually here for their stimulating requests and they therefore remain anonymous (some should prefer it that way, and others are too distant in the mists of time for anything else). We have also benefitted from many discussions with colleagues at Reading and elsewhere and particularly wish to thank Richard Coe, Robert Curnow, John Fenlon, Geoff Freeman, Peter Goos, Derek Pike, Roger Stern and Luzia Trinca, without whom this book would have had a more stunted growth.

Part I

Overture

1

Introduction

1.1 Why a statistical theory of design?

The need to develop statistical theory for designing experiments stems, like the need for statistical analysis of numerical information, from the inherent variability of experimental results. In the physical sciences, this variability is frequently small and, when thinking of experiments at school in physics and chemistry, it is usual to think of 'the correct result' from an experiment. However, practical experience of such experiments makes it obvious that the results are, to a limited extent, variable, this variation arising as much from the complexities of the measurement procedure as from the inherent variability of experimental material. As the complexity of the experiment increases, and the differences of interest become relatively smaller, then the precision of the experiment becomes more important. An important area of experimentation within the physical sciences, where precision of results and hence the statistical design of experiments is important, is the optimisation and control of industrial chemical processes.

Whereas the physical sciences are thought of as exact, it is quite obvious that biological sciences are not. Most experiments on plants or animals use many plants or animals because it is clear that the variation between plants, or between animals, is very large. It is impossible, for example, to predict quantitatively the exact characteristics of one plant from the corresponding characteristics of another plant of the same species, age and origin.

Thus, no medical research worker would make confident claims for the efficacy of a new drug merely because a single patient responded well to the drug. In the field of market research, no newspaper would publish an opinion poll based on interviews with only two people, but would require a sample of at least 500, together with information about the method of selection of the sample. In a drug trial, the sample of patients would often be quite small, possibly between 20 and 100, but for a final trial before the release of the drug, as many as 2000–3000 patients might be used to detect unexpected side effects of the drug. In psychological experiments, the number of subjects used might be only 8–12. In agricultural experiments, there may be 20–100 plots of land, each with a crop grown on it. In a laboratory experiment hundreds of plants may be treated and examined individually. Or just six cows may be examined while undergoing various diets, with measurements taken frequently and in great detail.

The size of an experiment will vary according to the type of experimental method and the objective of the experiment. One of the important statistical ideas of experimental design is the choice of the size of an experiment. Another is the control of the use of experimental material. It is of little value to use large numbers of patients in the comparison of two drugs, if all the patients given one drug are male, aged between 20 and 30, and all the patients given the other drug are female, aged 50–65. Any reasonably sceptical person would doubt claims made about the relative merits of the two drugs from such a trial. This example may seem trivially obvious, but the scientific literature in medicine and many other disciplines shows that many examples of badly planned (or unplanned) experiments occur.

And this is just the beginning of statistical design theory. From avoiding foolish experiments, we can go on to plan improvements in precision for experiments. We can consider the choice of experiments as part of research strategy and can, for example, discuss the relative merits of many small experiments or a few large experiments. We can consider how to design experiments when our experimental material is generally heterogeneous, but includes groups of similar experimental units. Thus, if we are considering the effects of applying different chemicals on the properties of different geological materials, then these may be influenced by the environment from which they are taken, as well as by the chemical treatment applied. However, we may have only two or three samples from some environments, but as many as ten samples from other environments; how then do we decide which chemicals to apply to different samples so that we can compare six different chemical treatments?

1.2 History, computers and mathematics

If we consider the history of statistical experimental design, then most of the developments have been in biological disciplines, in particular in agriculture, and also in medicine and psychology. There is therefore an inevitable agricultural bias to any discussion of experimental design. Many of the important principles of experimental design were developed in the 1920s and 1930s, in particular by R. A. Fisher. The practical manifestation of these principles was very much influenced by the calculating capacity then available. Had the computational facilities which we now enjoy been available when the main theory of experimental design was being developed it is quite possible that the whole subject of design would have developed very differently. Whether or not this belief is valid, it is certainly true that a view of experimental design today must differ from that of the 1930s, or even the 1970s. The principles have not changed, but the principles are often forgotten, and only the practical manifestation of the principles retained; these practical applications do require rethinking.

The influence of the computer is one stimulus to reassessing experimental design. Another cause for concern in the development of experimental design is the tendency for increasingly formal mathematical ideas to supplant the statistical ideas. Thus the fact that a particularly elegant piece of mathematics can be used to demonstrate the existence of groups of designs, allocating treatments to blocks of units in a particular way, begs the statistical question of whether such designs would ever be practically useful.

Although our backgrounds have been in mathematics, we believe that the presentation of statistical design theory has tended to be quite unnecessarily mathematical, and we will hope to demonstrate the important ideas of statistical design without excessive mathematical encumbrance. The language of statistical theory, like that of physics, is mathematical and there will be sections of the book where those with a mathematical education beyond school level will find a use for their mathematical expertise. However, even in these sections, which we believe should be included because they will improve the understanding of statistical theory of those readers able to appreciate the mathematical demonstrations, there are intuitive explanations of the theory at a less advanced mathematical level.

1.3 The influence of analysis on design

To write a book solely about the theory of experimental design, excluding all mention of the analysis of data, would be impossible. Any experimenter must know how he or she intends to analyse the experimental data before the experiment to yield the data is designed. If not, how can the experimenter know whether the form of information which is to be collected can be used to answer the questions which prompted the experiment? And just as crucially, whether the amount of information from the experiment is not only sufficient but is not excessive.

Thus, consider again the medical trial to compare two drugs. Suppose the experimenter failed to think about the analysis and argued that one of the drugs was well known, while the other was not; in the controlled experiment to compare them, there are available 40 patients. Since a lot is already known about drug A, let it be given to one patient, and let drug B be given to the other 39. When the data on the response to the drugs are obtained, the natural analysis is to compare the mean responses to the drugs, and to consider the difference $(\bar{y}_A - \bar{y}_B)$. To test the strength of evidence for a real difference in the effects of the two drugs, we need the standard error of $(\bar{y}_A - \bar{y}_B)$, which will be

$$\sigma(1/1 + 1/39)^{1/2} = 1.013\sigma.$$

However, if the experimenter had considered this analysis of data before designing the experiment, he would have realised that the effectiveness of his experiment depended on making the standard error of $(\bar{y}_A - \bar{y}_B)$ within his experiment as small as possible (to use the previous knowledge about the effects of drug A requires that the experimental conditions are identical to those in which the previous experience was gained). This is achieved by allocating 20 patients to drug A and 20 to drug B. This gives a standard error of

$$\sigma(1/20 + 1/20)^{1/2} = 0.316\sigma,$$

showing that equal allocation to the two groups reduces the standard error by a factor of more than 3. Of course, it would not be necessary to consider the standard error formally to guess that equal allocation of patients gives the most precise answer. In a sense, it would be intuitively surprising if any unequal division were to be more efficient than the equal division. Prior thought of what is to be done with the results of the experiment should lead to avoiding a foolish design.

Nevertheless, although analysis is integral to any consideration of design, this book is about the design of experiments, and will not be concerned with the analysis of data, except insofar as this is essential to the understanding of the design principles discussed. The general theory of the analysis of data from designed experiments, using the method of least squares estimation for linear models is developed in Chapter 4. Some particular examples of this general theory will appear in later chapters for some examples of experimental designs. There will also be some discussion, particularly in Chapters 9, 11 and 18, of other methods and models for analysis; these methods are related to, but different from, the general method described in Chapter 4.

We shall assume throughout the book that the analysis of data from experiments will be achieved through the use of computer programs. There is available a wide range of programs for the analysis of experimental data. Some of the programs apply the simplest forms of the general method which we describe in Chapter 4, but there are many other analysis programs available, and it is important that an experimenter should understand enough about different methods of analysis to be able to detect when the method used in a particular program is appropriate to his situation.

Thus, not only is it possible to derive the algebraic formulae necessary to define the calculations required for the analysis of data from any experiment, but the computer programs should make it possible for an experimenter, without the mathematical skills necessary to derive the algebraic formulae, to understand the statistical basis for the interpretation of the calculations. The need, in the earlier development of statistical methods, to be able to derive the algebraic form of analysis has made it easier to discuss some of the important design principles we shall discuss later. In this way, the earlier lack of advanced computing power may be regarded as a blessing, though there must be many erstwhile technical assistants to statisticians who would find that hard to accept! A caveat to the advice to use the general computer programs to analyse all experimental data must be to remember that much of our understanding of design theory was stimulated by algebraic necessity because of the lack of computers. Consequently, we should search for increased understanding both through the use of computers and through algebraic manipulation.

1.4 Separate consideration of units and treatments

Throughout this book, we shall emphasise the need to think separately about the properties of the experimental units available for the experiment, and about the choice of experimental treatments. Far too often, the two aspects are confused at an early stage, and an inefficient or useless experiment results. Thus, an experimenter considering a comparison of the effects of different diets on the growth of rats may observe that the rats available for experimentation come from litters with between five and ten animals per litter. Having heard about the randomised block design, in which each treatment must occur in each block, he decides that he must use litters as blocks, and therefore will have blocks of five units. He further decides that he should consequently have five treatments, and chooses his treatments on this premise, rather than considering how many treatments are required for the objectives of his experiment.

Similarly, a microbiologist investigating the growth of salmonella as affected by five different temperatures, four different media and two independent different chemical additives, may decide that a factorial set of 80 treatments is necessary. Recognising the possibility of day-to-day variation and also knowing about the randomised block design he tries to squeeze the preparation of 80 units into a single day, where only 40 can be efficiently managed, and consequently suffers a much higher error variance for his observations than would normally be expected.

In both of these examples, which though plausible are fictitious, we believe that separate consideration of properties of units and choice of treatments, together with an appreciation of the possibilities of modern statistical design, would lead to better experiments. Thus, in the rat growth experiment, on considering the experimental material available, the experimenter might recognise that each litter forms a natural block, and that he has one block of five units, one of six units, two of seven units, and one of ten units; each block could be reduced in size while still retaining its natural block quality. For treatments, he has a control and three additions to the basic diet, each of which he wishes to try at two concentrations; a total of seven treatments. It is quite simple to construct a sensible design to compare seven treatments in blocks of five, six, seven, seven and ten units, respectively, and this problem will be discussed further in Chapter 7.

For the microbiologist, again the solution of his problem requires only the application of standard statistical design theory, though because the problem does not fit into any neat mathematical classification, the reader of most books on experimental design might be forgiven for thinking that the problem was impossible or at best required very complex mathematical theory. The problem of using blocks of 40 units with a set of 80 treatments in a $5 \times 4 \times 2 \times 2$ factorial structure is a very simple example of confounding to be met again in Chapter 15.

In summary, to achieve good experimental design, the experimenter should think first about the experimental units, and should make sure that he understands the likely patterns of variation. He should also consider the set of treatments appropriate to the objectives of his experiment. If the set of treatments fits simply with the structure of units using a standard design, then the experiment is already designed. If not, then a design should be constructed to fit the requirements of units and treatments, either by the experimenter alone or with the assistance of a statistician. The natural pattern of units and treatments should not be deformed in a Procrustean fashion to fit a standard design. It may perhaps reassure the reader to point out that, for most experiments (80 or 90%), a standard design will be appropriate. To counterbalance the note of reassurance, it is obvious from the experimental scientific literature that frequently an inappropriate simple standard design has been used when an appropriate but slightly more complex standard design exists and should have been used.

1.5 The resource equation

A major principle of design arises from the consideration of degrees of freedom (df) for treatment comparisons. These will be defined more formally in Chapter 2, but in essence df count the number of independent comparisons which can be made. The total information in an experiment involving N experimental units may be represented by the total variation based on $(N - 1)$ df. In the general experimental situation, this total variation is divided into three components, each serving a different function:

(a) the treatment component, T, being the sum of df corresponding to the questions to be asked;
(b) the blocking component, B, representing all the environment effects that are to be included in the fitted model, usually $(b - 1)$ df for the b blocks, but other alternatives are $(r - 1) + (c - 1)$ for row and column designs, or more generally, all effects in multiple control or multiple covariate designs. It may also be desirable to allow a few df for unforeseen contingencies such as missing observations;
(c) the error component, E, being used to estimate σ^2 for the purpose of calculating standard errors for treatment comparison, and, more generally, for inference about treatments.

We can express this division of information as *the resource equation* in terms of the df used in each component T, B and E:

$$T + B + E = N - 1.$$

Now, to obtain a good estimate of error, it is necessary to have at least 10 df. Many statisticians would take 12 or 15 df as their preferred lower limit, although there are situations where statisticians have yielded to pressure from experimenters and agreed to a design with only 8 or even 6 df. The basis for the decision is most simply seen by examining the 5% point of the t-distribution, and using sufficient df so that increasing the error df makes very little difference to the significance point, and hence to the interpretation.

To return to the implications for design, it is necessary to have at least 10 df for error, to estimate σ^2. In choosing the number of treatment question df, we should not normally allow E to fall below 10. But equally, if E is allowed to be large, say greater than 20, then the experimenter is wasting resources by using too much resource on estimating s^2, and not asking enough questions. This is a failure of experiments which occurs much more frequently than allowing E to fall too low, and it is just as much a failure of design. Often, the way of increasing the number of treatment question df is through adding an additional treatment factor, so that the previously planned treatments are all applied at each of two levels of some other factor. The additional factor may be an additional interference treatment, or it could be a treatment factor representing different environments. Various approaches to the choice of treatment factors will be discussed in Chapter 12 and subsequent chapters. The basic economic advantage of factorial experiments arises from asking many questions simultaneously and to insist, as many scientists do, on asking only one or two questions in each experiment, is to waste resources on a grand scale.

One final comment on this question of efficient use of resources. It is, of course, possible to keep the value of E in the region of 10–20 by doing several small experiments, rather than one rather larger experiment in which the total number of experimental units is much less than the sum of the units in the separate small experiments. This is inefficient because of the need to estimate σ^2 for each separate experiment. Experimenters should identify clearly the questions they wish the experiment to cover, and they should also consider carefully if they are asking enough questions to use the experimental resources efficiently.

2

Elementary ideas of blocking: the randomised complete block design

2.1 Controlling variation between experimental units

In an experiment to compare different treatments, each treatment must be applied to several different units. This is because the responses from different units vary, even if the units are treated identically. If each treatment is only applied to a single unit, the results will be ambiguous – we will not be able to distinguish whether a difference in the responses from two units is caused by the different treatments, or is simply due to the inherent differences between the units.

The simplest experimental design to compare t treatments is that in which treatment 1 is applied to n_1 units, treatment 2 to n_2 units, ... and treatment t to n_t units. In many experiments the numbers of units per treatment, $n_1, n_2, \ldots, n_t$, will be equal, but this is not necessary, or even always desirable. Some treatments may be of greater importance than others, in which case more information will be needed about them, and this will be achieved by increasing the replication for these treatments. This design, in which the only recognisable difference between units is the treatments which are applied to those units, is called a *completely randomised design*.

However, the ambiguity, when each treatment is only applied to a single unit, is not always removed by applying each treatment to multiple units. One treatment might be 'lucky' in the selection of units to which it is to be applied, and another might be 'unlucky'. There is no foolproof method of overcoming the vagaries of allocating treatments to units, since the responses, if the same treatment were applied to all units, are not known before the experiment. However, most experimenters have some idea about which units are likely to behave similarly, and such ideas can be used to control the allocation of treatments to units in an attempt to make the allocation more 'fair'. Essentially each group of 'similar' units should include roughly equal numbers of units for each treatment. This control is referred to as *blocking*, and the simplest and most frequently used design resulting from the idea of blocking is the randomised block design, more correctly called the *randomised complete block design*, in which each block contains a single replicate of each of the treatments.

The term blocking arises from the introduction of a controlled treatment allocation in agricultural crop experiments. Here the experimental unit is a small area of land, often called a plot, typically between $2 \, \text{m}^2$ and $20 \, \text{m}^2$, depending on the crop and on the treatments. The set of units or plots for an experiment might be arranged as in Figure 2.1(a). In general we might expect that adjacent plots would behave more similarly than those further apart. Of course, the experimenter might have more specific knowledge about the plots than this. For example, there might be a slope from left to right, so that the plots down the left hand side might be

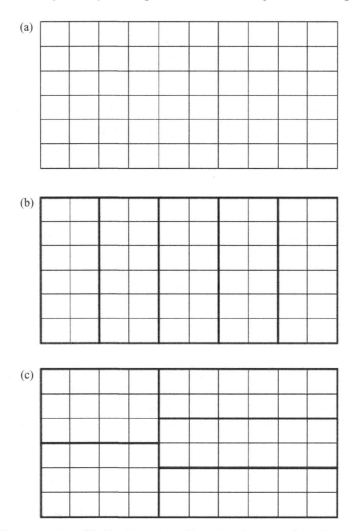

Figure 2.1 Possible blocking plans (b) and (c) for a set of 60 plots (a).

expected to behave similarly, and differently from those on the right hand side, with a trend in response from left to right. However, either from general or particular considerations, if the plots are grouped together in sets of, say, 12, so that the plots in a set might be expected to behave similarly, then the resulting grouping might look like Figure 2.1(b) for a trend from left to right or (c) for a trend from top to bottom. Such groupings are naturally called blocks of plots, and the term has become standard for groupings of units in disciplines other than agronomy.

The patterns of variation on which blocking systems can be based are very numerous. But the most important source of information in determining blocks of units must always be the specialist knowledge of the experimenter about his experimental material. Some examples of blocking systems are discussed here, to help the experimenter to identify the appropriate choice of blocks for their particular experiment.

For field trials concerned with the control of pests or diseases, there may be physical features, such as hedges or field edges, which might affect the direction in which the pests or disease will arrive in the experiment. For experiments with plants grown in pots on benches in glasshouses or laboratories, where different treatments are applied to different groups of pots (even individual pots), geographical blocks of similarly situated pots will usually prove useful. The major controllable sources of variation between responses from different pots are likely to be the position of the pot on the bench, or relative to other fixed features such as the edge of the room, the door, heating pipes or lighting. In experiments with perennial crops, such as fruit trees, the experimental unit will often be an individual tree, and geographical blocks may again be useful. Possibly more useful blocks can be based on the responses of the trees in previous years, or more generally based on the previous histories of the trees. Mushrooms are grown experimentally in trays arranged in a three-dimensional array, and here blocks might need to account for potential sources of variation in all three dimensions. When the experimental unit is an animal, then blocks can be based on genetic similarity, leading to the use of complete litters as blocks, or on the weights of the animals prior to the experiment, or on some other aspect of the previous history of the animals.

In medical trials, where the experimental unit is usually an individual person, the number of classifications which can be used as the basis for blocking is large. These include the age, sex, height, weight, social class, medical history, or racial characteristics of the person, or even the time at which the person becomes available for the trial. Many of the same classifications are relevant for psychological experiments. In such experiments, the experimental unit will often be a short time period in a person's life. If several observations are made for different periods for the same person, with different stress treatments being applied during the different periods, then the experimental unit may be defined to be each short period within the person's life, and the set of units for one person then constitutes a sensible block. A possible complication here is that time might be considered an important factor influencing response *within* the total experimental period, in which case the experimental design requires two forms of blocking – by persons and by times.

In industrial experiments, a small number of machines may each be observed at different times, working under different conditions. The set of observations for a single machine may then be a sensible block. Alternatively, if the environment changes diurnally or between days, the set of observations at the same time within a day, or within a week, may form a useful block. In industrial or other experiments, consistent differences may be suspected between different machine operators, observers or recorders, and the sets of observations for each operator or observer may conveniently be regarded as blocks.

In many experiments, as suggested above, there may be several different ways of blocking the experimental units. We shall return to this problem later in Chapters 7 and 8. For the present, we shall assume that for an experiment with t treatments the experimenter has considered the available units and has divided the units into blocks so that each block contains t units, which are expected to behave similarly, and has then allocated one unit to each treatment in each block. The resulting design in which there is exactly one observation for each treatment in each block is the randomised complete block design. The analysis of data from this design, and from the completely randomised design, are now considered in some detail at a fairly simple level.

Table 2.1. *Diameter (mm) of each affected area*

A	B	C
17.1	24.3	14.3
14.9	21.7	14.5
23.6	22.3	17.1
16.0	26.4	13.2
18.7	23.2	15.6
21.7	20.7	10.3
21.4	22.9	
23.0	19.6	
19.0	22.4	
18.6		

2.2 The analysis of variance identity

The initial technique in the analysis of most data from designed experiments is the *analysis of variance*. This has two purposes. First, it provides a subdivision of the total variation between the experimental units into separate components, each component representing a different source of variation, so that the relative importance of the different sources can be assessed. Second, and more important, it gives an estimate of the underlying variation between units, which provides a basis for inferences about the effects of the applied treatments. We develop the principles of the analysis of variance for the two designs described in the previous section, using two examples.

First we consider a laboratory experiment in which three different chemicals (A, B and C) were added to cultures of a virus growing within Petri dishes. The purpose of the experiment was to compare the effects of the chemicals on the growth of the virus, measured by the colony diameter. The experiment was arranged as a completely randomised design with a total of 25 units, divided unequally (10, 9, 6) between the treatments. The observed data are given in Table 2.1.

The second data set, shown in Table 2.2, is from an experiment to compare ten plant spacing treatments for rice, arranged as a randomised complete block design. The ten treatments comprised all possible combinations of pairs of inter-seedling distances (15 cm, 20 cm, 24 cm and 30 cm). The aims of the experiment were to investigate both the overall effect of changing the density of seedlings and the effect of spatial arrangement (measured by rectangularity). The area available for the experiment was sufficient for 40 experimental plots, each measuring 2.4 m × 3.6 m. The area was divided into four quarters, each containing ten plots, so that the ten plots in each quarter were hopefully similar, and the ten treatments were then randomly allocated, one to each plot in each of the four quarters, or blocks. The experiment, at Belle Vue Farm, Mauritius, was conducted during the 1970–71 season, using variety Taichung Native No. 1.

The basic philosophy of analysis of variance is to assume that each unit has an inherent response, possibly related to the position of the unit within the experiment, which is modified by the effect of the particular treatment applied to the unit. The response for each unit can

Table 2.2. *Yields of rice grain (kg per plot)*

Spacing	Block			
	I	II	III	IV
30 cm × 30 cm	5.95	5.30	6.50	6.35
30 cm × 24 cm	7.10	6.45	6.60	5.75
30 cm × 20 cm	7.00	6.50	6.35	8.90
30 cm × 15 cm	8.10	5.50	6.60	7.50
24 cm × 24 cm	8.85	7.65	7.00	7.90
24 cm × 20 cm	7.65	6.90	8.25	8.30
24 cm × 15 cm	7.80	6.75	8.20	7.25
20 cm × 20 cm	8.05	6.65	8.10	8.05
20 cm × 15 cm	9.30	8.75	8.75	8.00
15 cm × 15 cm	9.35	8.10	7.60	7.75

then be written as a sum of the relevant terms. For the completely randomised design the response for the kth unit to which treatment j is applied can be written as:

$$y_{jk} = \mu + t_j + \varepsilon_{jk}. \tag{2.1}$$

Here, μ represents the average response for the whole set of experimental units, and the ε_{jk} represent the deviations of the particular units from that average. The treatment effects, t_j, represent the deviations for each treatment from the average of the set of treatments included in the experiment.

If the units are grouped in blocks, such as in a randomised complete block design, then an additional set of terms is needed to represent the average differences between blocks. Hence, for the randomised complete block design, the response for treatment j in block i, y_{ij}, is written as

$$y_{ij} = \mu + b_i + t_j + \varepsilon_{ij}. \tag{2.2}$$

Here μ again represents the average response for the whole set of experimental units, b_i is the extra term representing the average deviation from μ of the set of units in block i, and the ε_{ij} again represent the deviations of the particular units from the average response, $\mu + b_i$, of the units in block i. The treatment effects, t_j, again represent the deviations for each treatment from the average of the set of treatments included in the experiment. This introduces the basic terminology which will be used throughout the book.

To complete the definitions we assume that there are t treatments and n_j units for treatment j, with a total of $n = \sum n_j$ units in the completely randomised design, whilst in the randomised complete block design there are b blocks and t treatments, and a total of $n = bt$ experimental units. The definitions of b_i and t_j further imply that

$$\sum_{j=1}^{t} t_j = 0, \qquad \sum_{i=1}^{b} b_i = 0.$$

The algebraic statements (2.1) and (2.2), with the accompanying verbal explanations, constitute the models for the sets of responses. Two basic assumptions implied by the use of these models must be remembered when calculating an analysis of variance. The first is that the individual unit variation, ε, is not affected by the particular treatment applied to the unit, and, second, that the effect of each treatment is additive. A consequence of these assumptions is that if two treatments applied to a pair of units had been swapped and applied each to the other unit of the pair, then the total response from the pair of units would have been unaltered, the loss for one response being exactly counterbalanced by the gain for the other. This is, of course, a purely hypothetical concept, since each unit can have only one treatment applied to it, but the consequence of the assumption is a powerful one which the experimenter must be prepared to accept as reasonable if the standard form of analysis is to be used. Of course, if treatments have different effects on different units, the whole idea of '*a* treatment effect' becomes meaningless.

The analysis of variance of data from an experimental design is simply the division of the total variation between the n observations into sensibly distinguishable components, and the subsequent interpretation of the relative size of the components.

In the completely randomised design, there are two distinguishable components of the total variation, the variation between the units receiving the same treatment, and the variation between units receiving different treatments.

The total variation between all n observations is

$$\sum_{jk} (y_{jk} - y_{..})^2, \tag{2.3}$$

where $y_{..}$ is the mean of all n observations (the use of a dot replacing a suffix indicates the mean over all possible values of the suffix).

Similarly, the variation between the responses for all units receiving treatment j is most simply written

$$\sum_{k} (y_{jk} - y_{j.})^2, \tag{2.4}$$

where

$$y_{j.} = \sum_{k} y_{jk}/n_j.$$

Summing (2.4) across all treatments then gives an expression for the total variation within treatments, i.e. the total variation not due to differences between treatments:

$$\sum_{j} \left\{ \sum_{k} (y_{jk} - y_{j.})^2 \right\}. \tag{2.5}$$

Now, both (2.3) and (2.5) can be written in extended form as

$$\sum_{jk} (y_{jk} - y_{..})^2 = \sum_{jk} y_{jk}^2 - \left(\sum_{jk} y_{jk} \right)^2 \Big/ n,$$

Table 2.3. *Analysis of variance for a completely randomised design*

Source	Sum of squares (SS)	df
Between treatments	$\sum_{j}\left(\sum_{k} y_{jk}\right)^{2}\bigg/ n_{j} - \left(\sum_{jk} y_{jk}\right)^{2}\bigg/ n$	$t-1$
Within treatments	$\sum_{j}\left\{\sum_{k} y_{jk}^{2} - \left(\sum_{k} y_{jk}\right)^{2}\bigg/ n_{j}\right\}$	$\sum_{j}(n_{j}-1)$
Total	$\sum_{jk} y_{jk}^{2}\left(\sum_{jk} y_{jk}\right)^{2}\bigg/ n$	$n-1$

and

$$\sum_{k}(y_{jk}-y_{j.})^{2} = \sum_{jk} y_{jk}^{2} - \sum_{j}\left(\sum_{k} y_{jk}\right)^{2}\bigg/ n_{j}.$$

We know that the total variation between all units is the variation between units receiving the same treatment, summed across treatments, and the variation between units receiving different treatments. Thus the difference between these two expressions gives us an expression for this latter term, initially in extended form:

$$\sum_{j}\left(\sum_{k} y_{jk}\right)^{2}\bigg/ n_{j} - \left(\sum_{jk} y_{jk}\right)^{2}\bigg/ n = \sum_{j}\sum_{k}(y_{j.}-y_{..})^{2}. \tag{2.6}$$

This allows us to deduce the analysis of variance identity:

$$\sum_{jk}(y_{jk}-y_{..})^{2} \equiv \sum_{j}\left\{\sum_{k}(y_{jk}-y_{j.})^{2}\right\} + \sum_{j}\sum_{k}(y_{j.}-y_{..})^{2}. \tag{2.7}$$

The three terms in the identity (2.7) represent the total variation in responses over all n observations, the sum of the within-treatment variations and the variation between treatment means. To transform these expressions into sample variances, (2.6) would be divided by the sum of the $(n_{j}-1)$ across all treatments, to give the total within-treatment variance, and (2.3) by $(n-1)$ to give the total variance across the set of observations. An analogous identity for these divisors, which are called df, is

$$(n-1) = \sum_{j}(n_{j}-1) + (t-1) \tag{2.8}$$

and the divisor $(t-1)$ is used to transform (2.6) into the between-treatment variance.

The analysis of variance structure is normally presented in tabular form, as shown in Table 2.3. For the analysis of the virus growth diameter data, we can calculate the total variation (usually called the total sum of squares (SS)) and the between-treatment variation

Table 2.4. *Analysis of variance for virus growth data*

Source	SS	df
Between treatments	256.88	2
Within treatments	136.28	22
Total	393.16	24

(usually called the between-treatment SS) by calculating three terms extracted from the expressions in extended form:

$$\sum_{jk} y_{jk}^2 = 9705.41,$$

$$\sum_j \left(\sum_k y_{jk} \right)^2 \Big/ n_j = 9569.13,$$

$$\left(\sum_{jk} y_{jk} \right)^2 \Big/ n = 9312.25.$$

The total SS is then obtained by subtracting the third term from the first, and the treatment SS is obtained by subtracting the third term from the second. The third term is often referred to as the *correction factor* as it modifies, or 'corrects' simple sums of squared quantities to be sums of squared deviations.

Calculating the within-treatment variation (within-treatment SS) by subtracting the between-treatment SS from the total SS completes the analysis of variance, which is summarised in Table 2.4.

Clearly the major component of the total variation is attributable to the treatment effects, almost two-thirds of the total variation representing between-treatment variation. The comparison of variances, obtained by dividing each of the SS by the appropriate df, is even more dramatic, the variances being 128, 6 and 16 for between treatments, within treatments and total, respectively.

The analysis of variance of data from a randomised complete block design is similarly based on an algebraic identity, but now with four component terms. The extra term, in comparison with the analysis of variance for the completely randomised design, is for the variation between blocks. The analysis of variance identity is

$$\sum_{ij} (y_{ij} - y_{..})^2 \equiv \sum_{ij} (y_{i.} - y_{..})^2 + \sum_{ij} (y_{.j} - y_{..})^2 + \sum_{ij} (y_{ij} - y_{i.} - y_{.j} + y_{..})^2$$

$$\text{total SS} \qquad \equiv \text{block SS} \qquad + \text{treatment SS} \; + \text{error SS}.$$

The second and third terms in the identity represent the variation between the block means, $y_{i.}$, and between the treatment means, $y_{.j}$, respectively. By analogy with the between-treatment variation in the previous analysis of variance, the corresponding df are $(b-1)$ and $(t-1)$, respectively. The final term in the identity, the error SS, corresponds to the ε terms in the

Table 2.5. *Analysis of variance for randomised complete block design*

Source	SS	df
Blocks	$\sum_i Y_{i.}^2/t - Y_{..}^2/n$	$b-1$
Treatments	$\sum_j Y_{.j}^2/b - Y_{..}^2/n$	$t-1$
Error	By subtraction	$(b-1)(t-1)$ by subtraction
Total	$\sum_{ij} y_{ij}^2 - Y_{..}^2/n$	$n-1$

model, and represents the inconsistency of treatment differences over the different blocks. The label 'error' is unfortunate but traditional – see later in this chapter for an alternative label.

The truth of the analysis of variance identity for the randomised complete block design may be verified simply by rewriting the various terms in extended form. So, the total SS can be written as

$$\sum_{ij}(y_{ij} - y_{..})^2 = \sum_{ij} y_{ij}^2 - \left(\sum_{ij} y_{ij}\right)^2 \bigg/ n = \sum_{ij} y_{ij}^2 - Y_{..}^2/n,$$

where we have introduced an extension of the dot suffix notation. The dot suffix with a capital letter represents the total response summed over the dotted suffix or suffices. Similarly, the block SS, treatment SS and error SS can be written as

$$\sum_{ij}(y_{i.} - y_{..})^2 = \sum_i Y_{i.}^2/t - Y_{..}^2/n,$$

$$\sum_{ij}(y_{.j} - y_{..})^2 = \sum_j Y_{.j}^2/b - Y_{..}^2/n,$$

$$\sum_{ij}(y_{ij} - y_{i.} - y_{.j} + y_{..})^2 = \sum_{ij} y_{ij}^2 - \sum_i Y_{i.}^2/t - \sum_j Y_{.j}^2/b + Y_{..}^2/n,$$

respectively. The term $Y_{..}^2/n$, which occurs in each expression, is again the *correction factor* we introduced earlier, though written in a slightly different form.

The analysis of variance structure for the randomised complete block design is also normally presented in tabular form as shown in Table 2.5.

For the rice spacing example given earlier, the calculations of the total SS, block SS and treatment SS are

$$\text{total SS} = 2252.61 - 2211.17 = 41.44,$$
$$\text{block SS} = 2217.05 - 2211.17 = 5.88,$$
$$\text{treatment SS} = 2234.31 - 2211.17 = 23.14,$$

with the error SS most easily calculated by subtracting the block SS and treatment SS from the total SS. The completed analysis of variance table is shown in Table 2.6.

Table 2.6. *Analysis of variance for rice spacing data*

Source	SS	df	Mean square (MS)
Blocks	5.88	3	1.96
Treatments	23.14	9	2.57
Error	12.42	27	0.46
Total	41.44		

For the final column in Table 2.6, each mean square (MS) is calculated by dividing the appropriate SS by the corresponding df. Each MS is then a variance, the method of calculation being such that the mean squares are directly comparable. Thus the treatment mean square (TMS) is

$$\text{TMS} = \left(\sum_j Y_{.j}^2 \Big/ b - Y_{..}^2/n \right) \Big/ (t-1)$$

$$= 1/b \left[\left\{ \sum_j y_{.j}^2 - \left(\sum_j Y_{.j} \right)^2 \Big/ t \right\} \Big/ (t-1) \right]$$

$$= 1/b \,(\text{variance of the } Y_{.j}\text{'s}),$$

which, since $Y_{.j}$ is the sum of b observations, is on the same scale as $\text{Var}(y_{ij})$. A similar algebraic manipulation is possible for the expression for the block MS. Thus the block, treatment and error MSs may be directly compared to assess the relative magnitude of the between-block variance, the between-treatment variance and the between-unit or error variance.

For the rice spacing experiment, the block variance is certainly larger than the error variance, whilst the treatment variance is considerably larger. Later, in Chapter 4, we show formally that the MSs can be compared using F tests if we assume normally distributed unit variation. For the present, our purpose is to consider the analysis of variance as an arithmetic subdivision of the total variance. We may reasonably conclude, however, for the rice spacing example that the different spacing treatments lead to different yields.

2.3 Estimation of variance and the comparison of treatment means

In the construction of the models (2.1) and (2.2), the terms ε_{jk} and ε_{ij} represented the deviations of unit responses from the average responses (in model (2.1)) and from the average response of the units in block i (model (2.2)), if the model assumptions are correct. The precision of comparisons between the mean responses for different treatments will depend only on the variability of these ε_{ij}, since in the completely randomised design there are no other effects, and in the randomised complete block design the block differences have been eliminated from treatment comparisons. To assess the precision of treatment comparisons, we require explicit assumptions about the mean and variance of the εs. Formally, we assume that the units, and hence the ε, are a random sample from a population, such that the expectation of ε, $E(\varepsilon)$, equals zero, and the variance $\text{Var}(\varepsilon) = \sigma^2$.

In the completely randomised design, the variance from the sample of observations for each treatment provides an estimate of σ^2. The pooled within-treatment MS provides a better estimate of σ^2 than any single treatment sample, because the variance of observations is assumed to be unaffected by the particular treatment considered. Hence, the estimate of σ^2 for a completely randomised design is the within-treatment MS:

$$s^2 = \sum_j \left\{ \sum_k y_{jk}^2 - \left(\sum_k y_{jk} \right)^2 \Big/ n_j \right\} \Big/ \sum_j (n_j - 1).$$

The variance of the difference between two treatment means is then

$$\text{Var}(y_{.j} - y_{.j'}) = \sigma^2 (1/n_j + 1/n_{j'})$$

and is estimated by $s^2 (1/n_j + 1/n_{j'})$.

In the randomised complete block design model (2.2), the block effects, b_i, sum to zero and the expected value of a treatment mean response is

$$E(Y_{.j}/b) = (\mu + t_j).$$

Additionally, as

$$Y_{.j} = b\mu + bt_j + \sum_i \varepsilon_{ij},$$

the variance of a treatment mean is

$$\text{Var}(Y_{.j}/b) = \text{Var}\left(\mu + t_j + \sum_i \varepsilon_{ij}/b \right) = \text{Var}\left(\sum_i \varepsilon_{ij}/b \right) = \sigma^2/b.$$

Similarly, the variance of the difference between two treatment means is

$$\text{Var}(y_{.j} - y_{.j'}) = 2\sigma^2/b.$$

The definition of σ^2 is clear, but it is not immediately clear how to estimate σ^2, since the observations, y_{ij}, all contain components due to both block and treatment differences. However, it is clear from the analysis of variance identity that the error SS is not affected by either block or treatment differences, and, in fact, the error MS provides an unbiased estimator of σ^2. To see that this is algebraically correct, consider the error SS:

$$\sum_{ij} (y_{ij} - y_{i.} - y_{.j} + y_{..})^2$$

$$= \sum_{ij} \left(y_{ij}^2 + y_{i.}^2 + y_{.j}^2 + y_{..}^2 - 2y_{ij}y_{i.} - 2y_{ij}y_{.j} + 2y_{ij}y_{..} + 2y_{i.}y_{.j} - 2y_{i.}y_{..} - 2y_{.j}y_{..} \right)$$

$$= \sum_{ij} y_{ij}^2 + t\sum_i y_{i.}^2 + b\sum_j y_{.j}^2 + ny_{..}^2 - 2t\sum_i y_{i.}^2 - 2b\sum_j y_{.j}^2$$

$$+ 2ny_{..}^2 + 2ny_{..}^2 - 2ny_{..}^2 - 2ny_{..}^2$$

$$= \sum_{ij} y_{ij}^2 - t\sum_i y_{i.}^2 - b\sum_j y_{.j}^2 + ny_{..}^2.$$

But, from model (2.2),

$$y_{..} = \mu + \varepsilon_{..},$$
$$y_{i.} = \mu + b_i + \varepsilon_{i.},$$
$$y_{.j} = \mu + t_j + \varepsilon_{.j},$$

and, since $E(\varepsilon_{ij}) = 0$ and $Var(\varepsilon_{ij}) = \sigma^2$,

$$E\left\{\sum_{ij}(y_{ij} - y_{i.} - y_{.j} + y_{..})^2\right\} = E\left(\sum_{ij} y_{ij}^2 - t\sum_i y_{i.}^2 - b\sum_j y_{.j}^2 + ny_{..}^2\right)$$

$$= \left(n\mu^2 + t\sum_i b_i^2 + b\sum_j t_j^2 + n\sigma^2\right) - t\left\{b\mu^2 + \sum_i b_i^2 + b(\sigma^2/t)\right\}$$

$$- b\left\{t\mu^2 + \sum_j t_j^2 + t(\sigma^2/b)\right\} + n(\mu^2 + \sigma^2/n)$$

$$= (n - t - b + 1)\sigma^2 = (b-1)(t-1)\sigma^2.$$

Hence the expected value of the error SS is $(b-1)(t-1)\sigma^2$, and the expected value of the error MS is σ^2. The error MS is therefore referred to as s^2 and provides an unbiased estimator of σ^2 and hence of the variances required for comparisons of treatment means.

In this chapter only simple comparisons of treatments are considered. The philosophy of choosing treatment comparisons to answer the questions which provoked the experiment is addressed in Chapter 12.

The obvious treatment comparisons to make are between two particular treatments. So, for example, the difference between treatments 1 and 2, $(t_1 - t_2)$, is estimated by $(y_{.1} - y_{.2})$, for which the standard error is $(2\sigma^2/b)^{1/2}$. An alternative comparison is between one treatment and the average of two other treatments, for example, $\{t_1 - (t_2 + t_3)/2\}$, for which the relevant standard error is

$$\{\sigma^2/b + \sigma^2/(2b)\}^{1/2} = (3\sigma^2/2b)^{1/2}.$$

There are two problems concerned with making too many comparisons between treatment means. First, the more tests we make of treatment comparisons, the more likely we are to find a result significant at the 5% level from the 1 in 20 chance inherent in the concept of such a significance test. In addition, it is natural, when making comparisons, to concentrate on those comparisons which look larger, which will be between the treatments giving the more extreme results and which are then subject to a selection bias. Both problems are discussed in more detail in Chapter 12, but both must be considered when comparing treatment means after any analysis of variance.

We first illustrate the testing of comparisons between treatment means for the experiment on viral growth. The three treatment means are

A	B	C
19.4	22.6	14.2,

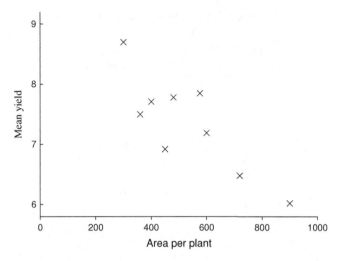

Figure 2.2 Relationship between mean yield and area per plant.

and the standard errors of the differences (SEDs) between each pair of means are

$$1.14 \quad \text{for} \quad A - B,$$
$$1.29 \quad \text{for} \quad A - C,$$
$$1.31 \quad \text{for} \quad B - C,$$

the different values being a result of the different replication levels for the three treatments.

Clearly chemical C reduces the growth of the virus compared with either of chemicals A and B. The difference between A and B is the smallest of the three differences but even this is quite large compared with the standard error of the difference, and we may reasonably conclude that the growth rate is different for all three chemicals, with B giving the greatest growth and C the smallest.

In the rice spacing example, there are various obvious simple comparisons, for example:

(i) between spacings with one dimension in common, for example $t_1 - t_2, t_1 - t_3, t_9 - t_{10}$;
(ii) between spacings which are square, for example $t_1 - t_5, t_5 - t_8, t_8 - t_{10}$;
(iii) between spacings at a similar overall density, for example $t_4 - t_8$.

In this instance, however, as in many experiments with quantitative treatments, a graphical presentation of the means is best. Figure 2.2 shows the mean yields plotted against the area per plant. The values of mean yield and area per plant for the ten treatments are given in Table 2.7.

The linear regression analysis of mean yield on area per plant gives

$$\text{mean yield} = 9.11 - 0.0034 \, \text{area},$$
$$\text{regression SS} = 4.20 \quad \text{on 1 df},$$
$$\text{residual SS} = 1.59 \quad \text{on 8 df},$$
$$\text{total SS} = 5.79 \quad \text{on 9 df}.$$

Table 2.7. *Summaries for the rice spacing experiment*

Mean yield (y)	Area per plant (x)
6.02	900
6.48	720
7.19	600
6.92	450
7.85	576
7.78	480
7.50	360
7.71	400
8.70	300
8.20	225

Table 2.8. *Residuals for rice spacing experiment*

Spacing	Block			
	I	II	III	IV
30 × 30	−0.55	−0.14	+0.52	+0.19
30 × 24	+0.14	+0.55	+0.16	−0.87
30 × 20	−0.66	−0.10	−0.79	+1.58
30 × 15	+0.70	−0.84	−0.28	+0.44
24 × 24	+0.52	+0.38	−0.81	−0.09
24 × 20	−0.61	−0.30	+0.51	+0.38
24 × 15	−0.18	−0.17	+0.74	−0.39
20 × 20	−0.14	−0.48	+0.43	+0.20
20 × 15	+0.12	+0.63	+0.09	−0.84
15 × 15	+0.67	+0.48	−0.56	−0.59

Clearly the linear regression accounts for most of the treatment variation, and the residual treatment variation is relatively small, suggesting that the interpretation of results should concentrate on the effect of space per plant. The analysis of this set of data is considered further in Chapter 12.

2.4 Residuals and the meaning of error

The error SS in the analysis of variance for a randomised complete block design is the SS of the quantities

$$r_{ij} = y_{ij} - y_{i.} - y_{.j} + y_{..}.$$

These are usually called *residuals*, because they represent the unexplained, or residual, variation for each observation after the block and treatment effects have been fitted. We will thus often refer to the error SS and error MS as the *residual SS* (RSS) and *residual MS*. The residuals for the rice spacing data are listed in Table 2.8 and show the pattern of the residuals

Table 2.9. *Observed yields and analysis of variance for rice spacing experiment: (a) yields; (b) analysis*

(a)

Block	Treatment 1	Treatment 2	Total	Difference $(1-2)$
I	5.95	7.10	13.05	−1.15
II	5.30	6.45	11.75	−1.15
III	6.50	6.60	13.10	−0.10
IV	6.35	5.75	12.10	+0.60
Total	24.10	25.90	50.00	−1.80

(b)

Source	SS	df	MS
Blocks	0.69	3	
Treatments	0.40	1	
Error	1.11	3	$s^2 = 0.37$
Total	2.20	7	

after the elimination of block and treatment differences. The residuals contain no element due to treatments and so the total of the residuals for each treatment is zero; similarly, the block totals of residuals are all zero. The pattern of the residuals displays the consistency of the treatment (or block) effects. A treatment with very small residuals, such as spacing treatment 8 (20×20), shows similar effects in all blocks, whereas one with large residuals, such as spacing treatment 3 (30×20), has rather discrepant effects in the different blocks.

The interpretation of error as consistency can be seen more clearly in the analysis of a subset of the data comprising just two treatments. For the first two treatments (30×30 and 30×24), the observed yields in each block and the analysis of variance table are shown in Table 2.9(a) and (b), respectively.

Calculating the mean of the treatment differences in the four blocks, and calculating the difference between the treatment means gives the same estimate for the difference between the two treatments, $(t_1 - t_2)$: -0.45. Based on the variance of the sample of four differences,

$$s_d^2 = \{(1.15)^2 + (1.15)^2 + (0.10)^2 + (0.60)^2 - (1.80)^2/4\}/3 = 0.735,$$

where 1.80 is the difference between the treatment totals, the standard error (SE) of this mean difference is

$$\left(s_d^2/4\right)^{1/2} = 0.43,$$

which is the same as the SE of the difference between the two spacing means as calculated from the error MS obtained from the analysis of variance,

$$(2s^2/4)^{1/2} = (0.74/4)^{1/2} = 0.43.$$

Thus, when only two treatments are considered, the error MS from the analysis of variance provides a direct measure of the consistency of the difference between the two treatments, over the different blocks.

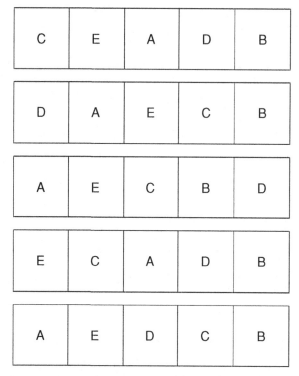

Figure 2.3 Random allocation of five treatments in five randomised blocks.

2.5 The random allocation of treatments to units

We have discussed the randomised complete block design in terms of blocking to control the effects of patterns of variation among the experimental units. The other aspect of this design is that, within each block, the treatments are allocated to units by a formal process of randomisation. This process ensures that each treatment is equally likely to be allocated to any unit within the block. This random allocation has two important consequences, the more important of which is discussed in detail in Chapter 10. The other consequence is quite simply that it removes the subjective element of choosing how to allocate treatments.

Subjective allocation is extremely difficult because it inevitably involves simultaneous avoidance of any tendency to include or exclude any particular patterns of allocation. If anyone is asked to allocate six treatments, A, B, C, D, E, F, to six units, 1, 2, 3, 4, 5, 6, in each of a number of blocks, it is commonly found that because of the subjective wish to avoid identical or similar patterns each letter tends to be spread evenly over the positions 1 to 6, which in itself produces a systematic pattern which may be very desirable but is also clearly a pattern and therefore not what is assumed. More simple failures of subjective allocations have occurred when doctors, having to allocate two drugs to pairs of patients, show a tendency to allocate one treatment to those patients more obviously in need.

Application of a random allocation procedure allows all possible patterns of treatment allocation to occur. For example, in a randomised complete block design, with five blocks of five units each, it allows the design shown in Figure 2.3. If this design seems undesirable in prospect, probably because treatment B occurs towards the right-hand end of each of the blocks, then this implies that the blocking system used is insufficient, and suggests that a further blocking restriction needs to be introduced. It is important to distinguish blocking, which controls the extent to which treatments may be allocated to similar units, from randomisation. Randomisation offers no control and allows all possible allocations to occur with equal probability, but it avoids subjective allocation with its inevitable, but unspecified, tendency towards particular patterns.

Many statistical design packages now provide tools for the randomisation of treatment allocations once the basic design structure has been specified. In the absence of access to such packages, the random allocation of treatments to units in each block may be achieved through the use of pseudo-random numbers, either generated by a computer package, or in printed versions to be found in many statistical textbooks and tables (or even from the final digits of telephone numbers in a telephone directory). For example, for a set of ten treatments, the units in each block are numbered 0 to 9. The first treatment is then allocated to the plot corresponding to the first digit in the random number sequence, the second to that corresponding to the second digit (if unique), and so on. Suppose the random number sequence is

$$3\ 9\ 3\ 8\ 4\ 3\ 1\ 9\ 4\ 3\ 7\ 7\ 1\ 5\ 5\ 4\ 0\ 0\ 4\ 7\ 7\ 0\ 5\ 3\ 2\ 1.$$

Then treatment A goes to plot 3, B to 9, C to 8 (not 3 which is already occupied), D to 4, E to 1, F to 7, G to 5, H to 0, I to 2 and, by elimination, J to 6:

$$0\ 1\ 2\ 3\ 4\ 5\ 6\ 7\ 8\ 9$$
$$H\ E\ I\ A\ D\ G\ J\ F\ C\ B.$$

If there are fewer than ten treatments, then digits $9, 8, 7 \ldots$ are ignored. If there are more than ten treatments, then two digit numbers can be used, but usually there are many fewer than 100 treatments, and to avoid 'wasting' digits each two-digit number may be replaced by the remainder after dividing by the number of treatments. To maintain equal probabilities for each treatment those two-digit numbers greater than or equal to the largest multiple of the number of treatments less than 100 are ignored. To illustrate this, suppose that, in the earlier example, there were 12 treatments. The two digit numbers formed from the sequence are

$$39\ 38\ 43\ 19\ 43\ 77\ 15\ 54\ 00\ 47\ 70\ 53\ 21.$$

To maintain equal probabilities, values greater than or equal to 96 would be ignored. After division by 12 these yield remainders

$$3\ 2\ 7\ 7\ 7\ 5\ 3\ 6\ 0\ 11\ 10\ 5\ 9.$$

The treatment allocation would now be A to unit 3, B to unit 2, C to 7, D to 5, E to 6, F to 0, G to 11, H to 10, with the remaining allocation of I, J, K and L to 1, 4, 8 and 9, determined by the continuation of the sequence.

The method of using the random numbers can be reversed, so that, instead of the digits representing units, they can represent the treatments: A = 0, B = 1, C = 2, up to J = 9. Now, starting at the beginning of the random number sequence and reverting to the ten-treatment case, the treatment corresponding to the first digit (3 = D) is allocated to the first plot, the next $(9 = J)$ to the second plot, and so on, giving the random allocation:

$$0\ 1\ 2\ 3\ 4\ 5\ 6\ 7\ 8\ 9$$
$$D\ J\ I\ E\ B\ H\ F\ A\ C\ G.$$

Note that each block must be randomised separately. For other designs involving more complicated blocking structures, the appropriate randomisation procedures will be discussed in Chapter 10. For the completely randomised design, there is no blocking, but simply a set of treatments, A, B, ..., to be allocated, each to a number of units. This may be achieved by numbering all units and selecting the set of units for each treatment in turn randomly from the complete set. Suppose six treatments are to be allocated to four units each; the units are numbered 1 to 24 and the previous sequence of two-digit random numbers is used. The remainders after division by 24, ignoring 96, 97, 98 and 99, are

$$15\ 14\ 19\ 19\ 19\ 5\ 6\ 0\ 23\ 22\ 5\ 21,$$

so treatment A is allocated to units 5, 14, 15 and 19 (the first four unique plot numbers), treatment B to units 6, 22, 23 and 24 (represented by 0) and treatment C to unit 21 and three other plots, with further numbers being required to complete the randomisation.

2.6 Practical choices of blocking patterns

In some situations, the choice of a blocking system for a set of experimental units may be obvious. In others, there may be two or more patterns of likely variation between units, each of which could usefully be identified with blocks. In yet other situations, there may be no obvious pattern on which to base a blocking system, but there may be advantages in using blocks as an 'insurance' against possible patterns not yet identified or simply for administrative reasons.

When there are alternative patterns on which to base a blocking system the experimenter can adopt one of two approaches. He or she may select one of the patterns, usually that suspected of accounting for more variation, and record for each unit information relevant to the other patterns, hoping to be able to use this information to remove 'nuisance' variability in the analysis, perhaps by the method of covariance discussed in Chapter 4. Alternatively, he or she may attempt to use two or more blocking patterns simultaneously, classifying each unit by two or more blocking factors and ideally arranging the treatments so that differences between the blocks of each system may be eliminated from treatment comparisons; this is the idea of multiple blocking systems, which are discussed in Chapter 8. An example where the first method might be appropriate is in experiments on animals, where both litter effects and initial size effects could be used as a basis for blocks. Using litters as blocks and adjusting for initial weight by covariance is a standard technique. An example of the second method would occur in industrial or laboratory experiments, where there are a limited number of machines or pieces of equipment which may be expected to show systematic differences, and where observations cannot all be taken at the same time, and time differences must be allowed for. Hence both machines and time periods would be included to define blocking systems.

The most common situation where blocks are used without any dominant pattern to define them is in field crop experiments. Typically, the experimenter is allocated an area of land,

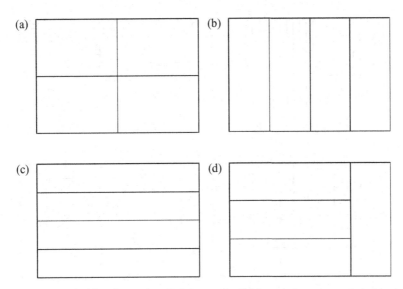

Figure 2.4 (a)–(d) Alternative divisions of a field experiment area into blocks depending on anticipated patterns of variability.

apparently fairly homogeneous, and has to choose the size and shape of his plots. Let us suppose that the experiment involves comparison of 12 treatments, and it is decided to have four plots for each treatment. Although the area is apparently uniform, it is a good precaution to decide to use four randomised blocks of 12 plots each, because there is likely to be some variation within the area, and plots close together are, in general, more likely to behave similarly. The experimenter is, however, left with several choices. Blocks could be defined in various ways, some of which are shown in Figure 2.4; within one of these arrangements of blocks, plots can then also be arranged within each block in many patterns, some of which are shown in Figure 2.5. How should the choice be made? Plots within each block should be as similar as possible. Hence, the arrangement of blocks and plots must be chosen to maximise the differences between plots in different blocks, and to minimise the differences between plots within blocks.

All other things being equal, and they rarely are, these criteria suggest that blocks should tend to be nearly square in shape, and that within each block the plots should be long and thin, thus sampling the remaining differences in the block equally. But this suggestion must be considered afresh in each situation. Suppose that there is suspected to be a fertility gradient from left to right within each block of Figure 2.5; then it may well be more effective to use Figure 2.5(b) rather than Figure 2.5(a). And long, thin plots may be quite inappropriate if only the centre portion of the plot is to be harvested as representative of plants in a large field, because the discarded edge area will be a very large proportion of the total plot. So ignore the rules, but remember the principles.

A more general example of the problem of choosing plots and blocks is illustrated in Figure 2.6. This shows a plan of about 200 rubber trees planted along the contours on a steeply sloping hill. Further difficulties are caused by the occurrence of drains, roads and rocks, all of which may influence production of neighbouring trees. At the beginning of the experiment some trees will have 'dried up' and cannot be used, whilst others will dry up

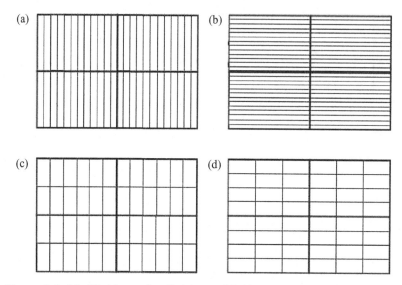

Figure 2.5 (a)–(d) Alternative divisions of field experiment blocks into plots depending on anticipated patterns of variability.

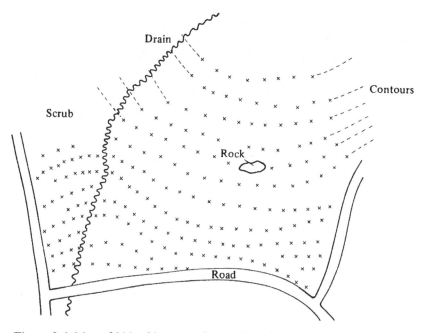

Figure 2.6 Map of 200 rubber trees from Sri Lanka, for a future experiment.

during the experiment. The minimum plot size is a set of four adjacent trees. How does the experimenter define blocks as groups of plots, each plot containing four trees, so that the plots within blocks will be as homogeneous as possible?

3

Elementary ideas of treatment structure

3.1 Choice of treatments

It might seem that the choice of treatments is not a statistical matter, being a question for the experimenter alone. However, there are very many examples of research programmes where the objectives of the programme either have been, or could have been, achieved much more efficiently by using statistical theory. In some situations, there are several intuitively appealing alternative sets of treatments, and the choice between these alternatives can be assisted by a statistical assessment of the efficiency with which the alternative sets of treatments satisfy the objectives of the experiment. In other situations statistical methods can allow more objectives to be achieved from the same set of experimental resources. In yet other situations statistical arguments can lead to adding experimental treatments which avoid the necessity of making possibly unjustifiable assumptions in the interpretation of results from the experimental data. The early theory of the design of experiments was less concerned with the choice of treatments than with the control of variation between units, but there is a large body of knowledge about treatment structure. Later, much of the statistical theory of experimental design was concerned with the optimal choice of treatments, and while some of this theory is very mathematical and some has little relevance to real experiments, the general principles are important and will be considered in Chapter 16.

The first point to clarify is that there are many different forms of objective for an experiment. Some experiments are concerned with the determination of optimal operating conditions for some chemical or biological process. Others are intended to make very precise comparisons between two or more methods of controlling or encouraging the growth of a biological organism. Some experiments are intended to provide information about the effects of increasing amounts of some stimulus on the output of the experimental units to which the stimulus is applied. And other experiments are planned to screen large numbers of possible contestants for a job, for example, drugs for treating cancer or athletes for a place in the Olympics. And most experiments have not one but several objectives.

In this chapter we consider some basic aspects of a particular pattern of treatments, namely factorial structure. Factorial structure is the single most important concept of treatment design, and much of the third part of the book is devoted to this concept.

3.2 Factorial structure

Suppose we are investigating the effects on the growth of young chicks of adding small levels of copper to their diet, and wish to assess and compare these effects for two basic diets, maize (M) and wheat (W). To find out whether the effects of the copper additive are similar for both wheat and maize based diets it is necessary to use as experimental treatments the same set of different copper levels, both for wheat based diets and for maize based diets. We might decide to use three levels of copper (0, 50 and 100 units), both with a wheat based diet and with a maize based diet. If the set of six experimental treatments is reduced, then it is not possible to assess whether the effects of copper on chick growth are different for the two basic diets. Consider the difficulties using the set of four treatments:

(1) W + 0 units copper,
(2) W + 50 units copper,
(3) W + 100 units copper,
(4) M + 0 units copper.

We can assess the maize–wheat difference from a comparison of treatments (1) and (4); we can find something about the effect of copper from treatments (1), (2) and (3). But there is no information on whether the copper effects will be the same for a maize based diet, and to predict the chick growth for M + 100, for example, we have to make strong assumptions which might be very wrong.

Other benefits of factorial structure, which are perhaps more surprising and important even than these, are examined in Section 3.6. Before looking at those advantages, we examine the standard form of analysis for factorial structure.

First we introduce some terminology: a *treatment factor* is a set of alternatives which can be applied to experimental units. A *level* of a factor is a particular one from the set of alternatives which constitute the factor. An *experimental treatment* is the description of the way in which a particular unit is treated, and comprises one level from each factor.

In the chick growth example, there were two factors (copper, basic diet): there were three levels of the copper factor (0, 50, 100) and two levels of the basic diet factor (maize, wheat). The set of experimental treatments numbers $3 \times 2 = 6$, being M + 0, M + 50, M + 100, W + 0, W + 50 and W + 100.

Other examples of factorial structure can be found in all areas of experimentation. In medical research, different drugs for treatment of a disorder could be one factor, different concentrations of the drug could be a second factor and different frequencies of administration a third factor. In experiments on metal fractures the thickness of material could be one factor, the form of alloy a second factor and the force of the stress a third factor. In microbiological investigations, different strains of a virus could be one factor, and different media in which the virus is immersed a second factor. In the chemical industry, the number of constituents for a process is large, and variation of each constituent would define a separate factor; for many processes it would be easy to think of a factorial structure with eight or more factors, and this produces different problems of design discussed in Chapter 14. In crop experiments the spacing of the crop, the crop variety, the nutrients added to the soil, and the timing of sowing offer a wide range of possible factors.

3.3 Models for main effects and interactions

Given a set of treatments with factorial structure, such as those for the chick growth experiment, there are several different approaches to interpreting the results. Consider the following weight gains, set out in a form which emphasises the structure:

	copper		
	0	50	100
wheat	20.6	24.1	25.6
maize	21.8	26.7	25.9

This presentation emphasises the overall levels of the response, whereas answering the experimenters' questions requires us to think about differences in response between treatments.

(i) We could consider the response to copper for each basal diet in turn. For a wheat diet, the effect of increasing copper appears to be to produce an increased weight for copper, increasing both from 0 to 50 and from 50 to 100, but with the second increase smaller than the first. In contrast, for a maize diet, increasing copper produces first an increase and then a decrease in weight. To complete the summary, we might note that weights are, on average, 1.37 greater for maize than for wheat.

(ii) Alternatively, and in this example rather less convincingly, we could consider the maize–wheat yield difference for each copper level; these are 1.2 at zero copper, 2.6 at 50 and 0.3 at 100. In addition, we should consider the average response to copper:

0	50	100
21.2	25.4	25.75

showing an initial increase disappearing almost totally between 50 and 100.

(iii) The third approach, which is that conventionally adopted for factorial structure, is to consider first the average maize–wheat difference, 1.37, then the average response to copper, 21.2 to 25.4 to 25.75, and then the way in which the overall pattern differs from a combination of these two effects. Qualitatively, this last component of interpretation could be expressed by saying either that the maize–wheat difference was largest for the middle copper level, or that the response to increasing copper declines more rapidly for maize than for wheat.

The philosophy of this third approach can be expressed in an algebraic model

$$t_{jk} = d_j + c_k + (dc)_{jk}, \tag{3.1}$$

where t_{jk} is the treatment effect for basal diet j, and copper level k; d_j is the average treatment effect for basal diet j; c_k is the average treatment effect for copper level k; and $(dc)_{jk}$ is the difference between t_{jk} and $(d_j + c_k)$.

The model (3.1) is superficially similar to (2.2) for the randomised block design in the previous chapter:

$$y_{ij} = \mu + b_i + t_j + \varepsilon_{ij},$$

but the interpretation of the terms is different, and the two models should not be thought of as being two particular cases of the same model. In particular, whereas ε_{ij} is a characteristic of the experimental units representing variation between the units within each block, and is quite independent of the treatment effects, t_j, the $(dc)_{jk}$ term represents deviations of the treatment effects relative to both d_j and c_k and is only meaningful in the context of the d_j and c_k effects.

Effects which involve comparisons between levels of only one factor are called *main effects* of that factor, and effects which involve comparisons for more than a single factor are called *interactions*. To make these concepts more precise, we define effects as follows.

Main effect of a factor: a comparison between the expected responses for different levels of one factor, averaging over all levels of all other factors. Algebraically, this is written

$$\sum_j l_j t_{j.},$$

where $\sum_j l_j = 0$ and $t_{j.}$ represents the average of t_{jk} over all possible levels of k.

Interaction between two factors: a comparison, of an effect comparison between levels of one factor, between levels of the second factor. Algebraically, and more simply than in words, this is written

$$\sum_k m_k \left(\sum_j l_j t_{jk} \right),$$

where $\sum_k m_k = 0$. Although defined non-symmetrically, the interaction effect is symmetric, as is apparent if the definition is expanded:

$$\sum_k m_k \left(\sum_j l_j t_{jk} \right) = \sum_j \sum_k l_j m_k t_{jk}.$$

The effects d_j, c_k and $(dc)_{jk}$ in model (3.1) can now be recognised as main effects and interactions if we define them as follows:

$$d_j = t_{j.} - t_{..},$$
$$c_k = t_{.k} - t_{..},$$
$$(dc)_{jk} = (t_{jk} - t_{j.}) - (t_{.k} - t_{..}) = (t_{jk} - t_{.k}) - (t_{j.} - t_{..}) = t_{jk} - t_{j.} - t_{.k} + t_{..}.$$

For the numerical values used earlier, the treatment effects t_{jk} are estimated by the deviations of treatment mean yields from the overall average 24.12 (so that $t_{..} = 0$):

$$\begin{array}{ccc} -3.52 & -0.02 & +1.48 \\ -2.32 & +2.58 & +1.78 \end{array}$$

and the effects are:

$$d_1 = (-3.52 - 0.02 + 1.48)/3 = -0.68,$$
$$d_2 = (-2.32 + 2.58 + 1.78)/3 = +0.68,$$
$$c_1 = (-3.52 - 2.32)/2 = -2.92,$$
$$c_2 = (-0.02 + 2.58)/2 = +1.28,$$
$$c_3 = (+1.48 + 1.78)/2 = +1.63,$$

$$(dc)_{11} = (-3.52) - (-0.68) - (-2.92) = +0.08 \quad (dc)_{21} = -0.08,$$
$$(dc)_{12} = (-0.02) - (-0.68) - (1.28) = -0.62 \quad (dc)_{22} = +0.62,$$
$$(dc)_{13} = (1.48) - (-0.68) - (1.63) = +0.53 \quad (dc)_{23} = -0.53.$$

These numerical values confirm the general pattern of conclusions discussed earlier. The average maize–wheat difference is 1.37. The major effect for the copper levels is the lower weight for zero copper; the difference between 50 and 100 is small. And the interaction pattern occurs mainly between the second and third copper levels, with wheat producing a positive additional difference: $(dc)_{13} - (dc)_{12} = +1.21$, to add to the main effect difference $c_3 - c_2 = +0.35$, giving $+1.56$, whereas maize produces a negative difference: $+0.35 + (-1.21) = -0.86$.

3.4 The analysis of variance identity

We return now to the formal representation of the models on which the analysis of data from designed experiments may be based. The model for the randomised block design,

$$y_{ij} = \mu + b_i + t_j + \varepsilon_{ij},$$

and the associated analysis of variance were discussed in Chapter 2. Now suppose that the treatments have a factorial structure so that the treatment effects are classified by two factors, and can be expressed in terms of two main effects and an interaction:

$$t_{jk} = p_j + q_k + (pq)_{jk}.$$

Here, p_j and q_k represent the main effects of factors P and Q and $(pq)_{jk}$ represents the interaction effects. As in the earlier definitions of b_i and t_j, the p_j and q_k are regarded as deviations from the average yield of all treatment combinations for factors P and Q, respectively. In addition, each $(pq)_{jk}$ is regarded as the deviation of the yield of the treatment combination for levels j and k of factors P and Q from the sum $(p_j + q_k)$. Consequently, there are implied relations within each set of effects:

$$\sum_{j=1}^{p} p_j = 0, \ \sum_{k=1}^{q} q_k = 0, \ \sum_{j=1}^{p} (pq)_{jk} = 0, \ \sum_{k=1}^{q} (pq)_{jk} = 0,$$

where p and q are the numbers of levels for factors P and Q respectively.

The definition of a model for the treatment effects for a factorial structure is reflected in a second part of the analysis of variance, dividing the treatment SS into constituent parts. Since two factors are used to classify the treatment effects, the experimental observations are classified by three suffixes giving an overall model:

$$y_{ijk} = \mu + b_i + p_j + q_k + (pq)_{jk} + \varepsilon_{ijk}.$$

The treatment SS becomes

$$\sum_{ijk} (y_{.jk} - y_{...})^2$$

and the analysis of variance identity for this SS is

$$\sum_{ijk} (y_{.jk} - y_{...})^2 \equiv \sum_{ijk} (y_{.j.} - y_{...})^2 + \sum_{ijk} (y_{..k} - y_{...})^2$$
$$+ \sum_{ijk} (y_{.jk} - y_{.j.} - y_{..k} + y_{...})^2.$$

Table 3.1. *Analysis of variance for a two-factor design*

Source	SS	df
Blocks	$\sum_i Y_{i..}^2/(pq) - Y_{...}^2/n$	$b - 1$
Main effect P	$\sum_j Y_{.j.}^2/(bq) - Y_{...}^2/n$	$p - 1$
Main effect Q	$\sum_k Y_{..k}^2/(bp) - Y_{...}^2/n$	$q - 1$
Interaction PQ	$\sum_{jk} Y_{.jk}^2/b - \sum_j Y_{.j.}^2/(bq) - \sum_k Y_{..k}^2/(bp) + Y_{...}^2/n$	$(p-1)(q-1)$
Error	By subtraction	$(b-1)(pq-1)$
Total	$\sum_{ijk} Y_{ijk}^2 - Y_{...}^2/n$	$n - 1$

The three terms on the right hand side may be identified as:

(i) the main effect SS for factor P

$$\sum_{ijk}(y_{.j.} - y_{...})^2 = \sum_j Y_{.j.}^2/(bq) - Y_{...}^2/n;$$

(ii) the main effect SS for factor Q

$$\sum_{ijk}(y_{..k} - y_{...})^2 = \sum_k Y_{..k}^2/(bp) - Y_{...}^2/n; \text{ and}$$

(iii) the interaction SS for factors P and Q

$$\sum_{ijk}(y_{.jk} - y_{.j.} - y_{..k} + y_{...})^2 = \sum_{jk} Y_{.jk}^2/b - Y_{...}^2/n - \text{SS(P)} - \text{SS(Q)}.$$

Combined with the earlier analysis of variance identity, this gives a full analysis of variance structure for a randomised block design with factorial treatment structure (see Table 3.1). The similarity of the structure of the analysis of variance identity for treatments to that for the randomised block design is clear. In each, we isolate two constituents, each representing variation over the different classes of one of the two classifications, and then consider the remaining variation as a third component. However, as with the similarity between models (2.2) and (3.1), this analogy should not be pursued too enthusiastically. In terms of designed experiments, block and treatment effects have quite different interpretations, and the error SS, although arithmetically equivalent to the block × treatment interaction, is more appropriately viewed as representing variation between units within a block.

Nonetheless, the similarity of the two analysis of variance identities does suggest a method of extending the analysis of variance to include the case of three factors in the treatment structure. The second factor Q may be replaced by two factors Q and R. The main effect SS for the earlier factor, Q, may be divided into constituent SS for the main effects of Q and R and for the QR interaction. Similarly, the original PQ interaction SS can be divided into

Table 3.2. *Data for experiment on amphibia*

Treatment	Results		Total
Toad wet control	+2.31	−1.59	+0.72
Toad wet hormone	+28.37	+14.16	+42.53
Toad dry control	+17.68	+25.23	+42.91
Toad dry hormone	+28.39	+27.94	+56.33
Frog wet control	+0.85	+2.90	+3.75
Frog wet hormone	+3.82	+2.86	+6.68
Frog dry control	+2.47	+17.72	+20.19
Frog dry hormone	+13.71	+7.38	+21.09

the interaction SS for PQ and PR and the remainder which can be interpreted as due to the interaction of all three factors. The resulting extended analysis of variance identity for the complete three-factor treatment structure becomes:

$$\sum_{ijkl}(y_{.jkl} - y_{....})^2 \equiv \sum_{ijkl}(y_{.j..} - y_{....})^2 + \sum_{ijkl}(y_{..k.} - y_{....})^2 + \sum_{ijkl}(y_{...l} - y_{....})^2$$

$$+ \sum_{ijkl}(y_{.jk.} - y_{.j..} - y_{..k.} + y_{....})^2 + \sum_{ijkl}(y_{.j.l} - y_{.j..} - y_{...l} + y_{....})^2$$

$$+ \sum_{ijkl}(y_{..kl} - y_{..k.} - y_{...l} + y_{....})^2$$

$$+ \sum_{ijkl}(y_{.jkl} - y_{.jk.} - y_{.j.l} - y_{..kl} + y_{.j..} + y_{..k.} + y_{...l} - y_{....})^2.$$

The further extension to four or more factors is conceptually simple though even more typographically painful!

Example 3.1 The analysis of variance calculation for a design with factorial treatment structure is illustrated for data from an experiment on the water uptake of amphibia. Frogs and toads were kept in moist or dry conditions prior to the experiment. Half of the animals were injected with a mammalian water balance hormone. There were thus three treatment factors: species (S), pre-experiment moisture condition (M) and hormone (H) – each with two levels. Two animals were observed for each of the eight treatment combinations, but there was no blocking of the 16 animals. The model for the analysis of variance is therefore an extension of that derived above but with no block effect term:

$$y_{jkli} = \mu + s_j + m_k + h_l + (sm)_{jk} + (sh)_{jl} + (mh)_{kl} + (smh)_{jkl} + \varepsilon_{jkli}.$$

The variable measured was the percentage increase in weight after immersion in water for two hours (see Table 3.2). The totals for the eight combinations of treatment levels can be further totalled in a set of two-way tables as in Table 3.3. The method of manual calculation

Table 3.3. *Totals for factor combinations*

	Wet	Dry	Total	Control	Hormone
Toad	43.25	99.24	142.49	43.63	98.86
Frog	10.43	41.28	51.71	23.94	27.77
Total	53.68	140.52	194.20	67.57	126.63
Control	4.47	63.10	67.57		
Hormone	49.21	77.42	126.63		

Table 3.4. *Analysis of variance for amphibia experiment*

Source	SS	df	MS
Species	515.06	1	
Moisture	471.33	1	
Hormone	218.01	1	
SM	39.50	1	
SH	165.12	1	
MH	57.73	1	
SMH	43.43	1	
Error	276.05	8	$s^2 = 34.51$
Total	1786.33	15	

of the SS is shown below:

$$\text{SS (species)} = (142.49^2 + 51.71^2)/8 - 194.20^2/16 = 515.06,$$
$$\text{SS (moisture)} = (53.68^2 + 140.52^2)/8 - 194.20^2/16 = 471.33,$$
$$\begin{aligned} \text{SS (SM)} = &(43.25^2 + 99.24^2 + 10.43^2 + 41.28^2)/4 \\ &- 194.20^2/16 - 515.06 - 471.33 = 39.50, \end{aligned}$$
$$\begin{aligned} \text{SS (SMH)} = &(0.72^2 + 42.91^2 + 42.53^2 + 56.33^2 + 3.75^2 + 20.19^2 \\ &+ 6.68^2 + 21.09^2)/2 - 194.20^2/16 - \text{SS(S)} - \text{SS(M)} \\ &- \text{SS(H)} - \text{SS(SM)} - \text{SS(SH)} - \text{SS(MH)} = 43.53. \end{aligned}$$

The analysis of variance is given in Table 3.4. Note that there is no SS for blocks, since the animals were not blocked. The error SS could therefore be calculated alternatively from the within treatment variation:

$$\begin{aligned} \text{error SS} = &(2.31^2 + 1.59^2 - 0.72^2/2) + (28.37^2 + 14.16^2 - 42.53^2/2) \\ &+ \cdots + (13.71^2 + 7.38^2 - 21.09^2/2) = 276.05. \end{aligned}$$

3.5 Interpretation of main effects and interactions

The analysis of variance includes a subdivision of the treatment variation into various components, and this division can be used to provide a structure for the interpretation of the results of the experiment. The relative sizes of the SS show the relative importance of the different sets of effects. As a further aid to interpretation, each MS = SS/df can be compared with the error MS, s^2, to assess which effects are important. A formal derivation of the procedure for assessing the significance of the component MS will be developed in Chapter 4. This procedure involves the calculation of the ratio of each MS to the error MS and the comparison of the ratio with the F distribution with the df of the two MS.

The possible patterns of large or small MS (and the corresponding F ratios) are many, and rules for the interpretation of sets of effects are vulnerable to counterexamples. However, the following principles will be useful in most situations:

(i) The definition of main effects and interactions implies a natural ordering of the effects for interpretation. Main effects should be examined first, then the two-factor interactions, then three-factor interactions, and so on. This is because interactions essentially represent modifications of the main effects, and have no sensible interpretation without consideration, at the same time, of the corresponding main effects.

(ii) If the two-factor and higher-order interactions appear negligible, then the results of the experiment should be interpreted in terms of the main effect mean yields only, ignoring the mean yields for the combinations of levels. If the three-factor and higher-order interactions appear negligible but the two-factor interactions are not negligible, then the mean yields for combinations of levels from pairs of factors should provide the basis for interpretation.

(iii) If a two-factor interaction is clearly important, then the interpretation of the effects of these two factors should normally be based on the mean yields for the combinations of levels for those two factors. If the main effect MS are of the same order of magnitude as the interaction MS, then the main effect yields will add little to an interpretation based on the mean yields for the two-factor combinations of levels. If, on the other hand, one or both of the main effect MS are large compared with the interaction MS, then the main effect comparison of mean yields will be meaningful, and the interpretation should be in terms of the main effect, or main effects, modified by the interaction effects. In other words mean yields should be presented for the two-way combinations of levels and also for the levels of the factor whose main effect is larger than the interaction.

For the example of the toads and frogs, the MS for species and moisture are large (515 and 471), that for the hormone treatment is not so large (218), but still substantially bigger than the error MS (34.5) and the species × hormone interaction (165) is certainly big enough for that two-factor interaction to be investigated. The other three interaction MS are small (39, 58 and 44) and those effects can be ignored. In interpreting the results, the first thing to notice is that the moisture effect appears not to be influenced by other factors, so that for the moisture factor only the main effect mean yields need to be presented. The species main effect MS is quite a bit larger than the species × hormone interaction MS and therefore we should present the species main effect yields as well as the species–hormone combination mean yields. However, the hormone main effect mean yields are unlikely to be helpful in arriving at an interpretation because the MS for the hormone main effect and species × hormone interaction are similar in size. The relevant mean yields are therefore:

moisture

	wet	dry	difference (D − W)
	6.72	17.56	10.84

$$\text{SE(difference (D − W))} = \{2(34.5)/8\}^{1/2} = 2.94.$$

species + hormone

	toads	frogs	
control	10.91	5.98	
hormone	24.72	6.94	difference (T − F)
mean	17.82	6.46	11.36 SE = 2.94
			difference of differences (H − C)
differences (H − C)	13.81	0.96	12.85

$$\text{SE(differences (H − C))} = \{2(34.5)/4\}^{1/2} = 4.15$$
$$\text{SE(difference of differences)} = \{4(34.5)/4\}^{1/2} = 5.87.$$

The interpretation is clear. Toads show a greater increase in weight after immersion in water than frogs. The effect of the hormone is to raise the weight increase for toads from 11% to 25%, but for frogs the hormone appears to have little effect. Finally, keeping the animals in dry conditions before the experiment leads to an additional 10% increase in weight, as compared with animals kept in moist conditions before the experiment.

This example, with two levels for each factor, is not difficult to interpret. The use of more levels for a factor allows a wider range of possible patterns of results, and hence of interpretations. Many such examples occur later in the book. However, the general principles enunciated in this section will usually provide a sound basis for interpretation.

3.6 Advantages of factorial structure

In introducing the ideas of factorial structure, we have emphasised the advantage of being able to examine interactions between factors. There are two other advantages. First, the conclusions about the effects of a particular factor have a broader validity because of the range of conditions under which that factor has been investigated. Thus, in the example of frogs and toads, the effect of the different pre-experimental conditions has been assessed for both frogs and toads, and with and without a hormone treatment. The estimate of the extra 10% weight increase might be expected to apply in other conditions as well, because of its consistency over levels of two other factors within this experiment.

The other advantage of factorial experiments is still more important, and is essentially that a factorial experiment allows several experiments to be done simultaneously. This advantage is seen most clearly in the situation where there are assumed to be no interactions between factors. Consider an experimenter who wants to investigate the effects of three factors, each at two levels, and who has the resources sufficient for 24 observations. Let the factors be P, Q and R with levels p_0 and p_1, q_0 and q_1, r_0 and r_1. Three different designs of experiment are considered:

(a) Three separate experiments, one for each factor, eight observations per experiment:
   (i) $(p_0q_0r_0, p_1q_0r_0)$ with four observations each,
   (ii) $(p_0q_0r_0, p_0q_1r_0)$ with four observations each,
   (iii) $(p_0q_0r_0, p_0q_0r_1)$ with four observations each.

This is the classical scientific experiment, isolating the effect of each factor in turn by controlling all other factors.

(b) Instead of 'wasting' resources by using $(p_0q_0r_0)$ in each subexperiment, the four distinct treatments may be replicated equally:

$(p_0q_0r_0)$, $(p_1q_0r_0)$, $(p_0q_1r_0)$, $(p_0q_0r_1)$ with six observations each.

(c) The factorial experiment with eight treatments:

$(p_0q_0r_0)$, $(p_0q_0r_1)$, $(p_0q_1r_0)$, $(p_0q_1r_1)$, $(p_1q_0r_0)$, $(p_1q_0r_1)$, $(p_1q_1r_0)$, $(p_1q_1r_1)$, each having three observations.

To compare the three designs, consider the variance of the comparison of mean yields for p_0 and p_1; obviously the comparison of q_0 and q_1, or of r_0 and r_1 will be equivalent. The three experiments give variances for $(p_1 - p_0)$ as follows:

(a) $2\sigma^2/4 = 0.5\sigma^2$,

(b) $2\sigma^2/6 = 0.33\sigma^2$,

(c) $2\sigma^2/12 = 0.17\sigma^2$.

The third experiment gives the smallest variance, and is therefore the most efficient design, because in the absence of interactions the difference

$$(p_1q_kr_l) - (p_0q_kr_l)$$

has the same expectation for all (k, l) pairs. Thus, the effective replication of the $(p_1 - p_0)$ comparisons is 12. Looked at from another angle, in comparing p_1 with p_0 in experiment (c), all 24 observations are used, whereas in (a) and (b) only 8 and 12 observations, respectively, are used. The factorial experiment is said to have 'hidden replication' for the $p_1 - p_0$ comparison.

If there are interactions between factors, then some of this hidden replication may disappear, and this advantage of factorial experiments is diminished. However, if there are interactions, then the other two forms of design, (a) and (b), are still inferior to the factorial, because they do not permit recognition that the size of the $(p_1 - p_0)$ effect depends on the particular combination of factor Q and factor R levels, and hence results from (a) and (b) may not be reproducible if the levels of Q and R are changed.

Only some of the hidden replication is lost when there are interactions between factors, and it is important to realise that hidden replication is still a benefit when some, but not all, interactions exist. Consider again the experiment for the frogs and toads. Because the moisture factor does not interact with the other factors, the effective replication for the wet–dry comparison is eight – two explicit replications times four hidden replications. The species and hormone interaction requires consideration of the mean yields for the four combinations. Each of these mean yields is based on four replications – two explicit replications times two hidden replications. So, the factorial structure has enabled us to discover the interaction effect of species and hormone, but also gives the benefit of some hidden replication for all the comparisons of means which are of importance.

The abandonment of one-factor-at-a-time experiments might seem to contradict everything some physical scientists have been taught about good experimental practice. However, this is not really so. The scientific principle often stated as 'hold all factors constant except the one whose effects we are investigating' is correctly stated as 'hold all factors constant except

Table 3.5. *Analysis of variance for simplified model*

Source	SS	df	MS
Species	515.06	1	
Moisture	471.33	1	
Hormone	218.01	1	
SH	165.12	1	
Error	416.81	11	$s^2 = 37.89$
Total	1786.33	15	

the *ones* whose effects we are investigating'. There is no contradiction between this and the recognition that we should investigate several input variables in one experiment, rather than in separate experiments.

3.7 Treatment effects and treatment models

The discussion above has been in terms of treatment *effects*, which are determined by the treatment structure and lead inevitably to a model including main effects and interactions. Sometimes, there will be a more specific model for the treatment means which can be assumed before conducting the experiment. For example, it might be that a model containing only main effects is expected to be adequate in advance of running the experiment, or, when the factors are levels of a continuous variable, some linear or non-linear functional form might be assumed. Such models are discussed in detail in Chapters 16 and 17. Here, we reflect a little more on the differences between treatment effects and treatment models.

The strength of belief in a model assumed before carrying out the experiment can vary. It might be simply a tentative assumption that a model will be adequate, made for the purposes of reducing the experiment to a size that can be afforded. Alternatively, there might be considerable previous experience with the treatment factors being studied, for example, on a pilot scale, which leads one to assume particular effects will be important or not. Finally, there might be some accepted scientific theory which demands that the model should take a particular form. Whatever the source of the model, we can never be sure that it will apply in a specific experiment and will usually want to at least check whether it holds. Even incontestable theoretical models only apply under the appropriate conditions and it is never guaranteed that these conditions will hold in our experiment.

A careless, purely model based, approach to analysing experimental data can fail to make the best use of the data. Consider again Example 3.1 on the water uptake of amphibia. If prior knowledge told us that the appropriate model for the treatments would have only one interaction, that between species and hormone, it might seem natural to fit this model. This would give the analysis of variance in Table 3.5. The SS for the treatment effects remaining are the same as in Table 3.4, as are the estimated treatment differences and the estimate of σ^2 is very similar. The conclusions do not change. Nevertheless, we have lost some information by this presentation, in particular the reassurance that our assumed model does actually fit the data well.

Table 3.6. *Full analysis for simplified model*

Source	SS	df	MS
Species	515.06	1	
Moisture	471.33	1	
Hormone	218.01	1	
SH	165.12	1	
Lack of fit	140.71	3	46.90
Error	276.05	8	$s^2 = 34.51$
Total	1786.33	15	

In situations like this, with sufficient replication, we would always recommend that the analysis of variance be presented as in Table 3.4, or as in Table 3.6, where interactions assumed negligible have been bundled together as lack of fit. Nothing is lost by this presentation and, especially with more complex treatment structures, it ensures that if different people analyse the data they will reach conclusions which are the same, or at least not conflicting. With unreplicated experiments, we might have to compromise on this and interpretation of these experiments will be discussed further in Chapter 13.

4

General principles of linear models for the analysis of experimental data

4.1 Introduction and some examples

In this chapter we consider the principles on which the classical analysis of experimental data is based. The explanation of these principles necessarily involves mathematical terminology, but in the main part of the chapter this will be reduced as far as possible so that the results will hopefully be appreciated by readers without a sophisticated mathematical background. The formal mathematical derivation of results is included in the appendix to this chapter, the numbering of sections in the appendix matching the section numbering in the main body of the chapter. The discussion of general principles will be illustrated by reference to five examples of data. Two of the examples have already been seen in the two previous chapters.

Example A is the randomised block design with ten spacing treatments for rice in four blocks of ten plots each, for which the yield data are given in Section 2.2. The interest in this example is in the relative magnitude of the treatment effects, the block effects and the residual variation, and also in the relationship between yield and spacing.

Example B is the experiment on the water uptake of amphibia, described in Section 3.4, in which eight treatment combinations were replicated twice each. The interest in this example is in the separation of the different components of treatment variation, using the factorial structure of the eight treatments.

Example C is an experiment on tomatoes in which five spray treatments using a chemical growth regulator were compared. Thirty plots were used, arranged in six rows and five columns. The results were as shown in Table 4.1. The analysis must allow for systematic differences between rows and between columns, while being principally concerned with the differences between the five treatments for which the questions are concerned with the general effect of spraying with the growth regulator, and the particular factorial comparisons of the first four treatments.

The final two examples are related to the analysis of covariance, and are considered in some detail in Section 4.8.

Example D is a strawberry variety trial in which eight varieties were compared in four blocks of eight plots. The blocks were chosen so that each block included eight plots in a row as shown in Figure 4.1. Examination of the yields, displayed in Figure 4.1, revealed a clear trend along the blocks, and on enquiry it was discovered that there had been a hedge at the right hand end of the trial. Clearly, the blocks have not been chosen appropriately and, in analysing the yields from the experiment, it is important to allow for the decline in yield towards the hedge.

Example E is a trial to compare three drugs whose purpose is to relieve pain, in which each patient was given a drug whenever the patient decided that his level of pain required

Table 4.1. *Data for tomato spray experiment*

A 3.72	B 3.39	C 2.95	D 2.92	E 1.68
C 3.50	D 2.73	E 2.99	A 3.08	B 1.72
D 2.30	A 4.36	B 4.18	E 1.27	C 0.81
E 3.98	C 5.45	D 5.48	B 5.26	A 4.85
B 6.40	E 3.25	A 8.44	C 4.90	D 2.53
B 7.69	C 5.97	E 6.91	D 5.41	A 6.92

The treatment definitions were as follows:
A: early spray 75 p.p.m.
B: early spray 150 p.p.m.
C: late spray 75 p.p.m.
D: late spray 150 p.p.m.
E: control.

Block									
I	G	V	Rl	F	Re	M	E	P	H
	5.8	6.3	4.9	6.5	4.5	5.2	6.5	3.8	
II	E	P	M	Re	G	V	F	Rl	E
	6.9	7.6	7.9	5.6	7.0	5.5	4.0	2.7	
III	V	F	Rl	G	P	E	Re	M	D
	7.6	6.4	5.0	6.9	7.4	5.3	5.2	3.2	
IV	E	Re	M	P	G	F	V	Rl	G
	7.5	7.0	6.1	7.2	6.5	5.6	5.8	1.4	
									E

Figure 4.1 Experimental plan for eight strawberry varieties in four blocks of eight plots, with hedge.

relief. The number of hours relief provided by each drug was recorded (time from the administration of the drug to the next request for a drug). For each patient the sequence of drugs to be provided was predefined (but unknown to the patient). There were six groups of patients with each group receiving a different combination of two drugs in a specified order. The number of patients in each group varied because of patients dropping out of the trial. For each patient the number of hours relief from pain for each drug provided was recorded, and the data are shown in Table 4.2.

The data are to be analysed to estimate differences between treatments, allowing for differences between patients and the effect of the order in which the drugs were administered. However, the timing of the administration of the second drug for each patient was determined by the length of the period of relief afforded by the first drug for the patient. It would be reasonable, in assessing the effect of the second drug, to try to allow for the variation in the length of the interval from the beginning of the trial before the second drug was administered.

4.2 The principle of least squares and least squares estimators

A general model for the linear dependence of yields, from a set of experimental units, on a set of parameters which may be expected to cause some of the variation observed in the

Table 4.2. *Hours of relief from pain for each drug for each patient*

1st drug T_1	2	6	4	13	5	8	4	
2nd drug T_2	10	8	4	0	5	12	4	
1st drug T_2	2	0	3	3	0			
2nd drug T_1	8	8	14	11	6			
1st drug T_1	6	7	6	8	12	4	4	
2nd drug T_3	6	3	0	11	13	13	14	
1st drug T_3	6	4	4	0	1	8	2	8
2nd drug T_1	14	4	13	9	6	12	6	12
1st drug T_3	12	1	5	2	1	4	6	5
2nd drug T_2	11	7	12	3	7	5	6	3
1st drug T_2	0	8	1	4	2	2	1	3
2nd drug T_3	8	7	10	3	12	0	12	5

yields may be expressed as follows:

$$y_i = \sum_{j=1}^{p} a_{ij}\theta_j + \varepsilon_i. \tag{4.1}$$

In this formulation y_i are the yields, θ_j the parameters, a_{ij} represent the structure of dependence of yields on parameters, and ε_i represent the random variation resulting from inherent variation between individual units. The models we have already used for examples A and B can be recognised as fitting within this general structure. For example A the model was

$$y_{ij} = \mu + b_i + t_j + \varepsilon_{ij}. \tag{4.2}$$

An alternative model representing only block averages and a dependence of yield on area per plant would be

$$y_{ij} = b_i + \beta a_j + \varepsilon_{ij}. \tag{4.3}$$

For example B a model for the observations on the eight treatments $(t_1, \ldots, t_8)$ could be written

$$y_{jk} = t_j + \varepsilon_{jk}. \tag{4.4}$$

In the general model (4.1) the parameters θ_j may represent block or treatment effects (b_i, t_j) or a mean effect (μ); in each case the a_{ij} term takes values 1 or 0 according to whether the particular effect (j) is relevant or not to yield i, respectively. For quantitative relationships, such as that between yield and area per plant, the θ_j parameter becomes the regression coefficient, β, in model (4.3) and the a_{ij} become the values of the quantitative variable a_j in model (4.3).

To obtain estimates of the parameters some assumptions are necessary about the properties of the variation among experimental units, represented by the ε_{ij}. We must also define the principle on which the estimation of effects is to be based. For the ε_{ij} the simple assumption is that the observations from different units are independent (apart from the patterns represented by the θ_j) and that the variability of the observations is homogeneous (unaffected by θ_j). Formally these assumptions may be written

$$E(\varepsilon_{ij}) = 0,$$
$$\text{Var}(\varepsilon_{ij}) = \sigma^2,$$
$$\text{Cov}(\varepsilon_{ij}, \varepsilon_{i'j'}) = 0 \text{ unless } (i, j) = (i', j').$$

The principle on which the estimation of parameters is based is the *least squares principle*. This requires that the estimates of parameters ($\hat{\boldsymbol{\theta}}$) are chosen so that the deviations of the observed yields from the fitted values ($\hat{\mathbf{y}}$), where

$$\hat{y}_i = \sum_{j=1}^{p} a_{ij}\hat{\theta}_j,$$

are minimised, in the sense of minimising the SS of deviations

$$S = \sum_{i=1}^{n} \left(y_i - \sum_{j=1}^{p} a_{ij}\theta_j \right)^2. \tag{4.5}$$

The minimisation of S with respect to any particular parameter, θ_k, involves setting

$$\partial S/\partial \theta_k = 0,$$

which leads to an equation

$$\sum_{i=1}^{n} \left\{ \left(y_i - \sum_{j=1}^{p} a_{ij}\theta_j \right) a_{ik} \right\} = 0,$$

which may be written

$$\sum_{j=1}^{p} \hat{\theta}_j \left(\sum_{i=1}^{n} a_{ik}a_{ij} \right) = \sum_{i=1}^{n} a_{ik}y_i, \tag{4.6}$$

where the 'hat' symbol represents an estimate ($\hat{\boldsymbol{\theta}}$) of the parameter $\boldsymbol{\theta}$.

For a set of p parameters a set of p equations of the form (4.6) is obtained and the parameter estimates, $\hat{\theta}_j$, satisfying these equations are termed *least squares estimates*.

4.2.1 *Some examples*

We illustrate these equations for models (4.3) and (4.4). First for (4.3) in which, with parameters representing the four average block values (b_i) and the regression coefficient (β), there are five parameters and 40 observations. The individual model equations are typically

$$y_{11} = b_1 + \beta(900) + \varepsilon_{11},$$
$$y_{12} = b_1 + \beta(720) + \varepsilon_{12},$$
$$\vdots \qquad \qquad \vdots$$
$$y_{4,10} = b_4 + \beta(225) + \varepsilon_{4,10},$$

and the least squares equations are

$$10\hat{b}_1 + 5011\hat{\beta} = 79.15,$$
$$10\hat{b}_2 + 5011\hat{\beta} = 68.55,$$
$$10\hat{b}_3 + 5011\hat{\beta} = 73.95,$$
$$10\hat{b}_4 + 5011\hat{\beta} = 75.75,$$
$$5011(\hat{b}_1 + \hat{b}_2 + \hat{b}_3 + \hat{b}_4) + 2883301\hat{\beta} = 144847.9,$$

which may be solved to give

$$\hat{\beta} = -0.0034, \hat{b}_1 = 9.62, \hat{b}_2 = 8.56, \hat{b}_3 = 9.10, \hat{b}_4 = 9.29.$$

For model (4.4) there are eight parameters and 16 observations. The individual equations are

$$y_{11} = t_1 + \varepsilon_{11},$$
$$y_{12} = t_1 + \varepsilon_{12},$$
$$\vdots \qquad \vdots$$
$$y_{82} = t_8 + \varepsilon_{82},$$

and the least squares equations are

$$2\hat{t}_1 = 0.72, \ 2\hat{t}_2 = 42.91, \ 2\hat{t}_3 = 42.53, \ 2\hat{t}_4 = 56.33,$$
$$2\hat{t}_5 = 3.75, \ 2\hat{t}_6 = 20.19, \ 2\hat{t}_7 = 6.68, \ 2\hat{t}_8 = 21.09,$$

from which the solutions are obtained immediately.

For most of the models considered in this book the least squares equations have a particularly simple form. Whenever the coefficients a_{ij} for a parameter θ_j are all 0 or 1, as in the case for b_i in (4.3) or t_j in (4.4) the corresponding least squares equation has on the right hand side the total of observations for which θ_j is part of the yield expression. The left hand side is the expected value of the total on the right hand side, written in terms of parameter estimates $\hat{\theta}$, instead of parameters θ.

4.3 Properties of least squares estimators

In this section we consider the properties of least squares estimators, and to do so we assume that the set of least squares equations (4.6) can be solved uniquely. In fact, for most of the experimental design models considered in this book, the equations do not have unique solutions and we consider how to obtain solutions in these cases in the next section. The properties of the modified least squares estimators of the next section are simply related to those of the simple least squares estimators for the case where (4.6) does yield unique solutions, and it is convenient to derive these properties for the simple case.

The most important properties for any estimator are the expected value of the estimator and the variance of the estimator. Since the variance of the estimator will inevitably be related to the variance of the original observations, σ^2, it is necessary to be able to obtain an estimate of σ^2 in order to make practical use of the information about the variance of estimators.

To demonstrate the argument required to derive the properties of least squares estimators it is helpful to reexpress the least squares equations in terms of matrices. The set of design coefficients (a_{ij}) is defined for n observations, referenced by suffix i, and p parameters, referenced by suffix j. The set can be considered in the form of a two-way array, or design matrix:

$$\mathbf{A} = \begin{bmatrix} a_{11} & a_{12} & \cdots & a_{1p} \\ a_{21} & a_{22} & \cdots & a_{2p} \\ \vdots & \vdots & & \vdots \\ a_{n1} & a_{n2} & \cdots & a_{np} \end{bmatrix}.$$

It is convenient to define

$$c_{jk} = \sum_{i=1}^{n} a_{ij} a_{ik}.$$

Essentially c_{jk} is the sum of products of pairs of values in columns j and k for each row of the matrix $\mathbf{A}$. The set $\{c_{jk}\}$ may also be written as a matrix:

$$\mathbf{C} = \begin{bmatrix} c_{11} & c_{12} & \cdots & c_{1p} \\ c_{21} & c_{22} & \cdots & c_{2p} \\ \vdots & \vdots & \cdots & \vdots \\ c_{p1} & c_{p2} & \cdots & c_{pp} \end{bmatrix}.$$

Notice that the definition of c_{jk} means that $c_{kj} = c_{jk}$, and the matrix $\mathbf{C}$ is said to be symmetric. Equation (4.6) may be rewritten in terms of the c_{jk} in the form

$$\sum_{j=1}^{p} c_{jk} \hat{\theta}_j = \sum_{i=1}^{n} a_{ik} y_i.$$

There is an equation of this form for each of the p values of k, and the set of such equations,

$$\sum_{j=1}^{p} c_{j1} \hat{\theta}_j = \sum_{i=1}^{n} a_{i1} y_i$$
$$\vdots \tag{4.7}$$
$$\sum_{j=1}^{p} c_{jp} \hat{\theta}_j = \sum_{i=1}^{n} a_{ip} y_i,$$

may be written in matrix form

$$\mathbf{C} \hat{\boldsymbol{\theta}} = \mathbf{A}' \mathbf{y}. \tag{4.8}$$

If the matrices $\mathbf{C}$, $\boldsymbol{\theta}$, $\mathbf{A}$, and $\mathbf{y}$ are written as arrays

$$\begin{bmatrix} c_{11} & c_{12} & \cdots & c_{1p} \\ c_{21} & c_{22} & \cdots & c_{2p} \\ \vdots & \vdots & \cdots & \vdots \\ c_{p1} & c_{p2} & \cdots & c_{pp} \end{bmatrix}, \begin{bmatrix} \hat{\theta}_1 \\ \hat{\theta}_2 \\ \vdots \\ \hat{\theta}_p \end{bmatrix}, \begin{bmatrix} a_{11} & a_{12} & \cdots & a_{1p} \\ a_{21} & a_{22} & \cdots & a_{2p} \\ \vdots & \vdots & & \vdots \\ a_{n1} & a_{n2} & \cdots & a_{np} \end{bmatrix}, \text{ and } \begin{bmatrix} y_1 \\ y_2 \\ \vdots \\ y_n \end{bmatrix},$$

then each sum of terms in (4.7) is a sum, over rows, of products of terms in $\hat{\boldsymbol{\theta}}$ and a column of $\mathbf{C}$, or in $\mathbf{y}$ and a column of $\mathbf{A}$.

Now associated with the matrix $\mathbf{C}$, for the situation when (4.8) has a unique solution there is a matrix $\mathbf{C}^{-1}$, the inverse matrix of $\mathbf{C}$, such that the solutions to (4.8) are given by

$$\hat{\boldsymbol{\theta}} = \mathbf{C}^{-1} (\mathbf{A}' \mathbf{y}). \tag{4.9}$$

where the multiplication of the matrix $\mathbf{C}^{-1}$ by $\mathbf{A}'\mathbf{y}$ is again a set of products obtained by multiplying a column of $\mathbf{C}^{-1}$ by the column of values represented by $\mathbf{A}'\mathbf{y}$.

The crucial characteristic about least squares estimators, which may be seen directly from (4.9) or may be inferred from (4.6), is that least squares estimators are linear combinations of the original observations, y_i. From this general characteristic and the particular form of (4.9) the following results may be deduced (proofs are in the appendix to this chapter).

(1) The expected value of a least squares estimator of a parameter in a general linear model is the true value of the parameter: i.e. least squares estimators are unbiased.
(2) The variance of a least squares estimator of a parameter is a simple multiple of σ^2, the multiplying factor being the corresponding diagonal element of the matrix $\mathbf{C}^{-1}$.
(3) The covariance of least squares estimators for a pair of parameters is also a simple multiple of σ^2, the multiplying factor being a non-diagonal element of $\mathbf{C}^{-1}$.
(4) For any linear comparison or combination of parameters, the corresponding contrast of the least squares estimators of the parameters is unbiased, and this estimator gives the most precise estimate that can be obtained using an unbiased linear combination of observations.

The implications of these results are that if we are considering linear models, i.e. models in which the parameters appear additively, and intend to use linear combinations of the observations to estimate parameters or combinations of parameters, then least squares estimates are the best available.

4.3.1 The estimation of σ^2

The other important question is how to estimate σ^2. The least squares principle requires that the parameter estimates minimise the sum of squared deviations (4.5). If we consider the resulting minimised SS

$$S_\mathrm{r} = \sum_{i=1}^{n} \left(y_i - \sum_{j=1}^{p} a_{ij}\hat{\theta}_j \right)^2 ,$$

then it can be shown that the expected value of S_r is $(n-p)\sigma^2$ (see the appendix).

The calculation of S_r is straightforward following an expansion of the squared terms of the summation as follows:

$$S_\mathrm{r} = \sum_{i=1}^{n} \left\{ y_i^2 - \sum_{k=1}^{p} \hat{\theta}_k a_{ik} \left(2y_i - \sum_{j=1}^{p} a_{ij}\hat{\theta}_j \right) \right\}$$

$$= \sum_{i=1}^{n} y_i^2 - \sum_{k=1}^{p} \hat{\theta}_k \left(2 \sum_{i=1}^{n} a_{ik}y_i - \sum_{j=1}^{p} \hat{\theta}_j \sum_{i=1}^{n} a_{ij}a_{ik} \right)$$

$$= \sum_{i=1}^{n} y_i^2 - \sum_{k=1}^{p} \hat{\theta}_k \left(\sum_{i=1}^{n} a_{ik}y_i \right) .$$

The first term is the SS of the original observations. The second is the sum of products of each $\hat{\theta}_k$ estimate and the corresponding right hand side total in the least squares equations (4.6).

In practice therefore we set up and solve the least squares equations and calculate the residual SS S_r by subtracting from the total SS the fitting SS, which is calculated as the sum of products of each least squares estimate times the corresponding total.

For the model (4.3) for example A, the total SS is

$$\text{total SS} = 5.95^2 + \cdots + 7.75^2 = 2252.61,$$

and the fitting SS is

$$\text{fitting SS} = (9.62)(79.15) + (8.56)(68.55) + (9.10)(73.95) + (9.29)(75.85)$$
$$+ (-0.0034)(144847.9) = 2233.85.$$

Hence the residual SS is

$$S_r = 2252.61 - 2233.85 = 18.76,$$

and the estimate of σ^2 is

$$18.76/(40 - 5) = 0.536.$$

For the model (4.4) for example B the SS are

$$\text{total SS} = 2.31^2 + \cdots + 7.38^2 = 17821.49,$$
$$\text{fitting SS} = (0.36)(0.72) + (21.46)(42.91) + \cdots + (10.54)(21.09)$$
$$= 12545.44.$$

Hence the residual SS is

$$S_r = 17821.49 - 17545.44 = 276.05,$$

and the estimate of σ^2 is 34.5, as in Section 3.4.

4.4 Overparameterisation, constraints and practical solution of least squares equations

The natural form in which most models for data from designed experiments are expressed includes more parameters than can be estimated from the information available. The cause of the overparameterisation is almost always the desire to retain symmetry for sets of effects. The most obvious example is the randomised block design model such as (4.2) where we include a complete set of block effects (b_i) and a complete set of treatment effects (t_j) in addition to the overall mean effect (μ). When this model was first developed in Chapter 2 the b_i represented deviations of each block from the experiment average, and the t_j represented the deviation for a specific treatment from the average of all treatments. It was recognised that this interpretation implied two restrictions

$$\sum_i b_i = 0 \quad \text{and} \quad \sum_j t_j = 0.$$

If we consider the predicted values from the fitted model

$$\hat{y}_{ij} = \hat{\mu} + \hat{b}_i + \hat{t}_j,$$

then it is clear that the model contains superfluous parameters. Suppose that we have a set of parameter estimates satisfying the least squares equations. Then clearly we would get the same fitted values, and therefore the same residual SS, if the value for $\hat{\mu}$ was increased by 1 and the values for $\hat{b}_i$ were each reduced by 1. Alternatively, we could decrease $\hat{\mu}$ by 2 and increase each $\hat{t}_j$ by 2. We cannot simply identify μ as superfluous and consider the model

$$y_{ij} = b_i + t_j + \varepsilon_{ij}, \tag{4.10}$$

since if each of the fitted $\hat{b}_i$ values is increased by 5 and each $\hat{t}_j$ decreased by 5 then again the fitted $\hat{y}$ values will be unchanged.

For the model (4.10) we could think of the b_i as representing average block yields and the t_j as deviations from the average of all treatments, implying $\sum_j t_j = 0$. Or, by the symmetry of the data structure, we could think of the t_j as representing average treatment yields and the b_i as deviations from the average of all blocks, implying $\sum_i b_i = 0$.

For a practical approach to models for experimental data, it is sensible to accept that models will be overparameterised and that we should obtain least squares estimates by recognising the inevitability of overparameterisation and using methods which allow us to deal with the resulting difficulty simply and directly. The theory of a general method of modified least squares estimation for overparameterised models is developed in the appendix to this chapter. We shall concentrate here on a practical approach.

4.4.1 Practical constraints on parameter values

Usually consideration of the practical interpretation of the terms in the model will identify sensible restrictions, or constraints, on the parameter values, and use of these constraints with the least squares equations will lead directly to parameter estimates. Consider the set of least squares equations for the model (4.2) for the randomised block rice spacing example:

$$40\hat{\mu} + 10\sum_i \hat{b}_i + 4\sum_j \hat{t}_j = \sum_i \sum_j y_{ij} = 297.40,$$

$$10\hat{\mu} + 10\hat{b}_1 + \sum_j \hat{t}_j = \sum_j y_{1j} = 79.15,$$

$$\vdots \qquad \vdots \qquad \vdots$$

$$10\hat{\mu} + 10\hat{b}_4 + \sum_j \hat{t}_j = \sum_j y_{4j} = 75.75,$$

$$4\hat{\mu} + \sum_i \hat{b}_i + 4\hat{t}_1 = \sum_i y_{i1} = 24.10,$$

$$\vdots \qquad \vdots \qquad \vdots$$

$$4\hat{\mu} + \sum_i \hat{b}_i + 4\hat{t}_{10} = \sum_i y_{i10} = 32.80.$$

If the restrictions $\sum_i b_i = 0$ and $\sum_j t_j = 0$ are imposed then obviously equivalent restrictions will apply to the parameter estimates, $\sum_i \hat{b}_i = 0$ and $\sum_j \hat{t}_j = 0$, and all summations on the left hand side of the equations disappear leaving

$$40\hat{\mu} \qquad\qquad = 297.40,$$
$$10\hat{\mu} + 10\hat{b}_1 = 79.15,$$
$$\vdots$$
$$4\hat{\mu} + 4t_{10} \qquad = 32.80,$$

from which least squares estimates can be obtained as

$$\hat{\mu} = 7.435,$$
$$\hat{b}_1 = 0.480,$$
$$\vdots$$
$$\hat{t}_{10} = 0.765.$$

We now consider briefly the effect of alternative constraints. We have already mentioned the possibility of setting $\mu = 0$ and $\sum_j t_j = 0$. The least squares estimates are then

$$\hat{b}_1 = 7.915$$
$$\vdots$$
$$\hat{b}_4 = 7.575$$

and the $\hat{t}_j$ remain unaltered.

If we consider all block effects relative to block 1 and all treatment effects relative to treatment 1 then we could use the constraints $b_1 = 0, t_1 = 0$. The solution of the least squares equations is more complex but can be achieved quite quickly to give

$$\hat{\mu} = 6.51 \quad \hat{b}_2 = -1.06 \qquad \hat{t}_2 = 0.45$$
$$\hat{b}_3 = -0.52 \qquad \hat{t}_3 = 1.16$$
$$\hat{b}_4 = -0.34 \qquad \vdots$$
$$\hat{t}_{10} = 2.18.$$

The fitted values are

$$\hat{y}_{11} = 6.51 \quad \hat{y}_{21} = 5.45 \quad \hat{y}_{31} = 5.99 \quad \hat{y}_{41} = 6.17$$
$$\hat{y}_{12} = 6.96 \quad \hat{y}_{22} = 5.90 \quad \hat{y}_{32} = 6.44 \quad \hat{y}_{42} = 6.62$$
$$\vdots \qquad\qquad \vdots \qquad\qquad \vdots$$
$$\hat{y}_{1,10} = 8.69 \quad \hat{y}_{2,10} = 7.63 \quad \hat{y}_{3,10} = 8.17 \quad \hat{y}_{4,10} = 8.35,$$

exactly the same as for any other set of constraints.

Because the fitted values are unchanged by the use of alternative constraint systems it follows that the residual SS is invariant over constraint systems. In particular if constraints were all of the form $\theta_j = 0$ for sufficient different j we can recognise that the number of parameters would be reduced to the number that can actually be estimated from the data. If the model is rewritten to exclude the parameters which are set to zero by the constraints, then the effective number of parameters, p', may be identified. From the results of the previous section the expected value of S_r is known to be $(n - p')\sigma^2$. This allows us to estimate σ^2, when using constraints of any form, p' being the number of parameters minus the number of constraints.

For the usual randomised block model (4.2) using constraints $\sum_i b_i = 0$, $\sum_j t_j = 0$ the least squares estimates have been obtained earlier in this section. The fitted SS is calculated as

$$\text{fitted SS} = \hat{\mu}\left(\sum_{ij} y_{ij}\right) + \hat{b}_1\left(\sum_j y_{1j}\right) + \cdots + \hat{t}_{10}\left(\sum_i y_{i10}\right)$$
$$= (7.435)(297.40) + (0.480)(79.15) + \cdots + (0.765)(32.80)$$
$$= 2240.19.$$

The residual SS is $2252.61 - 2240.19 = 12.42$. This residual SS is the same as would have been obtained if b_1 and t_1 had been set to zero, in which case the number of parameters in the model would have been

$$p' = 1 + 3 + 9 = 13.$$

Therefore the estimate of σ^2 is

$$S_r/(n - p') = 12.42/27 = 0.46.$$

Variances for parameters when constraints have been used can be obtained quite easily, but some care is necessary in interpreting variance information about parameters subject to constraints. If the model is

$$y_{ij} = \mu + b_i + t_j + \varepsilon_{ij}$$

and constraints on the b_i and on the t_j are used, then because there is an arbitrariness in the choice of constraints there is a corresponding arbitrariness about individual parameter values. Thus, as argued previously, the value for $\hat{b}_1$ is dependent on the values for $\hat{\mu}$ and $\hat{t}_j$, and therefore discussion about the absolute variance of $\hat{b}_1$ is meaningless. However, differences between the parameters within the set of block parameters or between treatment parameters are not affected by the choice of constraint.

To derive the variance of any comparison of parameters the estimate of the comparison must be expressed as a linear combination of the observations, y_{ij}. The variance of a linear combination of y values can then be evaluated directly. Thus for the rice spacing data the difference $(t_1 - t_2)$ is estimated by

$$\left(\sum_i y_{i1} - \sum_i y_{i2}\right)\bigg/4,$$

and the variance is obviously $2\sigma^2/4$.

4.4.2 A more complex example of the use of constraints

For a final illustration of the use of constraints and the calculation of least squares estimates we consider example C. The model must allow for row effects, column effects and treatment effects,

$$y_{ij} = \mu + r_i + c_j + t_{k(ij)} + \varepsilon_{ij},$$

where the actual treatment k occurring for the combination of row i and column j is determined by the design. The least squares equations are:

$$30\hat{\mu} + 5\sum_i \hat{r}_i + 6\sum_j \hat{c}_j + 6\sum_k \hat{t}_k = \sum_{ij} y_{ij} = 125.04,$$

$$5\hat{\mu} + 5\hat{r}_1 + \sum_j \hat{c}_j + \sum_k \hat{t}_k \qquad = \sum_j y_{1j} = 14.66,$$

$$\vdots \qquad\qquad \vdots \qquad \vdots$$

$$5\hat{\mu} + 5\hat{r}_6 + \sum_j \hat{c}_j + \sum_k \hat{t}_k \qquad = \sum_j y_{6j} = 32.90,$$

$$6\hat{\mu} + \sum_i \hat{r}_i + 6\hat{c}_1 + \sum_k \hat{t}_k + \hat{t}_B = \sum_i y_{i1} = 27.59,$$

$$\vdots \qquad\qquad \vdots \qquad \vdots$$

$$6\hat{\mu} + \sum_i \hat{r}_i + 6\hat{c}_5 + \sum_k \hat{t}_k + \hat{t}_A = \sum_i y_{i5} = 18.51,$$

$$6\hat{\mu} + \sum_i \hat{r}_i + \sum_j \hat{c}_j + \hat{c}_5 + 6\hat{t}_A = \sum_A y_{ij} = 31.37,$$

$$\vdots \qquad\qquad \vdots \qquad \vdots$$

$$6\hat{\mu} + \sum_i \hat{r}_i + \sum_j \hat{c}_j + \hat{c}_3 + 6\hat{t}_E = \sum_E y_{ij} = 20.08.$$

The natural constraints are $\sum_i r_i = 0$, $\sum_j c_j = 0$, $\sum_k t_k = 0$ and applying these gives simplified equations:

$$30\hat{\mu} \qquad\qquad = 125.04,$$

$$5\hat{\mu} + 5\hat{r}_1 \qquad = 14.66,$$

$$\vdots \qquad\qquad \vdots$$

$$5\hat{\mu} + 5\hat{r}_6 \qquad = 32.90,$$

$$6\hat{\mu} + 6\hat{c}_1 + \hat{t}_B = 27.59,$$

$$\vdots \qquad\qquad \vdots$$

$$6\hat{\mu} + 6\hat{c}_5 + \hat{t}_A = 18.51,$$

$$6\hat{\mu} + \hat{c}_5 + 6\hat{t}_A = 31.37,$$

$$\vdots \qquad\qquad \vdots$$

$$6\hat{\mu} + \hat{c}_3 + 6\hat{t}_E = 20.08.$$

The estimates for $\hat{\mu}$ and $\hat{r}_i$ are simple. Those for $\hat{c}_j$ and $\hat{t}_k$ occur in pairs of equations, of which one pair, by chance, is the two equations for $\hat{c}_5$ and $\hat{t}_A$ shown adjacently above. After calculating $\hat{\mu} = 4.168$ and eliminating $\hat{\mu}$ from the $\hat{c}_5$ and $\hat{t}_A$ equations we have

$$6\hat{c}_5 + \hat{t}_A = -6.50,$$

$$\hat{c}_5 + 6\hat{t}_A = +6.36$$

and hence

$$\hat{c}_5 = -1.296, \quad \hat{t}_A = +1.276.$$

Similarly

$$\hat{c}_1 = +0.339, \quad \hat{t}_B = +0.549,$$
$$\hat{c}_2 = +0.065, \quad \hat{t}_C = -0.249,$$
$$\hat{c}_3 = +1.159, \quad \hat{t}_E = -1.015,$$
$$\hat{c}_4 = -0.268, \quad \hat{t}_D = -0.562.$$

For completeness the other estimated effects are

$$\hat{\mu} = 4.168,$$
$$\hat{r}_1 = -1.234, \hat{r}_2 = -1.362, \hat{r}_3 = -1.585, \hat{r}_4 = +0.838, \hat{r}_5 = +0.938, \hat{r}_6 = +2.414.$$

To determine the variance of a difference between two treatments, for example t_A and t_E, we have to express $\hat{t}_A - \hat{t}_E$ as a linear combination of the y values. To do this we note that if we consider the simple differences $c_5 - c_3$ and $t_A - t_E$, then the equations can be written

$$6(\hat{c}_5 - \hat{c}_3) + (\hat{t}_A - \hat{t}_E) = 18.51 - 30.95 = C_5 - C_3,$$
$$(\hat{c}_5 - \hat{c}_3) + 6(\hat{t}_A - \hat{t}_E) = 31.37 - 20.09 = T_A - T_E,$$

where capital letters are used to denote yield totals. We can now solve for $(\hat{t}_A - \hat{t}_E)$:

$$\begin{aligned}
35(\hat{t}_A - \hat{t}_E) &= 6(T_A - T_E) - (C_5 - C_3) \\
&= 6y_{11} + y_{13} - 7y_{15} - 5y_{23} + 6y_{24} - y_{25} \\
&\quad + 6y_{32} + y_{33} - 6y_{34} - y_{35} - 6y_{41} + y_{43} \\
&\quad + 5y_{45} - 6y_{52} + 7y_{53} - y_{55} - 5y_{63} + 5y_{65}.
\end{aligned}$$

A simple diagrammatic method for writing out such linear combinations with minimum error is described in Section 7.1. The variance of $(\hat{t}_A - \hat{t}_E)$ can now be calculated as

$$\begin{aligned}
\text{Var}(\hat{t}_A - \hat{t}_E) &= \sigma^2\{6(36) + 2(49) + 4(25) + 6(1)\}/(35)^2 \\
&= 12\sigma^2/35.
\end{aligned}$$

This is, interestingly, very close to $2\sigma^2/6$, which would be the variance for a simple design with six blocks of five treatments. We consider designs of this pattern further in Chapters 7 and 8.

Finally for this example note that the residual SS can be calculated by first finding the total SS and fitting SS:

$$\begin{aligned}
\text{total SS} &= 3.72^2 + 3.39^2 + \cdots + 6.92^2 = 630.12, \\
\text{fitting SS} &= (4.168)(125.04) + (-1.234)(14.66) + \cdots + (-1.015)(20.08) \\
&= 622.00.
\end{aligned}$$

Hence, residual SS $= 630.12 - 622.07 = 8.05$ and

$$s^2 = S_r/(35 - 14) = 8.05/21 = 0.39.$$

4.5 Subdividing the parameters; extra SS

For many models the set of parameters consists of several subsets, and the interest in the parameters may be restricted to only some of the subsets. More generally we may think of the model as being built in stages by

(i) starting with parameters which are consequences of the initial structure of the experimental units,
(ii) adding parameters to represent the treatment effects,
(iii) trying, parsimoniously, to reduce the set of parameters required to represent the effects of the treatments, and possibly
(iv) adding further parameters to represent modification of the treatment effects by the environment.

For example in a randomised block design there are three subsets of parameters, μ, $\{b_i\}$ and $\{t_j\}$. The model must include μ and $\{b_i\}$ because they are implied by the structure of the units in the design. However, the real interest is in the $\{t_j\}$ and consequently in their estimates, standard errors and the SS for fitting the treatment effects. In the rice spacing example we have used two models representing the effects of treatments, first the full model (4.2)

$$y_{ij} = \mu + b_i + t_j + \varepsilon_{ij},$$

and secondly a model where the treatment effects are represented by a linear dependence on area per plant, model (4.3)

$$y_{ij} = \mu + b_i + \beta a_j + \varepsilon_{ij}.$$

The latter is a special case of the former assuming a special pattern among the t_j (and using one parameter instead of ten).

Factorial structure also provides a situation in which different subsets of treatments may be identified as being of different a priori importance. Main effects will almost certainly be important, two-factor interactions may be, but three-factor interactions are unlikely to be important, though we may wish to check this last presumption.

The fundamental results which allow us to analyse these situations are derived by considering a simple split of the vector of parameters θ into two subsets θ_1 and θ_2. There is a corresponding split of the design matrix $\mathbf{A}$ into those columns related to θ_1, denoted by $\mathbf{A}_1$, and the remainder related to θ_2, denoted by $\mathbf{A}_2$. There is a further, corresponding, split of the matrices $\mathbf{C}$ and $\mathbf{C}^{-1}$ so that $\mathbf{C}_{11}$ is the submatrix obtained from pairs of columns both in $\mathbf{A}_1$, $\mathbf{C}_{22}$ from pairs of columns both in $\mathbf{A}_2$, and $\mathbf{C}_{12}$ from pairs of columns, one from $\mathbf{A}_1$ and one from $\mathbf{A}_2$.

We assume that the interest is in the second set of parameters θ_2, while θ_1 represents parameters inevitably included in the model but not of direct interest. Explicit expressions may be obtained for estimates of θ_2, allowing for θ_1, without actually calculating $\hat{\theta}_1$, and these are derived in Appendix 4.A5. It is also important to consider the SS, which is the difference between the SS for fitting the full model with θ_1 and θ_2 and the SS for fitting the reduced model including only θ_1. If the complete set θ includes p parameters, and the subset of interest θ_2 includes q parameters, then we can identify an analysis of variance structure with three components as shown in Table 4.3. We have noted that it can be shown that the

Table 4.3. *Analysis of variance showing the extra SS*

Source	SS	df
Fitting $\boldsymbol{\theta}_1$ ignoring $\boldsymbol{\theta}_2$	S_{p-q}	$p-q$
Fitting $\boldsymbol{\theta}_2$ allowing for $\boldsymbol{\theta}_1$	S_q	q
Residual	S_r	$n-p$

expected value of S_r is $(n-p)\sigma^2$. It can also be shown that if all the parameters in the set $\boldsymbol{\theta}_2$ are zero then the expected value of S_q is $q\sigma^2$, making no assumptions about the values of the parameters in the set $\boldsymbol{\theta}_1$. In other words, we can use S_q as a basis for assessing the evidence about the parameters $\boldsymbol{\theta}_2$, while allowing for the effects of the parameters $\boldsymbol{\theta}_1$. The SS, S_q, is termed the 'extra sum of squares' attributable to $\boldsymbol{\theta}_2$.

There is one further important concept and this is whether S_{p-q} may be used to assess the importance of the parameters $\boldsymbol{\theta}_1$. Clearly the roles of $\boldsymbol{\theta}_1$ and $\boldsymbol{\theta}_2$ could be reversed and the model fitted first with $\boldsymbol{\theta}_2$ only so that the fitting SS for fitting $\boldsymbol{\theta}_1$, allowing for $\boldsymbol{\theta}_2$, may be calculated. This 'extra' SS for $\boldsymbol{\theta}_1$, allowing for $\boldsymbol{\theta}_2$, provides an assessment of $\boldsymbol{\theta}_1$ and may sometimes be the same as the SS for $\boldsymbol{\theta}_1$ ignoring $\boldsymbol{\theta}_2$. It can be shown (Appendix 4.A6) that, for the SS for fitting $\boldsymbol{\theta}_1$ ignoring $\boldsymbol{\theta}_2$ and the SS for fitting $\boldsymbol{\theta}_1$ allowing for $\boldsymbol{\theta}_2$ to be identical, it is necessary that the design matrix columns for $\boldsymbol{\theta}_1$ are orthogonal to those for $\boldsymbol{\theta}_2$. This orthogonality may be expressed in terms of the columns of the design matrix $\mathbf{A}$ in the following form: the sum (over rows) of products of pairs of values in columns j (from $\mathbf{A}_1$) and k (from $\mathbf{A}_2$) for each row must be zero for each possible pairing (j, k). Any constraints on parameters must be allowed for by rewriting the parameters when applying this concept of orthogonality.

We can interpret orthogonality practically as requiring that the estimates of comparisons between the parameters $\boldsymbol{\theta}_2$ are unaffected by the need to allow for the parameters $\boldsymbol{\theta}_1$. In this practical sense comparisons between treatments in a randomised complete block design are clearly unaffected by differences between blocks because all block–treatment combinations occur exactly once, and hence treatments are orthogonal to blocks. However, in example C the occurrence of treatment A twice in column 5 and treatment E twice in column 3 means that the apparent $(A - E)$ difference could be partially caused by differences between columns 5 and 3. Therefore in example C treatments are not orthogonal to columns.

The subdivision of parameters into two sets $\boldsymbol{\theta}_1$ and $\boldsymbol{\theta}_2$ generalises to more than two sets, and analyses of variance can be constructed with components:

(i) fitting θ_1, ignoring $\theta_2, \ldots, \theta_k$;
(ii) fitting θ_2, allowing for θ_1, ignoring $\theta_3, \ldots, \theta_k$;
(iii) fitting θ_3, allowing for θ_1 and θ_2, ignoring $\theta_4, \ldots, \theta_k$;
$\vdots$
(k) fitting θ_k, allowing for $\theta_1, \theta_2, \ldots, \theta_{k-1}$.

Unless all sets of parameters are mutually orthogonal, the SS for fitting a set of parameters will depend on the order of fitting which defines the other parameter sets allowed for or ignored, and the interpretation of each SS is specific to this context.

Table 4.4. *Analysis of variance for rice spacing experiment*

Source	SS	df
μ	2211.17	1
b_i	5.88	3
t_j	23.14	9
Residual	12.42	27
Total	2252.61	40

Consider again example A. For the model (4.2)

$$y_{ij} = \mu + b_i + t_j + \varepsilon_{ij},$$

we can define three sets of parameters $\theta_1 = \mu, \theta_2 = \{b_i\}$ and $\theta_3 = \{t_j\}$. Fitting $\theta_1 = \mu$ is only sensible as the first step since the interpretation of $\{b_i\}$ and $\{t_j\}$ as deviations from an overall average yield would not be valid if μ is not included in the model. We might consider therefore two orders of fitting terms for the analysis of variance, $\theta_1 \rightarrow \theta_2 \rightarrow \theta_3$ or $\theta_1 \rightarrow \theta_3 \rightarrow \theta_2$. In fact, block effects and treatment effects are orthogonal so the order of fitting is irrelevant and the resulting analysis of variance shown in Table 4.4, is essentially that calculated in Chapter 2. It is clear from the practical form of calculation of block and treatment SS in Chapter 2 that the SS are unaffected by the order of fitting. We could reinforce this by noting that the block effects are

$$\hat{b}_1 = +0.480, \hat{b}_2 = -0.580, \hat{b}_3 = -0.040, \hat{b}_4 = +0.140,$$

as calculated in the previous section whether or not the $\{t_i\}$ are included in the model.

4.5.1 *An example where different orders of fitting produce different SS*

Now consider example C with the model

$$y_{ij} = \mu + r_i + c_j + t_{k(ij)} + \varepsilon_{ij}$$

for which we can identify four sets of parameters $\theta_1 = \mu, \theta_2 = \{r_i\}, \theta_3 = \{c_j\}, \theta_4 = \{t_k\}$. Sets θ_2 and θ_3 are orthogonal since each row 'occurs' in each column and vice versa, and also θ_2 and θ_4 are orthogonal since each treatment occurs in each row. However, the pattern for θ_3 and θ_4 is not orthogonal, since the replication pattern of the five treatments varies between columns. The implications of this non-orthogonality can be identified by considering the situation where there are no difference effects between the columns. If the treatments give different values, then there will appear to be differences between the columns reflecting the treatments which appear in the columns. Thus suppose we assume for the model

$$y_{ij} = \mu + r_i + c_j + t_{k(ij)} + \varepsilon_{ij}$$

that all c_j are zero and, to make the pattern simple, that $\mu = 20$, all r_i and ε_{ij} are zero and the t_k are $(+5, +3, +1, -1, -8)$. Then the observed yields would be as in Table 4.5. Column 3

Table 4.5. *Artificial observed yields*

					Row totals
A 25	B 23	C 21	D 19	E 12	100
C 21	D 19	E 12	A 25	B 23	100
D 19	A 25	B 23	E 12	C 21	100
E 12	C 21	D 19	B 23	A 25	100
B 23	E 12	A 25	C 21	D 19	100
B 23	C 21	E 12	D 19	A 25	100
Column totals 123	121	112	119	125	

appears to give lower values than column 5, if possible treatment effects are ignored. That this effect does not disappear if the row effects or random effects are made non-zero can be confirmed by adding any set of numerical values for such effects to the observed yields.

In the previous section the least squares equations for example C allowed us to estimate the column and treatment sets of parameters, fitted together, giving

$$\hat{c}_1 = +0.339, \hat{c}_2 = +0.065, \hat{c}_3 = +1.159, \hat{c}_4 = -0.268, \hat{c}_5 = -1.296;$$
$$\hat{t}_A = +1.276, \hat{t}_B = +0.549, \hat{t}_C = -0.249, \hat{t}_D = -0.562, \hat{t}_E = -1.015.$$

If treatment effects are ignored the estimates for column effects are

$$\hat{c}_1 = +0.430, \hat{c}_2 = +0.024, \hat{c}_3 = +0.990, \hat{c}_4 = -0.361, \hat{c}_5 = -1.083.$$

Column 3 appears less high and column 5 less low because the extra occurrence in column 3 of the low value treatment E and the extra occurrence of the high value treatment A in column 5 are ignored.

Similarly if we ignore the column effects the treatment effect estimates are

$$\hat{t}_A = +1.060, \hat{t}_B = +0.605, \hat{t}_C = -0.238, \hat{t}_D = -0.606, \hat{t}_E = -0.821.$$

The SS for fitting column effects, treatment effects and both sets of effects are:

$$\text{SS (columns only)} = (0.430)(27.59) + \cdots = 14.82,$$
$$\text{SS (treatments only)} = (1.060)(31.37) + \cdots = 15.53,$$
$$\text{SS (both)} = (0.339)(27.59) + \cdots + (1.276)(31.37) + \cdots$$
$$= 34.24.$$

The two analyses of variance for the two orders $\theta_1 \to \theta_2 \to \theta_3 \to \theta_4$ and $\theta_1 \to \theta_2 \to \theta_4 \to \theta_3$ are given in Table 4.6(a) and (b).

The SS for column effects and for treatment effects are broadly similar in the two analyses, and in each case are markedly larger than the residual, so that the interpretation is going to be similar for treatment effects whether or not we allow for column effects. Nevertheless the change of 3.89 in the SS is sufficiently large to indicate that where an effect is less clearly substantial, or where the non-orthogonality or degree of interference between treatment and column effects is larger, then the conclusions from an analysis which failed to allow for other effects when assessing treatment effects could be severely misleading.

Table 4.6. *Analysis of variance for the tomato experiment, with two orders of fitting: (a) with* c *before* t; *(b) with* t *before* c

(a)

Source	SS	df
μ (ignoring all other terms)	521.17	1
r_i (ignoring c_j and t_k)	66.66	5
c_j (allowing for r_i, ignoring t_k)	14.82	4
t_k (allowing for r_i and c_j)	19.42	4
Residual	8.05	21

(b)

Source	SS	df
μ (ignoring all other terms)	521.17	1
r_i (ignoring c_j and t_k)	66.66	5
t_k (allowing for r_i, ignoring c_j)	15.53	4
c_i (allowing for r_i and t_k)	18.71	4
Residual	8.05	21

Table 4.7. *Analysis of variance for regression model for rice spacing experiment*

Source	SS	df
μ	2211.17	1
Block effect	5.88	3
Regression on area per plant	16.80	1
Residual	18.76	35

One further example of the use of the extra SS principle concerns example A again and the comparison of the two models considered for the treatment effects. In one form of model the ten separate treatment effects are included giving a SS of 23.14. Alternatively the effects of treatments are represented by a regression on area per plant. The full analysis of variance for this alternative model is given in Table 4.7. Now the regression model for treatment effects is a special case of the more general model allowing for ten different treatment effects, and we could write an extended model to include both regression and separate treatment effects:

$$y_{ij} = \mu + b_i + \beta a_j + t'_j + \varepsilon_{ij}.$$

We can calculate the SS for fitting four sets of parameters $\theta_1 = \mu, \theta_2 = \{b_i\}, \theta_3 = \beta, \theta_4 = \{t'_j\}$. However, since the above model is exactly equivalent to the earlier model which allowed for possible differences between all ten treatments

$$y_{ij} = \mu + b_i + t_j + \varepsilon_{ij},$$

Table 4.8. *Full analysis of variance for rice spacing experiment*

Source	SS	df
μ	2211.17	1
Blocks	5.88	3
Regression	16.80	1
Remaining treatment effects (t'_j)	6.34	8
Residual	12.42	27

we already have the SS for all submodels and for the full model and can construct the analysis of variance given in Table 4.8. We may now observe that the treatment SS after allowing for the regression effect is not very large so that we may be prepared to accept the regression model as an adequate summary of the treatments.

4.6 Distributional assumptions and inferences

All the results discussed so far in this chapter have ignored assumptions about the sampling distributions of the observations. It is important to notice that much of the analysis of experimental data does not rely on particular distributional assumptions. However, if we assume that the observations are taken from a normally distributed population, then we can develop methods for more detailed inferences.

The first consequence of the normal distribution assumption is that, since least squares estimates are linear combinations of the ys and since linear combinations of normally distributed variables are themselves normally distributed, then the least squares estimates are normally distributed. Least squares estimates are unbiased and variances of least squares estimates were derived in Section 4.3, so the information on the distribution of least squares estimates is complete. In particular, confidence intervals for parameters may be constructed, and tests of significance calculated using the t-distribution.

The second important consequence of the normal distribution assumption is that the distributional properties of SS in the analysis of variance are all described in terms of χ^2-distributions. First, quite generally, the residual SS, S_r, is distributed as σ^2 times a χ^2-distribution on $(n - p')$ df. Second, where the parameters are partitioned into two sets, θ_1 and θ_2, then, whenever the null hypothesis (that the parameters in the second set are zero) is true, the SS due to fitting θ_2 allowing for θ_1 is distributed as σ^2 times a χ^2-distribution on q df, where q is the number of effective parameters in θ_2. Hence, the null hypothesis that the parameters θ_2 are zero may be tested by calculating the ratio of the MS for θ_2 to the residual MS, and comparing that ratio with the F-distribution on q and $(n - p')$ df. Note that for the fitting order of θ_1 followed by θ_2 the F-distribution applies only to the MS for θ_2, and not to that for θ_1, unless θ_1 and θ_2 are orthogonal.

To illustrate the application of these distributional results we repeat several previous analyses of variance with added MS and F statistics.

Example A First we consider the simple analysis for block and treatment effects (see Table 4.9). The treatment MS with an F statistic on 9 and 27 df is clearly significant so that

Table 4.9. *Analysis of variance for rice spacing experiment*

Source	SS	df	MS	F
μ	2211.17	1		
Blocks	5.88	3	1.96	4.26
Treatments	23.14	9	2.57	5.59
Residual	12.42	27	0.46	

Table 4.10. *Analysis of variance for rice spacing experiment*

Source	SS	df	MS	F
μ	2211.17	1		
Blocks	5.88	3	1.96	
Regression	16.80	1	16.80	36.52
Remaining treatment effects	6.34	8	0.79	1.72
Residual	12.42	27	0.46	

the differences between treatments are well substantiated. The F statistic for blocks could also be tested because of the orthogonality between blocks and treatments, but the formal hypothesis that $b_i = 0$ for all i is neither credible nor important.

The secondary analysis with the regression effect is given in Table 4.10. The F tests provide confirmation, if any was needed, that there is strong regression effect. More importantly, we can examine the remaining treatment variation not explained by the regression. The F statistic of 1.72 indicates that there may be other important effects amongst the treatments but the evidence is not significant at the 5% level. We return to the interpretation of these data in Chapter 12.

Example C The analysis of variance (Table 4.11) includes the SS for fitting treatment effects allowing for column effects, and it is that SS for treatments that we should use to test the significance of the treatment effects. Plainly the treatment effects are very significant. Although we should not wish to test the significance of the MS for columns or rows, since the hypothesis of no row or column effects is not important to the objectives of the experiment, there are clearly substantial differences between rows and, to a lesser extent, between columns. The design of the experiment has thus been clearly successful in providing more precise information about treatments, and the degree of that success may be assessed by estimating the error MS that might have been obtained if the treatments had been allocated without restriction. The error MS might then have been expected to be about

$$(66.66 + 14.82 + 8.05)/(5 + 4 + 21) = 2.98$$

instead of 0.38.

Table 4.11. *Analysis of variance for tomato data*

Source	SS	df	MS	F
μ	521.17	1		
Rows	66.66	5	13.33	
Columns (ignoring treatments)	14.82	4	3.70	
Treatments (allowing for columns)	19.42	4	4.86	12.67
Residuals	8.05	21	0.38	

4.7 Contrasts, treatment comparisons and component SS

The main use of formally defined contrasts between observations is in the detailed examination of treatment comparisons in the analysis of experimental data. However, the definition and identification of contrasts between individual observations is also useful in the design of experiments, particularly in understanding the concepts of confounding in Chapter 15. We consider contrasts first in the context of treatment comparisons.

Assume a set of t treatments, and for simplicity assume initially that there are n observations for each treatment, and that each treatment effect estimate is orthogonal to all block effects and is in the form

$$\hat{t}_j = y_{.j} - y_{..}.$$

A general treatment comparison, L_k, is defined as a linear contrast of treatment effects

$$L_k = \sum_j l_{kj} t_j,$$

where $\sum_j l_{kj} = 0$. The least squares estimate of L_k is

$$\hat{L}_k = \sum_j l_{kj} \hat{t}_j$$

$$= \sum_j l_{kj} (y_{.j} - y_{..})$$

$$= \sum_j l_{kj} y_{.j}$$

since $\sum_j l_{kj} = 0$. The variance of $\hat{L}_k$ is

$$\text{Var}(\hat{L}_k) = \sigma^2 \sum_j l_{kj}^2 / n,$$

which may be estimated, using the residual MS, s^2, as

$$s^2 \sum_j l_{kj}^2 / n.$$

The estimate of a contrast and the estimated variance of that estimate provide all the necessary information about the contrast, but it is useful to develop the concept further to consider the relationship between different contrasts and the relationship between individual contrasts and the total treatment variation. First, two treatment comparisons, L_k and $L_{k'}$, are defined to be orthogonal if

$$\sum_j l_{kj} l_{k'j} = 0 \quad \text{if } k \neq k'.$$

It can be shown that we can define a component SS corresponding to a contrast which can be identified in the form

$$SS(L_k) = n \left(\sum_j l_{kj} y_{.j} \right)^2 \Big/ \left(\sum_j l_{kj}^2 \right)$$

(see the appendix to this chapter). Note that the divisor in this expression is a scaling factor. There are no restrictions on the size of coefficients in the definition of a comparison so that

$$t_1 - t_2 \text{ and } 20t_1 - 20t_2$$

are both valid forms of comparison. In calculating the component SS this arbitrariness of scale inherent in the definition of L_k must be taken into account.

These component SS for comparisons L_k have two important properties.

(i) If $L_k = 0$, i.e. the treatment comparison has no effect, the SS is distributed as σ^2 times a χ^2-distribution on 1 df.
(ii) The total of the SS for $(t - 1)$ orthogonal comparisons, from t treatments, is the total treatment SS.

The implications are that we can examine $(t-1)$ comparisons between the treatments independently provided the comparisons are constructed so as to be orthogonal, and together such a set of $(t - 1)$ orthogonal comparisons represents the total variation between the treatments.

4.7.1 Examples of the use of contrasts

The practical use of treatment comparisons can be illustrated for the first three of our examples. For example B on water uptake of amphibia the analysis of variance for main effects and interactions was considered in Chapter 3. The same analysis may be constructed in terms of treatment contrasts. Seven contrasts corresponding to the main effects and interactions may be defined as shown in Table 4.12. Contrasts L_1, L_2 and L_3 are main effect contrasts between the two levels of each factor. Contrasts L_4, L_5 and L_6 are two-factor interactions, in the form of differences of differences, and contrast L_7 is the three-factor interaction, a difference of differences of differences. It can be verified by inspection that any two contrasts are orthogonal. The component SS are calculated, using $n = 2$; for example

$$SS(L_1) = 2(0.360 + 21.265 + \cdots - 10.545)^2/(1^2 + 1^2 + \cdots + 1^2)$$
$$= 2(45.39)^2/8 = 515.06,$$

agreeing with the value for the species SS calculated in Chapter 3.

For example C there are five treatments and various alternative sets of contrasts may be considered. Meaningful contrasts that might be considered are defined in Table 4.13. The set (L_1, L_2, L_3, L_4) are orthogonal, and the contrasts represent first the effect of spray compared

Table 4.12. *Contrasts for water uptake experiment*

Treatment	L_1	L_2	L_3	L_4	L_5	L_6	L_7	Treatment mean
Toad wet control	+1	+1	+1	+1	+1	+1	+1	0.360
Toad wet hormone	+1	+1	−1	+1	−1	−1	−1	21.265
Toad dry control	+1	−1	+1	−1	+1	−1	−1	21.455
Toad dry hormone	+1	−1	−1	−1	−1	+1	+1	28.165
Frog wet control	−1	+1	+1	−1	−1	+1	−1	1.875
Frog wet hormone	−1	+1	−1	−1	+1	−1	+1	3.340
Frog dry control	−1	−1	+1	+1	−1	−1	+1	10.095
Frog dry hormone	−1	−1	−1	+1	+1	+1	−1	10.545

Table 4.13. *Contrasts for the tomato experiment*

Spray treatment	L_1	L_2	L_3	L_4	L_5	L_6	$\hat{t}_k$
Early 75 ppm	+1	−1	−1	+1	0	−3	+1.276
Early 150 ppm	+1	−1	+1	−1	+1	+2	+0.549
Late 75 ppm	+1	+1	−1	−1	0	−3	−0.249
Late 150 ppm	+1	+1	+1	+1	+1	+2	−0.562
Control	−4	0	0	0	−2	+2	−1.015

with control and then the main effects and interaction for the factorial structure of the four spray treatments. Alternatively, the control may be viewed as a zero level spray and (L_5, L_6) defined to represent two orthogonal contrasts between the three levels of spray, L_5 comparing the two extremes and representing linear trend while L_6 compares the middle level with the extremes, representing the curvature of the trend. A second complete set of four orthogonal contrasts is $(L_2, L_4, L_5$ and $L_6)$ but neither L_1 nor L_3 is orthogonal to either L_5 or L_6.

The calculation of the component SS for these contrasts introduces a problem not considered in the development of the component SS because the treatment effects are not orthogonal to the column effects. Consequently, the treatment effects are not in the simple form specified earlier in this section. We can now extend the results for component SS for the situation where, as in example C, treatment effects are not orthogonal to block effects but the precision of treatment differences is identical for all treatment differences. We shall return to this concept of equality of precision (balance) in Chapters 7 and 8. For the present it is sufficient to note that the symmetry of the design for example C, with each treatment occurring a second time in just one column, implies that the variances of all simple treatment differences should be equal.

In Section 4.4 the variance of a simple treatment difference was derived as $12\sigma^2/35$, which is larger than the variance for a treatment difference in a randomised block design with the same six-fold replication, by a factor of 36/35. It can be shown that the variance for any treatment contrast is similarly inflated by the factor 36/35, and that correspondingly the component SS must be reduced by the reciprocal factor 35/36. Essentially the non-orthogonality reduces the information on treatment differences by a factor 35/36.

Table 4.14. *Two analyses of variance for the tomato experiment: (a) with contrasts 1, 2, 3 and 4; (b) with contrasts 2, 4, 5 and 6*

(a)

Source	MS	df	F
L_1	7.51	1	19.59
L_2	10.13	1	26.43
L_3	1.58	1	4.12
L_4	0.25	1	0.65
Residual	0.38	21	

(b)

Source	MS	df	F
L_2	10.13	1	26.43
L_4	0.25	1	0.65
L_5	3.96	1	10.33
L_6	5.13	1	13.38
Residual	0.38	21	

The SS for the six components are therefore

$$
\begin{aligned}
SS(L_1) &= (35/36)6[1.276 + 0.549 + 0.249 + 0.562 - 4(1.015)]^2/20 = \; 7.51, \\
SS(L_2) &= (35/36)6[-1.276 - 0.549 + 0.249 + 0.562]^2/4 && = 10.13, \\
SS(L_3) &= (35/36)6[-1.276 + 0.549 - 0.249 + 0.562]^2/4 && = \; 1.58, \\
SS(L_4) &= (35/36)6[1.276 - 0.549 - 0.249 + 0.562]^2/4 && = \; 0.25, \\
SS(L_5) &= (35/36)6[0.549 + 0.562 - 2(1.015)]^2/6 && = \; 3.96, \\
SS(L_6) &= (35/36)6[-3(1.276) + 2(0.549) - 3(0.249) + 2(0.562) \\
&\quad + 2(1.015)]^2/30 && = \; 5.13.
\end{aligned}
$$

The two alternative analyses of treatment variation are given in Table 4.14(a) and (b). Clearly the timing of the spray (L_2) has the largest single effect but there is no evidence of an interaction (L_4) between timing and concentration. The most striking pattern of the effect of concentration is apparently provided by L_1, the simple contrast of spray with no spray. Alternatively the mean yields at the three concentrations should be presented

control	75 ppm	150 ppm
3.15	4.68	4.16.

As a final illustration of the use of contrasts note that the regression effect for example A could have been expressed as a treatment comparison. If the regression is written in the form

$$
E(y_{.j}) = \alpha + \beta(a_j - \bar{a}),
$$

Table 4.15. *Contrast for regression effect*

Area/plant for each treatment	Treatment contrast $(a_j - \bar{a})$	Treatment mean
900	+398.9	6.02
720	+218.9	6.48
600	+98.9	7.19
450	−51.1	6.92
576	+74.9	7.85
480	−21.1	7.78
360	−141.1	7.50
400	−101.1	7.71
300	−201.1	8.70
225	−276.1	8.20
Mean = 501.1		

then a regression comparison with coefficients $(a_j - \bar{a})$ may be defined as shown in Table 4.15. The component SS for the regression effect is therefore

$$SS \text{ (regression)} = 4\{(398.9)(6.02) + \cdots$$
$$- (276.1)18.20\}^2/(398.94^2 + \cdots + 276.1^2)$$
$$= 4(-1250.185)^2/372289 = 16.79.$$

4.7.2 Contrasts defined for a complete set of observations

At the beginning of this section the comment was made that the concept of contrasts may be applied more widely than within the set of treatments. We may more generally define contrasts within the complete set of n observations. A complete set of contrasts would lead to an analysis of variance in which each SS was based on a single df and would, of course, be quite useless. Nevertheless, the identification and interpretation of individual contrasts may sometimes be helpful in understanding the properties of designs. To illustrate this wider concept of contrasts we consider a trivial example of a design for four treatments in three randomised blocks. A set of possible contrasts is defined in Table 4.16.

Contrasts L_1 and L_2 are contrasts between blocks; contrasts L_3, L_4 and L_5 are contrasts between treatments; the remaining contrasts are for residual variation. Rather laboriously it may be verified that all pairs of contrasts are orthogonal. In fact the residual contrasts are formed as products of block and treatment contrasts ($L_6 = L_1 L_3, L_7 = L_1 L_4, \ldots, L_{11} = L_2 L_5$), which emphasises the interpretation of the residual variation both as block × treatment interaction and as a measure of the consistency of treatment effects over the different blocks.

4.8 Covariance – extension of linear design models

4.8.1 The use of additional information

Frequently in experiments, the experimenter has additional information about each experimental unit, which might be expected to be related to the yield of the unit. There are various

Table 4.16. *Contrasts for randomised complete block design*

Block	Treatment	L_1	L_2	L_3	L_4	L_5	L_6	L_7	L_8	L_9	L_{10}	L_{11}
1	1	-1	-1	-1	-1	$+1$	$+1$	$+1$	-1	$+1$	$+1$	-1
1	2	-1	-1	-1	$+1$	-1	$+1$	-1	$+1$	$+1$	-1	$+1$
1	3	-1	-1	$+1$	-1	-1	-1	$+1$	$+1$	-1	$+1$	$+1$
1	4	-1	-1	$+1$	$+1$	$+1$	-1	-1	-1	-1	-1	-1
2	1	$+1$	-1	-1	-1	$+1$	-1	-1	$+1$	$+1$	$+1$	-1
2	2	$+1$	-1	-1	$+1$	-1	-1	$+1$	-1	$+1$	-1	$+1$
2	3	$+1$	-1	$+1$	-1	-1	$+1$	-1	-1	-1	$+1$	$+1$
2	4	$+1$	-1	$+1$	$+1$	$+1$	$+1$	$+1$	$+1$	-1	-1	-1
3	1	0	$+2$	-1	-1	$+1$	0	0	0	-2	-2	$+2$
3	2	0	$+2$	-1	$+1$	-1	0	0	0	-2	$+2$	-2
3	3	0	$+2$	$+1$	-1	-1	0	0	0	$+2$	-2	-2
3	4	0	$+2$	$+1$	$+1$	$+1$	0	0	0	$+2$	$+2$	$+2$

reasons why such additional information may not have been used as the basis for choosing a blocking system for the units. The information may not have been available at the time when the experiment was being designed. The number of blocking factors that could be used may have been too large for all factors to be used in constructing blocks. Or, having thought about the possibility of using the methods described in this section, it might have been decided that it was not necessary to use blocking methods.

Examples in agricultural experiments of additional information which might be expected to be relevant to the yield comparisons being investigated include:

(i) In animal experiments, with a blocking system based on genetic characteristics of the animals (i.e. blocks may be litters), it is clear that the growth or other response of the animals may also be related to the initial weight of each animal.

(ii) In fruit tree experiments, blocking is frequently based on the geographical positions of the trees, but much of the variation in the future yields of individual trees may be predictable from previous yielding records.

(iii) In crop experiments, damage by birds, waterlogging or human error may be recognisably uneven between plots, and plot yield may be closely related to some simple quantitative score of damage.

(iv) In any experiment, where units remain in the same spatial positions, trends in yield variation may become apparent during the experiment, resulting from fertility or light trends.

In clinical trials, a great deal of information is available on each individual experimental unit, which usually is an individual patient. Each patient can be classified by age, sex, occupation, various physical characteristics and by historical or health factors appropriate to the particular experiment. There are many potential blocking factors, or covariates. Even if the clinical trial is planned with full information about all patients to be included in the trial available at the outset of the trial, then the use of blocking will not be simple. Patients will not occur in equal numbers for each class of each blocking factor. Typically, the pattern for four blocking factors each with two classes might be as shown in Table 4.17. In most clinical trials, moreover, patients enter the trial sequentially, so that it is impossible to designate a precise blocking structure in advance of the experiment. Methods of

Table 4.17. *Numbers of patients in 16 combinations from four blocking factors*

		X_1		X_2	
		W_1	W_2	W_1	W_2
Y_1	Z_1	3	5	4	4
	Z_2	6	4	8	1
Y_2	Z_1	3	7	2	5
	Z_2	8	0	3	1

sequential allocation intended to avoid extreme forms of non-orthogonality by using restricted, or partially determined, randomisation will be discussed in Section 11.7, but these have not yet been widely accepted. This situation has led some medical statisticians to advocate covariance adjustments as the principal control technique for achieving precision in clinical trials with blocking, or stratification, relegated to a minor, or even non-existent role.

In chemical experiments, obvious blocking factors include different times, different machines, and different sources of material. In addition, preliminary measurements on the experimental material before treatment may reveal other potential differences.

In all these examples, the general philosophy is the same. The object of the experiment is to compare the experimental treatments as precisely as possible. The precision of comparison is determined primarily by the background variation, represented by the variance, σ^2, of the error term, ε. If some of the error variation can be related to variation in additional variables, measured on each experimental unit and called covariates, then the effective background variance σ^2 will be reduced and the treatment comparisons, which may require adjustment to allow for uneven patterns of values of the additional variables, can be made more precise.

The dependence of the main variate on the covariate is represented by adding a regression term to the design model for yield variation. The covariate may be a quantitative variable as in linear regression with the covariance effect represented by a linear regression coefficient, or can be a qualitative variable taking values 0 or 1, in which case the covariance effect represents the difference between the two groups of units for which the covariate is 0 and 1 respectively.

4.8.2 *Covariance analysis for a randomised block design*

To demonstrate the ideas of covariance analysis in more detail, we shall consider the algebraic form of the calculations for a randomised block design with a single covariate. The initial design model is

$$y_{ij} = \mu + b_i + t_j + \varepsilon_{ij}$$

and the full model with a covariate, x, is

$$y_{ij} = \mu + b_i + t_j + \beta(x_{ij} - \bar{x}) + \varepsilon_{ij}.$$

As mentioned in the previous section, covariance analysis takes the form of a regression of the residuals of the principal variate, y, on the residuals of the covariate, x. The residuals represent the variation of y, or x, after allowing for the systematic effects of blocks and treatments. The covariance coefficient is calculated like any simple regression coefficient, as the ratio of the sum of products of residuals of x and y to the sum of squares of the residuals of x.

The SS of the residuals of y is the error SS

$$R_{yy} = \sum_{ij} y_{ij}^2 - \sum_i Y_{i.}^2/t - \sum_j Y_{.j}^2/b + Y_{..}^2/(bt)$$

and correspondingly the SS of the residuals of x is

$$R_{xx} = \sum_{ij} x_{ij}^2 - \sum_i X_{i.}^2/t - \sum_j X_{.j}^2/b + X_{..}^2/(bt).$$

The sum of products of residuals is $\sum_{ij}(y_{ij} - \hat{y}_{ij})(x_{ij} - \hat{x}_{ij})$, which can be written

$$R_{xy} = \sum_{ij} x_{ij}y_{ij} - \sum_i X_{i.}Y_{i.}/t - \sum_i X_{.j}Y_{.j}/b + X_{..}Y_{..}/(bt).$$

The covariance coefficient is

$$\hat{\beta} = R_{xy}/R_{xx},$$

the covariance SS is

$$(R_{xy})^2/R_{xx}$$

and the residual SS is

$$S_r = R_{yy} - R_{xy}^2/R_{xx}.$$

The treatment effects after adjustment for covariance are estimated by the adjusted treatment means:

$$z_j = y_{.j} - \hat{\beta}(x_{.j} - \bar{x}).$$

To examine the practical effect of this adjustment consider the diagram shown in Figure 4.2. If, for three treatments A, B and C, it is assumed that the covariance relationship has the same slope, then the three parallel lines in Figure 4.2 represent the covariance relationships for the three treatments. The range of values of x is not the same for the different treatments and the three different ranges are indicated by the horizontal spreads of the three lines. If the x values are ignored and the three mean y values are directly compared, then the real differences between the treatments will be misrepresented and the treatment means may occur in the wrong order. Covariance analysis standardises the mean y yields so that they are compared at a constant x value. Traditionally the constant x value is usually taken as $\bar{x}$, but since the covariance slope is assumed invariant over treatments the comparison of adjusted means is unaltered by any choice of the standard value of x.

If the covariance slope is positive, then yields for treatments with above average x values should be adjusted downwards. In Figure 4.2 treatment B has an unusually high set of x values and would be expected to produce correspondingly higher y values by virtue of the positive covariance whether it is a superior treatment (in terms of y yields) or not.

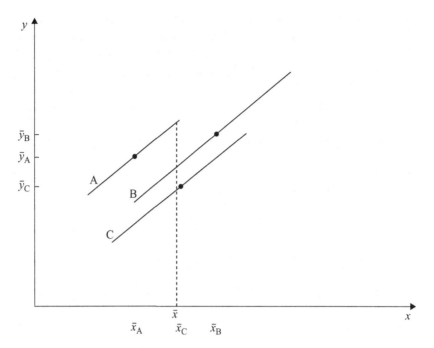

Figure 4.2 Covariance relationships for three treatments with different ranges for the x-variable.

The difference between two adjusted treatment means is

$$z_j - z_{j'} = (y_{.j} - y_{.j'}) - \hat{\beta}(x_{.j} - x_{.j'}).$$

The variance of this difference is obtained from the variances of its two components. The variance of $(y_{.j} - y_{.j'})$ is $2\sigma^2/b$, and the variance of $\hat{\beta}$ is σ^2/R_{xx}. The covariance between the two components is zero because $\hat{\beta}$ is a linear combination of residuals of y and the linear coefficients (which are residuals of x) sum to zero over blocks for each particular treatment. Hence, the variance of $z_j - z_{j'}$ is

$$\mathrm{Var}(z_j - z_{j'}) = 2\sigma^2/b + \sigma^2(x_{.j} - x_{.j'})^2/R_{xx},$$

where σ^2 is estimated from the residual SS,

$$s^2 = S_\mathrm{r}/\{(b-1)(t-1) - 1\}.$$

In practice, this variance, although different for each pair of treatments, will often vary only slightly, and in the interests of simplifying the presentation of results it may be reasonable to present the adjusted treatment means, with a single 'average' variance

$$(2\sigma^2/b)(1 + \mathrm{T}_{xx}/\mathrm{R}_{xx}),$$

where T_{xx} is the treatment MS in the analysis of variance for x,

$$\mathrm{T}_{xx} = \left\{ \sum_j X_{.j}^2/b - X_{..}^2/(bt) \right\} \bigg/ (t-1).$$

To assess the overall significance of the treatment effects allowing for the covariance adjustment, the covariance effect is calculated for the model omitting the treatment effects to obtain the adjusted residual SS omitting treatments, as follows:

$$R'_{yy} = \sum_{ij} y_{ij}^2 - \sum_i Y_{i.}^2 / t,$$

$$R'_{xx} = \sum_{ij} x_{ij}^2 - \sum_i X_{i.}^2 / t,$$

$$R'_{xy} = \sum_{ij} x_{ij} y_{ij} - \sum_i X_{i.} Y_{i.} / t,$$

$$S'_r = R'_{yy} - (R'_{xy})^2 / R'_{xx}.$$

The adjusted treatment SS, allowing for the effect of covariance, is

$$S'_r - S_r \quad \text{on} \quad (t-1)\,\mathrm{df},$$

and the test of the null hypothesis of no treatment effects is

$$F = \frac{(S'_r - S_r)/(t-1)}{S_r/\{(b-1)(t-1)-1\}}.$$

4.8.3 Examples of the use of covariance analysis

Example 4.1 For the first example we consider the trial of pain-relieving drugs for which the data are given in example E, in the introduction to this chapter; these data are also used for illustration in Chapters 6, 8 and 9.

It seems possible that the level of pain might be expected to diminish with time after the operation and that as a result a treatment applied later might be expected to achieve more hours of pain relief. This would mean that the performance of a drug following a less effective drug might appear to be worse than that of the same drug following a more effective drug. A possible model for the results could be

$$y_{ij} = \mu + p_i + o_j + d_{k(ij)} + \beta(x_{ij} - \bar{x}) + \varepsilon_{ij},$$

where p_i, o_j, d_k are patient, order and drug effects and x_{ij} is the hours of relief afforded by the previous drug (x_{ij} will be zero for the first drug for each patient). The residual SS after fitting various models are given in Table 4.18, and the two resulting analyses of variance are given in Table 4.19(a) and (b).

In this instance the use of a covariance term in the model appears to account for most of the apparent differences between drugs. The F statistic to test for differences between the effects of the drugs allowing for the effect of the covariate is $13/8.3$ and is clearly unconvincing. However, when the fitted model is examined the fitted value for $\hat{\beta}$ is found to be -0.77, which is clearly not compatible with the assumption made of an increase in the hours of relief when the drug administration is delayed.

The explanation of the results appears to be that because the two drugs which each patient receives are always different the covariance effect is really describing the negative correlation within each pair of drugs. If one drug is better than average the other will tend to be worse

Table 4.18. *Sums of squares for pain relief data*

Fitting	Residual SS	df
μ	1417	85
μ, p_i	810	43
μ, p_i, o_j	548	42
μ, p_i, o_j, d_k	432	40
$\mu, p_i, o_j, d_k, \beta$	325	39
μ, p_i, o_j, β	351	41

Table 4.19. *Analysis of variance for pain relief data: (a) with drugs before covariance; (b) with covariance before drugs*

(a)

Source	SS	df	MS
Patients	607	42	
Order	262	1	
Drugs	116	2	
Covariance	107	1	107
Residual	325	39	8.3
Total	1417	85	

(b)

Patients	607	42	
Order	262	1	
Covariance	197	1	
Drugs	26	2	13
Residual	325	39	8.3
Total	1417	85	

than average. The use of a trial in which each patient experiences two different drugs and the artificial environment which that design imposes on the comparison of drugs will produce uninformative results and the negative covariance could be an artefact of the design.

If the relation between the hours relief from the first and second drugs is examined separately for each of the six groups of patients receiving a particular ordered pair of drugs, then the six regression relationships shown in Figure 4.3 are not strong or consistent and there is clearly very little to be gained from any form of covariance in the analysis of these data. The conclusions from the trial should be those from the analysis ignoring covariance (given in Chapter 8) with most of the information coming from the analysis of the effects of the first drug only (given in Chapter 6).

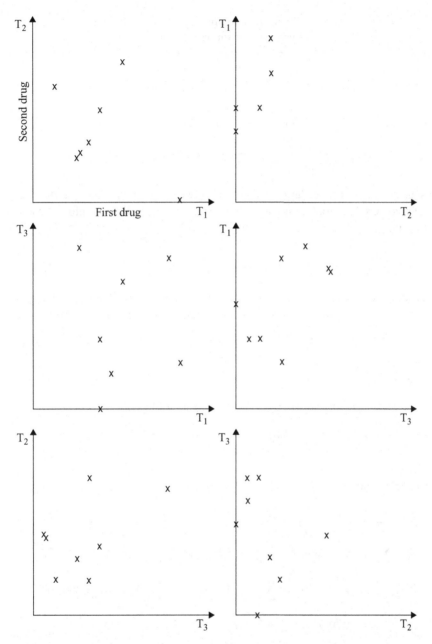

Figure 4.3 Relationship between relief from second drug and relief from first drug for each pair of drugs.

Example 4.2 An illustration of how covariance analysis can improve the precision of information from an experiment is provided by the strawberry experiment data in example D at the beginning of this chapter. The initial analysis of variance of yield, ignoring the hedge effect, is shown in Table 4.20.

Table 4.20. *Analysis of variance for strawberry data*

	SS	df	MS
Blocks	1.22	3	
Varieties	29.08	7	4.15
Error	42.70	21	2.03
Total	73.00	31	

This suggests that there are large differences between varieties. However, the hedge effect also appears to be substantial, as is shown by the total yields of the four plots at each distance from the hedge:

distance	8	7	6	5	4	3	2	1
total yield	27.8	27.1	23.9	26.2	25.4	21.6	21.5	11.1

The most reasonable simple covariance model would seem to be that the effect of the hedge on yields diminishes as the reciprocal of the distance from the hedge. We therefore try a covariate

$$x = 1/\text{distance}$$

giving eight x values:

$$0.125, \ 0.143, \ 0.167, \ 0.200, \ 0.250, \ 0.333, \ 0.500, \ 1.000.$$

The graph of individual yields against x shown in Figure 4.4 suggests that the model is a reasonable one.

The block and variety totals for y and x are given in Table 4.21. To calculate the analysis of covariance, we need

$$\sum_{ij} y_{ij}^2 = 1140.22, \ \sum_{ij} x_{ij}^2 = 6.1094, \ \sum_{ij} x_{ij} y_{ij} = 51.974,$$

$$\sum_{i} Y_{i.}^2/8 = 1068.44, \ \sum_{i} X_{i.}^2/8 = 3.6938, \ \sum_{i} X_{i.} Y_{i.}/8 = 62.786,$$

$$\sum_{j} Y_{.j}^2/4 = 1096.30, \ \sum_{j} X_{.j}^2/4 = 4.1019, \ \sum_{j} X_{.j} Y_{.j}/4 = 60.001,$$

$$R_{yy} = 42.70, \ R_{xx} = 2.0075, \ R_{xy} = -8.027,$$

$$\hat{\beta} = -8.027/2.0075 = -4.186,$$

covariance SS $= (-8.027)^2/2.0075 = 32.096,$

residual SS $= 10.60.$

The analysis of covariance is shown in Table 4.22, and the adjusted treatment means are calculated as shown in Table 4.23. Note that in Table 4.23 we have adjusted the yields to an x value of 0.125, i.e. to the level of x at the furthest point from the hedge within the experiment. The predicted yields for an 'average' hedge effect are not appropriate because the hedge has clearly caused a reduction from the 'normal' yield, and the prediction should discount the

Table 4.21. *Block and variety totals*

	y	x
Block 1	43.5	2.718
Block 2	47.2	2.718
Block 3	47.0	2.718
Block 4	47.1	2.718
Variety G	26.2	0.825
Variety V	25.2	1.101
Variety Rl	14.0	2.334
Variety F	22.5	1.176
Variety Re	22.3	1.093
Variety M	22.4	1.667
Variety E	26.2	1.083
Variety P	26.0	1.593
Total	184.8	10.872

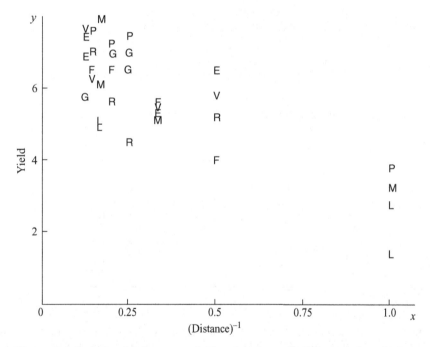

Figure 4.4 Relationship between yield and reciprocal of distance from hedge. Varieties represented by letter, with L for Rl and R for Re.

hedge effect. The complete set of standard errors for comparing the eight varieties is shown in Table 4.24. Clearly the standard errors are very similar, and it is sensible to use the average variance to calculate a common standard error of differences:

$$\sqrt{2(0.53)/4(1 + 0.0583/2.0075)} = 0.52.$$

Table 4.22. *Analysis of covariance*

	SS	df	MS
Blocks	1.22	3	
Treatments	29.08	7	
Covariance	32.10	1	32.10
Error	10.60	20	0.53
Total	73.00	31	

Table 4.23. *Adjusted treatment means*

Variety	$y_{.j}$	$x_{.j}$	$y_{.j} + 4.186(x_{.j} - 0.125)$
G	6.55	0.206	6.89
V	6.30	0.275	6.93
Rl	3.50	0.584	5.42
F	5.62	0.294	6.37
Re	5.58	0.273	6.20
M	5.60	0.417	6.82
E	6.55	0.271	7.16
P	6.50	0.398	7.64
Mean		0.340	

Table 4.24. *Standard errors for comparing pairs of treatments*

	V	Rl	F	Re	M	E	P
G	0.52	0.55	0.52	0.52	0.53	0.52	0.52
V		0.54	0.51	0.51	0.52	0.51	0.52
Re			0.54	0.54	0.52	0.54	0.52
F				0.51	0.52	0.51	0.52
Rl					0.52	0.51	0.52
M						0.52	0.51
E							0.52

This may be compared with the SED from the analysis without covariance, which is $\sqrt{2(2.03)}/4 = 1.01$. To make an overall test of significance of variety differences, allowing for the hedge covariate effect, we calculate

$$R'_{yy} = 71.78,$$
$$R'_{xx} = 2.4156,$$
$$R'_{xy} = -10.812,$$
$$S'_r = 71.78 - (10.812)^2/2.4156 = 23.39,$$

Table 4.25. *Analysis of variance for strawberry data, fitting covariance before varieties*

	SS	df	MS	F
Blocks	1.22	3		
Covariance (ignoring varieties)	48.39	1		
Varieties (allowing for covariance)	12.79	7	1.83	3.45
Error	10.60	20	0.53	
Total	73.00	31		

Table 4.26. *Analysis of variance for strawberry data, splitting the variety variation into two components*

	SS	df	MS	F
Blocks	1.22	3		
Covariance	31.81	1		
Rl v. rest	22.16	1		
Remaining variety variation	7.21	6	1.20	2.26
Error	10.60	20	0.53	
Total	73.00	31		

and the adjusted treatment SS

$$S'_r - S_r = 12.79 \text{ on } 7 \text{ df},$$

which gives a second analysis of variance in Table 4.25.

The conclusions from the analysis are:

(i) The hedge effect is very substantial, and allowing for it reduces the SE of estimated differences between variety effects by a factor of $0.522/1.008 = 0.52$.

(ii) The principal difference between varieties is that variety Rl gives the lowest yield, although the hedge adjustment reduces the amount by which Rl yields less than the other varieties. The variation between the remaining varieties can be tested by splitting the variety SS into one component for the comparison between Rl and the rest, and the remaining variation between the other seven. If this is done, and the latter variation assessed, allowing for covariance, the resulting analysis of variance is given in Table 4.26. The analysis suggests that there is still some variation between the other seven varieties, though it is not significant at the 5% level. We believe the best conclusion is that there is a range of yield variation amongst the eight varieties. The most nearly clear-cut difference is the poor yielding of Rl. The best variety is P with an advantage of about one standard error over its nearest rival, E.

(iii) Experimenters should be a lot more careful in choosing blocking structures for their experiments!

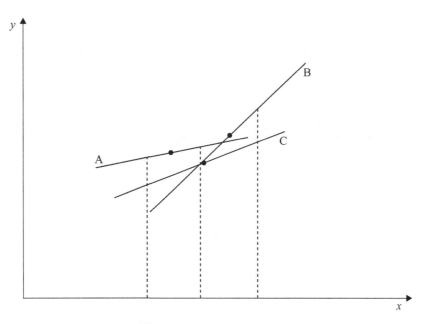

Figure 4.5 Differing relationships for three treatments.

4.8.4 Assumptions and implications of covariance analysis

To examine the additional assumptions required for a covariance analysis, consider the general form of a covariance model:

$$\text{yield} = \text{design model} + \text{regression model} + \text{error}$$

or, in algebraic notation,

$$\mathbf{y} = \mathbf{A}\boldsymbol{\theta} + \mathbf{X}\boldsymbol{\beta} + \boldsymbol{\varepsilon}.$$

The regression component of the model is clearly assumed to be independent of the design component. In particular, the regression relationship of y on a particular covariate is assumed to have the same slope for all the experimental treatments. This is a very strong assumption, and it is important for the statistician and experimeter to consider the assumption, and to make a positive decision that the assumption is reasonable. Writing the model in the above, simple, form also makes it clear that it would be possible to consider also models in which the regression coefficients were affected by the experimental treatments. Such models, allowing 'interaction' between treatments and covariates, can be fitted with ease using modern statistical computer packages. However, while fitting such models may be useful in checking the assumptions of covariance analysis, the results are of limited value in interpreting comparisons between treatments, which is the original purpose of the experiment.

To appreciate this, consider the situation for three treatments illustrated in Figure 4.5. The three treatments are assumed to have different rates of dependence of yield (y) on the covariate (x) shown by the three different lines. If the treatment means are adjusted to any specified level of the covariate the comparison between the adjusted treatment means depends critically on the level of x which is chosen.

If the parallel line assumption of the covariance model is correct, then comparisons between treatments are independent of the value of the covariate at which the comparison is made. It is, however, important to determine the level of the covariate used for adjustment for each treatment, so that the interpretation of individual adjusted treatment means is practically relevant. Adjustment to the overall mean value of x for the whole experiment is obviously suitable for a covariate such as the initial body weight of animals in a growth experiment, where the average initial weight is a typical value that would be relevant in practice. However, as clearly noted for the strawberry experiment, $\bar{x}$ is not always a meaningful level to use in standardising the adjusted yields, and in general the adjustment will be

$$z_j = y_{.j} - \beta(x_{.j} - x_0)$$

for an appropriately chosen x_0.

It is sometimes asserted that for a covariance analysis to be valid it is necessary that the covariate be not affected by the experimental treatments. It will often be true that the form of covariate makes it impossible for there to be systematic difference in the value of the covariate *caused* by treatments. However, the model used in covariance analysis does not require that there are no systematic treatment differences for the covariate. There are situations where the covariate is affected by treatments and yet adjustment of treatment comparisons to allow for the variation in the covariate is sensible.

An example occurs in the comparison of onion varieties in a large breeding trial. When the onions are grown from seed, the field germination may fluctuate wildly between different varieties or breeding lines. Although germination may be tested in the laboratory, this does not provide a reliable prediction of field germination, even when the laboratory testing is extensive, which would not be possible for a large variety trial. It is known that onion yields on a per acre basis are substantially affected by crop density. Also, when varieties are selected for large-scale use, their germination rates will be investigated more intensively so that, in future use, any selected variety will be grown at about the optimal density for that variety. In these circumstances, it is essential that varieties in the trial be compared, eliminating the effects of the variation in the achieved field crop density. This comparison may be achieved by using crop density as a covariate in the analysis of the variety trial.

As with any parallel regression model, if the values of the covariates vary too much between treatments, then it is difficult to assess whether the regression slope is consistent for the different treatments. In the context of an experimental comparison of treatments, this means that, if the values of the covariate are very different for different treatments, then it will be difficult to detect treatment differences after allowing for the covariance effect.

4.9 Computers for analysing experimental data

4.9.1 Introduction and some history

There is now a wide range of computer packages for the analysis of data from designed experiments and it is clear that the number and diversity of programs and packages will continue to increase. We will not give precise advice on which packages to use. Such advice would rapidly become out of date, might be irrelevant to many readers immediately because of a lack of availability of some programs and would be clearly biased by the personal preference of the writers. Instead we shall explain the principles on which different types of program for the analysis of experimental data may be based and attempt to advise on the characteristics which should be sought in selecting appropriate computer packages.

It is informative to trace some of the developments of computer programs for the analysis of experimental data. At the time of writing the entire history is only about half a century. When computers first became generally available for statistical use, the methods for analysing experimental data were already well developed. There was a wide range of designs for which the algebraic definition for the calculation of the analysis of variance was well defined. The natural way to attempt to develop statistical programs was to aim to provide programs to perform the calculations, which were then laboriously achieved by manual calculation. There was, at that time, no concept of the increases in computer speed and size that would become available so that thoughts about the possible generality of computer programs were inevitably not considered.

Initially, then, many statisticians wrote programs to analyse randomised complete block designs, Latin square designs and factorial designs, and the complexity of design catered for by these programs gradually increased. Typically, institutions with large experimental programmes developed suites of programs for different designs. The different programs at a single institution tended to have some standardisation of input and output. New programs were developed mainly through modification of existing programs. Special routines for missing values or for covariance were devised and added to many programs in each suite.

Gradually aspirations to generality led to attempts to replace the suites of programs by general packages which would provide analyses for most of the frequently used designs. Initially the dominant characteristic of designs which the general programs were constructed to accept was orthogonality. Because orthogonality allowed the separate calculation of SS and treatment means for orthogonal 'factors', general programs could allow completely unrestricted orders for the introduction and definition of the factors to be included in the model. Programs used two basic principles. Either the SS and effects were calculated directly from the original yields, or the calculations for each set of levels were performed on the residuals from the previous sets of calculations using the philosophy of the sweeping methods developed by Kuipers (1952), Corsten (1958) and Wilkinson (1970). The latter approach was one of the first indications that computers would use different forms of calculations from those in general use before computers.

The major change which encouraged thoughts of computer programs which could handle a very much wider range of designs, far wider than was conceivable in pre-computer days, was the development of computer methods for handling matrices. After the introduction of analysis programs and packages based on the matrix algebra of the earlier sections of this chapter, the scope of the programs and packages that could be made available was determined by the objectives of the programmer or the 'package philosopher'.

As packages become more general there is a tendency for the package to become more complicated to use, and this aspect of packages is considered briefly in the next subsection. In the third subsection we look at the requirements which we believe a package should satisfy. The two main alternative philosophies for the calculations in a package are the factor and the regression approaches and these will be briefly outlined in further subsections. Finally the implications of the computer revolution for the design of experiments, and for the remainder of this book, will be discussed. Because the computer packages include the possibility of analysing data from any of the designs discussed in this book, this section includes some references to concepts which are introduced formally in chapters after this one.

4.9.2 How general, how friendly?

The programs now available range from the simple program written by an experimenter to analyse an experiment with four randomised complete blocks for five treatments, to packages which can handle almost any designed experiment. How should the user choose the packages that should be available in the institution where he or she works, and how should the program to be used in any particular situation be chosen?

There is one very important benefit which is implied by the decision to implement a general program. The limits on available experimental designs have been primarily imposed by the available computational facilities and the mathematical knowledge necessary for devising more complex methods of analysis within the limitation of those facilities. The early history of the development of new experimental designs has reflected the increasing mathematical ingenuity of statisticians in finding design patterns for which the analysis of data could be performed using only the limited computing facilities offered by manually operated calculating machines. Now that packages exist which can provide the analysis for any experiment, including experimental structures deformed by accidents, experimenters and statisticians should avoid choosing to equip themselves only with programs which perform, much more quickly, the analysis of designs available in the precomputer era. Correspondingly the limits on available experimental designs must change.

There are, of course, advantages in using specific programs. If the data for analysis come from a randomised block design with three blocks of a 2^3 factorial set of treatment combinations and a program is available for just that design, then the program will require nothing more than the set of data, perhaps with identification of the factor names, and a definition of the order in which the data are to be presented. A more general program may require preliminary information about the numbers of blocks and of factors and factor levels, and a definition of the pattern of treatment allocation to blocks. It may be tempting to retain some of the simpler, more specific, programs and thereby reduce the need to come to terms with the more general programs. However, research resources will tend not to be used with maximum efficiency if we voluntarily prescribe the scope of experimental designs by relying on specific programs.

One problem with general programs, already implied in the previous discussion, is the knowledge which the user requires before using the program. The difficulty arises directly from the generality of the program. The more general the program, the more information that is required by the program before it can provide the analysis of any particular data set, and the more complex the language required for communication between user and program. The description of input and output information requires the development of a communication system, and this is a difficult skill with a strong educational requirement. Far too little attention has been paid to this aspect of the development of general statistical packages. The need to understand the particular form in which information must be provided for programs is obvious. Input to general programs will usually involve various forms of coded information devised to minimise the actual amount of typed information required to activate the program.

4.9.3 *Requirements of packages for the analysis of experimental data*

The requirements for a general package for the analysis of data from designed experiments are not very different from those for data from observational studies. Nevertheless this discussion will be restricted to experimental data and it will be left to others to make comparisons with observational data. The requirements set out below are an attempt to include everything that is important.

1. *Variables* An initial requirement which applies to all analysis programs is that the program can handle many variables for the same design structure, without repetition of information about that structure. This must include the facility to derive new variables as functions or combinations of the original variables.

2. *Structure* The program should accept *any* structure of blocking and treatment factors. Either explicitly, or implicitly, the structure is defined by the set of experimental units and an indexing of each unit within the set of levels for each factor.

3. *Multiple levels of variation* The program must be able to analyse, in a single analysis, multiple error designs, i.e. designs in which information is available about treatment effects in different levels of the design structure. The program should be capable of performing all the calculations for each level separately and should present results, with correct information about precision, for separate and combined analyses. Further discussion of the concept of combining information from different levels of a design is included in Chapters 9 and 18.

4. *Fitting order* The program must be able to include terms representing sets of effects in the fitted model in any order as specified by the user, and the definition of the order of fitting terms must be simple. The strong implication is that the program must be usable interactively. The order of fitting is important in the interpretation of results, and the determination of the order or, in many cases, orders, in which terms are fitted, must be directly controllable by the user. Default settings for order are not appropriate and positive action should be required of the user. However, there should be restrictions on the order to avoid the possibility of fitting an interaction term before the corresponding main effects are fitted.

5. *Treatment contrasts* The program should allow, and generally require, the user to specify contrasts of levels of treatment factors to represent the questions which the experiment is intended to answer (see Chapter 12 for discussion about objectives).

6. *Quantitative variables* The program should allow the inclusion of terms representing regression dependence on quantitative variables in the fitted model. The quantitative variables may be covariates, or may represent levels of a treatment factor to allow regression of the analysed variable on quantitative treatment factor levels. Within the model quantitative and qualitative terms should have equal status.

7. *Multivariate* The program should allow the joint analysis of several variables, providing a bivariate analysis, or more generally a multivariate analysis of variance.

8. *Subset analysis* The program should allow the analysis of a subset of the experimental units, the subset being definable in terms of the defined structure of the experimental units or through intervention of the user. Correspondingly the program should allow some variables to be defined for only a subset of units.

9. *Non-normal error* The program should permit the definition of non-normal error structures. Essentially the whole set of ideas relating to generalised linear models (for a basic

exposition see McCullagh and Nelder (1989)), generalised linear mixed models and hierarchical generalised linear models (see Lee, Nelder and Pawitan (2006)) should be implementable within the program. This extension of classical linear models should be considered a basic tool in the statistical analysis of experimental data.

10. *User specified additions* The program must present and identify the results from any analysis in such a way that the user can define additional calculations and specify the form of presentation of information within the operation of the program. This will include an extensive provision of graphical facilities, simple arithmetical operations and matrix manipulation.

11. *Presentation of information* The program must provide the user with all information necessary to interpret the results. This information must be clear and succinct without superfluous detail. The essential requirements must include:
    (i) the initial subdivision of information in an analysis of variance including alternative orders of fitting terms, and
    (ii) tables of fitted mean values with the SED between means for all important effects (and not for unimportant effects).

12. *Interactive education* Most important of all, the program must be so written that the facilities provided can be described and presented in such a manner that each user may obtain as much information as is necessary to interpret the results. This will require an initial definition of the general level of statistical knowledge expected of the user, and thereafter the program must allow the possibility of the user requesting information about the methods used. Indeed, the program should provoke the user to assess whether he or she does fully understand the procedures and methods which are being used in the program. Of course, the initial definition could be set at a sufficiently high level of statistical sophistication that no further information could possibly be required. This represents an extreme of user-unfriendliness and should not be considered appropriate in an age when many computer programs are very easily available. Users tend to use user-friendly programs regardless of whether they are statistically valid.

13. *Database information* Packages should have facilities for consulting and summarising the results of analyses from similar experiments and experimental material. For any experiment there will always have been other experiments using similar sets of experimental units, and related treatments. The results for these experiments and the consequent analyses should be available in a data bank in a form which can be accessed by the program being used to analyse the current data. The program should be capable of extracting either general information from previous analyses (error MS, general mean, coefficient of variation) or specific information about mean yields and SEs for treatments or factor levels common to both the current and previous data sets. More generally it should be easy to consider combining the results from several different experiments (this aspect of the analysis of experimental data is considered in more depth in Chapter 19).

4.9.4 The factor philosophy of analysis programs

As mentioned previously the earlier programs for the analysis of experimental data were all based on the pattern of calculations used before computers. The underlying concept was that the units were classified according to several factors, such as blocks, treatments, rows, columns, levels of factor A, levels of factor B and so on. The classes, or levels, of each classification, or factor, are viewed symmetrically as expressed in the model

$$y_{ijk} = \mu + a_i + b_j + c_k + \varepsilon_{ijk}.$$

The analysis of variance provides information about the sets of parameters (a_i), (b_j), (c_k), and an estimate of σ^2, the variance of the ε_{ijk}.

There are two computational problems in the factor based programs. The first arises from the interrelationships between the different factors, the second from the non-absoluteness of the parameter values.

The majority of simple programs for the analysis of variance handle only designs in which all factors are orthogonal to each other. Most of the classical experimental designs, and most of the designs actually used in experiments, assume orthogonal relationships between factors. The consequences of orthogonality are that the SS for the different factors are calculated independently. For orthogonal designs there is no need to consider the order of fitting sets of parameters, and the actual calculation of SS can be achieved in various ways, either directly from the raw data, or successively from the residuals from previous stages in the fitting.

When factors are not orthogonal the interpretation of the analysis of variance depends in varying degrees on the pattern of interrelationships between the different factors. The basic theoretical results required for the analysis of non-orthogonal factors are explained in Section 4.5 on partitioning. The crucial characteristic, computationally, is the pattern of the matrix $(\mathbf{C}^{-1})_{22}$. If the off-diagonal terms of this matrix are all equal, then the parameters in the set θ_2 are said to be *balanced*. The concept of balance is examined in Chapter 7. More generally any pattern of the matrix $(\mathbf{C}^{-1})_{22}$ allows a more or less simple form of analysis, and general analysis programs can be defined to allow different degrees of complexity of this matrix. The degree of complexity permitted within the program defines the class of designs which the program can analyse.

The second problem is that of constraints, discussed in Section 4.4 and arising from the fact that the sets of parameters corresponding to the levels of each factor are not uniquely defined if there are two or more factors. Consider the model

$$\mathrm{E}(y_{ij}) = a_i + b_j.$$

It is not possible from any set of observations on (i, j) combinations to deduce absolute values for either set of parameters without imposing an arbitrary restriction on the other set of parameters. Inferences are possible only for differences within a single set of parameters. This restriction may be managed in various ways either by imposing an arbitrary, non-symmetric constraint on the set of parameters (for example $a_1 = 0$) or by adopting a symmetric constraint which is suitable in the context of the generality of the $(\mathbf{C}^{-1})_{22}$ matrix which the program is designed to handle.

A further difficulty arises from the attempt to retain the symmetry of levels within a factor. If particular treatment contrasts are of interest, then these will, almost invariably, be non-symmetrical in nature. To include a subdivision of treatment SS corresponding to particular contrasts will require a different philosophy from the symmetry of the factor approach and will often be difficult or impossible in the context of a general analysis of variance program based on the factor approach.

The advantages of a general analysis of variance program based on a linear model using a factor approach to models are that the computational methods may be simple and efficient within the degree of generality permitted by the program. It is possible to retain the symmetry of factor levels. The disadvantages are that there is a restriction on the designs which can be analysed, and non-symmetric questions about the factor levels have to be considered separately from the main pattern of the program.

There is an alternative approach to analysing data using the factor approach, the 'sweeping' methods mentioned early in this section. The fundamental concept involves estimating the effects of each level of a factor by sweeping through the data, attributing each yield to the relevant level of the factor, calculating a mean effect for each level and subtracting the mean for each level effect from all yields corresponding to that level. The reduction in the SS of the residuals represents the SS due to the factor used in the sweep. For orthogonal designs the set of SS produced immediately is exactly the same as those produced by the more familiar methods. For non-orthogonal designs the sweeping has to be applied in an iterative scheme. This produces the correct SS for sets of factorial effects and enables any form of non-orthogonal design to be handled. Few analysis of variance packages have adopted the sweeping approach, but more recently interest in the method has revived and the method could be the basis for new packages.

4.9.5 The regression model for analysis programs

The alternative to a factor based approach for the construction of analysis programs is to view all models as essentially multiple linear regression models. Instead of a set of parameters (a_i) we now think of the parameters individually, $a_1, a_2, \ldots, a_m$. This approach immediately removes the question of which classes of design can be handled by the program since, for any design which actually contains information about all the required parameters, the appropriate model can be fitted and analysed. The constraint problem must be addressed directly by the elimination of one parameter from each set (or several parameters from sets of parameters for interaction effects). Essentially if a_1 is omitted or set equal to zero, then the remaining parameters in the set represent deviations from a_1.

In experiments with multiple levels of information, discussed in Chapter 9, it is necessary to go beyond linear regression models to linear mixed models, which have more than one random (error) term. Many packages now include routines for fitting linear mixed models and some also allow generalised linear mixed models, or hierarchical generalised linear models, and non-linear mixed models to be fitted. In principle, this covers all classes of models that experimenters might want to fit to experimental data. However, we must recognise that the usual methods of estimating the parameters of these models, based on residual maximum likelihood (REML) are less well behaved than linear models and can suffer from convergence problems, grossly underestimated SEs and overdependence on distributional assumptions. This continues to be an active area of research, with continual improvements in available software. Some of these issues are discussed further in Chapter 9.

The early history of the analysis of data from designed experiments was dominated by the factor philosophy of analysis because of the easy form of calculations for orthogonal designs. The benefits of the regression approach have been less appreciated because the computational problems of fitting multiple regression models were prohibitive before computers were available. There are three main advantages of the regression approach.

(i) Any data structure can be accommodated, and this includes any experimental design from which some observations are not available. The implications for design will be considered further in the next section.
(ii) The parameters corresponding to treatments can be written in the form most appropriate to the questions which the treatments were chosen to answer. In other words, the model may be expressed directly in terms of treatment contrasts.
(iii) The models may be generalised to allow a much wider class of distributional assumptions.

The advantage of choosing parameters to represent treatment comparisons directly should be considerable. For example, if three equally spaced levels of a quantitative treatment factor are chosen, the pattern of yield response to the factor can be assessed by using a quadratic response relationship. This can be directly represented either by replacing the model

$$y_{ij} = a_i + b_j + \varepsilon_{ij} \quad i = 1, 2, 3$$

by the contrast model

$$y_{ij} = a_L x_L + a_Q x_Q + b_j + \varepsilon_{ij},$$

where x_L and x_Q have values $(-1, 0, 1)$ and $(1, -2, 1)$ for $i = (1, 2, 3)$, or equivalently by the more obvious (and equivalent) regression model

$$y_{ij} = \beta (x_i - \bar{x}) + \gamma (x_i - \bar{x})^2 + b_j + \varepsilon_{ij}.$$

Models allowing the quantitative response for levels of factor A to vary over levels of factor B can be written similarly.

More generally, any set of treatment contrasts may be represented directly as model parameters. This should have the considerable advantage of encouraging the experimenter to think of the analysis of data directly in relation to the original questions instead of going through two stages: first obtaining a measure of the overall variation among the experimental treatments, and subsequently defining a meaningful set of treatment comparisons.

A major development has been that of hierarchical generalised linear models in which the error structure can be modelled quite separately from the model for the expected value of y. The use of hierarchical generalised linear models was mentioned in the subsection on requirements for packages and is extremely important. A full account is given in the monograph by Lee, Nelder and Pawitan (2006).

4.9.6 Implications for design

It is extremely important to realise that the possible scope for the subject of experimental design has been completely changed by the advent of the computer. The chronological development of the theory of experimental design can be seen, with the benefit of hindsight, to consist of a sequence of designs of increasing complexity of analysis. The early orthogonal designs were extended to include some non-orthogonal designs as mathematical devices were discovered to allow the simple computational analysis of data. It is certainly possible to argue that the statistical considerations of the efficiency of designs for providing answers to questions became of secondary importance to the consideration of the possibility of the analysis of the data resulting from the design.

Not only was the theoretical development of design theory constricted by the necessity of a practical form of analysis, and incidentally a method of analysis restricted to the algebraically manipulable theory of least squares estimation, but experimental research developed a very limited catalogue of designs deemed suitable for practical experiments. For the vast majority of experiments all the statistical ideas of blocking became identified with the randomised complete block design with the result that many practical experiments have been and still are designed in the randomised complete block format using blocks which are a complete travesty of the principle of blocking.

It is time for a complete reappraisal of the principles of experimental design without the restraints imposed by the limits of computational practice which existed when such admirable books as Cochran and Cox (1957), Cox (1958), or Kempthorne (1952) were written. In the rest of this book we examine the important concepts of the statistical philosophy of experimental design assuming the availability of general packages for the analysis of experimental data which allow the analysis of data from any experimental design. The mathematical theory of general linear models is necessary to permit a comprehensive assessment of the statistical concepts, but these concepts only rarely require mathematical theory for their development; consequently the following chapters employ very little mathematics.

Appendix to Chapter 4

The sections in the appendix are numbered to correspond to the relevant sections in the main part of the chapter.

4.A2 Least squares estimators for linear models

The general linear model, relating n observations to p parameters, may be written

$$E(\mathbf{y}) = \mathbf{A}\boldsymbol{\theta} \quad D(\mathbf{y}) = \sigma^2 \mathbf{I}.$$

The SS of deviations of y_i from its expected value is

$$S = (\mathbf{y} - \mathbf{A}\boldsymbol{\theta})'(\mathbf{y} - \mathbf{A}\boldsymbol{\theta})$$

and is minimised by

$$\frac{\partial S}{\partial \boldsymbol{\theta}} = 2\mathbf{A}'(\mathbf{y} - \mathbf{A}\boldsymbol{\theta}) = \mathbf{0},$$

giving least squares equations

$$\mathbf{A}'\mathbf{A}\hat{\boldsymbol{\theta}} = \mathbf{A}'\mathbf{y}$$

or

$$\mathbf{C}\hat{\boldsymbol{\theta}} = \mathbf{A}'\mathbf{y},$$

where

$$\mathbf{C} = \mathbf{A}'\mathbf{A}.$$

If the rank of $\mathbf{C}$ is p, $\mathbf{C}^{-1}$ exists and

$$\hat{\boldsymbol{\theta}} = \mathbf{C}^{-1}\mathbf{A}'\mathbf{y}.$$

4.A3 Properties of least squares estimators

The expected value and mean of $\hat{\boldsymbol{\theta}}$ may be obtained as follows:

$$\begin{aligned}
\mathrm{E}(\hat{\boldsymbol{\theta}}) &= \mathrm{E}(\mathbf{C}^{-1}\mathbf{A}'\mathbf{y}) \\
&= \mathbf{C}^{-1}\mathbf{A}'\mathrm{E}(\mathbf{y}) \\
&= \mathbf{C}^{-1}\mathbf{A}'(\mathbf{A}\boldsymbol{\theta}) \\
&= (\mathbf{A}'\mathbf{A})^{-1}\mathbf{A}'\mathbf{A}\boldsymbol{\theta} \\
&= \boldsymbol{\theta}.
\end{aligned}$$

Hence $\hat{\boldsymbol{\theta}}$ is unbiased. Also

$$\begin{aligned}
\hat{\boldsymbol{\theta}} - \mathrm{E}(\hat{\boldsymbol{\theta}}) = \hat{\boldsymbol{\theta}} - \boldsymbol{\theta} &= \mathbf{C}^{-1}\mathbf{A}'\mathbf{y} - \mathbf{C}^{-1}\mathbf{A}'\mathbf{A}\boldsymbol{\theta} \\
&= \mathbf{C}^{-1}\mathbf{A}'(\mathbf{y} - \mathbf{A}\boldsymbol{\theta}).
\end{aligned}$$

Hence

$$\begin{aligned}
\mathrm{Var}(\hat{\boldsymbol{\theta}}) &= \mathrm{E}\left[\{\hat{\boldsymbol{\theta}} - \mathrm{E}(\hat{\boldsymbol{\theta}})\}\{\hat{\boldsymbol{\theta}} - \mathrm{E}(\hat{\boldsymbol{\theta}})\}'\right] \\
&= \mathrm{E}\{\mathbf{C}^{-1}\mathbf{A}'(\mathbf{y} - \mathbf{A}\boldsymbol{\theta})(\mathbf{y} - \mathbf{A}\boldsymbol{\theta})'\mathbf{A}\mathbf{C}^{-1}\} \\
&= \mathbf{C}^{-1}\mathbf{A}'\mathrm{E}\{(\mathbf{y} - \mathbf{A}\boldsymbol{\theta})(\mathbf{y} - \mathbf{A}\boldsymbol{\theta})'\}\mathbf{A}\mathbf{C}^{-1} \\
&= \sigma^2\mathbf{C}^{-1}\mathbf{A}'\mathbf{A}\mathbf{C}^{-1} \\
&= \sigma^2\mathbf{C}^{-1},
\end{aligned}$$

since

$$\mathbf{A}'\mathbf{A} = \mathbf{C}.$$

An important result about least squares estimators is contained in the following theorem proved originally by Gauss.

Theorem 4.3 *Consider estimation of a linear contrast* $\mathbf{l}'\boldsymbol{\theta} = \sum_j \mathrm{l}_j\boldsymbol{\theta}_j$ *by a linear combination of the observations* $\mathbf{m}'\mathbf{y} = \sum_i \mathrm{m}_i y_i$. *Then the unbiased estimator with the minimum variance is the corresponding contrast of the least squares estimators* $\mathbf{l}'\hat{\boldsymbol{\theta}}$.

Proof The first stage of the proof concerns the requirement that the estimator be unbiased, for which we require the condition

$$\mathrm{E}(\mathbf{m}'\mathbf{y}) = \mathbf{l}'\boldsymbol{\theta} \text{ for all } \boldsymbol{\theta},$$

or equivalently

$$\mathbf{m}'\mathbf{A}\boldsymbol{\theta} = \mathbf{l}'\boldsymbol{\theta} \text{ for all } \boldsymbol{\theta},$$

whence

$$\mathbf{m}'\mathbf{A} = \mathbf{l}'.$$

The variance of the estimator is

$$\mathrm{Var}(\mathbf{m'y}) = \mathbf{m'}\,\mathrm{Var}(\mathbf{y})\mathbf{m} = \mathbf{m'm}\sigma^2.$$

Now

$$\begin{aligned}
\mathbf{m'm} &= (\mathbf{m'} - \mathbf{l'C}^{-1}\mathbf{A'} + \mathbf{l'C}^{-1}\mathbf{A'})(\mathbf{m} - \mathbf{AC}^{-1}\mathbf{l} + \mathbf{AC}^{-1}\mathbf{l}) \\
&= (\mathbf{m'} - \mathbf{l'C}^{-1}\mathbf{A'})(\mathbf{m} - \mathbf{AC}^{-1}\mathbf{l}) + \mathbf{l'C}^{-1}\mathbf{A'AC}^{-1}\mathbf{l} \\
&\quad + \mathbf{l'C}^{-1}\mathbf{A'}(\mathbf{m} - \mathbf{AC}^{-1}\mathbf{l}) + (\mathbf{m'} - \mathbf{l'C}^{-1}\mathbf{A'})\mathbf{AC}^{-1}\mathbf{l} \\
&= (\mathbf{m'} - \mathbf{l'C}^{-1}\mathbf{A'})(\mathbf{m} - \mathbf{AC}^{-1}\mathbf{l}) + \mathbf{l'C}^{-1}\mathbf{A'AC}^{-1}\mathbf{l},
\end{aligned}$$

since

$$\mathbf{m'A} = \mathbf{l'C}^{-1}\mathbf{A'A} = \mathbf{l'},$$

and hence $\mathbf{m'm}$ is minimised when $\mathbf{m'} = \mathbf{l'C}^{-1}\mathbf{A'}$, which is the required result. Further, the minimised form is $\mathbf{l'C}^{-1}\mathbf{l}$, which could of course have been deduced from the form of the variance–covariance matrix of $\boldsymbol{\theta}$.

This completes the properties of $\hat{\boldsymbol{\theta}}$. Now consider the estimation of σ^2 and the minimised or residual SS:

$$S_\mathrm{r} = (\mathbf{y} - \mathbf{A}\hat{\boldsymbol{\theta}})'(\mathbf{y} - \mathbf{A}\hat{\boldsymbol{\theta}}).$$

Theorem 4.4 $\mathrm{E}(S_\mathrm{r}) = (n-p)\sigma^2$, *and hence* σ^2 *may be estimated by* $S_\mathrm{r}/(n-p)$.

Proof First note that

$$\begin{aligned}
(\mathbf{y} - \mathbf{A}\boldsymbol{\theta})'(\mathbf{y} - \mathbf{A}\boldsymbol{\theta}) &= \{\mathbf{y} - \mathbf{A}\hat{\boldsymbol{\theta}} + \mathbf{A}(\hat{\boldsymbol{\theta}} - \boldsymbol{\theta})\}'\{\mathbf{y} - \mathbf{A}\hat{\boldsymbol{\theta}} + \mathbf{A}(\hat{\boldsymbol{\theta}} - \boldsymbol{\theta})\} \\
&= (\mathbf{y} - \mathbf{A}\hat{\boldsymbol{\theta}})'(\mathbf{y} - \mathbf{A}\hat{\boldsymbol{\theta}}) + (\hat{\boldsymbol{\theta}} - \boldsymbol{\theta})'\mathbf{A'A}(\hat{\boldsymbol{\theta}} - \boldsymbol{\theta}),
\end{aligned}$$

the other terms being zero since $\mathbf{A'y} = \mathbf{A'A}\hat{\boldsymbol{\theta}}$. Now

$$\mathrm{E}\{(\mathbf{y} - \mathbf{A}\boldsymbol{\theta})'(\mathbf{y} - \mathbf{A}\boldsymbol{\theta})\} = \sum_i \mathrm{E}\{y_i - \mathrm{E}(y_i)\}^2 = n\sigma^2$$

and

$$\mathrm{E}(\hat{\boldsymbol{\theta}} - \boldsymbol{\theta})'\mathbf{A'A}(\hat{\boldsymbol{\theta}} - \boldsymbol{\theta})\} = \sum_{jk} c_{jk}\mathrm{E}\{(\hat{\theta}_j - \theta_j)(\hat{\theta}_k - \theta_k)\}.$$

But

$$\mathrm{E}\{(\hat{\theta}_j - \theta_j)(\hat{\theta}_k - \theta_k)\} = \mathrm{Cov}(\hat{\theta}_j, \hat{\theta}_k) = (\mathbf{C}^{-1})_{jk}\sigma^2.$$

Hence

$$\begin{aligned}
\mathrm{E}\{(\hat{\theta}_j - \theta_j)(\hat{\theta}_k - \theta_k)\} &= \sum_{jk} c_{jk}(\mathbf{C}^{-1})_{jk}\sigma^2 \\
&= \sum_{jk} (\mathbf{C})_{kj}(\mathbf{C}^{-1})_{jk}\sigma^2
\end{aligned}$$

since $\mathbf{C}$ is symmetric and $c_{jk} = c_{kj} = (\mathbf{C})_{kj}$. Hence,

$$
\begin{aligned}
\mathrm{E}\{(\hat{\theta}_j - \theta_j)(\hat{\theta}_k - \theta_k)\} &= \sum_k \left\{ \sum_j (\mathbf{C})_{kj}(\mathbf{C}^{-1})_{jk} \right\} \sigma^2 \\
&= \sigma^2 \operatorname{trace}(\mathbf{C}\mathbf{C}^{-1}) \\
&= \sigma^2 p.
\end{aligned}
$$

Hence

$$
\begin{aligned}
\mathrm{E}(S_r) &= \mathrm{E}\{(\mathbf{y} - \mathbf{A}\theta)'(\mathbf{y} - \mathbf{A}\theta)\} - \mathrm{E}\{(\hat{\theta} - \theta)\mathbf{A}'\mathbf{A}(\hat{\theta} - \hat{\theta})\} \\
&= n\sigma^2 - p\sigma^2
\end{aligned}
$$

as required. Finally, the practically most useful form of S_r is

$$
\begin{aligned}
S_r &= (\mathbf{y} - \mathbf{A}\theta)'(\mathbf{y} - \mathbf{A}\hat{\theta}) \\
&= \mathbf{y}'\mathbf{y} - \mathbf{y}'\mathbf{A}\hat{\theta},
\end{aligned}
$$

since

$$
\mathbf{A}'\mathbf{y} = \mathbf{A}'\mathbf{A}\hat{\theta}.
$$

This relationship is expressed verbally as

$$
\text{residual SS} = \text{total SS} - \text{fitting SS},
$$

where the fitting SS, $\mathbf{y}'\mathbf{A}\hat{\theta}$ or $(\mathbf{A}'\mathbf{y})'\hat{\theta}$, is the SS attributable to the fitted parameters, $\hat{\theta}$.

4.A4 Overparameterisation and constraints

Now consider the general model

$$
\mathbf{y} = \mathbf{A}\theta + \varepsilon,
$$

where the model contains more parameters than are necessary. Formally this is expressed by the statements that

  (i) the matrix $\mathbf{A}$ is singular, or
 (ii) the p columns of $\mathbf{A}$ are linearly dependent, or
(iii) the rank of $\mathbf{A}$ is p', less than p.

It follows that the rank of $\mathbf{C} = \mathbf{A}'\mathbf{A}$ is less than p and that the least squares equations $\mathbf{A}'\mathbf{A}\hat{\theta} = \mathbf{A}'\mathbf{y}$ cannot be directly solved. To solve the equations formally, additional equations defining the relationships between the parameters, θ_j, are needed. These additional equations are called *constraints*, and because there are many possible sets of constraint equations it is necessary to determine the conditions which the constraint equations must satisfy so that there will be unique solutions to the least squares equations. The properties of the resulting modified least squares parameter estimates must also be examined.

Theorem 4.5 *If the rank of $\mathbf{A}$ is p' and the least squares equations are considered without additional constraint equations, then*

(i) there is no set of estimates **My** of the set of parameters, $\boldsymbol{\theta}$, such that the set of estimators are all unbiased for all possible values of $\boldsymbol{\theta}$, but

(ii) an unbiased estimator **m'y** of a linear combination of parameters **l'**$\boldsymbol{\theta}$ does exist provided the set of coefficients **l** satisfies the same linear dependence relation as the columns of the matrix **A**.

Proof (i) The linear relationship between the columns of **A** can be expressed in the form

$$\mathbf{AD} = \mathbf{0},$$

where **D** is a $p \times (p - p')$ matrix of rank $(p - p')$. Now

$$\mathrm{E}(\mathbf{My}) = \mathbf{MA}\boldsymbol{\theta}$$

and hence for **My** to be a set of unbiased estimators for $\boldsymbol{\theta}$ we require

$$\boldsymbol{\theta} = \mathbf{MA}\boldsymbol{\theta} \text{ for all } \boldsymbol{\theta},$$

which would require

$$\mathbf{MA} = \mathbf{1}.$$

Consequently,

$$\mathbf{MAD} = \mathbf{D}$$

and hence

$$\mathbf{D} = \mathbf{0},$$

since $\mathbf{AD} = \mathbf{0}$. But as **D** is of rank $(p - p') > 0$ this cannot be true, and therefore there is no unbiased set of estimators of $\boldsymbol{\theta}$.

(ii) For a linear combination **m'y** to provide an unbiased estimator of **l'**$\boldsymbol{\theta}$ we require

$$\mathbf{l'}\boldsymbol{\theta} = \mathbf{m'A}\boldsymbol{\theta} \text{ for all } \boldsymbol{\theta},$$

whence

$$\mathbf{l'} = \mathbf{m'A}$$

and

$$\mathbf{l'D} = \mathbf{m'AD} = \mathbf{0}.$$

Therefore, if $\mathbf{l'D} = \mathbf{0}$, or equivalently if the linear relations between the columns of **A** hold also for the coefficients of the linear combination, **l**, then there is an unbiased estimator of **l'**$\boldsymbol{\theta}$.

Theorem 4.6 *If a set of* $(p - p')$ *constraints* $\mathbf{B}\boldsymbol{\theta} = \mathbf{0}$ *is imposed, subject to the condition that* **BD** *has full rank* $(p - p')$, *then the modified least squares estimates are given by the least squares equations and the constraint equations, and the minimised SS, S_r, is independent of the particular form of constraint.*

Proof The constraints $\mathbf{B}\boldsymbol{\theta} = \mathbf{0}$ satisfy the condition that **BD** has full rank, which is equivalent to the requirement that the linear dependence relations for the columns of **A** do not hold for the columns of **B**.

Modified least squares estimators are those which minimise

$$S = (\mathbf{y} - \mathbf{A}\boldsymbol{\theta})'(\mathbf{y} - \mathbf{A}\boldsymbol{\theta})$$

subject to the constraint

$$\mathbf{B}\boldsymbol{\theta} = \mathbf{0}.$$

By introducing a vector, $\boldsymbol{\lambda}$, of undetermined multipliers, the least squares estimators of $\boldsymbol{\theta}$ must minimise

$$S' = (\mathbf{y} - \mathbf{A}\boldsymbol{\theta})'(\mathbf{y} - \mathbf{A}\boldsymbol{\theta}) + \boldsymbol{\lambda}\mathbf{B}\boldsymbol{\theta}.$$

Differentiating with respect to $\boldsymbol{\theta}$

$$2(\mathbf{y} - \mathbf{A}\hat{\boldsymbol{\theta}})'\mathbf{A} - \boldsymbol{\lambda}\mathbf{B} = \mathbf{0}.$$

Postmultiplying by $\mathbf{D}$ gives

$$\boldsymbol{\lambda}\mathbf{B}\mathbf{D} = \mathbf{0} \text{ since } \mathbf{A}\mathbf{D} = \mathbf{0},$$

whence

$$\boldsymbol{\lambda} = \mathbf{0},$$

since $\mathbf{B}\mathbf{D}$ is of full rank. Hence the absolute minimum of S is the same as the minimum conditional on $\mathbf{B}\boldsymbol{\theta} = \mathbf{0}$ for any set of constraints for which $\mathbf{B}\mathbf{D}$ has full rank. And the equations satisfied by the modified least squares estimates are

$$\mathbf{A}'\mathbf{A}\hat{\boldsymbol{\theta}} = \mathbf{A}'\mathbf{y},$$
$$\mathbf{B}\hat{\boldsymbol{\theta}} = \mathbf{0}.$$

General solutions may be obtained by using the partitioned matrix

$$\begin{bmatrix} \mathbf{A} \\ \mathbf{B} \end{bmatrix} = \mathbf{A}^*,$$

which is of full rank because the linear dependencies for the columns of $\mathbf{A}$ do not apply to the columns of $\mathbf{B}$. Hence,

$$\mathbf{A}'\mathbf{A} + \mathbf{B}'\mathbf{B} = \mathbf{A}^{*\prime}\mathbf{A}^*$$

is non-singular. Thus the modified least squares equations can be written

$$\mathbf{A}^{*\prime}\mathbf{A}^*\hat{\boldsymbol{\theta}} = \mathbf{A}'\mathbf{y},$$

with the solution

$$\hat{\boldsymbol{\theta}} = (\mathbf{A}^{*\prime}\mathbf{A}^*)^{-1}\mathbf{A}'\mathbf{y} = (\mathbf{A}'\mathbf{A} + \mathbf{B}'\mathbf{B})^{-1}\mathbf{A}'\mathbf{y}.$$

Theorem 4.7 (This theorem extends the earlier Gauss' theorem.) *The minimum-variance unbiased linear estimator of a linear contrast $\mathbf{l}'\boldsymbol{\theta}$ is the corresponding contrast of the modified least squares estimators $\mathbf{l}'\hat{\boldsymbol{\theta}} = \mathbf{l}'(\mathbf{A}'\mathbf{A} + \mathbf{B}'\mathbf{B})^{-1}\mathbf{A}'\mathbf{y}$.*

Proof   A linear estimator, $\mathbf{m}'\mathbf{y}$, will be unbiased if

$$\mathbf{l}'\boldsymbol{\theta} = \mathbf{m}'\mathbf{A}\boldsymbol{\theta}$$

whenenever $\mathbf{B}\boldsymbol{\theta} = \mathbf{0}$ or equivalently

$$\mathbf{l}'\boldsymbol{\theta} = \mathbf{m}'\mathbf{A}\boldsymbol{\theta} + \mathbf{n}'\mathbf{B}\boldsymbol{\theta} \text{ for all } \boldsymbol{\theta},$$

which requires

$$\mathbf{l}' = \mathbf{m}'\mathbf{A} + \mathbf{n}'\mathbf{B}.$$

Here $\mathbf{n}$ is an arbitrary column vector, which may be eliminated by postmultiplying by $\mathbf{D}$:

$$\mathbf{l}'\mathbf{D} = \mathbf{m}'\mathbf{A}\mathbf{D} + \mathbf{n}'\mathbf{B}\mathbf{D} = \mathbf{n}'\mathbf{B}\mathbf{D},$$

whence

$$\mathbf{n}' = \mathbf{n}'\mathbf{D}(\mathbf{B}\mathbf{D})^{-1}$$

and the required condition for $\mathbf{m}'\mathbf{y}$ to be unbiased becomes

$$\mathbf{m}'\mathbf{A} = \mathbf{l}' - \mathbf{l}'\mathbf{D}(\mathbf{B}\mathbf{D})^{-1}\mathbf{B}.$$

But

$$(\mathbf{A}'\mathbf{A} + \mathbf{B}'\mathbf{B})\mathbf{D} = \mathbf{B}'\mathbf{B}\mathbf{D},$$

since $\mathbf{A}\mathbf{D} = \mathbf{0}$ and hence

$$\mathbf{D}(\mathbf{B}\mathbf{D})^{-1} = (\mathbf{A}'\mathbf{A} + \mathbf{B}'\mathbf{B})^{-1}\mathbf{B}',$$

giving the final form of the unbiasedness condition

$$\mathbf{m}'\mathbf{A} = \mathbf{l}' - \mathbf{l}'(\mathbf{A}'\mathbf{A} + \mathbf{B}'\mathbf{B})^{-1}\mathbf{B}'\mathbf{B}$$
$$= \mathbf{l}'(\mathbf{A}'\mathbf{A} + \mathbf{B}'\mathbf{B})^{-1}\mathbf{A}'\mathbf{A}.$$

Now consider the variance of $\mathbf{m}'\mathbf{y}$,

$$\begin{aligned}
\operatorname{Var}(\mathbf{m}'\mathbf{y}) &= \sigma^2 \mathbf{m}'\mathbf{m} \\
&= \sigma^2 \{\mathbf{m}' - \mathbf{l}'(\mathbf{A}'\mathbf{A} + \mathbf{B}'\mathbf{B})^{-1}\mathbf{A}' + \mathbf{l}'(\mathbf{A}'\mathbf{A} + \mathbf{B}'\mathbf{B})^{-1}\mathbf{A}'\} \\
&\quad \times \{\mathbf{m} - \mathbf{A}(\mathbf{A}'\mathbf{A} + \mathbf{B}'\mathbf{B})^{-1}\mathbf{l} + \mathbf{A}(\mathbf{A}'\mathbf{A} + \mathbf{B}'\mathbf{B})^{-1}\mathbf{l}\} \\
&= \sigma^2 [\{\mathbf{m}' - \mathbf{l}'(\mathbf{A}'\mathbf{A} + \mathbf{B}'\mathbf{B})^{-1}\mathbf{A}'\}\{\mathbf{m} - \mathbf{A}(\mathbf{A}'\mathbf{A} + \mathbf{B}'\mathbf{B})^{-1}\mathbf{l}\} \\
&\quad + \mathbf{l}'(\mathbf{A}'\mathbf{A} + \mathbf{B}'\mathbf{B})^{-1}\mathbf{A}'\mathbf{A}(\mathbf{A}'\mathbf{A} + \mathbf{B}'\mathbf{B})^{-1}\mathbf{l}].
\end{aligned}$$

Hence $\mathbf{m}'\mathbf{y}$ has minimum variance when $\mathbf{m}' = \mathbf{l}'(\mathbf{A}'\mathbf{A} + \mathbf{B}'\mathbf{B})^{-1}\mathbf{A}'$, and

$$\operatorname{Var}(\hat{\boldsymbol{\theta}}) = \sigma^2 (\mathbf{A}'\mathbf{A} + \mathbf{B}'\mathbf{B})^{-1}\mathbf{A}'\mathbf{A}(\mathbf{A}'\mathbf{A} + \mathbf{B}'\mathbf{B})^{-1}. \tag{4.11}$$

Since it is still true that $\mathbf{A}'\mathbf{A}\hat{\boldsymbol{\theta}} = \mathbf{A}'\mathbf{y}$ the expression for the minimised residual SS is unchanged

$$S_{\mathrm{r}} = \mathbf{y}'\mathbf{y} - \mathbf{y}'\mathbf{A}\hat{\boldsymbol{\theta}}.$$

Further, since the above results hold for any set of constraints for which $\mathbf{B}\mathbf{D}$ is of rank $(p - p')$ they hold for the special set of constraints which set $(p - p')$ of the parameters of $\boldsymbol{\theta}$ equal to zero. In this case the model has no surplus parameters and the results of the previous section hold. In particular

$$\mathrm{E}(S_{\mathrm{r}}) = (n - p')\sigma^2,$$

since the reduced model has precisely p' parameters.

Before leaving the subject of constraints for overparameterised models it must be emphasised that the theory of this section does not give the usual form of calculations for practical analysis of data. For most analyses of experimental data, the sensible approach is to write down the least squares equations, use sensible constraint equations to simplify the least squares equations as far as possible and then solve the least squares equations directly with the aid of the constraints. The benefit of the theory is in the assurance that it provides that modified least squares estimators can be obtained and that the form of constraint does not affect any practically important estimator. In particular, the variances of estimators are, in practice, not calculated from (4.11) but through considering each estimator as a linear combination of y values.

4.A5 Partitioning the parameter vector and the extra SS principle

Now consider the partition of the p parameter vector $\boldsymbol{\theta}$ into two parameter vectors:

$$\boldsymbol{\theta}_1 \text{ containing } p - q \text{ parameters}$$

and

$$\boldsymbol{\theta}_2 \text{ containing } q \text{ parameters.}$$

Our primary interest will be in $\boldsymbol{\theta}_2$. The model for the observations, $\mathbf{y}$, will be

$$\mathrm{E}(\mathbf{y}) = \mathbf{A}_1 \boldsymbol{\theta}_1 + \mathbf{A}_2 \boldsymbol{\theta}_2,$$

where $\mathbf{A}$ is partitioned into two matrices $\mathbf{A}_1 \{n \times (p-q)\}$ and $\mathbf{A}_2 (n \times q)$. The matrix $\mathbf{C} = \mathbf{A}'\mathbf{A}$ is similarly partitioned:

$$\mathbf{C} = \begin{bmatrix} \mathbf{C}_{11} & \mathbf{C}_{12} \\ \mathbf{C}_{21} & \mathbf{C}_{22} \end{bmatrix},$$

where $\mathbf{C}_{11}$ is $(p - q) \times (p - q)$ and $\mathbf{C}_{22}$ is $q \times q$.

From Section 4.4 (and Section 4.A4), any model for which the rank of $\mathbf{A}$ is less than p, may be reparameterised to produce an effective model for which the least squares matrix, $\mathbf{C}$, can be inverted. Without losing any generality therefore we discuss the theory for partitioned parameter vectors for a model in which the rank of $\mathbf{A}$ is p. Hence $\mathbf{C}^{-1}$ exists and is partitioned according to the same pattern as $\mathbf{C}$:

$$\mathbf{C}^{-1} = \begin{bmatrix} (\mathbf{C}^{-1})_{11} & (\mathbf{C}^{-1})_{12} \\ (\mathbf{C}^{-1})_{21} & (\mathbf{C}^{-1})_{22} \end{bmatrix},$$

where $(\mathbf{C}^{-1})_{11}$ is $(p - q) \times (p - q)$ and $(\mathbf{C}^{-1})_{22}$ is $q \times q$. The least squares equations are, in partitioned form,

$$\mathbf{C}_{11}\hat{\boldsymbol{\theta}}_1 + \mathbf{C}_{12}\hat{\boldsymbol{\theta}}_2 = \mathbf{A}_1'\mathbf{y},$$
$$\mathbf{C}_{21}\hat{\boldsymbol{\theta}}_1 + \mathbf{C}_{22}\hat{\boldsymbol{\theta}}_2 = \mathbf{A}_2'\mathbf{y}.$$

The fitting procedure is taken in two stages. First ignore $\boldsymbol{\theta}_2$ and fit only $\boldsymbol{\theta}_1$, obtaining

$$\boldsymbol{\theta}_1^* = (\mathbf{C}_{11})^{-1}\mathbf{A}_1'\mathbf{y}.$$

Second, fit $\boldsymbol{\theta}_2$ allowing for $\boldsymbol{\theta}_1$ by eliminating $\boldsymbol{\theta}_1$ from the partitioned equations

$$\hat{\boldsymbol{\theta}}_1 = \mathbf{C}_{11}^{-1}\mathbf{A}_1'\mathbf{y} - \mathbf{C}_{11}^{-1}\mathbf{C}_{12}\hat{\boldsymbol{\theta}}_2,$$

whence the second set of least squares equations becomes

$$(\mathbf{C}_{22} - \mathbf{C}_{21}\mathbf{C}_{11}^{-1}\mathbf{C}_{12})\hat{\boldsymbol{\theta}}_2 = (\mathbf{A}_2' - \mathbf{C}_{21}\mathbf{C}_{11}^{-1}\mathbf{A}_1')\mathbf{y}.$$

By consideration of the similar equations arising from the equation $\mathbf{CC}^{-1} = \mathbf{I}$

$$\mathbf{C}_{11}(\mathbf{C}^{-1})_{12} + \mathbf{C}_{12}(\mathbf{C}^{-1})_{22} = \mathbf{0},$$
$$\mathbf{C}_{21}(\mathbf{C}^{-1})_{12} + \mathbf{C}_{22}(\mathbf{C}^{-1})_{22} = \mathbf{I},$$

we can deduce that $(\mathbf{C}_{22} - \mathbf{C}_{21}\mathbf{C}_{11}^{-1}\mathbf{C}_{12})(\mathbf{C}^{-1})_{22} = \mathbf{I}$, and hence

$$\hat{\boldsymbol{\theta}}_2 = (\mathbf{C}^{-1})_{22}(\mathbf{A}_2' - \mathbf{C}_{21}\mathbf{C}_{11}^{-1}\mathbf{A}_1')\mathbf{y}.$$

This result for $\hat{\boldsymbol{\theta}}_2$ is not in any way different from that obtained directly from the original least squares equations. However, use of this form to consider the SS for fitting sets of parameters leads to a particularly neat form for the SS.

Theorem 4.8 *If the SS for fitting the full parameter vector is $S_p = \mathbf{y}'\mathbf{A}\hat{\boldsymbol{\theta}}$, and the SS for fitting only the subset of parameters $\boldsymbol{\theta}_1$ is $S_{p-q} = \mathbf{y}'\mathbf{A}_1'\boldsymbol{\theta}_1$, then the difference $S_q = S_p - S_{p-q}$ is the fitting SS for $\boldsymbol{\theta}_2$ allowing for the effects of $\boldsymbol{\theta}_1$ and is*

$$S_q = \hat{\boldsymbol{\theta}}_2'(\mathbf{C}^{-1})_{22}^{-1}\hat{\boldsymbol{\theta}}_2.$$

Proof

$$\begin{aligned}
S_q &= S_p - S_{p-q} \\
&= \mathbf{y}'\mathbf{A}\hat{\boldsymbol{\theta}} - \mathbf{y}'\mathbf{A}_1\boldsymbol{\theta}_1^* \\
&= \mathbf{y}'(\mathbf{A}_1\hat{\boldsymbol{\theta}}_1 + \mathbf{A}_2\hat{\boldsymbol{\theta}}_2) - \mathbf{y}'\mathbf{A}_1\boldsymbol{\theta}_1^* \\
&= \mathbf{y}'\mathbf{A}_1\big(\mathbf{C}_{11}^{-1}\mathbf{A}_1'\mathbf{y} - \mathbf{C}_{11}^{-1}\mathbf{C}_{12}\hat{\boldsymbol{\theta}}_2\big) + \mathbf{y}'\mathbf{A}_2\hat{\boldsymbol{\theta}}_2 - \mathbf{y}'\mathbf{A}_1\mathbf{C}_{11}^{-1}\mathbf{A}_1'\mathbf{y} \\
&= \mathbf{y}'\big(\mathbf{A}_2 - \mathbf{A}_1\mathbf{C}_{11}^{-1}\mathbf{C}_{12}\big)\hat{\boldsymbol{\theta}}_2 \\
&= \hat{\boldsymbol{\theta}}_2'(\mathbf{C}^{-1})_{22}^{-1}\hat{\boldsymbol{\theta}}_2.
\end{aligned}$$

This SS for fitting $\boldsymbol{\theta}_2$, allowing for the effects of $\boldsymbol{\theta}_1$, is known as 'the extra SS' and the extra SS principle is fundamental to almost all linear models analysis.

Note that the SS for fitting $\boldsymbol{\theta}_2$, ignoring $\boldsymbol{\theta}_1$, is not usually the same as the extra SS for $\boldsymbol{\theta}_2$ after fitting $\boldsymbol{\theta}_1$. The reason for this can be seen simply as follows. The SS for fitting $\boldsymbol{\theta}_2$, without $\boldsymbol{\theta}_1$, is

$$\mathbf{y}'\mathbf{A}_2(\mathbf{C}_{22})^{-1}\mathbf{A}_2'\mathbf{y},$$

whereas the extra SS for $\boldsymbol{\theta}_2$, allowing for $\boldsymbol{\theta}_1$, is

$$\mathbf{y}'\big(\mathbf{A}_2 - \mathbf{A}_1\mathbf{C}_{11}^{-1}\mathbf{C}_{12}\big)\big(\mathbf{C}_{22} - \mathbf{C}_{21}\mathbf{C}_{11}^{-1}\mathbf{C}_{12}\big)^{-1}\big(\mathbf{A}_2' - \mathbf{C}_{21}\mathbf{C}_{11}^{-1}\mathbf{A}_1'\big)\mathbf{y}.$$

It can be seen that a sufficient condition for the equality of these two expressions is $\mathbf{C}_{12} = \mathbf{0}$, implying also $\mathbf{C}_{21} = \mathbf{0}$ by the symmetry of $\mathbf{C}$.

The condition $\mathbf{C}_{12} = \mathbf{0}$ implies that the product of any two column vectors from $\mathbf{A}$, one from $\mathbf{A}_1$ and the other from $\mathbf{A}_2$, is zero. This is simply expressed by saying that the column vectors of $\mathbf{A}_1$ are orthogonal to those of $\mathbf{A}_2$, or equivalently the estimation of the parameters $\boldsymbol{\theta}_1$ and $\boldsymbol{\theta}_2$ is orthogonal.

4.A6 Distributional assumptions and inferences

Theorem 4.9 *Assuming that the $\boldsymbol{\varepsilon}$ are normally distributed the residual SS, S_r, is distributed as $\sigma^2 \chi^2_{n-p}$, independently of $S_p = \mathbf{y}'\mathbf{A}\hat{\boldsymbol{\theta}}$, the fitted SS. Further, if $\boldsymbol{\theta} = \mathbf{0}$, then S_p is itself distributed as $\sigma^2 \chi^2_p$.*

Proof This can be developed from a purely algebraic approach. There is also a geometric interpretation which may assist understanding.

Consider the model

$$\mathbf{y} = \mathbf{A}\boldsymbol{\theta} + \boldsymbol{\varepsilon},$$

and an orthogonal transformation,

$$\mathbf{z} = \mathbf{L}\mathbf{y},$$

where $\mathbf{L}$ is an orthogonal matrix such that

$$\mathbf{H} = \mathbf{L}\mathbf{A} = \begin{bmatrix} t_{11} & t_{12} & t_{13} & t_{14} & \cdots & t_{1p} \\ 0 & t_{22} & t_{23} & & \cdots & t_{2p} \\ 0 & 0 & t_{33} & & \cdots & t_{3p} \\ \vdots & \vdots & & \ddots & & \vdots \\ 0 & 0 & \cdots & & 0 & t_{pp} \\ 0 & 0 & \cdots & & 0 & 0 \\ \vdots & \vdots & & & \vdots & \vdots \\ 0 & 0 & \cdots & & 0 & 0 \end{bmatrix} = \begin{bmatrix} \mathbf{T} \\ \mathbf{0} \end{bmatrix},$$

where $\mathbf{T}$ is an upper triangular matrix $(p \times p)$ and $\mathbf{0}$ is a $(n - p) \times p$ matrix of zero elements.

Define also $\boldsymbol{\eta} = \mathbf{L}\boldsymbol{\varepsilon}$ so that the original model

$$\mathbf{y} = \mathbf{A}\boldsymbol{\theta} + \boldsymbol{\varepsilon}$$

transforms to

$$\mathbf{z} = \mathbf{H}\boldsymbol{\theta} + \boldsymbol{\eta}.$$

(The geometric interpretation thus far is as follows. The design matrix $\mathbf{A}$ is considered as a set of vectors $\mathbf{a}_1, \ldots, \mathbf{a}_p$, where $\mathbf{A} = \{\mathbf{a}_1, \mathbf{a}_2, \ldots, \mathbf{a}_p\}$, and the vectors $\mathbf{y}, \mathbf{a}_1, \mathbf{a}_2, \ldots, \mathbf{a}_p$ are considered in the n-dimensional space. Since the rank of $\mathbf{A}$ is p, the vectors $\mathbf{a}_1, \ldots, \mathbf{a}_p$ span a subspace of dimension p. The orthogonal transformation $\mathbf{z} = \mathbf{L}\mathbf{y}$ effectively rotates the axes so that the first axis lies along $\mathbf{a}_1$, the second in the plane $\{\mathbf{a}_1, \mathbf{a}_2\}$, the third in the hyperplane $\{\mathbf{a}_1, \mathbf{a}_2, \mathbf{a}_3\}$ and so on.)

Since $\mathbf{L}$ is orthogonal

$$\boldsymbol{\eta}'\boldsymbol{\eta} = \boldsymbol{\varepsilon}'\mathbf{L}'\mathbf{L}\boldsymbol{\varepsilon} = \boldsymbol{\varepsilon}'\boldsymbol{\varepsilon}.$$

Also $\boldsymbol{\eta} = \mathbf{L}\boldsymbol{\varepsilon}$ so that the $\boldsymbol{\eta}$ are normally distributed with common variance σ^2. Also

$$\begin{aligned} \hat{\boldsymbol{\theta}} &= (\mathbf{H}'\mathbf{H})^{-1}\mathbf{H}'\mathbf{z} \\ &= (\mathbf{A}'\mathbf{L}'\mathbf{LA})^{-1}\mathbf{A}'\mathbf{L}'\mathbf{y} \\ &= (\mathbf{A}'\mathbf{A})^{-1}\mathbf{A}'\mathbf{y}, \end{aligned}$$

$$S_p = \mathbf{z}'\mathbf{H}\hat{\boldsymbol{\theta}} = \mathbf{y}'\mathbf{L}'\mathbf{LA}\hat{\boldsymbol{\theta}} = \mathbf{y}'\mathbf{A}\hat{\boldsymbol{\theta}},$$

and finally

$$\text{total SS} = \mathbf{z}'\mathbf{z} = \mathbf{y}'\mathbf{L}'\mathbf{Ly} = \mathbf{y}'\mathbf{y}.$$

Hence $\hat{\boldsymbol{\theta}}$, S_r and S_p are unchanged by the transformation $\mathbf{z} = \mathbf{Ly}$. (Because the transformation is simply a rotation of axes the SS, which are distances in the geometrical interpretation, will inevitably be unchanged. The relationship of the fitted model and the data, through the fitted parameter values, also must be unchanged by a rotation of axes.) Now

$$(\mathbf{z} - \mathbf{H}\boldsymbol{\theta})'(\mathbf{z} - \mathbf{H}\boldsymbol{\theta}) = \sum_{i=1}^{p} \left(z_i - \sum_{j=1}^{p} \mathbf{t}_{ij}\theta_j \right)^2 + \sum_{i=p+1}^{n} z_i^2.$$

The parameter estimates are given uniquely by the equations

$$z_i = \sum_{j=1}^{p} \mathbf{t}_{ij}\hat{\theta}_j, \quad \text{or} \quad \mathbf{z}_1 = \mathbf{T}\hat{\boldsymbol{\theta}},$$

where $\mathbf{z}_1$ includes the first p elements of $\mathbf{z}$. These equations can be solved directly and uniquely. But

$$\mathrm{E}(z_i) = 0, \quad i > p \text{ since } z_i = \eta_i.$$

Hence S_r is the sum of $(n - p)$ squared normal variables, η_i, each with zero expectation and variance σ^2 and therefore S_r is $\sigma^2 \chi^2_{n-p}$, independent of $S_p = \sum_{i=1}^{p} z_i^2$.

Further, if $\boldsymbol{\theta} = \mathbf{0}$, $E(z_i) = 0$, $i \le p$ and so S_p is $\sigma^2 \chi^2_p$.

One direct consequence of the theorem is that to test the rather useless hypothesis that $\boldsymbol{\theta} = \mathbf{0}$ the appropriate statistic is

$$F = \frac{S_p/p}{S_r/(n - p)},$$

which is compared with the F-distribution for p and $(n - p)$ df.

One further result for the entire vector of parameters relates to the joint distribution of the parameter estimates. Since

$$(\mathbf{y} - \mathbf{A}\boldsymbol{\theta})'(\mathbf{y} - \mathbf{A}\boldsymbol{\theta}) \text{ is } \sigma^2\chi^2_n$$

and

$$(\mathbf{y} - \mathbf{A}\hat{\boldsymbol{\theta}})'(\mathbf{y} - \mathbf{A}\hat{\boldsymbol{\theta}}) \text{ is } \sigma^2\chi^2_{n-p},$$

and from Section 4.2

$$(\mathbf{y} - \mathbf{A}\boldsymbol{\theta})'(\mathbf{y} - \mathbf{A}\boldsymbol{\theta}) = (\mathbf{y} - \mathbf{A}\hat{\boldsymbol{\theta}})'(\mathbf{y} - \mathbf{A}\hat{\boldsymbol{\theta}}) + (\hat{\boldsymbol{\theta}} - \boldsymbol{\theta})'\mathbf{A}'\mathbf{A}(\hat{\boldsymbol{\theta}} - \boldsymbol{\theta}),$$

then

$$(\hat{\boldsymbol{\theta}} - \boldsymbol{\theta})' \mathbf{A}' \mathbf{A} (\hat{\boldsymbol{\theta}} - \boldsymbol{\theta}) \text{ is } \sigma^2 \chi_p^2,$$

independent of the true values of $\boldsymbol{\theta}$.

Hence, confidence intervals and confidence regions for the parameters can be constructed. These distributional results extend to the partitioned parameter vector as follows.

Theorem 4.10 *If S_q is the fitting SS for $\boldsymbol{\theta}_2$, the extra SS, for which the algebraic expression derived earlier was*

$$S_q = \hat{\boldsymbol{\theta}}_2' (\mathbf{C}^{-1})_{22}^{-1} \hat{\boldsymbol{\theta}}_2,$$

then on the hypothesis that $\boldsymbol{\theta}_2 = \mathbf{0}$, S_q is distributed as $\sigma^2 \chi_q^2$, and is independent of S_r.

Proof Again, use the orthogonal transformation, $\mathbf{L}$, such that

$$\mathbf{L}\mathbf{A} = \begin{pmatrix} \mathbf{T} \\ \mathbf{0} \end{pmatrix}$$

and consider $\mathbf{W} = \mathbf{T}^{-1}$ and the partitioned forms of $\mathbf{T}$ and $\mathbf{W}$:

$$\mathbf{T} = \begin{bmatrix} \mathbf{T}_{11} & \mathbf{T}_{12} \\ \mathbf{0} & \mathbf{T}_{22} \end{bmatrix} \quad \mathbf{W} = \begin{bmatrix} \mathbf{W}_{11} & \mathbf{W}_{12} \\ \mathbf{0} & \mathbf{W}_{22} \end{bmatrix}.$$

Now

$$\mathbf{C}^{-1} = (\mathbf{A}'\mathbf{A})^{-1} = (\mathbf{A}'\mathbf{L}'\mathbf{L}\mathbf{A})^{-1} = (\mathbf{T}'\mathbf{T})^{-1} = \mathbf{W}\mathbf{W}'.$$

Therefore

$$(\mathbf{C}^{-1})_{22} = \mathbf{W}_{22}\mathbf{W}_{22}^{-1}.$$

Since

$$\mathbf{T}_{22}\mathbf{W}_{22} = \mathbf{I},$$
$$(\mathbf{C}^{-1})_{22}^{-1} = \mathbf{T}_{22}'\mathbf{T}_{22}$$

and therefore

$$\hat{\boldsymbol{\theta}}_2' (\mathbf{C}^{-1})_{22}^{-1} \hat{\boldsymbol{\theta}}_2 = \hat{\boldsymbol{\theta}}_2' \mathbf{T}_{22}' \mathbf{T}_{22} \hat{\boldsymbol{\theta}}_2$$

$$= \sum_{i=p-q+1}^{p} \left(\sum_{j=1}^{p} t_{ij} \hat{\theta}_j \right)^2$$

$$= \sum_{i=p-q+1}^{p} z_i^2$$

and, on the hypothesis $\boldsymbol{\theta}_2 = \mathbf{0}$, $E(z_i) = 0$, $p - q + 1 \le i \le p$ and S_q is $\sigma^2 \chi_{n-p}^2$.

This result provides the basic theory for interpreting analyses of variance. The first consequence is that, on the hypothesis that $\boldsymbol{\theta}_2 = \mathbf{0}$,

$$\frac{S_q/q}{S_r/(n-p)}$$

Table 4.27. *General form of analysis of variance with extra sum of squares*

Source	SS	df
Due to θ_1 (ignoring θ_2)	S_{p-q}	$p - q$
Due to θ_2 (allowing for θ_1)	S_q	q
Residual	S_r	$n - p$

is distributed as F for q and $n - p$ df and so the extra SS can be used formally to assess the parameters θ_2.

The χ^2-distributional results also allow df to be added to the analysis of variance structure derived in Sections 4.5 and 4.A5 (see Table 4.27).

It is important to emphasise that, when θ is partitioned into θ_1 and θ_2, then, in the second theorem of the previous section, the two subsets of parameters have quite distinct interpretations. We cannot simply reverse S_q and S_{p-q} and use S_{p-q} to test the hypothesis that $\theta_1 = 0$. In most practical situations, this would not even be a practically desirable procedure. Thus, if θ_1 consists of the set of block effects in a randomised block design and θ_2 is the set of treatment effects, we should not be interested in testing whether $\theta_1 = 0$ because the experiment is designed to allow for block differences, and therefore the analysis of experimental data must reflect this aspect of design. In other words, we should not consider obtaining estimates of treatment effects without simultaneously allowing for the block effects.

However, suppose that we were interested in inferences about θ_1 in addition to those about θ_2 and that we have fitted θ_1 first ignoring θ_2 and subsequently fitted θ_2 allowing for the effects of θ_1. Consider again the expected values of the z_i,

$$E(z_i) = \sum_{j=1}^{p} t_{ij}\theta_j$$

$$= \sum_{j=1}^{p-q}(t_{ij}\theta_j) + \sum_{j=p-q+1}^{p} (t_{ij}\theta_j).$$

The crucial difference between the expressions for the expected values for the first $(p-q)z_i$s and for the next q is that, for the latter z_is, the expected values do not involve θ_1 because of the triangular form of $\mathbf{T}$, whereas the first $(p-q)z_i$s do involve θ_2.

The necessary conditions for $S_{p-q} = \sum_{i=1}^{p-q} z_i^2$ to be distributed as σ^2 times χ^2_{p-q} are not only that $\theta_1 = 0$, which would set the first sum equal to zero, but also that

$$t_{ij} = 0 \quad \text{for} \quad 1 < i < p - q, p - q + 1 < j < p.$$

If the t_{ij} coefficients of θ_2 in the expected values of the first $(p-q)$ z_is are zero, then hypotheses about the θ_1 parameters may be tested independently of the values of θ_2. The condition can be written

$$\mathbf{T}_{12} = \mathbf{0}.$$

This is equivalent to $\mathbf{C}_{12} = \mathbf{0}$, so that $\mathbf{C}$ partitions in the pattern

$$\mathbf{C} = \begin{bmatrix} \mathbf{C}_{11} & \mathbf{0} \\ \mathbf{0} & \mathbf{C}_{22} \end{bmatrix}$$

and the first $p - q$ columns of the design matrix $\mathbf{A}$ are orthogonal to the last q columns. Thus the condition $\mathbf{C}_{12} = \mathbf{0}$, shown in Section 4.A5 to be sufficient for independent interpretation of the two SS S_{p-q} and S_q, is here shown to be necessary also.

4.A7 Treatment comparisons and component SS

A general treatment comparison L_k is defined as a linear contrast of treatment effects

$$L_k = \sum_j l_{kj} t_j,$$

where

$$\sum_j l_{kj} = 0.$$

Two treatment comparisons, L_k, $L_{k'}$, are orthogonal if

$$\sum_j l_{kj} l_{k'j} = 0, \quad k \neq k'.$$

A scale-standardised form of comparison is defined as

$$c_{kj} = l_{kj} \bigg/ \left(\sum_j l_{kj}^2 \right)^{1/2}.$$

Now the coefficients c_{kj} satisfy

$$\sum_j c_{kj} = 0,$$

$$\sum_j c_{kj}^2 = 1,$$

$$\sum_j c_{kj} c_{k'j} = 0, \quad k' \neq k.$$

Consider a set of $(t - 1)$ orthogonal comparisons. We can write

$$\mathbf{x} = \mathbf{C}\hat{\mathbf{t}},$$

where the matrix $\mathbf{C}$ has elements c_{kj} ($k = 1, 2, \ldots, t - 1, j = 1, 2, \ldots, t$), and the vector $\mathbf{x}$ is the set of estimates of the $(t - 1)$ orthogonal standardised contrasts for $L_1, L_2, \ldots, L_{t-1}$. The orthogonality of the contrasts means that the x_ks are independent. If the observations are assumed to be normally distributed, then $\mathbf{x}$ is a vector of normally and independently distributed variables, with each element having zero mean and unit variance.

Define $\mathbf{C}^*$ to be the matrix $\mathbf{C}$ with an added row vector consisting of $t^{-1/2}$ in each column, so that

$$\mathbf{C}^* = \begin{bmatrix} \mathbf{C} \\ t^{-1/2}, t^{-1/2}, \ldots, t^{-1/2} \end{bmatrix};$$

then $\mathbf{C^* C^{*\prime}} = \mathbf{I}$ and $\mathbf{C^*}$ is orthogonal. Now

$$
\begin{aligned}
x_k^2 &= \left(\sum_j c_{kj} \hat{t}_j \right)^2 \\
&= \left(\sum_j l_{kj} y_{.j} \right)^2 \Big/ \left(\sum_j l_{kj}^2 \right) \\
&= \left[\sum_j l_{kj} \left(\mu + t_j + \sum_i \varepsilon_{ij} \Big/ n \right)^2 \right] \Big/ \left(\sum_j l_{kj}^2 \right),
\end{aligned}
$$

where n is the number of observations per treatment (or the effective replication when treatments are not orthogonal to blocks but treatment differences are all equally precise). Since

$$
\sum_j l_{kj} = 0,
$$

$$
x_k^2 = \left\{ \sum_j l_{kj} \left(t_j + \sum_i \varepsilon_{kj} \Big/ n \right)^2 \right\} \Big/ \left(\sum_j l_{kj}^2 \right).
$$

Hence

$$
\begin{aligned}
\mathrm{E}(x_k^2) &= \left(\sum_j l_{kj} t_j \right)^2 \Big/ \left(\sum_j l_{kj}^2 \right) + \left(\sum_j l_{kj}^2 \sigma^2 \right) \Big/ \left(\sum_j l_{kj}^2 n \right) \\
&= \left(\sum_j l_{kj} t_j \right)^2 \Big/ \left(\sum_j l_{kj}^2 \right) + \sigma^2 / n.
\end{aligned}
$$

But if $L_k = 0$, $\mathrm{E}(x_k) = 0$ and, since x_k is normally distributed, then on the hypothesis $H_k : L_k = 0$,

$$
nx_k^2 \text{ is } \sigma^2 \chi_1^2.
$$

Now consider $\mathbf{x'x} = \hat{\mathbf{t}}' \mathbf{C'C} \hat{\mathbf{t}}$. Since

$$
\mathbf{C^{*\prime} C^*} = \mathbf{C'C} + \mathbf{J}(1/t),
$$

where $\mathbf{J}$ is the matrix whose elements are all 1, then

$$
\hat{\mathbf{t}}' \mathbf{C^{*\prime} C^*} \hat{\mathbf{t}} = \hat{\mathbf{t}}' \mathbf{C'C} \hat{\mathbf{t}} + (1/t) \hat{\mathbf{t}}' \mathbf{J} \hat{\mathbf{t}}.
$$

However, $\mathbf{C^*}$ is orthogonal and $\mathbf{J}\hat{\mathbf{t}} = \mathbf{0}$, because of the constraints $\sum_j t_j = 0$. Hence

$$
\hat{\mathbf{t}}' \mathbf{C'C} \hat{\mathbf{t}} = \hat{\mathbf{t}}' \hat{\mathbf{t}}.
$$

Consequently

$$\mathbf{x}'\mathbf{x} = \hat{\mathbf{t}}'\hat{\mathbf{t}}$$
$$= \sum_j (\mathrm{y}_{.j} - \mathrm{y}_{..})^2$$
$$= \sum_j \{\mathrm{Y}_{.j}/n - \mathrm{Y}_{..}/(nt)\}^2$$
$$= (\text{treatment SS})/n.$$

Hence the SS for comparison L_k is identified in the form

$$\mathrm{SS}(\mathrm{L}_k) = nx_k^2 = \left(\sum_j \mathrm{l}_{kj}\mathrm{Y}_{.j}\right)^2 \bigg/ \left(n\sum_j \mathrm{l}_{kj}^2\right),$$

with the properties that

(i) if $\mathrm{L}_k = 0$, the SS is σ^2 times a χ^2 variable on 1 df, and
(ii) the sum of the SS for $(t-1)$ orthogonal comparisons is the total treatment SS.

4.A8 The general theory of covariance analysis

The theory supporting the analysis of covariance is essentially an application of the results for partitioned models in Section 4.5. The theory presented here is for a general design with many covariates.

The covariance model assumes that the yields of experimental units can be modelled in two stages. First, the usual component of the model representing the experimental design with terms for block effects and treatment effects, both sets of effects being possibly structured, and also the block and treatment effects being combined according to the design. The second component is a set of regression terms expressing the dependence of yield on additional measured or descriptive variables.

If the design model is written

$$\mathbf{y} = \mathbf{A}\boldsymbol{\theta} + \boldsymbol{\eta}$$

and the regression model is

$$\boldsymbol{\eta} = \mathbf{X}\boldsymbol{\beta} + \boldsymbol{\varepsilon},$$

then the full model is

$$\mathbf{y} = \mathbf{A}\boldsymbol{\theta} + \mathbf{X}\boldsymbol{\beta} + \boldsymbol{\varepsilon}.$$

Assuming the use of necessary constraints, the parameter estimates for the design model are

$$\hat{\boldsymbol{\theta}} = (\mathbf{A}'\mathbf{A})^{-1}\mathbf{A}'\mathbf{y} = \mathbf{C}^{-1}\mathbf{A}'\mathbf{y}$$

and the residual SS is

$$\mathbf{y}'\mathbf{y} - \mathbf{y}'\mathbf{A}\hat{\boldsymbol{\theta}} = \mathbf{y}'(\mathbf{I} - \mathbf{A}\mathbf{C}^{-1}\mathbf{A}')\mathbf{y}.$$

The quantities $(\mathbf{I} - \mathbf{AC}^{-1}\mathbf{A}')\mathbf{y}$ may be recognised as the residuals after the fitting of the design model. The interpretation of the residual SS and the following results is clarified by the recognition that the matrix

$$\mathbf{R} = \mathbf{I} - \mathbf{AC}^{-1}\mathbf{A}'$$

is symmetric and idempotent. Thus

$$\mathbf{R}' = \mathbf{R}$$

and

$$\mathbf{RR} = (\mathbf{I} - \mathbf{AC}^{-1}\mathbf{A}')(\mathbf{I} - \mathbf{AC}^{-1}\mathbf{A}') = \mathbf{I} - \mathbf{AC}^{-1}\mathbf{A}'.$$

For the full model, the least squares equations are

$$\begin{pmatrix} \mathbf{A}'\mathbf{A} & \mathbf{A}'\mathbf{X} \\ \mathbf{X}'\mathbf{A} & \mathbf{X}'\mathbf{X} \end{pmatrix} \begin{pmatrix} \hat{\boldsymbol{\theta}} \\ \hat{\boldsymbol{\beta}} \end{pmatrix} = \begin{pmatrix} \mathbf{A}'\mathbf{y} \\ \mathbf{X}'\mathbf{y} \end{pmatrix} \tag{4.12}$$

and eliminating $\hat{\boldsymbol{\theta}}$ we obtain equations for the estimation of the covariance effects

$$(\mathbf{X}'\mathbf{X} - \mathbf{X}'\mathbf{AC}^{-1}\mathbf{A}'\mathbf{X})\hat{\boldsymbol{\beta}} = \mathbf{X}'\mathbf{y} - \mathbf{X}'\mathbf{AC}^{-1}\mathbf{A}'\mathbf{y}, \tag{4.13}$$

which we can rewrite in the form

$$\mathbf{X}'(\mathbf{I} - \mathbf{AC}^{-1}\mathbf{A}')\mathbf{X}\hat{\boldsymbol{\beta}} = \mathbf{X}'(\mathbf{I} - \mathbf{AC}^{-1}\mathbf{A}')\mathbf{y}$$

or

$$\mathbf{X}'\mathbf{R}'\mathbf{R}\mathbf{X}\hat{\boldsymbol{\beta}} = \mathbf{X}'\mathbf{R}'\mathbf{R}\mathbf{y}.$$

All the elements in this set of equations are SS and sums of products of the residuals of $\mathbf{y}$ and the residuals of the covariates $\mathbf{x}_i$. Hence we can identify the underlying methodology of covariance analysis as a multiple regression of the residuals of the yield variable, y, on the residuals of the covariates, all residuals being calculated in terms of the basic experimental design represented by the design matrix, $\mathbf{A}$.

The reduction in the residual SS due to fitting $\boldsymbol{\beta}$ is

$$\hat{\boldsymbol{\beta}}'(\mathbf{X}'\mathbf{R}'\mathbf{R}\mathbf{y}),$$

where

$$\hat{\boldsymbol{\beta}} = (\mathbf{X}'\mathbf{R}'\mathbf{R}\mathbf{X})^{-1}(\mathbf{X}'\mathbf{R}'\mathbf{R}\mathbf{y}). \tag{4.14}$$

The covariance analysis has been developed in terms of fitting $\hat{\boldsymbol{\theta}}$ first and $\hat{\boldsymbol{\beta}}$ subsequently so that the analysis of variance may be written as in Table 4.28.

However, the main purpose in introducing the covariates is to improve the precision of estimation of treatment parameters. The adjusted treatment effect estimates may be obtained from the full set of least squares equations (4.14):

$$\hat{\boldsymbol{\theta}} = (\mathbf{A}'\mathbf{A})^{-1}(\mathbf{A}'\mathbf{y} - \mathbf{A}'\mathbf{X}\hat{\boldsymbol{\beta}}),$$

when $\hat{\boldsymbol{\beta}}$ is given by (4.14), and the variance–covariance matrix of these adjusted effects is

$$\mathbf{V}(\hat{\boldsymbol{\theta}}) = \{\mathbf{A}'\mathbf{A} - \mathbf{A}'\mathbf{X}(\mathbf{X}'\mathbf{X})^{-1}\mathbf{X}'\mathbf{A}\}^{-1}\sigma^2.$$

Table 4.28. *Analysis of covariance*

Fitting	SS
$\hat{\theta}$ (ignoring $\hat{\beta}$)	$\mathbf{y'AC^{-1}A'y}$
$\hat{\beta}$ (allowing for $\hat{\theta}$)	$(\mathbf{X'R'Ry})'(\mathbf{X'R'RX})^{-1}(\mathbf{X'R'Ry})$
Residual	S_r

Whether or not the block and treatment effects were orthogonal in the original design model, it will usually not be true that they are orthogonal after adjustments for the effects of covariance. Hence, to assess the variation due to treatment effects, the reduced model including only the block effects (θ_1) and the covariance effects β is fitted. The fitting SS are:

(i) for $\hat{\theta}_1$ $\mathbf{y'A_1C_{11}^{-1}A_1'y}$,

(ii) for $\hat{\beta}$ in addition $(\mathbf{X'R_1'R_1y})'(\mathbf{X'R_1'R_1X})^{-1}(\mathbf{X'R_1'R_1y})$, where $\mathbf{R_1 = I - A_1C_{11}^{-1}A_1'}$.

The new residual SS, S_r', will of course be greater than the original S_r. The difference $S_r' - S_r$ is the SS for fitting the treatment effects $\hat{\theta}_2$, allowing for block effects and covariance effects.

The calculation of SS for any subset of treatment effects adjusted for the covariance effect is performed in the same manner as for the full treatment SS. That is, a design model is fitted omitting the subset of treatment effects to be tested, and any related effects which derive directly from those effects (for example, interaction effects when main effects of a factor are being considered); the covariance effects are then added to the fitted model, and the difference between the two RSS after using the full and partial design models represents the variation due to the omitted terms, allowing for the covariance effect.

Degrees of freedom for the various fitted SS are calculated in the usual way for a partitioned model. Effectively, this means that the df for any set of design parameters from the original model, θ, will be unchanged when covariance terms are included in the model.

Part II

First subject

5

Experimental units

5.0 Preliminary examples

(*a*) Gene expression studies using spotted microarray technologies allow the comparison of gene transcription responses for different experimental samples, such as plant samples taken from different plant lines (a wild-type and lines with different genetic mutations), having been exposed to different environmental conditions or inoculated with a pathogen, and possibly collected at different times after exposure or incoluation. Each microarray contains probes (spots) for a large number (many thousands) of genes, with these probes arranged in a rectangular grid within the microarray. Some microarray technologies only allow one sample to be hybridised to each array, but others (multichannel systems) allow a mixture of two (or more) samples to be hybridised to the array, with the samples being differentially labelled (using fluorescent dyes) prior to being mixed, and separate responses being measured for each of the fluorescent labels. Scientific interest is in both the patterns of gene expression measured for each probe across experimental samples, and in the relationships between gene expression responses measured for different probes within and between experimental samples. For multichannel systems, generally each combination of array and channel might be considered as the experimental unit, though considering the probe (gene) as a 'treatment' (as some analysis approaches do) suggests that each spot should be considered as the experimental unit. An added complication in most gene expression microarray studies is that the experimental samples will have been obtained from a field, glasshouse or controlled environment experiment, so that the design of this earlier experiment, and the various processing steps between initial experimental sample and microarray sample, should be considered in determining the structure of the microarray experiment. Similar issues arise with other 'high throughput' technologies used in molecular biology studies.

(*b*) A clinical trial was proposed to investigate the properties of a gel for rubbing on breasts to alleviate monthly breast pain in young women. The trial was designed to run for twelve months, and it was proposed to have two treatments, a new gel, and a placebo ointment. It was decided that there should be twice as many observations with the new gel as with the placebo, and it was proposed that this should be achieved by having three groups of women. The first group would use the new gel for all twelve months, the second group would use the placebo for the first six months and the new gel for the last six months, and the third group would use the new gel for the first six months, and the placebo for the last six months. This design implies that the experimental unit is not each woman in the trial, but is a period of six months for each woman. The design efficiency, having two groups with changing treatment and one group with no changes can be considerably improved, and this derives from the correct recognition of the experimental unit. Because the trial regime would involve considerable discomfort for the participants, it was important to design the trial as efficiently as possible, thereby involving as few women as possible while having enough participants to achieve sufficient precision.

(*c*) A large-scale experiment on the effectiveness of the control of tuberculosis in badgers, and the impact on the wider ecological environment, established a small number of very large areas, within each of which one of three 'treatments' was applied; the three treatments were: (a) no action, (b) total cull and (c) partial cull of badgers based on detection of tuberculosis incidence. Within each area detailed records of badger and other populations were made at various trapping and recording sites.

(*d*) A chemical plant has a large number of input materials, the amounts and constituents of which can all be varied, and operates under a set of conditions (temperatures, pressures, time periods for component activities), many of which can be varied. Therefore, a detailed description of how the plant operates can include many components, each of which could be at a range of different levels. An experiment to determine the optimal conditions for running the plant, where the criterion for optimality is clearly defined, will vary the levels of perhaps as many as 20 factors. This may be achieved by changing the settings for inputs and running conditions and running the plant for a period, which might be a day, or might be much shorter, for example half an hour. After the full set of experimental conditions has been completed the results are analysed using a mathematical model and an estimate of the optimum conditions obtained.

(*e*) The production of wine has two distinct stages where experimental research can take place, and sometimes an experiment can involve both. Different varieties of the vine, and grape, can be grown in the field, and the juice from the different field plots of grapes can be transferred to the laboratory, where the wine production process can be varied in the same kind of way as described in the previous example. The experimental design problem is to combine the designs that would be used if two separate trials were being planned, recognising, for example, where the blocking systems of the field trial should be relevant in the laboratory trial, and also deciding how the experimental units for the two stages are related.

The final two examples come from situations where the experimenter designed an experiment without consulting a statistician. When the experimenter came, with the experimental data, to consult the statistician the first problem was to identify the experimental units used and the structure of the design and then to provide a suitable analysis.

(*f*) The first experiment was concerned with the influences on the production of glasshouse tomatoes of differing air and soil temperatures. Eight glasshouse compartments were available, and these were paired in four 'blocks'. Each compartment contained two large troughs in which the tomatoes were grown and in each half of each trough the soil temperature could be heated to a required level or left unheated. In one compartment of each block the minimum air temperature was kept at 55 °F and in the other the minimum air temperature was 60 °F. In each trough, one half was maintained at the control soil temperature while the other half was at an increased temperature, the increased temperature being 65 °F for one trough, and 75 °F for the other trough. The design layout for one pair of compartments is shown in Figure 5.1. Yields of tomatoes from each half trough were recorded.

(*g*) The second experiment was also concerned with heating, this time the heating characteristics of different forms of plastic pot situated in cold frames and of different forms of

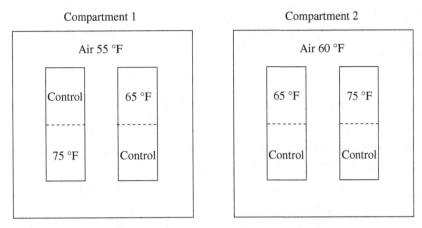

Figure 5.1 Design layout for one pair of compartments for the tomato experiment on the effects of air and soil heating.

covers for the frames. Six wooden frames were divided into four quarters and four different forms of pot were allocated to the four quarters randomly in each frame. The frames were covered either by glass or by polythene, three frames having glass covers and three having polythene covers. All the heat was derived from solar energy. After one week of observation, the covers were exchanged so that each frame that had previously been covered by glass was now covered by polythene and vice versa. The pot positions were unaltered. The design is represented in Figure 5.2. For each pot, in each week, the soil temperature was recorded.

5.1 Different forms of basic experimental units

It is extremely important when planning an experiment to define precisely what the experimental unit is. This may be thought to be too obvious to require discussion, but frequently there are several possible options. The analysis of information and the conclusions from results can be misleading if the experimental unit is not precisely defined. Part of the definition of an experimental unit involves both the 'treatments' that are to be applied and the kinds of measurements which are to be taken. We also need to have a very clear concept of the population from which the sample of experimental units has been taken, and for which the inferences from the experiment will apply.

The simplest possible situations, for agricultural, and for medical experimental units, are the following:

(i) In an agricultural trial different varieties of potatoes are grown on different plots, each consisting of four rows of 8 m. The experimental units are the individual plots, with treatments being the different varieties (which may result from a breeding programme), the measurements are the total yields of potatoes from each plot, and the populations which the results will represent, are the sets of potato plants from each of the different varieties, or possibly from all the possible varieties which might result from such a breeding programme.

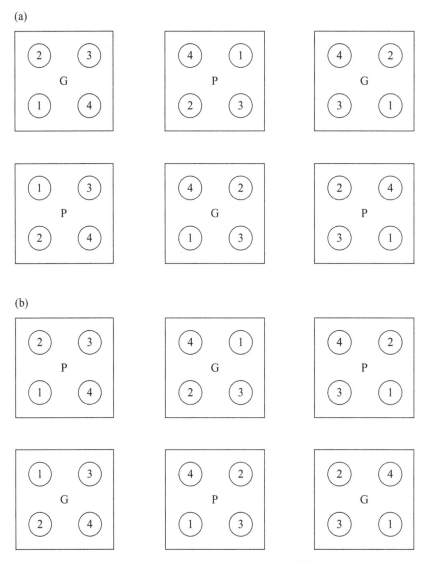

Figure 5.2 Design layout for experiment on effects of different materials on soil heating in pots: (a) week 1; (b) week 2. P = polythene, G = glass.

(ii) In a clinical trial to compare two dosage levels of a new drug, with the best available standard drug treatment, quite large numbers of patients (perhaps as many as 100 per drug) may be needed to detect with near-certainty whether there is a clear benefit of one dosage level compared with the other, and also with the placebo, or standard treatment. The experimental units are the individual patients, the measurements are appropriate measures of quality of life, and the population which the results will represent is the population from which the selection of patients was obtained.

More generally, experimental units need to be defined for a very wide range of situations, of which the following are examples.

(a) An individual person will often be the natural experimental unit when different drugs are to be compared. Each person receives one drug, and measurements of the response to the drug are made on the whole person, or on samples of blood or tissue, or on the reactions or opinions of the person, and are recorded. Sometimes the length of life is the relevant measurement or some measure of quality of life. Each measurement is taken in the same way for all experimental units. There are other forms of treatment than drugs, including various forms of surgery, different possibilities of physiotherapy, or different diets. The conclusions would apply to populations of people which the experimental sample could be argued to fairly represent.

(b) Alternatively there may be no physical interference treatment, but the treatment may be psychological, with different environments, and the measurements being based on responses to a variety of tasks set by the experimenter. The treatments may even be different classes of a classification, such as male or female, age, physical condition or mental capacity. Again the conclusions would apply to all the people who are in a similar condition to those who could have been included in the experimental sample.

(c) An individual food-production animal may similarly be an experimental unit. There are many different forms of feeding treatments. Measurements would be growth, meat or egg production. The conclusions would apply to all the animals in the same condition and with the same genetic background as the experimental sample.

(d) A single plant may be an experimental unit, treated as a separate entity, and observed throughout its life, measuring productivity (for example tomato yield), physical sizes of plant components, chemical measures of plant material, or taste of fruits. Treatments may be applied in the form of chemicals, or may be different forms of soil in which the plants are grown, or may be different environments. The conclusions would apply to all plants with the same background of growth and genetic history.

(e) A tree may be a unit, possibly observed over a lifetime, or at least over a lengthy period, for growth in terms of size, or of shape, or productivity of fruits or of quality. The conclusions would apply to all trees of the same genetic and environmental background.

(f) A single seed can be a unit, with the measurements being whether the seed germinates, and how the resulting plant grows. Treatments could be different chemicals or environments. Conclusions would apply to the population of seeds obtained in a manner similar to that for the experiment.

(g) A sample of soil taken from a defined area may be a unit, which can be treated chemically or through changes of environment, and measurements made on the physical characteristics of the soil, or of the growth of plants when grown in the soil. Conclusions could be assumed to apply to soils of the area and of similar areas, with similar histories.

(h) A batch of liquid in an industrial process may be a unit, with the treatments being variations of the applied chemical process, and the measurements being on the output at the end of the process. The conclusions would apply to the liquids in that processing plant and, arguably, to similar plants operated in similar conditions.

(i) A batch of material in a food-making process (bread, cake, meat, wine, vegetable) can be a unit, treated in various ways, including forms of heating, and chemical applications, with measurements on quality (taste, texture) or quantity. The conclusions would apply to all batches of material prepared in the same manner, within the same production plant, and, arguably, to other similar plants.

(j) A sample of liquid from a person, such as blood or plasma, or from a plant, may be a unit, being treated in different ways. Samples may be very small, and there may be very large numbers of samples; the samples may be contained in test tubes or in small holes in a large array on a single sheet. The sample of liquid in a single test tube or small hole is the unit. Different chemicals may be injected into the test tubes or the holes, and changes in the composition of the liquids observed. The relevant population is that from which the sample of sources for the original liquid samples was drawn.

(k) A physical construct such as a light bulb, a silicon chip, a car, or part of a car, or a component of an oil rig, can be a unit, tested under various conditions, with measurements of performance or strength, reliability or deterioration. The population to which the results will apply is all constructs produced in the same manner as those considered as available for the experimental sample.

(l) Pieces of genetic material may be units, treated with different chemicals, and observed for their resulting contents. The relevant population is defined by the organisms from which the genetic material was obtained, and the manner in which the genetic material was extracted.

(m) In an industrial process where experimentation is on one industrial plant, a day (or some shorter length of time) may define the unit, with different treatments being applied on different days, and the output observed on each day. Conclusions would apply to the performance of that particular plant.

5.1.1 *Variation between units*

Because in an experiment, different treatments will be applied to different units, and the resulting responses compared, we must think about the variation that we would expect to occur between the responses from different units if they had all been treated identically (we may think of this variation as the 'natural' variation innate within the population). This natural variation will be the basis for our deciding whether the different treatments have produced real differences, or whether the apparent differences between units treated differently are only those which would have been expected from the variation inherent in the population of units.

Obviously it is beneficial for the experimental objectives if the natural variation between units in the sample used for the experiment is small, because that will make it easier to identify when a treatment has produced a real difference. In general, variation between biological units ((a)–(f)) is often found to be relatively large, while variation in chemical or physical units ((g)–(k)) is often relatively small. This may explain why a lot of the work of developing statistical design concepts occurred originally in agricultural research and in clinical trials.

One approach to trying to make the variation between units smaller is to identify subpopulations of units such that the units in the subpopulation are particularly similar. For patients who may be subjects for a clinical trial, this could be achieved by identifying and separating the female patients from the male patients, separating different age groups, and using different histories of previous social or medical conditions. The experiment would then be designed so that the patients of a particular group would be split evenly between the different treatments. This would ensure that differences between the treatments and differences between the groups could both be estimated, and the important estimates of the treatment differences would not be biased by differences between groups. An additional benefit would be that the treatment differences could be estimated separately for the different groups allowing us to have information on whether there could be differences in the treatment differences depending on the group of patients. The ideas of control of variation between units, often referred to as blocking because of the original development of the ideas in agricultural research, and the implications for the design of the experiment are developed in Chapters 7 and 8.

Another consideration arising from thinking about the variation between units is the question of how many experimental units are required for each treatment. This is a crucial part of the design of the experiment. The experiment must include enough units to give a good chance that the experimental results will provide sufficient information about the treatment differences for the decisions for which the experiment was conducted to be made with confidence. Equally, since the experimental units are costly, we should avoid using more units than is necessary to obtain a result of adequate precision. These ideas, which are collected under the heading of replication are discussed in Chapter 6.

The third aspect of the experimental units which has a chapter to itself is randomisation. This is a more complex concept, and there is much confusion among experimenters as to its purpose. The way in which treatments are allocated to experimental units should be governed by the idea that the allocation should be random so that no-one would be able to predict in advance which units get which treatment, and ideally most or all people involved in the assessment and measurement of the progress of the units should not know which treatment any particular unit is getting.

There are two main reasons for using random allocation of treatments to units. One is to avoid using any more subjective system of allocation, which might give rise to suspicions of bias in the results. If the treatment allocation is random and unpredictable, then there can be no reason to suspect any bias. If the allocation system is subjective, then although the results may be an entirely fair reflection of the true differences between treatments, the experimenter has to justify the allocation method as providing unbiased estimates. Without a transparently random allocation system, this may be difficult. The second reason is that to estimate the variability, and hence the precision of the experimental results, it is necessary to have a valid estimate of the variance of the unit variation. The usual requirement for a statistically valid estimate of the variance is that the sample of units from which the variance is estimated should be, in some sense, a random sample from an identifiable population. Randomisation gives an indirect way of ensuring this. For a valid estimate of variance it is not necessary that the randomisation procedure should be a complete randomisation. Most of the block designs we discuss in later chapters require restricted randomisation, but this does provide a valid random allocation of treatments to experimental units. The subject of randomisation is discussed further in Section 5.6, and is developed thoroughly in Chapter 10, with some modifications considered in Chapter 11.

5.2 Experimental units as collections

In most agricultural crop trials, it is not realistic to treat and observe individual plants, and the conventional practice is to grow groups of plants on small areas of land, referred to as 'plots'. Plot size can vary, according to the size of the individual plants, and the scale of the agricultural practices being applied to the plots. A plot must be large enough for the results from observing plots treated differently to provide relevant information for the conclusions to be translated to the large fields in which many crops are grown commercially. Typically field plots may be as small as 3 metres square or as large as 20 metres square. They may be square or may consist of single rows of the crop, or may be rectangular.

Plots must be manageable in agricultural terms, but also need to be kept small because the natural, spatial variation between the performances of crops on different plots will be smaller when the plots are closer together. Obviously smaller plots can be arranged to be closer together than can larger plots. For similar reasons, long thin plots may be preferable because they can be arranged so that the plots are closer than would be the case for square plots.

In an experiment using crop plots, with many plants grown on each plot, the plot becomes the unit. It is the plot which is treated, each plot may be treated differently, and the measurements will be made on the whole plot. The population to which the results may be hoped to apply will be fields of plants, grown in the same way as the best treatment. The variation between individual plant performance may be of interest, but is not relevant to the comparison of the performances of different treatments which are observed on different plots. This is a particular example of the need to be very clear as to precisely what is the experimental unit.

There are other situations where a collection of what might be thought to be the natural experimental units is used as the actual experimental unit. Using collection units may be desirable because of the way the treatments are applied, or may be desirable for the making of appropriate measurements, or may be necessary for effective translation of the conclusions to real life. In some situations the collection unit may be the only possible form of experimental unit.

Groups of trees will sometimes be appropriate as experimental units for much the same reasons as using plots of agricultural plants. However, it will often be the case that individual trees can be separately treated, and the main question which will determine the form of the experimental unit is whether the results from individual trees can reasonably represent the results that would be obtained within a large array of trees.

Frequently the natural way of feeding and managing the growth of animals for agricultural produce will involve grouping the animals. Pigs are kept in group pens, grazing animals are kept in herds, chickens are kept in houses or in small yards. In all these cases it may be unnatural to manage and feed the animals individually, and the natural experimental unit becomes a group of animals. As with groups of plants, the variation between animals will often be of interest because of the interest in variation in itself, but this will not be relevant to the comparison of dietary or management treatments which are applied to whole groups. In all cases the population to which the results should apply is of animals managed in the same structure as in the experiment.

In education, children are sometimes taught as isolated individuals, but more frequently groups of children are taught together. Different teaching methods are sometimes tried on different classes, and when this is the case the whole class becomes the experimental unit. The measurements which are the basis for the comparison of different teaching method treatments will be those recorded for whole classes, and the population to which the results will refer is that of similar sized and socially based classes.

In any trial where groups of people are treated, as groups, in different ways, the experimental unit may be the whole group. This is usually called cluster randomisation. On the other hand, the treatment of an individual may be the involvement in a group therapy, and it may then be appropriate to regard the individual as the unit. It should be becoming clear that these situations require careful thought. Perhaps the best indicator is whether different groups are treated as groups in different ways, in which case it is likely that it is the group that is the unit, whereas if groups are treated similarly, and in particular if some individuals are in groups, whereas others are treated separately, then the group is part of the treatment but not of the unit. Take care!

In food processing, groups of food material may be treated together (cakes in an oven, peas in a batch for freezing, beer or wine in vats) in which case it is likely that a batch or group or large quantity of liquid will be the experimental unit, rather an individual cake, pea or bottle.

In ecological experiments it will usually be necessary to define a large area, or ecological zone, as the experimental unit, and this often causes real problems in having proper replication. Identifying and defining many equivalent ecological zones, which can be treated in different ways, but which would, apart from the different treatments be expected to behave in a broadly similar way, is difficult because of the natural variation of the environment.

One unusual and interesting example of individuals being treated in groups comes from the testing of large numbers of individuals for a particular form of defect or characteristic, when the probability of an individual having the characteristic is small. The original example was when blood samples from a large population were being tested for a rare disease. If subsamples from a group of samples are pooled and the test applied to the pooled sample, the presence of the disease in the pooled sample would indicate that at least one of the group had the disease; the individual samples can then be tested to identify which individual(s) possessed the disease. If the pooled sample shows no disease then all members of the group are clear. The efficiency of the design derives from eliminating many members of the group from a single test. Here the experimental unit is the pooled sample, but also the individual samples, if the pooled test proves positive and individual samples have to be tested.

5.3 A part as the unit and sequences of treatments

There are some situations where it is possible to treat and observe parts of a person, animal or plant as the experimental unit. For example if the experiment is concerned with the eyes of people, it may well be possible to treat the two eyes of each person separately and differently. The advantage of using each eye as a separate unit is that the two eyes of an individual are in general likely to behave more similarly than eyes of different people, and thus a more precise comparison of different treatments will be obtained if two treatments are applied to the two eyes of each person participating in the experiment than if an equivalent number of eyes all from different people were used. Different treatments here might involve variations of a basic surgical procedure, or different forms of lenses in glasses. And the measurement might be the improvement in focal power, or the rapidity of the eyes adapting to new glasses.

In the same way treating the two leaves in a pair on a plant differently will give a more precise comparison than using leaves from different plants. For laboratory experiments on the response of leaf material to a variety of different fungicide treatments it may be appropriate to use small circular discs cut from a leaf, with the different fungicide treatments being applied to individual discs.

Different fruits from a fruit tree may be used as units in an experiment on different methods of storage or treatment with chemicals to improve the ripening or storage properties of the fruit.

Different parts of a cake mix may be separated and treated in different ways such as cooking temperatures, or different variations of the latter parts of the preparation process.

All of these uses of a part of an original unit, with different parts being treated differently, instead of maintaining the original unit intact for each treatment, are motivated by the desire to improve the precision of the comparison because the parts of a single unit are likely to be more similar than are samples from the original different units. The part becomes the new unit, and the improved precision for comparing treatments enables the experimenter to use fewer units in total.

5.3.1 Sequences of treatments in patients in clinical trials

The most important example of using a part of a basic unit as the experimental unit is when different treatments are applied in sequence to people participating in a clinical trial. This means that each person provides several experimental units, with each time period for each person being a separate experimental unit. This is a very important aspect of clinical trials, and has also been a concept involving considerable controversy, so we shall consider it in some depth. The design of trials involving the application of sequences of treatments is also discussed at several other places in this book.

When human patients are used for trials to compare different drugs, or different concentrations of the same drug, or different physical or psychological treatments, it is normal to test the drug for a relatively short period. Certainly the period of testing will almost always be much shorter than a lifetime, and will often be as short as a year, a month, or even a few days. Because of the relatively short time periods for testing each drug, and because variation between patients, even those of similar age, sex, and environment, is large, whereas the variation between observations made on the same patient at intervals of a month or even a year tends to be much smaller, the idea of using each patient as both a control unit and also a test unit was developed.

In essence the idea is simple. Each patient is treated and observed for two periods, during one of which the patient receives the test drug, and during the other of which the patient receives either no treatment or a placebo made up to look identical to the test drug, so that neither patient nor experimenter and doctor will know which treatment is given in which period. If each patient receives both treatments, at different times, then each patient provides a measure of the difference between the effects of the two treatments, and the overall information from a group of patients gives the average difference between the two treatments, which because of the general consistency of most measurements within (at different times) a patient, will be much more precise than a comparable difference between the averages of two groups of patients, one group treated with the test drug, and the other treated with the placebo.

We now have the situation where the experimental unit is a patient for a period of time. Different units may be different times for the same patient, or may be periods for different patients.

As we mentioned, this idea of trials in which each patient receives more than one treatment, in different time periods, has caused great controversy, and for a long time the results of such trials were not accepted as providing valid information for the legislative processes dealing with drugs. The arguments against such trials are many and we shall consider them briefly.

(i) A trial in which a patient receives a sequence of different drugs cannot provide legitimate information for future treatment in which a patient will receive only a single drug.
(ii) Once a patient has received one drug, that patient is changed, and therefore does not provide the same condition for the second drug.
(iii) Treating a patient with a time-cocktail of drugs may provide a more dangerous situation for the patient, for which the dangers are difficult to predict.
(iv) The order in which treatments are presented may affect the apparent benefits of different treatments.
(v) It may be difficult to prevent the patient making judgements about which treatment he or she is receiving at a particular time and this may bias the results (whether or not the patient's judgement is correct or wrong).

The fact that such clinical trials, generally described as 'cross-over' trials are now accepted as a legitimate design for an investigation does not mean that these arguments are unimportant. On the contrary, each argument, and possibly others, must be carefully considered before the design of each particular trial is approved. To some extent there are standard ways of managing the design of a cross-over trial so that the difficulties represented by the above arguments are overcome, or at least minimised. Thus periods will be sufficiently separated by recovery periods to protect against argument (ii). The design as a whole will be pharmacologically assessed for potential dangers to protect against argument (iii). The order of treatments will be arranged to be systematically different for different patients to protect against argument (iv) (this will be considered in some detail in Chapter 8). Randomisation schemes to determine which treatment is applied in which period will be constructed so as to make 'guessing the treatment' as difficult as possible (this will be considered in Chapter 10). And the fundamental argument (i) must always be justified for each trial.

The overwhelming argument in favour of cross-over trials is, of course, the improved precision which can result from a such trials, allowing an adequately precise trial to be constructed using very considerably fewer patients than would be required for a conventional trial. Using each patient as their own control is essentially an exercise in controlling the variation between experimental units, and will be considered in some depth in Chapter 7. There are also aspects of using within-patient information, possibly in addition to using between-patient information which may require the ideas of Chapters 9 and 18.

Cross-over trials may be designed for more than two periods for each patient, and in general we should think in terms of sequences of treatments being presented to different patients. Even for just four treatments the number of possible sequences is 24 so that it will be clear that the design of each trial when several different treatments are to be compared can become quite complex. This is particularly true if it is thought that the effect of treatments might possibly be affected by the position in the order of the sequence. This aspect of design is a particular concern of Chapter 8.

5.3.2 *Other situations involving sequences of treatments*

The ideas of cross-over designs for clinical trials were also developed for feeding trials for animals. Again, it was realised that because variation of measurements between animals could often be very large, while variation within a single animal tends to be much smaller, an experiment could be much more precise if each animal was used for several periods, each period being a separate experimental unit. For feeding trials the time periods will usually be the same for each animal, and there will be changes of environment which make allowing for differences between time periods, as well as for differences between animals, essential. The resulting double control designs are considered in detail in Chapter 8.

Trees are another example of a natural experimental unit, which may be divided through time into several experimental units, with each tree receiving different treatments in successive years. Again as with animal feeding trials, the differences between the different years (periods) will usually be considerable and the design must allow for this.

One other major area of experimentation where sequences of treatments are applied through time is industrial experiments where the number of replicate industrial plants is small, or even in the most extreme case, where only a single plant is available. All the experimental units then consist of periods of use of the single plant, or of a period for one of the small number of plants. The sequence of treatment periods will then be long, and may well be spread over several days; there may be the possibility of substantial variation between days, or between times within a day if there are regular diurnal patterns of performance. Each period is a separate experimental unit, with all the potential problems for interdependence between units previously discussed.

5.4 Multiple levels of experimental units

We have considered alternative choices of the experimental units, which should always be appropriate to the objectives of the experiment, and the form of treatments and measurements being used. We can recognise that in some situations the appropriate form of experimental unit may consist of a part of what would be the natural experimental unit in a different form of experiment. In others the appropriate form of experimental unit may be a collection or group of what would in other circumstances be the obvious experimental units. There are also experimental situations where there are several different kinds of experimental treatments being investigated. We might be looking at various combinations of drugs and surgery, of grazing treatments and milking treatments, at large-scale and small-scale agricultural treatments, or at combinations of treatments which are applied at different stages of an entire process, such as different methods of growing vines and grapes for wine, and different methods of processing the liquid from the grapes to produce the wine.

In many investigations in which two or more treatment factors are being investigated, the size of experimental unit which is appropriate for one treatment factor is different from that appropriate to another treatment factor. In the chemical industry, a process may have several stages, starting with a large mixing stage from which the material is subdivided into smaller quantities for the later chemical interactions; if the constituents of the original mixture are varied, the natural experimental unit is the large container used for the initial stage, but for treatment factors involving changing the conditions for the subsequent stages of the process, the unit may be a subdivision of the large initial experimental unit.

In experiments on dairy cattle, the experimental treatments may include different treatments of the pasture and variation of milking conditions. Pasture treatments must be applied to groups of animals if they are to be representative of farming practice, but milking method treatments may be applied to individual animals. This division of treatments into those necessarily applied to groups of subjects and those applied to individuals also occurs in psychological experiments. In field crop experiments, or glasshouse experiments, some treatments such as cultivation or irrigation methods or electrical heating conditions must be applied to large areas for the results to have practical relevance, but treatments such as the variety of the crop or involving chemical treatment of individual plants or parts of plants may be advantageously applied to smaller units, such as individual plants or even individual leaves.

When the levels of one treatment factor are naturally applied to large units and the levels of a second treatment factor may be applied to smaller units within the larger units, then there are two possibilities open to the experimenter. The first is simply to use the large experimental units. This will almost inevitably place a severe restriction on the total number of units that can be available for the experiment, and will lead to few replicate units for each treatment combination. The second possibility is to use units of both sizes in the same experiment. We then have a relatively small number of large amalgamated units to which one set of treatments is applied, and within each large unit, a number of small units to which a second set of treatments is applied. In terms of the total set of factorial treatment combinations we effectively have restrictions on the allocation of individual treatment combinations to units such that treatment combinations are grouped by the treatment levels for the large units.

There is no problem in principle with using more than one size of experimental unit within a single overall experiment. We must obviously ensure that each set of alternative treatments is applied to the relevant size of experimental unit, and we must also ensure that the management and control of experimental units is consistent throughout the overall experiment. The detailed discussion of the design of multiple level experiments, sometimes referred to as split-plot or split-unit designs, is in Chapter 18. There is also a general discussion of the problems of analysing data from multiple level experiments in Chapter 9, though this covers aspects of designs other than the split-plot designs.

When we have a multistage experiment as with the wine growing/processing experiment, the forms of the experimental units may be very different. Field plots, with a structure to control the patterns of variation within the field, are the experimental units for the first stage. Containers of grape juice provide the experimental units for the second stage, and these will be processed using a variety of treatments; the variation between containers/times/processing plants may also need control. The two patterns of methods to control the variation may be related or not; it will certainly be necessary to record the relationship between experimental units in the two stages, for use in analysing the results, and decisions about restrictions on the randomisation must be carefully planned before the first stage of the experiment.

5.5 Time as a factor and repeated measurements

There is another situation where it is sometimes assumed that the approach of using two levels of unit is appropriate, which leads to the so-called 'repeated measurements' experimental design. This occurs when measurements are taken at repeated time points for each patient or animal, but where the experimental units, to which the treatments are applied are the patients or animals (a single level). Essentially it is argued that time is treated as a factor, resulting in a second level of experimental units. The repeated measurements design is sometimes treated as if it were the same as a split-unit design, and the analysis performed assuming two levels of units. We discuss the concept briefly here since it is a good example of the need to think clearly about exactly what constitutes an experimental unit.

In many agricultural and other biological experiments with both animals and plants, it is common to take measurements on each individual unit at several different times, and to regard time as a split-unit factor. The reason for this form of experiment is that the experimenter is interested in how the effects of treatments vary with time. This is the classic expression of interaction and therefore the investigation of treatment × time interaction within a split-unit stratum of analysis seems an obvious approach. There are, however, some major difficulties in this use of the split-unit concept.

First, there is an important philosophical difficulty that multiple measurements are different variables and not different experimental units. The number of units in the experiment cannot be increased by taking more measurements. By taking more observations on each unit more information is obtained about the set of units originally included in the experiment, but because the number of units is not increased, we do not have more information about the population of units. Even if we are prepared to confuse the fundamental concepts of experimental units and observed variables there are several technical difficulties about the practice of regarding time as a split-unit factor.

Two rather different situations in which time is used as a split-unit factor must be distinguished. In some experiments, usually with plants, the sample of experimental material at each time is different. Measurement of the variables of principal interest for plants usually involves destruction of the plants. Thus, weight measurements require the plant to be removed from its growth environment, and measurement of chemical constituents require further destruction. Plant material is usually relatively cheap so that the use of repeated 'harvests' is economically viable. Rather than using different experimental plots for each treatment × time combination it is often practically convenient to use more substantial plots and to subsample at the various times. The positions of the subplots to be used at the different times can be selected from the plots either randomly or systematically. Random sampling makes a split-unit form of analysis more reasonable than does systematic sampling, but the objections discussed below still require consideration.

Alternatively the repeated measurements at the different times may be taken on exactly the same experimental units. This will typically be the case in animal experiments and in clinical trials, and occurs sufficiently frequently for such experiments to be distinguished as a special form, the repeated measurements experiments already mentioned. With the rapid development of technology producing machinery capable of taking measurements increasingly frequently for human, animal and plant experimentation, the repeated measurements experiment is becoming extremely important, and there is a flourishing large literature on the subject.

Similar questions arise in many laboratory experiments which involve running (bio)chemical reactions. Sometimes the laboratory measurement of the output from the reaction at a specific time requires stopping the reactions, so that different times are represented by different experimental units. In other cases, contnuous monitoring of the reaction is possible, in which case repeated measurements over time will be taken from the same experimental unit.

In some industrial experiments, experimental units can have 'repeated measurements' in space, rather than time. For example, in assessing the effect on the constancy of a plasma layer on microchips, each experimental unit (run) will produce one or several chips and the thickness of the plasma layer will be assessed at several points on the chip. Although this introduces an extra dimension in the response, the basic idea is exactly the same as with repeated measurements in time.

The most valid approach to repeated measurements experiments is to define the experimental units as those units to which different treatments are applied, or for which different environmental conditions are identified. The multiple measurements, through time, are just measurements, not separate units, and the analysis must investigate the pattern of results at different times over a defined period, with the pattern being characterised in the same way for each and every experimental unit. In practice, this will require a suitable response model to be defined (this could be as simple as the difference between the first measurement and the last measurement), and then parameters from the response model should be calculated for all experimental units, and the resulting values analysed in the same fashion as any other single measurement made for each experimental unit. This approach clearly retains the integrity of the experimental units, and makes the identification of the populations to which the results apply clear and unambiguous.

When the measurements to be made at different times require destructive harvesting of different plants, it is necessary to define clearly the way in which the different plant samples are obtained at the different times. Usually the samples for the different times will all be obtained from within a single, relatively large, plot, taking either individual plants or rows of plants. There are two possible ways of selecting the plants, or rows, for each harvesting occasion. Either plants, or rows, may be selected at random from all the plants/rows in the plot, or they may be selected systematically, starting at one end of the plot. The practical benefits of a systematic selection, in reducing the change to the growing environment, should usually outweigh the theoretical benefit of a random selection of rows/plants, since there will usually be other randomnesses resulting from the random selection of seeds/seedlings in the different plant/row positions.

5.6 Protection of units, randomisation restrictions

All the units in an experiment are assumed to provide independent information about the effects of the treatments applied to those units. We know that some units may be expected to give more similar results than other units, and units may be grouped so that the units in a group are expected to give similar results and the design modified accordingly, but in general we still expect that the results from different units will not be affected by what is happening on other units.

There are two situations where this independence assumption may be particularly doubtful, and where some particular precautions may be adopted to protect units from each other. The two situations can be characterised as time dependence and spatial dependence.

When treatments are applied to plots in an array, such as occurs in many agricultural experiments, we have to consider carefully the size of plots and the nearness of different plots. Plot size must be sufficiently large that the results from a plot may be regarded as providing useful information for how an equivalent crop in a much larger area, such as an entire field, might be expected to respond to the different treatments. However, there will also be a need to make plots as small as is realistic so that the plots will be close enough that the plot-to-plot variation should be small enough to give acceptably precise results for the comparison between treatments. However, the closer together the plots are, the greater the possibility that the results from one plot may be affected by what happens on adjacent plots.

The effects of one plot on adjacent plots can result either from the treatments 'overflowing' from one plot to neighbouring plots or from the success of one plot depressing yields on the adjacent plot, through competition. There may also be effects on plot yield because the plot is surrounded by space to define the plots and to allow the management of the plots. In some cases these potential interference effects can be reduced, or eliminated, by using 'guard' areas around the central part of a plot on which the records are to be made. The guard areas will be treated like the central plot, but no records will be taken from the guard areas. they are there simply to protect the central areas, and to ensure as far as is possible that the results from the central, guarded area will be properly representative of plants grown in a field-sized array.

There are similar problems when treatments are applied in a sequence, as in cross-over designs for clinical trials. After each treatment there will almost always be a resting period, or wash-out time, when the short-term effects of the treatment can be assumed to be eliminated from the system, and after which the patient will be back to the state they were in before the first treatment. The same considerations are necessary for feeding trials with animals, or for any form of trial in which treatments are applied in sequence, and for which we would like to assume that the subject will be in exactly the same state before each treatment in the sequence. Essentially we would like to assume that each subject provides a system to which treatments may be applied, and is such that 'the system' returns to its original state fairly quickly.

If we have to protect the subjects by having extended periods between the treatment periods it becomes much more difficult, and expensive, to organise the experiment. As with the interference between plots, any residual effects from one treatment period to the next can be caused by the treatment itself, or by the effect on the subject (beneficial or deleterious) of the treatment. Of course, when treatments are applied to patients in clinical trials, there may sometimes be a hope that some treatment may produce a more permanent effect. This would produce problems when treatments are applied in a sequence, because the experimental unit would be being changed, and it would usually be wrong to use a cross-over design with sequences of treatments for such situations.

For both forms of unit protection, there are possibilities of modifying the method of analysing the results from an experiment to attempt to take account of the possible interference between plots or the residual effects from one treatment lingering on to the next treatment period. Even where such forms of analysis are not thought to be necessary, they may be used to check that there are no effects between plots or between time periods. These more complex methods of analysis are considered in Chapter 8.

Sometimes the procedures for allocating treatments to experimental units may be modified to recognise and protect against problems of interference between units. The normal randomisation methods of allocating treatments in principle allow any two treatments to occur on any two units and in particular allow any two treatments to occur on units adjacent in time or space. Blocking or control restricts the possibilities so that the same treatment will not occur twice in the same block or sequence, and each pair of treatments will occur adjacently at most once in a block or sequence. However, there are various situations where it may be desirable to impose additional restrictions.

Such restrictions may require that patterns of neighbouring adjacency in spatial arrays meet some requirement of similar frequency of occurrence of pairs of treatments. Similarly for temporal sequences it may be desired that the pattern of treatment succession is as diverse as possible. These would have benefits for the analysis of the results allowing for spatial or temporal dependence. Some examples of this form of restricted randomisation occur in Chapters 8 and 11.

There may also be restrictions on the patterns of adjacent treatments for practical reasons. In experiments on industrial plants where the experimental treatments are combinations from a factorial structure, it may be practicably desirable that the levels of some factors should change between successive treatment combinations as rarely as possible.

All these devices for protecting units from effects from other units, or restrictions on randomisation procedures are intended to improve the precision of comparisons between treatments. They should never be allowed simply because it becomes part of the standard procedure. The wide range of possibilities which have been discussed very briefly here illustrate how complicated the choice of sets of experimental units can be, and should demonstrate how much thought is required when designing an experiment.

6

Replication

6.0 Preliminary example

The student explained that his was a simple problem. He had done an experiment with eight treatments, each treatment replicated ten times. The analysis of variance seemed straightforward, the salient points being as follows:

	df	F
treatments	7	500
error	72	
total	79	

He wanted to be assured simply that his analysis was correct. Apart from a feeling that $F = 500$ suggested that ten replicates constituted overkill, there was nothing apparently suspicious. Requests for more information elicited a description of the units as small discs cut from leaves of a particular plant species, innoculated with spores of a fungus. The treatment consisted of injecting one of two fungicides into either the upper or lower sides of the leaf, using old or young leaves, the total set of treatments being old/young × upper/lower × two fungicides. Suspicions and questions now begin to emerge in the statistician's mind. What was the design? The discs for the old and young leaves must be from different leaves. The analysis implies that all the comparisons are at the same level of variation and hence that all 80 discs came from 80 different leaves. However, many alternative designs are possible. To determine whether the analysis was correct the statistician must identify precisely the design of experiment. How many leaves per plant? How many discs per leaf? How exactly were the leaves chosen?

6.1 The need for replication

In this chapter, we consider various aspects of replication, which is one of the fundamental concepts in experimental design. We start with the fundamental argument for replication. Consider the situation when farmer A uses variety X in one field and gets a yield of 24.1, while farmer B uses variety Y in one field and gets a yield of 21.4. Farmer A claims that his cultivation method is better than farmer B's; the seed merchant producing variety X claims it is better than variety Y. But, without any further information, the difference could be simply due to different fertility levels in the two fields. Suppose farmer A has also sown a second field also with variety X, and gets a yield of 24.2. The whole situation is now changed. The difference of 21.4 from 24.1 now seems notably large compared with the difference between 24.1 and 24.2. The smallness of the latter difference may lead to an unreasonable belief in the validity of the first difference. And it is still not possible to deduce whether the important difference is A − B or X − Y. But a second difference is now available against which to measure the first, and this is the crucial benefit of replication.

Suppose further that the two varieties are each grown in several fields, giving yields of 24.1, 24.2, 23.0, 25.4 and 22.8 for X and 21.4, 20.9, 23.1 and 22.6 for Y. The reality of a difference between X and Y (or A and B) is becoming more credible and, provided the allocation of fields to X and Y is not done in any way that favours X, the reality of the difference should by now be fairly convincing; the *t* statistic for comparing the two sample means is 2.5 on 7 df.

It is important to ensure that the replication is appropriate for the comparison that is to be made. Different varieties are used in different fields, and consequently replication of fields with the same variety is needed to provide a measure of variation against which the difference between fields with different varieties can be measured. Sometimes there is a temptation to take a short cut to achieve replication by measuring sample plots within fields. Suppose that each farmer applied his chosen variety to a single field but, within their fields, farmer A marks out five plots and farmer B marks out four plots and the yields from these plots are recorded. Now, if the yields obtained are the same as those discussed in the previous paragraph, the same confidence that the apparent difference between X and Y is real cannot be sustained. One field has been selected for X and one for Y and assessment of the variation that would have been found between two fields treated identically is necessary before the difference between the yields in the two fields can be interpreted. The variation between small areas within a field does not, in general, give information about differences between fields. A given field may be very homogeneous or it may be very variable. The small areas within a field represent a different scale of unit from that of complete fields, and information about variation between units for one size of unit may not be a reliable guide to the variation to be expected between units of another size.

6.2 The completely randomised design

When the experiment consists of several replicate observations for each of a number of units, and the available units are allocated quite arbitrarily to the required treatments, subject only to the restrictions on the total number of units for each treatment, then the experimental design is called a completely randomised design. A full discussion of the crucial concept of randomisation is deferred until Chapter 10, but it will be assumed here that random allocation of treatments to units is essential in experiments. The completely randomised design was considered briefly when introducing the concept of the analysis of variance in Chapter 2, and is now re-examined to illustrate the application of the formal models and methods of analysis developed in Chapter 4.

We assume, as before, that there are *t* treatments with replications $n_1, n_2, \ldots, n_t$ and a total number of observations N. The model is written formally as follows:

$$y_{jk} = \mu + t_j + \varepsilon_{jk} \quad j = 1, \ldots, t \quad k = 1, \ldots, n_j,$$

Table 6.1. *Analysis of variance for a completely randomised design*

Source	SS	df	MS
Mean (μ)	$Y_{..}^2/N$	1	
Treatments (t_j)	$\sum_j Y_{j.}^2/n_j - Y_{..}^2/N$	$t-1$	
Residual	S_r	$N-t$	s^2
Total	$\sum_{jk} Y_{jk}^2$	N	

where y_{jk} represents the observation for the kth unit to which treatment j is applied. The least squares equations are

$$\hat{\mu} \sum_j n_j + \sum_j \hat{t}_j n_j = Y_{..},$$

$$\hat{\mu} n_j + \hat{t}_j n_j = Y_{j.}, \qquad j = 1, 2, \ldots, t.$$

The constraint $\sum_j n_j t_j = 0$ results from the general concept of t_j as deviations from an overall mean and expresses the interdependence of the t_j in a form which allows easy solution of the least squares equations. The resulting parameter estimates are

$$\hat{\mu} = Y_{..}/N,$$

$$\hat{t}_j = Y_{j.}/n_j - Y_{..}/N.$$

The fitting SS is

$$\mathbf{y}' \mathbf{A} \hat{\boldsymbol{\theta}} = \hat{\mu} Y_{..} + \sum_j Y_{j.} \hat{t}_j$$

$$= Y_{..}^2/N + \sum_j \left(Y_{j.}^2/n_j \right) - Y_{..}^2/N.$$

The analysis of variance structure is shown in Table 6.1.

We have seen already, in Chapter 2, how a comparison of the TMS with the error MS provides an assessment of the overall strength of treatment effects. In Chapter 4, the formal justification for the use of the F-distribution to assess the statistic

$$F = \text{TMS}/\text{error MS}$$

was developed for the general design model. It is also instructive to examine, for this simple design, the expected value of the TMS.

For the completely randomised design model

$$\text{treatment SS} = \sum_j Y_{j.}^2/n_j - Y_{..}^2/N$$

$$= \sum_j n_j \left\{ \mu + t_j + \left(\sum_k \varepsilon_{jk}/n_j \right) \right\}^2 - N \left(\mu + \sum_{jk} \varepsilon_{jk}/N \right)^2,$$

since $\sum_j n_j t_j = 0$. Hence

$$\text{E(treatment SS)} = \sum_j \left(n_j \mu^2 + n_j t_j^2 + \sigma^2 \right) - (N\mu^2 + \sigma^2),$$

since $E(\epsilon_{jk}) = 0$, and so

$$\text{E(treatment SS)} = (t-1)\sigma^2 + \sum_j n_j t_j^2.$$

Hence the expected value of the TMS is

$$\sigma^2 + \sum_j n_j t_j^2/(t-1),$$

which simplifies to

$$\sigma^2 + n \sum_t t_j^2/(t-1)$$

for the case where all treatments have equal replication, n. Now from the general result in Section 4.2

$$\text{E(residual MS)} = \sigma^2$$

and therefore the power of the F test to detect treatment differences depends on the average of the squares of treatment deviations. This suggests that it is perfectly possible for one substantial treatment effect to be lost because other treatment deviations are much smaller and suggests that absolute rules such as 'Only examine detailed treatment differences if the overall F statistic is significant' can lead to failing to detect a real difference.

6.2.1 *Treatment means and SE*

The estimated means for the different treatments are formally $\hat{\mu} + \hat{t}_j$ estimated by $Y_{j.}/n_j$, the sample mean from the n_j observations for treatment j. Equally obviously the variance of a treatment mean is

$$\sigma^2/n_j,$$

and the variance of a difference between two treatment means, based on n_j and $n_{j'}$ observations, is

$$\sigma^2(1/n_j + 1/n_{j'}).$$

To illustrate further the use of the completely randomised design we consider the data from Example E in Chapter 4, the experiment to compare three drugs whose purpose is to relieve pain. The basic format of the experiment was such that each patient was given a drug whenever the patient decided that their level of pain required relief. The number of hours relief provided by each drug was recorded (time from administration of drug to the next request for a drug). For each patient the sequence of drugs to be provided was predefined (but unknown to the patient).

Table 6.2. *Analysis of variance for pain relief data*

Source	SS	df	MS
Mean	833	1	833
Treatment	96	2	48
Residual	337	40	8.4
Total	1265	43	

Table 6.3. *Treatment means and SE of differences for pain relief data*

Treatment means			SE (40 df)		
T_1	T_2	T_3	$T_1 - T_2$	$T_1 - T_3$	$T_2 - T_3$
6.4	2.2	4.3	1.12	1.06	1.08

The variable considered here is the number of hours relief afforded by the first drug given to each patient. The three drugs were labelled T_1, T_2 and T_3, T_1 being a form of control. The numbers of patients were 14 for T_1, 13 for T_2 and 16 for T_3, and the results were:

$$T_1 \quad 2, 6, 4, 13, 5, 8, 4, 6, 7, 6, 8, 12, 4, 4$$
$$T_2 \quad 2, 0, 3, 3, 0, 0, 8, 1, 4, 2, 2, 1, 3$$
$$T_3 \quad 6, 4, 4, 0, 1, 8, 2, 8, 12, 1, 5, 2, 1, 4, 6, 5$$

The two stages of model fitting are:

 (i) fitting the general mean (residual variation $= 433$);
(ii) fitting the treatment effects (residual variation $= 337$).

The analysis of variance is shown in Table 6.2.

The treatment means and standard errors of differences between these are in Table 6.3. Clearly the three drugs have substantially different effects with T_1 providing more hours relief than T_3, and T_3 providing more hours relief than T_2.

Note that there might be some concern whether the variances within separate drug samples are sufficiently similar to justify the assumption of a common within-drug variance. The three sample variances for the three drug samples are 9.6 for T_1, 4.7 for T_2 and 10.4 for T_3. The extreme variances are different by a factor of more than 2, but not significantly different by an F-test, and the assumption does not seem unreasonable.

6.3 Different levels of variation

We have already referred briefly to the need to ensure that replication is relevant to the treatment comparisons being made. There are many very tempting ways of apparently increasing replication, which in many cases do not provide genuine replication, and which lead very frequently to misleading results. In the previous example, the repetition of the drug for each patient may be appropriate to assess the real effect of each drug but, as mentioned previously, the results from the two successive applications do not provide useful replication and to use the variation between the two successive relief times as a measure of precision would be misleading. Some examples of other situations in which inappropriate forms of replication are tempting are:

(i) Different diets are applied to litters of pigs. Because of the shortage of litters, there are only two litters for each treatment. To overcome this lack of replication, it is proposed to record and analyse the weights of individual pigs. Each pig is an identifiable unit, and the variation between pigs is clearly important in assessing the difference between litters. Unfortunately, in most such experiments it is not practicable to feed and rear pigs individually, and whole litters are housed together. The variation between pigs within a litter is possibly similar to the variation between pigs in different litters with the same treatment. But pigs in the same litter might be expected to show less variation because of genetic similarity. Or to show more variation because of competitive feeding. There is no way of telling which pattern is occurring (or whether some other pattern is present) and so the within-litter variation cannot be used for treatment comparison.

(ii) In crop experiments it is often possible to record yields for each individual plant within a plot, or for each of several rows within a plot. Again, if all plots were treated identically, we might expect that the variation between plants within a plot might be less than the variation between plants in different plots because of locally homogeneous soil fertility. Alternatively it might be argued that competition between individual plants within a plot might lead to more variation of plant performance within plots than between plots. The crucial implication of these different possible expected patterns is that it is not valid to use within-plot variation between plants to assess the differences between treatments applied to whole plots.

(iii) In many scientific disciplines it is now possible to make very frequent observations automatically through the use of automatic data loggers. Such observations may provide very useful information about the detailed behaviour of the experimental unit which is being monitored. But they do not provide any basis for the comparison of treatments applied to different units. In extreme situations observations could be made so frequently that each observation must be almost identical with the previous one. If enough almost identical (false) replications are observed, then any difference between different experimental units could be made to appear relatively large and thereby be assessed as significant.

To demonstrate how misleading the use of multiple observations can be we consider some artificial data. Consider an educational experiment, in which the assessment results are influenced by two very different processes: first, a short-time-scale effect of the time within a day when the assessment is made (diurnal); second, a longer-term learning effect over several months. Three different treatments (forms of teaching) are applied to three classes each. The testing of the subject material takes one hour, and, because different tests must involve slightly different material, there is inevitably some variation of the recorded result over and above

Table 6.4. *Averages from single tests*

Treatment 1			Treatment 2			Treatment 3		
A	B	C	D	E	F	G	H	I
27	43	38	41	30	47	46	34	50

Table 6.5. *Averages from tests on four consecutive days*

Treatment 1			Treatment 2			Treatment 3		
A	B	C	D	E	F	G	H	I
27	43	38	41	30	47	46	34	50
25	43	36	43	35	42	48	37	44
30	46	37	44	31	46	46	38	52
31	44	41	45	35	48	45	35	49

the diurnal and long-term trend effects. Similar effects can be suggested for other fields of applications.

Consider first the average result of a single test for each class shown in Table 6.4. The analysis of variance for the completely randomised design is:

$$\text{between-treatments SS} = 81 \text{ on } 2 \text{ df, MS} = 40;$$
$$\text{within-treatments SS} = 447 \text{ on } 6 \text{ df, MS} = 74.$$

Clearly the evidence that the treatments have different effects is weak ($F = 0.54$). There is wide variation between classes for each treatment.

Next, suppose that each class is tested on four consecutive days, the test being at the same time each day to eliminate diurnal variation. The results are shown in Table 6.5.

The analysis of variance if the data are treated as 12 observations per treatment, ignoring the possible average differences between days, is

$$\text{between-treatments SS} = 288 \text{ on } 2 \text{ df, MS} = 144;$$
$$\text{within-treatments SS} = 1394 \text{ on } 33 \text{ df, MS} = 42.$$

The evidence for differential treatment effects now seems much stronger ($F = 3.4$), reflecting the apparent consistency of observations, which essentially is due to the fact that we are using (almost) the same observation four times for each class. With more such replicate observations for each class we would obtain results for the three treatments which were even more 'convincingly different'.

Now suppose that, on each day, only one class can be tested at a time, that all nine classes are tested each day, and that there is a diurnal pattern of test performance as shown in Figure 6.1. The pattern of the arrangement for testing the nine classes on the four days has been arranged, for this example, so that the set of times for each class is spread evenly over

Table 6.6. *Test results for four days with diurnal trend*

Treatment 1			Treatment 2			Treatment 3		
A	B	C	D	E	F	G	H	I
21	51	28	40	38	41	50	38	49
33	37	40	42	34	50	38	31	48
34	45	36	52	21	50	54	32	46
21	43	49	39	39	42	44	43	53

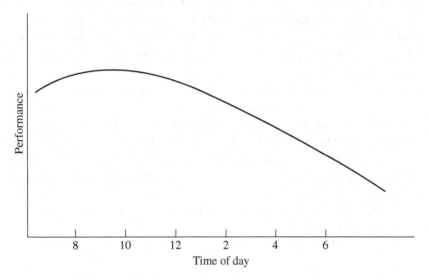

Figure 6.1 Diurnal pattern for test performance.

Time

		1	2	3	4	5	6	7	8	9
Day	1	A	D	G	E	B	H	I	F	C
	2	H	E	I	A	F	C	D	B	G
	3	I	C	F	G	D	A	B	H	E
	4	F	B	E	H	C	I	G	D	A

Figure 6.2 Arrangement of treatments within days and times.

the available time as shown in Figure 6.2. The effect of this is that the estimates of the average treatment differences are not much altered by the diurnal effects.

The test results for the four days might now be as in Table 6.6.

Now the analysis of variance for the three sets of 12 observations, ignoring the day and diurnal time effects, is

$$\text{between-treatments SS} = 322 \text{ on } 2 \text{ df, } \text{MS} = 162;$$

$$\text{within-treatments SS} = 2464 \text{ on } 33 \text{ df, } \text{MS} = 75.$$

The evidence for the existence of different effects of the three treatments is now apparently slightly weaker ($F = 2.2$) because the diurnal effects mask the treatment differences. Essentially, the false replication obtained by repeated observations on the same students has been diluted by the diurnal variation against which the treatments are evenly spaced. This kind of 'evening out' of effects is a form of blocking. It is clearly sensible to try to spread treatments evenly over the different times, but this control of the variation should also be reflected in the analysis of data, the result of which would be to reduce the random, within-treatment SS. In this example, of course, the use of replication of observations for each unit is wrong, and modifications of the analysis cannot correct this fundamental defect.

In this example, the treatment effects must be measured in terms of the variation between classes since each treatment can only be applied to the experimental unit of a class. As the conditions of the experiment have been changed while preserving the same form of analysis, the apparent strength of evidence for the existence of treatment differences has altered, although the real information available has not changed. The use of more complex analyses of data can also not change the real information available and, though alternative analyses should be preferred because they reflect the structure of observations, there is a danger that using the more appropriate analyses may blind the experimenter to the fact that the wrong form of replication is being used. The only proper measure of the differences between the teaching methods (applied to whole classes) is derived from the class averages using just one measurement per class. The learning effect is measured in terms of the variation over time of observations for the same class. The idea of information at several different levels of unit occurs in many later chapters, most notably in Chapter 18. In this present section, the important principle is that, in an experiment, the replication of observations must be exactly appropriate to the comparison of treatments.

6.4 Identifying and allowing for different levels of variation

Sometimes it is not clear from the description of the form of data whether the replication is at the same level of experimental unit as the application of treatments. Sometimes it is a matter of judgement whether the replication is appropriate. And sometimes it is necessary and desirable to use within-unit replication for the comparison of treatments applied to whole units, even though this may not be formally valid.

The lack of clarity frequently stems from an inadequate description of the experiment, and of the form of data. If results are presented properly, with the relevant df quoted for any SE, then it should be possible to deduce when within-unit replication has been used. For the example of the previous section, reference to the use of SEs based on 33 df combined with the use of a total of nine classes should immediately indicate what has happened. Less directly, if a randomised block design with three blocks of two treatments has been used, but no information on the analysis of variance or on SEs and their df is given, and effects for many variables are declared to be significant, then the results should be treated with suspicion, particularly if there is any implication that t-values of the order of 2 are significant, since the design gives only 2 df for error, and with 2 df a value of 4.3 is required for significance at the 5% level.

When experiments are not properly described, as in the example at the beginning of this chapter, then the statistician, or the experimenter considering another experimenter's results, is right to be suspicious of unspecified replication, since experience, particularly from reading papers in applied journals of all disciplines, shows that such suspicion is all too frequently justified. Let us return to the initial example of this chapter. The reader with a suspicious mind will recognise that the replication used to calculate the F ratio of 500 might have come from sampling different plants, different leaves from the same plant, or different discs from the same leaf. Also, the different treatments might have been applied to discs from different plants, from different leaves or from the same leaves. In this instance, the truth when it finally emerged was the saddest of the possibilities. The ten-fold replication for each treatment consisted of ten discs from a single leaf. And each different treatment was applied to discs from a leaf from a different plant. Thus, the value of 500 could have been due to the treatments but could equally well be attributable to the fact that the leaves come from different plants. The sad conclusion in this example was that there was no valid replication on which comparisons of the treatments could be based, and that the experiment had been so designed as to be completely useless in comparing the treatments.

There are some situations where the relevant form of replication is not entirely clear. For example, when comparing different types of soil it is necessary to use several samples of each type of soil as replicates in order to assess the extent to which apparent differences between soil types are meaningful. However, each soil type will occur over a relatively compact area, and therefore the different samples for each soil type will necessarily come from that area. In most investigations it would be proper to be suspicious if replicates of the different 'treatments' were bunched together within small areas. However, such replication provides the only genuine basis for comparison of soil types, since the area *is* the soil type and differences between soil types can only be meaningfully assessed in terms of the variation within the soil type. It is of course essential to keep each soil sample separate throughout the analysis procedure. The common laboratory procedure of mixing the soil to obtain a fully representative sample and then subsampling from the mixture is a disaster, and simply produces the situation where the replication is at a different level from that at which comparison between treatments is required.

Another situation where the replication must be carefully considered is when records of animal progeny, for example sheep, are obtained from a number of different farms, the progeny being classified by breed and farm. One of the purposes of such investigations is to compare performances of different breeds. However, each farm will usually have sheep from at most two or three out of the ten or so breeds for which comparison is required. The question then is whether the records for individual sheep at each farm can be used as replicate observations, or whether only the averages for each farm/breed combination can be used for assessing breed differences. Here, it is a matter of judgement, related to the purpose of the comparison, whether the variation between sheep within a farm/breed combination is relevant to breed comparison. The argument in favour is that breeds are collections of individual sheep, and that therefore breed comparisons should be made relative to variation between sheep within each breed. The opposite argument is that of the previous section, that variation between sheep within a farm may be increased by competition between individuals or decreased by being within a particularly homogeneous environment.

Table 6.7. *Analysis of variance for birthweights of lambs*

Source	df
Farms (ignoring breeds)	24
Breeds (adjusting for farms)	9
Residual between farm/breed combinations	$18 (= 51 - 24 - 9)$
Within farm/breed combinations	1966
Total	2017

The alternative estimators of replicate variance may be demonstrated by considering the analysis of variance structure (Table 6.7).

In this study there were observations (on the birthweight of twin lambs) from 2018 ewes from 25 farms, involving ten breeds but a total of only 52 farm breed combinations (of the 250 possible combinations).

If the interest is in predicting breed differences over a large population of farms, then the residual MS between farm/breed combinations forms the basis of our estimation of the SEs. However, if the objective is to estimate the variability within a breed for a single farm, then the within-farm/breed combinations MS should be used. The ideas of multiple levels of variation involved in these judgements are developed extensively in Chapter 9, and we return to the problem of determining the appropriate variance for comparing treatments in the final chapter of the book.

One extension of the idea of within-unit replication involves the effects of treatments on replication. Consider an experiment to compare four hormone treatments for pigs. Sixteen animals are available in four pens of four animals each. An initial design for the experiment might be to use hormone A on the four pigs in pen 1, hormone B for those in pen 2 and so on, as represented in Figure 6.3(a). However, the replication would be based on variation between animals within a pen, whereas the treatment is applied to a whole pen, a classic case of within-unit replication and one which we have hopefully learnt to avoid. An alternative design, illustrated in Figure 6.3(b) is to treat the four pigs in pen 1 with hormones A, B, C and D, one pig per hormone, and to repeat the procedure in each pen. This design resembles a randomised block design and at first seems appropriate.

However, in each pen the animals live and feed together competitively, the only applied differences being the different hormones. If we consider the final difference in, say, growth rate for individual pigs, then the observed differences would be due not only to the direct effects of the hormone, but also to the results of competition arising from the effects of the hormones. Thus, if hormone A makes pigs aggressive, then this will tend to make the hormone A pigs prosper in face of the lesser competition from the more docile hormone B pigs! Therefore, the experiment measures not only the effects of the different hormones, but also the effects of rearing pigs with different hormones together. This latter effect is not practically relevant since when one hormone is chosen for use, it will be applied to all the pigs in a pen. The search for appropriate replication has led, in this case, to inappropriate treatment of the individual pig units. The only solutions to the replication problem are to use replication of whole pens or to house the pigs individually (which may be inappropriate as a guide to behaviour of the hormone treated pigs in a normal farming environment).

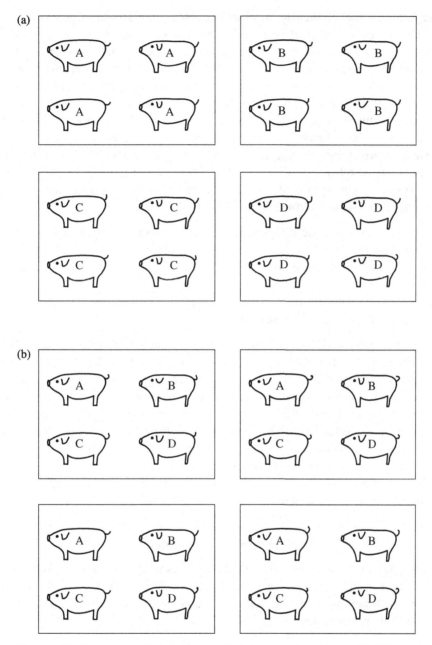

Figure 6.3 Experimental plans for the pig-hormone experiment: (a) different hormones in different pens, (b) each hormone in each pen.

Sometimes it is simply not practical to achieve sufficient real replication at the level appropriate to compare treatments and it is necessary to consider whether within-unit replication can be used to provide at least some idea of precision. This question really concerns the sources of the within-unit and between-unit variation. One subject of research which often leads to this problem is the growth of plants in controlled or artificial environments. Suppose an experimenter wishes to include, in an experiment on the growth of greenhouse plants, the investigation of the effect of heating the greenhouse. Distinct heating levels are only possible if large sections of the greenhouse are used and such sections will usually be few in number. Typically, there might be six greenhouse sections capable of independent heating: if three heating treatments are to be compared, then each may be applied to only two sections. Within each section, it will be possible to have a number of experimental units, probably with different treatments applied to different units. Can the within-section variation between units be used as the basis for comparison between heating treatments?

The strictly correct answer, as discussed earlier in the chapter, is 'No'. But since the experimenter has no other way of investigating the effects of heating, we should consider when the answer might be 'Yes, if . . . '. To use the within-section replication to compare treatments applied to whole sections, it is necessary to assume

(i) that there are no intrinsic differences between the sections which might affect plant growth and production;
(ii) that the variation between units both within and between sections is essentially due to plant-to-plant variation;
(iii) that there is no competitive variation between units within a section, induced by the treatment or the effects of the treatments.

The crucial requirement is that the experimenter could believe that the variation between experimental units is so dominated by plant-to-plant variation that other sources of variation can be ignored. If this is credible, then the experimenter *may* feel justified in using the within-section variation to compare the heating treatments. Note, however, that in presenting results for the heating treatments using SEs based on within-section replication, the experimenter no longer has the protection of a properly randomised experiment with genuine replication. The experimenter must demand an act of faith from the reader in accepting the experimenter's assertion that within-section variation between units would have been the same as between-section variation between units. The experimenter must make the assertion explicitly and has no statistical basis for arguing against the rejection of the assertion.

6.5 How much replication?

The question which is most frequently asked of a statistician when the design of an experiment is being considered is how many replicates should be used. It must be emphasised immediately that there is no simple rule for the amount of replication, which depends on many considerations, including:

(*a*) the resources available,
(*b*) the variability of experimental units,
(*c*) the treatment structure,
(*d*) the size of effect which is of importance, and
(*e*) the relative importance of different comparisons.

It may seem superfluous to suggest consideration of the available resources, but it is not unknown for lengthy discussion of the design of an experiment to lead to a decision that, for example, to detect treatment differences of a size which has important practical implications, it is necessary to include six replicate units for each of eight treatments, only to discover that a maximum of 25 units can be made available. The first question, then, for any experimenter is, how many experimental units can be made available? This may sometimes lead to consideration of replication of experiments in time in order to increase the scale of the experiment.

Suppose we have a total of N units available, and it is proposed that t treatments be compared. Then, assuming equal replication of treatments, the replication per treatment is $n = N/t$, and the SED between two treatment means is

$$(2\sigma^2/n)^{1/2}.$$

To assess whether this is sufficiently precise, we must ask first what the likely size of σ^2 is, and secondly what size of treatment difference is important. Suppose that σ is expected to be about 10% of the mean yield. Then, if resources allow $n = 4$, the SED between two treatment means will be approximately 7% of the mean yield. This means that an observed difference of 14% between two treatment means will be assessed as significant at the 5% level. However, if the true difference between two treatment effects is 14% of the mean yield, then the probability that the observed difference between the means for those two treatments will be 14% or greater is only 0.5. Hence, the probability of detecting as significant, at a 5% significance level, a true difference of 21% (or three times the SE) is 0.83; the probability of detecting a true difference of 28% (four times) is about 0.975.

We have assumed that a trustworthy estimate of σ^2 is available. Without any estimate of σ^2 it is of course quite impossible to assess whether the replication proposed in an experiment is adequate. Fortunately, for any proposed experiment, there are almost always previous experiments of sufficient similarity in which the same measurements have been made, so that a reasonable guess of the likely value of σ^2 for those measurements in the new experiment can be made from the analysis of previous data. For any predetermined level of resources in terms of the number of available experimental units, the replication is determined by the number of treatments, and, given an estimate of σ^2, the experimenter can assess the precision and hence the detectable size of any treatment difference. It is then possible to decide whether it is worth proceeding with the experiment.

If the resources are not, apparently, limiting, then to determine the appropriate level of replication the experimenter needs an estimate of σ^2 and also an estimate of the size of difference which it is important to detect. Returning to the previous discussion of the probability of detecting at a given significance level, α, the observed difference when the true difference is a specified size, d, the necessary replication, n, will be such that

$$d/(2\sigma^2/n)^{1/2} = t_\alpha + \Phi^{-1}(p),$$

where Φ is the normal distribution function and p is the probability of obtaining an observed difference which will be detected as significant. A realistic rule-of-thumb for most situations is to require that the replication r be sufficient to make the SED between two treatment means no bigger than $d/3$. This is equivalent to a probability, p, of 0.83 for a significance level, α, of 0.05.

For example, if it is important to detect true differences of $d = 5\%$ of the mean, then still using $\sigma = 10\%$ of the mean, and assuming that for $\alpha = 0.05$ t_α is approximately 2, we would require

$$\{2(10)^2\}/n \le (5/3)^2$$

or $n \ge 72$. If d is 20% of the mean, then a level of replication, n, of 5 should be adequate.

The choice of replication may not be determined by the precision required for a simple comparison between two treatment means. More generally, the crucial precision may be that for a treatment comparison $\sum_j l_j t_j$. The variance for such a comparison is $\sigma^2 \sum_j l_j^2/n$, and the previous argument holds for this general form of variance in place of the previously considered special case with $\sum_j l_j^2 = 2$.

This discussion of the necessary replication has ignored blocking and factorial structure. If blocks are used and the treatments are orthogonal to blocks (see Chapter 7 for discussion of orthogonality), then the replication arguments are unchanged. If the treatments proposed for the experiment have a factorial structure, then the comparisons considered in determining the required precision of the experiment may be main effect comparisons, in which case the hidden replication of the factorial structure (see Chapter 3) should be included in the calculation of replication.

6.5.1 When can unequal replication be useful?

So far, we have assumed that all treatments in the proposed experiment should be equally replicated. It is true that for a given total of observations, the precision of comparison of treatments will be optimised when each treatment has the same replication. However the loss of precision when unequal replication is used is surprisingly small if the replication is not too different. Suppose that we have a total of twelve observations to be divided between two treatments. If we use six observations for each treatment, then the variance of the comparison between two treatment means will be

$$\sigma^2(2/6).$$

If we use four observations for one treatment and eight for the other, then the variance for the comparison of the two treatment means will be

$$\sigma^2(3/8).$$

In terms of the SED between the two means, the unequal replication increases the SED by only 6%. This is not a large increase, and may well be a cost that an experimenter would be willing to pay for being able to use replications differing by a factor of 2 compared with equal replication. Obviously more extreme differences in the levels of replication for two treatments would lead to larger losses of precision.

In general, the use of equal replication will be sensible when all treatment comparisons are of equal importance, and there are no restrictions on the possible replication for any of the treatments. However, although these are the assumptions that are usually made, they are not necessary and, in some experimental situations, they are not sensible. A classic situation where they may not be reasonable is where one treatment is a standard or control. For this situation it may be that the more important comparisons are of each new treatment with the standard, the comparisons between the new treatments being relatively less important.

Table 6.8. *SED between new and standard treatments*

t	Unequal replication		Equal replication
	New–standard	New–new	Any comparisons
2	5.8	6.8	6.0
4	9.0	12.0	10.0
8	14.7	21.7	18.0
16	25.0	40.0	34.0
32	44.3	75.3	66.0

For the particular case of t new treatments, a total number of N available units and replication n for the new treatments and n_s for the standard, then

$$tn + n_s = N.$$

The variance for the comparison between the standard and one new treatment is

$$\sigma^2(1/n + 1/n_s).$$

If we write $n_s = kn$, then $(k + t)n = N$, and the variance becomes

$$\sigma^2(k + 1)(k + t)/Nk.$$

It can be shown quite simply that this variance is a minimum when $k = t^{1/2}$. Using this optimal form of unequal replication the variances for (a) comparing a new treatment with the standard, or (b) comparing two new treatments, can be substantially different from the variances that would be achieved if equal replication is adopted. The variance formulae for unequal replication are

$$\sigma^2 \left(t^{1/2} + 1 \right)^2 / N$$

when comparing a new treatment with the standard, and

$$2\sigma^2 t^{1/2} \left(t^{1/2} + 1 \right) / N$$

when comparing two new treatments. The formula for equal replication is

$$2\sigma^2(t + 1)/N$$

for both cases. Numerical values, for different values of t, of the variances (divided by σ^2/N) are given in Table 6.8. Thus the ratio of variances for the comparisons between a new treatment and the standard relative to the comparison of two new treatments tends downwards to a limiting value of $1/2$ as t increases. The ratio of variances for the new versus standard comparison using the unequal replication compared with the equal replication also tends to $1/2$, but considerably more slowly.

It is important to emphasise again that equal replication for all experimental treatments is not always necessary nor even desirable. If the replication of one or more treatments is limited by resources, as could be the case if, in a plant breeding trial, there are only small numbers of seeds for some breeding lines, then to some extent precision can be improved

Table 6.9. *Critical values from
t-distribution*

df	5% significance t-value
1	12.71
2	4.30
3	3.18
4	2.78
5	2.57
6	2.45
7	2.36
8	2.31
9	2.26
10	2.23
12	2.18
15	2.13
20	2.09
30	2.04
60	2.00
120	1.98

by using more replicates of other lines. In other words, since the variance of a comparison between two treatments with n_1 and n_2 replications is

$$\sigma^2(1/n_1 + 1/n_2),$$

then if $n_1 + n_2 = N$ is limited of course the variance is minimised when $n_1 = n_2 = N/2$. However, if n_2 is limited, the variance can be reduced by increasing n_1.

One final consideration when determining how much replication is needed is the precision of estimation of σ^2. The estimate of σ^2 is obtained from the residual MS, and the precision of this estimate is determined by the df for the RSS. As the residual df are increased, so the precision of the estimate of σ^2 is increased, and this may be most clearly seen by examining the set of values of the t statistic required for significance at 5%. The values given in Table 6.9 show how the critical t-values decrease substantially as the degrees of freedom increase up to about 10 df. After 10 df the improvement in the precision of estimating σ^2 tails off, and after 20 df the benefits of additional df in reducing the 5% t-values are negligible.

This aspect of replication was previously considered in Chapter 1, when the different uses of df within the resource equation were considered. It is quite clear, when determining the amount of replication that it is desirable that there should always be at least 10 df for the RSS but that no real advantage is gained by having more than 20 df.

Exercises

6.1 Examine critically the information about replication in an applied science journal. Assess the level of replication used in experiments from which results are reported, and judge whether the form of replication used to provide the estimated SE is appropriate to the treatment comparisons.

(We should be surprised if you do not find examples of inappropriate replication, and also if the necessary information is always provided.)

6.2 For a completely randomised design to compare three treatments, two experimenters have different views on the relative importance of the three comparisons:

(i) t_1-t_2,

(ii) t_1-t_3,

(iii) t_2-t_3.

They suggest that the SEs for these comparisons should be in the ratios

(*a*) 4:5:6, or

(*b*) 3:3:4.

Determine the relative replication needed for each suggestion. Assuming a total of 60 observations is available, compare the expected precision of treatment comparisons achieved using the designs appropriate to (*a*), (*b*) with the precision for the equal replicate design and present your advice to the experimenters.

7

Blocking and control

7.0 Preliminary examples

(*a*) An experiment is to be designed to investigate the effects of supplementary heating, supplementary lighting and carbon dioxide on the growth of peppers in glasshouses. The number of possible treatment combinations is eight, comprising all combinations of two levels of heating (standard or supplementary), two levels of lighting (standard or supplementary) and two levels of CO_2 (background or added). Each observation on a treatment combination requires a separate glasshouse compartment and there are 12 glasshouse compartments available, arranged in two linear runs of six compartments (each running east–west, one with a northerly aspect, the other with a southerly aspect). Compartments in different runs are therefore likely to produce rather different yields. The problem is to design an experiment in two blocks each of six units to compare the eight factorial treatment combinations. It is intended to have a second experiment, using the same glasshouse compartments, in the next year, after which the most effective treatment combination will be recommended for use.

(*b*) In a screening study 14 fungal isolates of a potential biological control agent of onion white rot (*Sclerotiorum cepivorum*) are to be produced using three standard methods and tested by assessing germination of sclerotia in the laboratory. The aim of the study is to compare the 42 treatment combinations and select the best six for more detailed research. The available resources allow the production of three replicate batches of each treatment combination, but at most 18 batches can be processed in a single production run, and it is known that there are likely to be substantial differences between production runs (each takes several weeks). Seven runs of 18 batches will accommodate the three replications of the 42 treatment combinations. How should the treatment combinations be allocated to runs?

(*c*) A microarray experiment using a two-colour (channel) technology is to be used to identify genes showing differential expression during leaf senescence in the model plant *Arabidopsis thaliana*. Leaf samples are collected from four replicate plants grown in a controlled environment on 13 alternate days following full leaf opening, resulting in 52 separate leaf samples. As well as wanting to identify genes showing differential expression over the time course, the experimenter is also interested in measuring the variability between plants, and estimating the mean response for each plant for use in subsequent analyses. The microarray technology allows two samples to be directly compared on each array, and resources allow about 100 arrays to be run. How should the samples be allocated to arrays?

Block

	I		II
	A		B
	C		C
	D		D
	E		E
	F		F

Figure 7.1 Experimental plan for comparing six treatments in two blocks of five units.

7.1 Design and analysis for very simple blocked experiments

The primary idea of blocking is that by identifying blocks of homogeneous units, such that there may be quite large differences between units allocated to different blocks, we are able to make more precise comparisons of the treatments by eliminating these potentially large block differences from the calculation of treatment differences. It follows that information from blocked experiments is predominantly based on the comparisons that can be made between treatment observations in the same block. If two treatments do not occur together in a block, then it will still be possible to make a valid comparison between the two treatments if each occurs in a block with a common third treatment. Essentially, a comparison between two treatments, A and B, within a block, has variance $2\sigma^2$, where σ^2 is the variance of units within a block. If each of A and B occur in separate blocks with each block also containing a third treatment C, then the A − B comparison can be calculated as $(A - C) + (C - B)$, each of the two-component comparisons being made within a block, and the variance of the resulting A − B comparison is $2(2\sigma^2)$.

If the indirect comparison of A and B can be made through several different intermediaries, then the precision of the A − B comparison is improved, depending on the number of intermediaries. Suppose we have two blocks of five units each, and wish to compare six treatments, A, B, C, D, E and F. With each of A and B appearing only once and in separate blocks, each block must then contain the remaining four treatments (C, D, E and F). This simple design is shown in Figure 7.1. To compare any two of C, D, E and F, we obviously use the mean response from the two observations for each of the two compared treatments. To compare A and B, we use all four other treatments as intermediaries to estimate the difference between the blocks, and compare each of A and B with the mean yields of the other four treatments in the block in which A, or B, appears. The variance of $A - (C + D + E + F)/4$ is $5\sigma^2/4$ and hence the variance of A − B is $5\sigma^2/2$. This may be compared with the variance for the difference between two treatments each replicated once, and occurring together in a block, $2\sigma^2$. As the number of intermediaries increases, the variance of the comparison between A and B in separate blocks moves closer to that for the comparison within blocks.

We can also consider the comparison of A, or B, with one of the other four treatments. A first thought might suggest that, since A only occurs in block I and since C also occurs there, the comparison A − C will be based only on observations from block I. However, it should be apparent that the C response in block II also contributes information to the comparison. To use the comparison between A in block I and C in block II, we have to estimate the difference between blocks I and II, and this is done through the treatments

(C, D, E and F) occurring in both blocks. However, this means that the estimate of the difference between A and C from A in block I and C in block II involves the differences between C in blocks I and II, and hence involves the A − C difference in block I. Thus, although the principle of the comparison is quite simple, the practical details are more difficult. These details can be formalised through consideration of the algebraic solution of the least squares equations, which also demonstrates the benefits of this general approach to designing experiments.

The model for this simple example is

$$y_{ij} = \mu + b_i + t_j + \varepsilon_{ij}$$

for $i = 1, 2$ and $j = A, B, C, D, E, F$, except $(i, j) = (1, B), (2, A)$. The constraints are

$$b_1 + b_2 = 0 \qquad \sum t_j = 0.$$

The least squares equations are:

$$10\,\hat{\mu} \;-\; \hat{t}_A - \hat{t}_B = Y_{..}, \tag{7.1}$$

$$5\,\hat{\mu} \;+\; 5\hat{b}_1 - \hat{t}_B = Y_{1.}, \tag{7.2}$$

$$5\,\hat{\mu} \;+\; 5\hat{b}_2 - \hat{t}_A = Y_{2.}, \tag{7.3}$$

$$\hat{\mu} \;+\; \hat{t}_A - \hat{b}_2 = Y_{.A}, \tag{7.4}$$

$$\hat{\mu} \;+\; \hat{t}_B - \hat{b}_1 = Y_{.B}, \tag{7.5}$$

$$2\,\hat{\mu} \;+\; 2\hat{t}_C \quad\;\; = Y_{.C}, \tag{7.6}$$

with (7.6) repeated for treatments D, E and F. These equations are solved by utilising the symmetry for $(\hat{t}_A, \hat{t}_B)$ and $(\hat{b}_1, \hat{b}_2)$ and considering $(\hat{t}_A - \hat{t}_B)$ and $(\hat{t}_A + \hat{t}_B)$. The steps in the algebraic manipulation are as follows.

Subtracting (7.3) from (7.2) gives

$$5(\hat{b}_1 - \hat{b}_2) + (\hat{t}_A - \hat{t}_B) = Y_{1.} - Y_{2.} \tag{7.7}$$

with the $\hat{\mu}$ cancelling out. Subtracting (7.5) from (7.4) gives

$$(\hat{t}_A - \hat{t}_B) + (\hat{b}_1 - \hat{b}_2) = Y_{.A} - Y_{.B} \tag{7.8}$$

with the $\hat{\mu}$ again cancelling out. Then subtracting (7.7) from five times (7.8) gives an expression for a multiple of the difference between treatments A and B in terms of the Y sums

$$4(\hat{t}_A - \hat{t}_B) = 5(Y_{.A} - Y_{.B}) - (Y_{1.} - Y_{2.}), \tag{7.9}$$

cancelling out the block terms, $\hat{b}_1$ and $\hat{b}_2$. Adding (7.4) and (7.5) gives

$$2\hat{\mu} + (\hat{t}_A + \hat{t}_B) = Y_{.A} + Y_{.B}, \tag{7.10}$$

again removing the block terms. Then subtracting (7.1) from five times (7.10) gives an expression for a multiple of the sum of treatments A and B in terms of the Y sums

$$6(\hat{t}_A + \hat{t}_B) = 5(Y_{.A} + Y_{.B}) - Y_{..} \tag{7.11}$$

eliminating the $\hat{\mu}$. Adding three times (7.9) to two times (7.11) eliminates $\hat{t}_B$ to leave an expression for $\hat{t}_A$ in terms of the Y sums

$$24\hat{t}_A = 25Y_{.A} - 5Y_{.B} - 3Y_{1.} + 3Y_{2.} - 2Y_{..}. \tag{7.12}$$

Adding (7.1) and (7.10) eliminates $\hat{t}_A$ and $\hat{t}_B$, leaving an expression for $\hat{\mu}$ in terms of the overall sum and treatment sums

$$12\hat{\mu} = Y_{..} + (Y_{.A} + Y_{.B}). \tag{7.13}$$

Multiplying (7.6) by 12, substituting for $\hat{\mu}$ from (7.13), and rearranging gives an expression for $\hat{t}_C$:

$$24\hat{t}_C = 12Y_{.C} - 2Y_{..} - 2Y_{1.} + 2Y_{2.} \tag{7.14}$$

Subtracting (7.14) from (7.13) then provides an expression for a multiple of the difference between treatments A and C in terms of the Y sums

$$24(\hat{t}_A - \hat{t}_C) = 27Y_{.A} - 3Y_{.B} - 12Y_{.C} - 3Y_{1.} + 3Y_{2.}. \tag{7.15}$$

Hence, (7.9) can be simplified to give

$$\hat{t}_A - \hat{t}_B = 1/4\{5(Y_{.A} - Y_{.B}) - (Y_{1.} - Y_{2.})\} \tag{7.16}$$

and (7.15) to give

$$\hat{t}_A - \hat{t}_C = 1/8\{9Y_{.A} - Y_{.B} - 4Y_{.C} - (Y_{1.} - Y_{2.})\}. \tag{7.17}$$

In each case, the estimate of the difference clearly involves the correction for the difference between the two blocks. The variances of the two estimated differences can be derived in various ways. The simplest is by considering the estimates as linear combinations of responses by expanding the Y sum expressions above:

$$
\begin{aligned}
\hat{t}_A - \hat{t}_B &= 1/4(4y_{1A} - y_{1C} - y_{1D} - y_{1E} - y_{1F} - 4y_{2B} + y_{2C} + y_{2D} + y_{2E} + y_{2F}), \\
\hat{t}_A - \hat{t}_C &= 1/8(8y_{1A} - 5y_{1C} - y_{1D} - y_{1E} - y_{1F} - 3y_{2C} + y_{2D} + y_{2E} + y_{2F}),
\end{aligned}
\tag{7.18}
$$

whence

$$\mathrm{Var}(\hat{t}_A - \hat{t}_B) = (\sigma^2/16)(2 \times 16 + 8 \times 1) = 5\sigma^2/2 \tag{7.19}$$

confirming the earlier result, and

$$\mathrm{Var}(\hat{t}_A - \hat{t}_C) = (\sigma^2/64)(64 + 25 + 9 + 6 \times 1) = 13\sigma^2/8. \tag{7.20}$$

These variances should be compared with those for an unblocked experiment with the same replication, $\sigma^2(1 + 1) = 2\sigma^2$, and $\sigma^2(1 + 1/2) = 3\sigma^2/2$, but we must remember that the effect of blocking should also be to reduce σ^2.

A diagrammatic method of obtaining the linear combination of responses for complex least squares estimates is illustrated in Figure 7.2. The method is essentially just writing out each total in (7.15) as the sum of the individual responses, but the diagram provides a structure, reducing the scope for mistakes. The separate figures in each 'unit' derive from the totals in (7.15) – top left for the treatment coefficients, top right for the block coefficients; the circled figures then represent the total coefficient for each unit (as in (7.16)).

Clearly this derivation of least squares estimates and their variances, although straightforward and useful in providing understanding, is too lengthy to be used in each particular case. General results are needed and these are derived in Section 7.3. Moreover, the analysis of experimental data can always be achieved through the use of general linear model algorithms

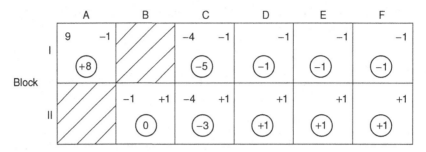

Figure 7.2 Diagrammatic representation of estimate of $t_A - t_C$.

```
                    Block

              I    II   III   IV

              A    A    A    A
              B    B    B    B
              C    C    C    C
              D    D    D    D
              E    E    E    E
              F    F    F    F
```

Figure 7.3 Experimental plan for comparing six treatments in four blocks of six units.

in statistical computer packages. However, the most important conclusion from this example concerns the simplicity of the design argument. The implications for analysis are not simple but, given the problem of comparing six treatments in two blocks of five units each, the design solution is inevitable, the only choice being which two treatments should appear only once.

7.2 Design principles in blocked experiments

In this section the thesis that the intuitive ideas of block designs are extremely simple is developed through considering a series of examples. In each example, the treatments are assumed to have no structure, and all treatment comparisons are assumed to be equally important.

Example 7.1 Six treatments, A–F, are to be compared using 24 units for which a natural blocking system gives four blocks of six units each. How should the six treatments be allocated to the units?

This is an easy problem. All six treatments can and should appear in each block leading to a randomised complete block design. Any other allocation will involve at least one treatment occurring twice in a block, and consequently some treatment comparisons being not directly estimated in that block, which must lead to less efficient comparison. The treatment allocation to blocks is shown in Figure 7.3 (throughout this chapter treatment allocations to blocks will be shown without randomisation).

Block

I	II	III	IV	V	VI
A	A	A	A	A	B
B	B	B	B	C	C
C	C	C	D	D	D
D	D	E	E	E	E
E	F	F	F	F	F

Figure 7.4 Experimental plan for comparing six treatments in six blocks of five units.

Example 7.2 Again, six treatments, A–F, are to be compared, this time using 30 units grouped in six natural blocks of five units each. How should the treatments be allocated to units?

Since all treatments are assumed equally important, each treatment should occur five times. The number of direct comparisons of different treatments within a block is maximised by avoiding repetition of any treatment within a block; thus, each block will contain five, different, treatments. The problem then reduces to choosing which treatment should be omitted from each block. Since there are six blocks and six treatments, the symmetrical arrangement of omitting a different treatment in each block, as shown in Figure 7.4, has an intuitive appeal, and the intuition is correct. The overall average precision of comparisons cannot be improved. The assumption of equal importance for all treatment comparisons provides the justification for the choice of the symmetric design.

Example 7.3 Again six treatments, A–F, are to be compared, this time using 24 units grouped in six natural blocks of four units each.

The equal importance of treatments leads to the first decision to have four units for each treatment. The primacy of direct comparisons requires that no treatment occurs twice in a block. In each block two treatments must be omitted, and correspondingly each treatment must be omitted from two blocks. This means that symmetry cannot be retained. Treatment A must be omitted twice and, to make direct comparisons with other treatments as even as possible, two different treatments will be omitted with treatment A. Suppose that in block I treatments A and B are omitted, and in block II treatments A and C are omitted. In the remaining four blocks, A occurs in each, B and C in three, and D, E and F in two each. Hence, there are three direct comparisons of A with B or with C, whereas there are only two with each of D, E and F. By considering the total possible direct comparisons for A, it can be recognised that no more even arrangement of direct comparisons can be achieved. Treatment A occurs in four blocks, in each of which there are three other units giving three direct comparisons in each block, and a total of 12 direct comparisons with A. Since there are five other treatments, and 12 is not a multiple of five, an uneven division of direct comparisons is inevitable.

An allocation in which each pair of omitted treatments is different seems intuitively attractive. There are two possible systems for the set of omitted pairs:

(AB), (BC), (CA), (DE), (EF), (FD),

or

$$(AB), (BD), (DF), (FE), (EC), (CA).$$

The ordering of the omitted pairs makes it obvious that the second system is based on one cycle, (ABDFEC), rather than two cycles for the first system, (ABC) (DEF), and this second system seems intuitively to offer a more even distribution of comparisons. This is confirmed by calculating the variances of the estimated treatment differences.

For the design with pairs of missing treatments based on two cycles there are two variances for treatment differences:

$$0.7303\sigma^2 \text{ for (AB), (BC), (CA), (DE), (EF), (FD),}$$
$$0.7601\sigma^2 \text{ for the other nine treatment differences.}$$

For the design with pairs of missing treatments based on one cycle there are three variances for treatment differences:

$$0.7321\sigma^2 \text{ for (AB) (BD) (DF) (FE) (EC) (CA),}$$
$$0.7579\sigma^2 \text{ for (AD) (DE) (EA) (BF) (FC) (CB),}$$
$$0.7596\sigma^2 \text{ for (AF) (BE) (CD).}$$

The difference is indeed slight with weighted mean variances of $0.7482\sigma^2$ and $0.7479\sigma^2$ respectively, though the second design also has the benefit of slightly more homogeneous variances. It would, of course, be ridiculous to argue that the first design is inefficient in any practically meaningful sense.

Example 7.4 Once again, six treatments, A–F, are to be compared, once again using 30 units, but this time in ten natural blocks of three units each. Again, the equal importance of the treatments implies five occurrences of each treatment, with the primacy of direct comparisons requiring that no treatment occurs twice in a block. With treatment A therefore appearing in five blocks, with two further units per block, there are ten possible direct comparisons between treatment A and the other treatments. With five other treatments this suggests that there may be a design offering equal frequency (2) of direct comparisons for all treatment pairs. Such a design, if it in fact exists, can be constructed logically as follows:

 (i) A must occur in five blocks (I, II, III, IV, V).
 (ii) B must occur in two of these blocks (I, II) and in three others (VI, VII, VIII).
(iii) C must occur twice with A, twice with B and a total of five times. Suppose that C occurs in block I, then the remaining four occurrences must include only one with A, and one with B. Since there are only two blocks (IX, X) available in which C can occur without A or B it follows that C must occur in both of these 'empty' blocks, once in a block with A alone (III), and once in a block with B alone (VI).
(iv) D must occur twice with A, twice with B, twice with C, and five times in all. Since all blocks now have at least one of A, B or C, it follows that D will occur with two of A, B or C in one block, and with one of them in each of its other four appearances. Without loss of generality we can add D to block II since one of D, E or F, must occur in this block. D then occurs once more with each of A (block IV) and B (block VII), and twice with C (in blocks IX and X).

Block

I	II	III	IV	V	VI	VII	VIII	IX	X
A	A	A	A	A	B	B	B	C	C
B	B	C	D	E	C	D	E	D	D
C	D	E	F	F	F	E	F	E	F

Figure 7.5 Experimental plan for comparing six treatments in ten blocks
of three units.

Block

I	II	III	IV	V	VI
A	A	A	A	A	A
B	B	B	B	B	B
C	C	C	C	C	C
D	D	D	D	D	D
E	E	E	E	E	E
F	F	F	F	F	F
A	B	C	D	E	F

Figure 7.6 Experimental plan for comparing six treatments in six blocks
of seven units.

(v) The allocation of E and F can be considered together. Both must occur in blocks V
and VIII as there are two spaces in each. Allocation to the remaining blocks can be
considered in pairs. One of E and F appears in blocks III and VII (with each of the other
four treatments once), and the other in blocks IV and VI (again with each of the other
four treatments once).

This experiment has been considered in some detail, and it should be clear that there is
no real choice in the construction of the design – the allocation process just follows a series
of logical and 'forced' steps. Any other design fitting the limitations of the design will be
identical (except for permutation of the letters) with the one we have constructed, shown in
Figure 7.5.

Example 7.5 Again six treatments, A–F, are to be compared, this time with 42 units grouped
in six blocks of seven units. The symmetry principles advanced in Example 7.2 apply again
here. The natural design includes all six treatments in each block, with an additional treatment
in each block, the additional treatment being different in each block. The resulting design is
shown in Figure 7.6.

Example 7.6 For the last time, six treatments, A–F, are to be compared, now in 25 units
grouped in five blocks of five units each. Again, comparison with Example 7.2 is useful.
Each block must contain five out of the six treatments, and a different treatment should
be omitted from each block. One treatment will inevitably occur in each block and the
comparisons involving that treatment will be more precisely determined than those of the
other treatments. The choice of this treatment with higher replication identifies the block
to be omitted from Example 7.2 (the block omitting this treatment), and the best design,

Block

I	II	III	IV	V
A	A	A	A	A
B	B	B	B	C
C	C	C	D	D
D	D	E	E	E
E	F	F	F	F

Figure 7.7 Experimental plan for comparing six treatments in five blocks of five units.

Block

I	II	III	IV	V
A	A	A	A	A
B	B	B	B	B
C	C	C	C	C
D	D	D	D	D
E	F	E	E	E
	G	F	F	F
		G	G	G
				E
				F
				G

Figure 7.8 Experimental plan for comparing seven treatments in five blocks with five, six, seven, seven and ten units.

offering most nearly even precision of comparisons will be identical (except for permutation of letters) with that shown in Figure 7.7.

Example 7.7 Finally in this section we consider an example discussed initially in Chapter 1, in which seven treatments affecting the growth of rats are to be compared using 35 rats from five litters, or blocks, containing five, six, seven, seven and ten rats, respectively. The two blocks of seven units can be dealt with first, by allocating all seven treatments to each block. The block of ten units must also contain all seven treatments, with three of the treatments being duplicated. These three treatments must then be omitted from one or other of the two smaller blocks. The design is shown in Figure 7.8. Clearly, the precision of different comparisons will vary slightly but the design is both obvious and efficient.

The examples presented in this section are intended to demonstrate that, when block size is not equal to the number of treatments, then the construction of a sensible design is not difficult. Two particular principles deserve special emphasis:

(i) if a design is possible with a complete, uniform, set of pairwise occurrences of treatments, then simple logical argument will produce that design (as for Example 7.4). If equal pairwise occurrence cannot be achieved, then simple logical argument will produce the most nearly equal pattern of occurrences;

Table 7.1. *Comparison of variances for treatment comparisons from incomplete block designs in Examples 7.2–7.7, with those for complete blocks and for direct comparisons*

Design	Variance for treatment differences	Complete block variance	Direct block variance
Table 7.2	$0.4167\sigma^2$	$0.4000\sigma^2$	$0.5000\sigma^2$
Table 7.3 AB	$0.5359\sigma^2$	$0.5000\sigma^2$	$0.6667\sigma^2$
AD	$0.5744\sigma^2$	$0.5000\sigma^2$	$1.0000\sigma^2$
AF	$0.5769\sigma^2$		
Table 7.4	$0.5000\sigma^2$	$0.4000\sigma^2$	$1.0000\sigma^2$
Table 7.5	$0.2917\sigma^2$	$0.2857\sigma^2$	$0.3056\sigma^2$
Table 7.6 AB	$0.4605\sigma^2$	$0.4500\sigma^2$	$0.5000\sigma^2$
BC	$0.5262\sigma^2$	$0.5000\sigma^2$	$0.6667\sigma^2$
Table 7.7 AB	$0.4000\sigma^2$	$0.4000\sigma^2$	$0.4000\sigma^2$
AE	$0.4112\sigma^2$	$0.4000\sigma^2$	$0.4688\sigma^2$
AF	$0.4130\sigma^2$	$0.4000\sigma^2$	$0.4688\sigma^2$
EF	$0.4156\sigma^2$	$0.4000\sigma^2$	$0.5000\sigma^2$
FG	$0.4000\sigma^2$	$0.4000\sigma^2$	$0.4000\sigma^2$

(ii) recognising that the randomised complete design is the most efficient possible blocked design, small modifications of the complete block design, such as omitting a different treatment from each block, will provide designs with high efficiency.

In the previous section we discussed the precision of treatment comparisons when direct and indirect comparisons are combined. The variance for any treatment comparison in an incomplete block design will include a multiplying factor larger than that for the same comparison in a complete block design, assuming the value of σ^2 were the same in both designs. However, the multiplying factors in the variances of treatment comparisons in incomplete block designs are usually not much larger than those for the equivalent comparisons (with the same treatment replication) in complete block designs (or equivalently in unblocked designs). Table 7.1 lists the variances for all the different forms of pairwise comparison between treatments for the various example designs considered in this section. To assess the impact of considering incomplete block designs, the variances for pairwise comparisons between treatments with the same replication levels in complete blocks, and the variances which would be obtained if only direct comparisons (from joint occurrence of the treatments in a block) were considered in the incomplete block designs, are also listed. All variances are written in terms of σ^2, and we should remember that if the blocks for incomplete block designs are 'better' blocks than those for complete block designs, then σ^2 will be smaller for the incomplete block design – sometimes very much smaller. It is clear from Table 7.1 that the increase in the multiplying factor from using incomplete blocks is usually small, the greatest loss in precision occurring for Example 7.4, when the smallest block size (half the number of treatments) is used. It should also be noted that the (direct plus indirect comparison) incomplete block variances are invariably much closer to the complete block variances than to the direct comparison variances. Thus the effective replication of treatment

Block

I	II	III	IV	V	VI
A	A	A	A	B	B
B	B	C	C	C	D
C	E	D	E	D	E
D	F	F	G	F	F
E	G	G	O	O	G
O	O	O	O	O	O

Figure 7.9 Experimental plan for comparing eight treatments in six blocks of six units.

comparisons, though less than the actual replication, is much greater than suggested by the direct comparison replication.

There is one final issue to consider before we leave this discussion of design principles for incomplete designs. In discussing these designs we have assumed the equal importance of all treatment comparisons. This will often be true, or at least the relative importance of different comparisons will be not clearly identified and it will be reasonable to proceed as if all comparisons are equally important. However, sometimes it will be clear that some treatment comparisons require greater precision than others. In such cases a sensible procedure is to consider first the desired relative precision of different comparisons and to choose the replication for each treatment, n_j, in such a way that the variances of treatment differences for an unblocked experiment, $\sigma^2(n_j^{-1} + n_{j'}^{-1})$, take appropriate relative values. Then the allocation of the treatments, with n_j observations for treatment j, should be chosen so as to achieve joint occurrences of pairs of treatments in numbers proportional to the product of their replications. Note that wide variation in precision is not a practical option (as was illustrated in Exercise 6.2).

Example 7.8 In an experiment to compare seven new treatments with an untreated control, it has been decided that the replication of the control treatment should be twice that of other treatments. Resources allow 36 units in total, so that the experiment should include eight observations of the control treatment, O, and four observations of each of the other seven treatments (A–G). The appropriate blocking system is believed to be six blocks of six units each. How should we then allocate the treaments to blocks? Each block must have at least one untreated control observation, with two blocks having two such observations. In these two blocks (IV and V in Figure 7.9) it would seem obvious to include all seven other treatments, so that one of the other treatments (C in the example) must occur in both of these blocks. In the remaining four blocks, treatment C must occur twice and the other six treatments three times each. With a total of 15 comparisons possible between any non-control treatment and the other six non-control treatments, our aim is to produce a design for which pairs of non-control treatments occur together two or three times in a block. A suitable design is shown in Figure 7.9. The resulting variances of treatment differences are tabulated in Table 7.2, and again it can be seen that the irregular pattern of the design is not reflected in the variances which are very homogeneous among the non-control treatments (the different levels indicating whether pairs occur together in two or three blocks) and between the control and non-control treatments.

Table 7.2. *Variances of treatment differences for Example 7.8*

	A	B	C	D	E	F	G
B	0.5477						
C	0.5236	0.5475					
D	0.5477	0.5228	0.5236				
E	0.5228	0.5239	0.5475	0.5477			
F	0.5477	0.5228	0.5475	0.5228	0.5467		
G	0.5228	0.5417	0.5475	0.5477	0.5228	0.5239	
O	0.3978	0.3983	0.3866	0.3978	0.3983	0.3983	0.3983

7.3 The analysis of block-treatment designs

Consider the general design with n_{ij} observations for treatment j in block i. For most designs considered in the previous section, and in the rest of this chapter, n_{ij} will be either 0 or 1. We write the model for the analysis of such designs in the form

$$y_{ijk} = \mathsf{b}_i + \mathsf{t}_j + \varepsilon_{ijk}, \quad k = 1, \ldots, n_{ij}, \tag{7.21}$$

where the b_is represent block mean responses and the t_js are deviations of the response for particular treatments from the mean treatment response. The least squares equations are

$$\hat{\mathsf{b}}_i N_{i.} + \sum_j \hat{\mathsf{t}}_j n_{ij} = Y_{i..}, \tag{7.22}$$

$$\sum_i \hat{\mathsf{b}}_i n_{ij} + \hat{\mathsf{t}}_j N_{.j} = Y_{.j.}, \tag{7.23}$$

where the dot notation introduced previously for totals is used for sums of both n_{ij} and y_{ijk}. Rearranging (7.22) gives an expression for the $\hat{\mathsf{b}}_i$:

$$\hat{\mathsf{b}}_i = \left(Y_{i..} - \sum_j \hat{\mathsf{t}}_j n_{ij} \right) / N_{i.}. \tag{7.24}$$

Using this to eliminate the $\hat{\mathsf{b}}_i$ from (7.23), we obtain

$$\hat{\mathsf{t}}_j \left(N_{.j} - \sum_i n_{ij}^2 / N_{i.} \right) - \sum_{j' \neq j} \mathsf{t}_{j'} \left(\sum_i n_{ij} n_{ij'} / N_{i.} \right) = Y_{.j} - \sum_i n_{ij} Y_{i..} / N_{i.}. \tag{7.25}$$

These equations are not independent, as can be seen by checking whether the sum of the coefficients of any particular t_j over the set of t equations is zero. The constraint $\sum_j \mathsf{t}_j = 0$ is therefore needed to solve the equations.

A more important deduction from the set of equations concerns the condition necessary for these equations to be symmetrical for all t treatments. If we consider the coefficient of $\mathsf{t}_{j'}$ in the equation for t_j, then we can see that all such coefficients will be equal if

$$\sum_i n_{ij} n_{ij'} / N_{i.} = p_{jj'} \text{ is constant for all } (j, j'). \tag{7.26}$$

This result was recognised first by Pearce (1963), who called the $p_{jj'}$ the 'sum of weighted concurrences' for treatments j and j'. Pearce's result is that, if all $p_{jj'}$ are equal, then all simple treatment comparisons are equally precise, and the design is said to be *balanced*. This is a wider definition of balance than the classical special case of balanced incomplete block designs, which will be considered later in this chapter.

A simple example, taken from Pearce's paper, demonstrates the general principle of balance and provides a further example of the detailed manipulation of the least squares equations to obtain variances for treatment comparisons.

Example 7.9 Consider a design for three treatments (O, A and B) in two blocks of four units and one of two units:

$$
\begin{array}{lcccc}
\text{block I} & \text{O} & \text{O} & \text{A} & \text{B} \\
\text{block II} & \text{O} & \text{O} & \text{A} & \text{B} \\
\text{block III} & \text{A} & \text{B} & &
\end{array}
$$

The sums of weighted concurrences are then

$$
\begin{array}{ll}
\text{O and A} & \dfrac{2 \times 1}{4} + \dfrac{2 \times 1}{4} = 1 \\[2mm]
\text{O and B} & \\[2mm]
\text{A and B} & \dfrac{1 \times 1}{4} + \dfrac{1 \times 1}{4} + \dfrac{1 \times 1}{2} = 1.
\end{array}
$$

The least squares equations (from (7.22) and (7.23)) for the three block effects and three treatment effects are

$$
\begin{array}{llll}
4\hat{b}_1 & + 2\hat{t}_O + \hat{t}_A + \hat{t}_B & = Y_{1.} \\
\quad 4\hat{b}_2 & + 2\hat{t}_O + \hat{t}_A + \hat{t}_B & = Y_{2.} \\
\quad\quad 2\hat{b}_3 & + \hat{t}_A + \hat{t}_B & = Y_{3.} \\
2\hat{b}_1 + 2\hat{b}_2 & + 4\hat{t}_O & = Y_{.O} \\
\hat{b}_1 + \hat{b}_2 + \hat{b}_3 & +3\hat{t}_A & = Y_{.A} \\
\hat{b}_1 + \hat{b}_2 + \hat{b}_3 & +3\hat{t}_B & = Y_{.B}.
\end{array}
$$

Eliminating $\hat{b}_1, \hat{b}_2$ and $\hat{b}_3$ leads to the set of equations

$$
\begin{bmatrix} 2 & -1 & -1 \\ -1 & 2 & -1 \\ -1 & -1 & 2 \end{bmatrix}
\begin{bmatrix} \hat{t}_O \\ \hat{t}_A \\ \hat{t}_B \end{bmatrix}
=
\begin{bmatrix} Y_{.O} - 1/2(Y_{1.} + Y_{2.}) \\ Y_{.A} - 1/4(Y_{1.} + Y_{2.} + 2Y_{3.}) \\ Y_{.B} - 1/4(Y_{1.} + Y_{2.} + 2Y_{3.}) \end{bmatrix}.
$$

These can be simply solved with the constraint

$$t_O + t_A + t_B = 0. \tag{7.27}$$

Alternatively, if we apply the constraint directly to the original least squares equations for the block sums, we have each block effect expressed in terms of the block sums and $\hat{t}_O$:

$$
\begin{aligned}
4\hat{b}_1 + \hat{t}_O &= Y_{1.}, \\
4\hat{b}_2 + \hat{t}_O &= Y_{2.}, \\
2\hat{b}_3 - \hat{t}_O &= Y_{3.}.
\end{aligned}
$$

Substitution for the block effects in the least squares equation relating $\hat{t}_O$ to the treatment O sum, and subsequently in the least squares equations for $\hat{t}_A$ and $\hat{t}_B$ for the treatment A and treatment B sums, provides direct estimates of the treatment effects. Either method leads to the same estimates:

$$3\hat{t}_O = Y_{.O} - 1/2(Y_{1.} + Y_{2.}),$$
$$3\hat{t}_A = Y_{.A} - 1/4(Y_{1.} + Y_{2.} + 2Y_{3.}),$$
$$3\hat{t}_B = Y_{.B} - 1/4(Y_{1.} + Y_{2.} + 2Y_{3.}),$$

from which we can construct equations for the differences between treatment effects:

$$3(\hat{t}_O - \hat{t}_A) = Y_{.O} - Y_{.A} - (Y_{1.} + Y_{2.})/4 + Y_{3.}/2,$$
$$3(\hat{t}_O - \hat{t}_B) = Y_{.O} - Y_{.B} - (Y_{1.} + Y_{2.})/4 + Y_{3.}/2,$$
$$3(\hat{t}_A - \hat{t}_B) = Y_{.A} - Y_{.B}.$$

Hence $\mathrm{Var}(\hat{t}_A - \hat{t}_B) = 2\sigma^2/3$, which, since each of A and B appears in each block, is an obvious result. And either directly or using the diagrammatic method of Figure 7.2,

$$12(\hat{t}_O - \hat{t}_A) = 3(y_{101} + y_{102} + y_{201} + y_{202}) - 5(y_{1A} + y_{2A})$$
$$- (y_{1B} + y_{2B}) - 2(y_{3A} + y_{3B}),$$

so that

$$\mathrm{Var}(\hat{t}_O - \hat{t}_A) = \sigma^2(4 \times 9 + 2 \times 25 + 2 \times 1 + 2 \times 4)/144$$
$$= 2\sigma^2/3,$$

with an identical result for the variance of the difference between treatments O and B.

Consider now the fitting SS for the analysis of the general block-treatment model:

$$\text{fitting SS} = \sum_i \hat{b}_i Y_{i..} + \sum_j \hat{t}_j Y_{.j.}. \tag{7.28}$$

Substituting for $\hat{b}_i$ from (7.24) gives

$$\text{fitting SS} = \sum_i Y_{i..} \left\{ (Y_{i..}/N_{i.}) - \sum_j (n_{ij}\hat{t}_j/N_{i.}) \right\} + \sum_j \hat{t}_j Y_{.j.}$$
$$= \sum_i Y_{i..}^2/N_{i.} + \sum_j \hat{t}_j \left\{ Y_{.j.} - \sum_i (n_{ij}Y_{i..}/N_{i.}) \right\}$$

and the two terms can be recognised as: (*a*) the SS for blocks ignoring treatment effects, and (*b*) the SS for treatments allowing for block effects. The second term is a particular example of the general result in Section 4.A5:

$$\theta_2'(\mathbf{A}_2' - \mathbf{C}_{21}\mathbf{C}_{11}^{-1}\mathbf{A}_1')\mathbf{y}. \tag{7.29}$$

7.3.1 *Examples of the analysis of data from incomplete block designs*

The practical analysis of data from incomplete block designs is illustrated in two examples.

Example 7.10 An experiment to examine preferences of cabbage root flies for six different brassica species on which to lay their eggs involved the use of ten cages of flies with plants of

Table 7.3. *Number of eggs laid*

Cage	Cabbage	Cauliflower	Broccoli	Sprouts	Kale	Romanesco
1	452	69	83			
2	802	143		53		
3	699		32		4	
4	1207			19		32
5	958				8	8
6		328	147			53
7		314		264	223	
8		158			36	5
9			117	14	115	
10			23	16		2

three brassica species available in each cage. The number of eggs laid on the various plants were as shown in Table 7.3.

The design may be recognised as that of Example 7.4. There are some difficulties in drawing inferences from these data. The eggs laid in each cage will be distributed between the plants of the three brassica species as a result of the 'competition' between the three plants for the flies' attention. But also, the overall rate of egg laying within each cage may be stimulated or depressed by the set of three brassica species in that cage. Therefore, inferences about differences between egg laying rates for the three brassica species may be unreliable if extrapolated to external conditions where only a single one of the species will be available. Nevertheless, as a basis for identifying the relative preferences of flies for the different brassica species, in the context of the experimental situation, the analysis is informative.

There are large overall differences between cages and, to try to satisfy the assumption that all observations are subject to the same random unit variance, σ^2, a square-root transformation of the egg counts seems appropriate. The square roots of observed counts are shown in Table 7.4.

Cage and species totals are:

Cage 1	38.7	Species	Cabbage (1)	141.8
2	47.6		Cauliflower (2)	68.7
3	34.1		Broccoli (3)	42.5
4	44.9		Sprouts (4)	35.6
5	36.6		Kale (5)	36.4
6	37.5		Romanesco (6)	19.4
7	48.8	Total		344.4
8	20.8			
9	25.2			
10	10.2			

Table 7.4. *Square roots of observed counts*

			Species			
Cage	Cabbage	Cauliflower	Broccoli	Sprouts	Kale	Romanesco
1	21.3	8.3	9.1			
2	28.3	12.0		7.3		
3	26.4		5.7		2.0	
4	34.8			4.4		5.7
5	31.0				2.8	2.8
6		18.1	12.1			7.3
7		17.7		16.2	14.9	
8		12.6			6.0	2.2
9			10.8	3.7	10.7	
10			4.8	4.0		1.4

Table 7.5. *Analysis of variance for brassica data*

	SS	df	MS
Cage effects (ignoring species)	458	9	51
Species effects (allowing for cages)	1859	5	372
Error	114	15	7.6
Total	2431	29	
or			
Species effects (ignoring cages)	1964	5	393
Cage effects (allowing for species)	353	9	39
Error	114	15	7.6
Total	2431	29	

Fitting cage and species effects using a general linear model algorithm gives the following residual SS for different models:

$$\text{total variation about the mean} = 2431 \ (29 \ \text{df}),$$
$$\text{SS (fitting cage effects only)} = 1973 \ (20 \ \text{df}),$$
$$\text{SS (fitting species effects only)} = 467 \ (24 \ \text{df}),$$
$$\text{SS (fitting cage and species effects)} = 114 \ (15 \ \text{df}).$$

The analysis of variance can be written in two forms shown in Table 7.5. The first of the analyses in Table 7.5 corresponds to the algebraic form of the fitting SS derived earlier in this section.

The least squares estimates of cage and species effects from the full model are given in Table 7.6. Allowing for differences between cages, the F ratio to test the barely credible null hypothesis of no difference between species is 49 on 5 and 15 df. The conclusions about the species are that cabbage is very much preferred by cabbage root flies, with cauliflower

Table 7.6. *Least squares estimates*

		Effect
Cage	1	−4.64
	2	−0.23
	3	−4.25
	4	1.77
	5	−1.37
	6	3.72
	7	7.94
	8	−0.78
	9	0.93
	10	−3.08
Species	Cabbage	18.62
	Cauliflower	1.05
	Broccoli	−1.52
	Sprouts	−5.83
	Kale	−4.70
	Romanesco	−7.65

SED between two species is 1.95.

Table 7.7. *Data from peppers experiment*

Heating	0	0	0	0	1	1	1	1
Lighting	0	0	1	1	0	0	1	1
CO_2	0	1	0	1	0	1	0	1
Year 1 Block 1	11.4	13.2	10.4		13.7		12.0	12.5
Year 1 Block 2		8.4	6.5	6.1	10.8	9.4		9.1
Year 2 Block 1		13.7			14.6	16.5 / 15.4	12.8	12.9
Year 2 Block 2		10.7			10.9	10.9	9.0 / 10.1	10.2

and broccoli next in the order of preference. The differences among the species other than cabbage although clearly not negligible do not show a clear grouping, the order of preference being cauliflower, broccoli, kale, sprouts, romanesco. As mentioned earlier, more detailed numerical interpretations are probably not capable of valid interpretation.

Example 7.11 The experiment on peppers discussed in the preliminary section of this chapter was the first part of a two-year programme to find the best treatment. The results (yield − costs) from 24 observations over two years of experiments were as shown in Table 7.7.

The philosophy behind the design is simple. In the first year eight treatments can be arranged in two blocks of six by having four treatments in both blocks and the other four

Table 7.8. *Analysis of variance for peppers experiment*

	SS	df	MS	F
Block effects (ignoring treatments)	116.5	3	38.8	
Treatment effects (allowing for blocks)	30.7	7	4.4	9.6
Error	6.0	13	0.46	
Total	153.2	23		

treatments in one of the two blocks. The four treatments to be used in both blocks may be chosen arbitrarily or selected to contain each level of each factor exactly twice as here (a concept we return to in Chapter 13). In the second year, the five most successful treatments are retained, with overall replication levels for these treatments made as even as possible. The complete set of 24 observations can be analysed considering the years and blocks as blocking factors and the treatment combinations simply as eight treatments. The resulting residual SS are

$$\text{total variation about the overall mean} = 153.2 \ (23 \ df),$$
$$\text{fitting block effects only} = 36.7 \ (20 \ df),$$
$$\text{fitting treatment effects only} = 91.3 \ (16 \ df),$$
$$\text{fitting block and treatment effects} = 6.0 \ (13 \ df).$$

The analysis of variance is shown in Table 7.8, and the least squares estimation of the treatment means (allowing for block effects) is shown in Table 7.9(a). The SEDs between treatment means are tabulated in Table 7.9(b).

The variation in replication of the treatments causes some variation of precision but treatment comparisons for treatments 2, 5, 6, 7 and 8 are similarly precise. The SE of a treatment difference using four replicates with a σ^2 value of 0.46 (as achieved by blocking) would be 0.48, almost the same as those shown in Table 7.9(b). However, if the variance ignoring block effects ($91.3/16 = 5.7$) is used then the SE would be 1.69. Thus the non-orthogonality of the block-treatment structure hardly affects the SE but the use of blocking reduces the SE by a factor greater than 3. Note that the analysis has not used the treatment structure. Further analysis would examine the main effect and interaction information in the data. It is, however, immediately obvious that the dominant effect is the main effect of heating.

7.4 Balanced incomplete block designs and classes of less balanced designs

There is a vast literature on the theory and indexing of classes of incomplete block designs. The original reason for the investigation of possible incomplete block designs arose from the wish to use practically relevant sets of treatments and practically sensible block sizes. Moreover, the interest in such designs began many years before the advent of computers, and so an overriding consideration in developing incomplete block designs was the need to be able to analyse the results using only a manually operated calculator. The main problem in devising designs was therefore to recognise mathematical structures enabling the least

Table 7.9. *Least squares estimates and standard errors for peppers experiment: (a) least squares estimates; (b) SEDs*

(a)

	H	L	CO_2	Estimate of mean
1	0	0	0	9.23
2	0	0	1	10.72
3	0	1	0	8.25
4	0	1	1	7.88
5	1	0	0	11.72
6	1	0	1	12.01
7	1	1	0	9.86
8	1	1	1	10.39

(b)

	1	2	3	4	5	6	7
2	0.80						
3	0.86	0.61					
4	1.06	0.80	0.86				
5	0.80	0.48	0.61	0.80			
6	0.84	0.49	0.64	0.81	0.49		
7	0.81	0.49	0.64	0.84	0.49	0.50	
8	0.80	0.48	0.61	0.80	0.48	0.49	0.49

squares equations to be solved in a way that involved simple arithmetical calculations. These mathematical structures essentially define restricted classes of designs. Consequently the history of the theory of incomplete block designs is primarily a sequence of increasingly intricate classes of designs, with occasional generalisations to combine several previous classes.

Clearly the computational facilities are now quite different, and much of the theory of particular classes of design is now irrelevant. Nevertheless we can learn something about the principles of experimental design by examining some of the major developments of the incomplete block design.

The general Pearce (1963) criterion of balance defines the condition required for the estimation of treatment differences to be equally precise for each pair of treatments. However, it was recognised many years earlier by Yates (1936) that balanced designs could be constructed through symmetry considerations. Essentially, if in a set of blocks of the same size, each pair of treatments occurs together in a block the same number of times, then, by symmetry, all treatment differences will be equally precise. The resulting designs are known as balanced incomplete block (BIB) designs.

We start by introducing some terminology for these designs. A BIB design has t treatments and each block has k units. There are b blocks and r replicates of each treatment. Each pair of treatments occurs λ times together in a block, with each of these five parameters taking integer values.

Trivial examples of BIBs are those which include all possible groups of k out of t treatments. These are known as unreduced designs and the appropriate values of b, r and λ may be found by simple combinatorial arguments:

$$b_u = \frac{t!}{k!(t-k)!},$$

$$r_u = \frac{(t-1)!}{(k-1)!(t-k)!},$$

$$\lambda_u = \frac{(t-2)!}{(k-2)!(t-k)!}.$$

More generally, by considering the total number of experimental units it can be seen that

$$bk = tr. \tag{7.30}$$

Similarly, by considering the number of units included in the set of blocks containing a particular treatment, we can see that

$$r(k-1) = (t-1)\lambda. \tag{7.31}$$

Hence, it can be seen that for any BIB design

$$b : r : \lambda = (t-1)t : (t-1)k : (k-1)k, \tag{7.32}$$

so that, for any pair of values of t and k, the ratios $b : r : \lambda$ are invariant over all possible designs, and in particular $b : r : \lambda$ will be the same as $b_u : r_u : \lambda_u$. It follows that reduced BIB designs can only exist when (b_u, r_u, λ_u) have a common factor greater than 1, and that in this case BIB designs may exist for all triples of values:

$$b = (m/f)b_u, \quad r = (m/f)r_u, \quad \lambda = (m/f)\lambda_u, \tag{7.33}$$

where f is the highest common factor of (b, r, λ) and m is any integer less than f. As might perhaps be inferred from the ease with which we constructed Example 7.4, designs are easily found, at least for relatively small numbers of treatments, replication levels and block sizes, whenever existence has been demonstrated to be possible. The set of (t, k) pairs for which designs have been shown to exist with a block size not bigger than ten, and a total number of units not exceeding 100, is listed in Table 7.10 and the sparseness of the available designs is clearly apparent. It is notable that the number of designs with a small number of replicates (≤ 5) such as experimenters would expect to use is extremely limited. More detailed design information is given in Fisher and Yates (1963).

One characteristic of incomplete block designs that is sometimes considered to be important is whether the blocks can be grouped so that each group of blocks constitutes a complete replicate of the set of treatments. When this division of an experiment into replicate groups of blocks is possible the design is said to be resolvable. None of the incomplete block examples considered so far in this chapter is resolvable and the concept of resolvable designs is chiefly relevant to designs for large numbers of treatments, where it is feared that mistakes in allocation may occur more easily in non-resolvable designs.

Table 7.10. *Pairs of values of* (k, t) *for which a BIB design is known to exist for* $k \leq 10$ *and* $bk < 100$.

k	t	b	r	λ	k	t	b	r	λ
3	4	4	3	2	4	13	13	4	1
3	5	10	6	3	4	16	20	5	1
3	6	10	5	2	5	6	6	5	4
3	6	20	10	4	5	9	18	10	5
3	6	10	5	2	5	10	18	9	4
3	7	7	3	1	5	11	11	5	2
3	9	12	4	1	6	7	7	6	5
3	10	30	9	2	6	9	12	8	5
3	13	26	6	1	6	10	15	9	5
4	5	5	4	3	6	11	11	6	3
4	6	15	10	6	6	16	16	6	2
4	7	7	4	2	7	8	8	7	6
4	8	14	7	3	8	9	9	8	7
4	9	18	8	3	9	10	10	9	8
4	10	15	6	2					

One important class of BIB designs for large numbers of treatments is the set of lattice designs. A simple lattice design is based on a square array containing the set of treatment identifiers. For a set of 25 treatments the square array would be

$$
\begin{array}{ccccc}
A & B & C & D & E \\
F & G & H & I & J \\
K & L & M & N & O \\
P & Q & R & S & T \\
U & V & W & X & Y
\end{array}
$$

A lattice design for 25 treatments uses blocks of five units in groups of five blocks. The first group consists of the five blocks given by the rows of the array. The second group consists of the five blocks given by the columns of the array. Subsequent groups are defined by superimposing on the 5×5 square, arrays whose elements are the digits 1, 2, 3, 4 and 5, arranged so that each digit occurs once in each row and once in each column. Thus the array

$$
\begin{array}{ccccc}
1 & 2 & 3 & 4 & 5 \\
2 & 3 & 4 & 5 & 1 \\
3 & 4 & 5 & 1 & 2 \\
4 & 5 & 1 & 2 & 3 \\
5 & 1 & 2 & 3 & 4
\end{array}
$$

can be used to generate a group of five blocks with each block defined by the treatments corresponding to a particular digit. For digit 1 we get the block (AJNRV). By discovering four such arrays of digits such that each pair of arrays is orthogonal in the sense that each digit in one array occurs with each digit in the second array we can construct four groups of five blocks which can be combined with the two groups for rows and columns to give a balanced design. We will see later (Chapter 8) that these four orthogonal arrays are Latin

Blocks

I	II	III	IV	V	VI	VII	VIII	IX	X
A	F	K	P	U	A	B	C	D	E
B	G	L	Q	V	F	G	H	I	J
C	H	M	R	W	K	L	M	N	O
D	I	N	S	X	P	Q	R	S	T
E	J	O	T	Y	U	V	W	X	Y

XI	XII	XIII	XIV	XV	XVI	XVII	XVIII	XIX	XX
A	B	C	D	E	A	B	C	D	E
J	F	G	H	I	I	J	F	G	H
N	O	K	L	M	L	M	N	O	K
R	S	T	P	Q	T	P	Q	R	S
V	W	X	Y	U	W	X	Y	U	V

XXI	XXII	XXIII	XXIV	XXV	XXVI	XXVII	XXVIII	XXIX	XXX
A	B	C	D	E	A	B	C	D	E
H	I	J	F	G	G	H	I	J	F
O	K	L	M	N	M	N	O	K	L
Q	R	S	T	P	S	T	P	Q	R
X	Y	U	V	W	Y	U	V	W	X

Figure 7.10 Experimental plan for comparing 25 treatments in a lattice design with 30 blocks of five units.

squares for five treatments, and that only four orthogonal arrays are possible. The contents of the 30 blocks are listed in Figure 7.10, with each set of five blocks providing a complete replicate of the treatment set. Simple lattice designs are therefore clearly resolvable but they are also very clearly restrictive, being limited in the possible numbers of treatments ($t = n^2$), block size $(t)^{1/2}$ and number of replications $(1 + t^{1/2})$. The last restriction may be relaxed by considering designs omitting one or more groups of blocks (though at least two groups of blocks need to be retained to provide information on treatment differences). The lack of balance in an incomplete lattice design is not extreme and the precision of treatment differences is at two levels depending on whether the two treatments to be compared occur together in a block. Such a design may be described as possessing partial balance with two classes of association, or, equivalently, two levels of precision. The class of partially balanced incomplete block designs includes many other families of designs and has been the source of an enormous literature. While this literature, and knowledge of particular families of partially balanced incomplete block designs, is now only of 'archaeological' interest, the use of such designs can still be valuable for scenarios where the particular design constraints are met, as a simple analysis can often be constructed using algorithms able to cope with generally balanced designs. In particular, such designs provide a useful base for constructing incomplete row–column designs (see Chapter 8).

A final class of designs which deserves brief mention is that of cyclic designs. In the basic form of such designs blocks are generated by cyclic rotation through the treatment letters, starting from an initial block. The initial block must be carefully chosen. Consider the problem of designing an experiment to compare seven treatments in seven blocks of three.

Table 7.11. *Analysis of variance for rice spacing experiment*

	SS	df	MS	F
Blocks	5.88	3	1.96	4.26
Spacing	23.14	9	2.57	5.59
Error	12.42	27	0.46	
Total	41.44	39		

We could start with the block (ABC) and by cyclic generation produce the design

$$(ABC), (BCD), (CDE), (DEF), (EFG), (FGA), (GAB).$$

This is obviously a poor design with treatment A occurring twice with B and G, once with C and F and not at all with D or E. However, starting with the block (ABD) we obtain

$$(ABD), (BCE), (CDF), (DEG), (EFA), (FGB), (GAC),$$

which is a BIB design and the best available design. Cyclic design structure has produced an extensive literature (see, for example, John and Williams (1995)), and may be generalised to include sets of treatments with a factorial structure. It may be recognised as the most sophisticated form of classical mathematical design theory. An important application of the cyclic design approach has been in the design of trials for large numbers of treatments (typically, varieties in large-scale breeding studies). We consider this particular application in more detail in Section 7.6.

7.5 Orthogonality, balance and the practical choice of design

We have introduced two particular characteristics of block-treatment designs: orthogonality and balance. These two concepts have been extremely important in the development of design theory and we should assess their importance and relevance for designing experiments in the present day.

Consider again the terminology of the general block-treatment design introduced in Section 7.2, with n_{ij} observations of treatment j in block i. An orthogonal design is such that treatment differences are estimated independently of block differences. A necessary and sufficient condition for this is that

$$n_{ij} = N_{i.} N_{.j} / N_{..} \quad \text{for all } i, j \text{ combinations,}$$

where $N_{i.}, N_{.j}$ and $N_{..}$ are, respectively, the total numbers of observations in block i, for treatment j, and in the complete experiment. Note that treatments do not have to be equally replicated in each block and that blocks may be of different sizes. The crucial characteristic is that the ratios of treatment replications must be constant over blocks.

The principal consequence of orthogonality is that treatment effects can be interpreted without simultaneously having to consider inferences about block effects. Consider again the analysis of variance for a randomised complete block design, the rice spacing experiment of Chapter 2 (see Table 7.11).

The structure of the design means that the SS for blocks and for treatments (spacing) are independent. Hence, the ratio of the spacing MS to the error MS can be used to test the hypothesis that all the treatment (spacing) effects, t_j, are zero, independent of any block effects, which have no impact on the observed differences between spacing treatments. Correspondingly, the ratio of the block MS to the error MS could be used to test the hypothesis that all the block effects, b_i, are zero, independent of the values of the treatment (spacing) effects. However, because the same divisor is used in both test statistics, the two test results are not independent, though it should also be emphasised that there is rarely any interest in drawing formal inferences about block effects. All we are usually interested in is the extent to which blocking has succeeded in reducing the variation between units (within blocks) so that we can decide whether to use similar blocking criteria (structures) in future experiments. Nevertheless, the orthogonality of treatments and blocks enables the effects of treatments and of blocks to be considered quite separately.

The benefits of orthogonality extend to the interpretation of orthogonal treatment contrasts, the philosophy for which was developed in Chapter 4, and the application of which will be considered in more detail in Chapter 12. If the block-treatment structure is orthogonal, then inferences about orthogonal treatment contrasts will also be independent of blocks, providing a simple interpretation of those orthogonal treatment contrasts. All other things being equal, designs should be arranged to be orthogonal and thus have the benefit of allowing independent interpretation of effects. However, non-orthogonality should not be considered a major defect of a design. Rather it produces a (usually) small loss of information. In all the non-orthogonal designs in this chapter (and book) the non-orthogonality of treatments and blocks is very much less than is typically found between variables in a multiple regression study. Consider the two analyses of variance in Example 7.10. The variation of SS according to the order of fitting is relatively slight. Certainly the interpretation of the strength of the substance effects, and of the cage effects, is unchanged whether or not we allow for the other set of effects, and this is typical of the impact of any non-orthogonality in well-designed experiments.

A second consequence of orthogonality, which was extremely important in the era before the powerful computers we have access to today, is the simplicity of the calculations both for the analysis of variance and for the comparison of treatment effects. The advent of powerful computers, and the development of analysis approaches, such as REML (Patterson and Thompson, 1971), that are able to take advantage of this computing power to estimate variance components and the sizes of treatment effects within non-orthogonal designs, means that this benefit of orthogonality is now greatly reduced. While there might previously have been a major conflict over whether to use the natural blocking system, or to use a blocking system which allowed orthogonal estimation of treatment and block effects, the decision now should be dominated by the choice of the most effective (natural) blocking system. This should imply a reduction in the propensity to use the randomised complete block design since, if natural blocking systems are being properly sought and recognised, it is improbable that the size of block will always be exactly equal to the number of treatments to be compared.

Balance, as defined in Section 7.3, requires that

$$p_{jj'} = \sum_i n_{ij} n_{ij'} / N_{i.}$$

is constant for all (j, j') pairs. When block size, N_i, is constant this reduces to the requirement that the total number of joint occurrences of two treatments in a block, across all blocks, should be the same for all treatment pairs, and this forms the basis for the construction of BIB designs. The constraints imposed by this property of balance mean that the class of BIB designs which actually exist is extremely limited. Hence it is usual to find that, given a block size of k units and a set of t treatments, the only BIB design which satisfies the requirements uses far more replicates, and hence resources, than are practical or available. Since designs which are not balanced but which are sensibly constructed to make the pairwise treatment occurrence as nearly equal as possible achieve very little variation of the precision of treatment differences, balance must be viewed as a pleasing but unimportant luxury. The theoretical developments of incomplete block design theory have led to the definition and tabulation of many classes of designs with mathematically neat patterns of treatment occurrence in blocks. The philosophy of such tabulation is, presumably, that the prospective experimenter, with a statistical consultant, should attempt to match the conditions for the desired experiment to that design in the available list which requires the smallest modification of the desired conditions.

While this philosophy might have been justified before the advent of powerful computers and new analysis algorithms, we believe that this is no longer a practically acceptable approach. Of course, where the actual constraints of treatments and blocks match those in the tabulated classes, such designs may provide a useful solution to the design problem. But it is then important to check that the imposed pattern of joint treatment occurrences fits the hypotheses of interest. More generally, it is important to first recognise and control potential variation in experimental units by using blocking in the most effective way possible. It is also important to choose the treatments (and treatment structures) for comparison in the experiment without constraint from considerations of the blocking structures. The construction of a design should then consist of choosing the allocation of treatments in such a way that treatment comparisons of interest are made as efficiently as possible, while minimising the degree of non-orthogonality between block and treatment effects. If an orthogonal design is possible it should be used. A small level of non-orthogonality, such as occurs when designs are constructed to make pairwise treatment occurrence as nearly equal as possible, will rarely cause ambiguity in interpreting results. Further, if the blocking requirements and the number of treatments are compatible with a balanced design, then clearly such a design should be used. However, such designs will automatically be obtained, when they are possible, if in each case the design is constructed to satisfy the general principles discussed in this chapter.

The relative precision of comparisons between different pairs of treatments is determined primarily by the number of joint occurrences of the treatment pairs within blocks, and secondarily by the number of second-order comparisons when the two treatments occur in different blocks but with a number of common, or linking, treatments. For incomplete blocks it follows that: (i) in each block there should be no repeated treatments; (ii) the pairwise occurrence of treatments should be as nearly even as possible; and (iii) if the joint occurrences of AB and of AC are relatively high, then that of BC need not also be relatively high. An implication and extension of this last idea is that the 'chains' of pairs with relatively high joint occurrence should be long, not short. Example 7.3 provides a simple illustration of the complement of this idea, with the second alternative, based on a single chain of omitted treatment pairs, giving a more homogenous set of variances for treatment differences than the first alternative, based on two chains, each of three omitted pairs.

For a particular design, a formal assessment of these concepts can be based on the matrix of block-treatment occurrences, $\mathbf{N}$, with elements n_{ij} as previously defined:

$$n_{ij} = \text{number of occurrences of treatment } j \text{ in block } i.$$

The total joint occurrences p_{jk} of treatments j and k in blocks are the elements of the matrix $\mathbf{P} = \mathbf{N}'\mathbf{N}$. The second-order links, s_{jk}, for treatments j and k are then the elements of the matrix $\mathbf{S} = \mathbf{P}'\mathbf{P}$. For a good design, if p_{jk} is relatively small, then s_{jk} should be relatively large, and vice versa.

In practice, the achievement of an 'optimal' design (achieving equal precision for treatment comparisons of equal interest) is not usually an absolute objective, since in most design problems there are many candidate designs of almost equal efficiency. It is therefore sufficient to devise methods of design construction which are simple but result in reasonably efficient designs. For relatively small numbers of treatments the philosophy evolved in Section 7.2 is adequate. Treatments should be allocated to sets of blocks sequentially, with all the treatment A allocations first, followed by all the treatment B allocations, and so on. Each allocation of the observations for a particular treatment should be arranged to achieve as nearly as possible the mean joint occurrence with the treatments already allocated, and to minimise restrictions on later treatments.

At the other extreme are designs for large numbers of treatments with small numbers of replications and block sizes such that the joint occurrences of treatment pairs are all zero or one. A substantial amount of work on the practical requirements for designs for national trials of large numbers of varieties of a particular crop has been produced by Patterson and coworkers. We consider the design of such large variety trials in detail in Section 7.6. The designs used in these trials are required to be resolvable, i.e. that a given number of blocks constitutes a complete replicate. Whilst having certain advantages for these large, often multisite, trials the requirement of resolvability should not be considered a crucial requirement for such designs, as it may impose unnecessary constraints on block size. Instead it should be seen as a useful property in the same way as orthogonality and balance. In the second example at the beginning of this chapter, the capacity to handle 18 units in a batch is more important than the resolvability concept which would use 14 units in a batch and thereby waste resources by requiring nine batches instead of seven. However, in constructing designs for large numbers of treatments it is sensible to consider each replicate in turn and within each replicate allocate treatments to complete as many blocks as possible, thus achieving near resolvability.

A helpful concept for the construction of designs for fairly large numbers of treatments is the general idea of lattices. If the blocks of the first replicate are written out as rows, then a second replicate should employ blocks which 'run across' the rows (are orthogonal to the rows), and the blocks of the third and subsequent replicates should make use of diagonals.

7.5.1 Some practical examples of designs for larger experiments

Our two general philosophies of design construction, first of allocating treatments sequentially when the number of treatments is relatively small, and second of allocating replicates

Blocks

	I	II	III	IV	V	VI	VII
VII	1	7					
VIII	2	8	13	19	25	31	
IX	3	9	14	20	26	32	
X	4	10	15	21	27		37
XI	5	11	16	22		33	38
XII	6		17	23	28	34	39
XIII		12	18	24	29	35	40
XIV					30	36	

	XIV	XV	XVI	XVII	XVIII	XIX	XX
		25	26	27	28	29	30
		32	33	34	35	36	31
	5	6	1	7	2	3	4
	8	18	12	13	9	10	11
	14	21	19	20	15	16	17
	37	38	39	40	22	23	24

Figure 7.11 Experimental plan for comparing 40 treatments in 20 blocks of six units.

sequentially when the number of treatments is large, inevitably imply an intermediate area and the designer of experiments has to learn how to select the appropriate philosophy for each individual problem. We conclude this section by first discussing possible solutions to the first two problems introduced at the start of this chapter we then provide a simple example as an introduction to the interesting problem posed by the two-channel microarray scenario, before suggesting possible ways to approach the more complex microarray scenario introduced at the start of the chapter.

Example 7.12 A trial is to compare 40 treatments, with three replicates of each treatment, using blocks of six units each. The first replicate is obtained by allocating the treatments in order to blocks I–VI, plus the first four units in block VII. Each block of the second replicate is constructed by taking one treatment from blocks I–VI, with treatments 1 and 7 first added to complete block VII (from the end of the first replicate), and a careful choice of which blocks to skip in order to include all but two treatments (30, 36) in blocks VIII–XIII. These two treatments are allocated to start block XIV. Given the joint treatment occurrences in the blocks for the first two replicates shown in Figure 7.11, then the construction of the third replicate must be planned to avoid repeating pairwise occurence of treatments in the remaining blocks. Since treatments 30 and 36 have occurred in the overflow into block XIV the treatments in each of blocks V and VI must occur one per block in the last six blocks, avoiding any pairwise occurences from blocks VIII–XIV. The remaining 28 treatments are allocated to blocks XIV–XX mainly by using diagonal transects across the two-way array of the blocks for the first two replicates (for example, treatments 2, 9, 15 and 22 are allocated to block XVIII, and treatments 4, 11, 17 and 24 to block XX), remembering that we have already allocated the treatments from blocks V and VI. Of course, there are a number of

Block

I	II	III	IV	V	VI	VII
1	19	37	7	13	4	10
2	20	38	8	14	5	11
3	21	39	9	15	6	12
4	22	40	10	16	28	19
5	23	41	11	17	29	20
6	24	42	12	18	30	21
7	25	1	25	31	7	16
8	26	2	26	32	8	17
9	27	3	27	33	9	18
10	28	4	28	34	22	40
11	29	5	29	35	23	41
12	30	6	30	36	24	42
13	31	19	37	1	13	31
14	32	20	38	2	14	32
15	33	21	39	3	15	33
16	34	22	40	25	37	34
17	35	23	41	26	38	35
18	36	24	42	27	39	36

Figure 7.12 Experimental plan for comparing 42 treatments in seven blocks of 18 units.

options in this last step, but the details of a possible set of final blocks to complete the design are shown in Figure 7.11.

Example 7.13 The second example in the problems outlined at the beginning of this chapter required three replicates of 42 treatments to be arranged in seven blocks of 18. With 51 comparisons possible within blocks for each treatment, clearly an ideal design would have each pair of treatments occurring together once or twice. However, it rapidly becomes clear that with the rather large blocks some pairs of treatments must occur together three times, i.e. for all three replicates. The first 18 treatments are allocated to the first block and the next 18 to the second block. The third block contains the remaining six treatments to complete the first replicate, and the remaining spaces in this block are arbitrarily allocated to the first six treatments from block I and the first six from block II. Since each treatment in these sets of six must occur once more, and there are only four more blocks to be filled, it follows that some of the treatments in each of these sets must occur together in a block again, and so in each of the three blocks in which they occur. The occurrence of new pairs is maximised by constructing blocks IV and V to each contain a remaining group of six treatments from each of blocks I and II, also allocating the last six treatments (previously allocated to start block III) to block IV. Triplets of treatments from previous blocks are then allocated together to the remaining blocks, and the resulting design is shown in Figure 7.12. It would be possible to arrange that some of the treatments occurred with all the treatments with which they had not previously occurred but this increases the number of treatment pairs occurring together three times. A clearer understanding of the structure of the comparisons in this design can be obtained by constructing the matrix of pairwise occurence of treatments in blocks (grouping the treaments in consecutive sets of six) – this exercise is left to the interested reader!

For these two examples it may be possible to find better solutions but the improvement in overall precision will be small. We should also note at this point that we have not considered how to allocate treatments to treatment identifiers; this is discussed in Chapter 9.

7.5.2 *Designs for two-channel microarray experiments*

Example 7.14 We first consider a relatively simple problem for allocating treatments to blocks of size 2 (as is the natural blocking structure for the two-channel microarray experiment introduced at the start of the chapter), for which all possible designs can be evaluated. We can therefore assess the performance of the designs derived through the mathematical classification methods mentioned in Section 7.4.

Suppose we wish to compare nine treatments (A, B, . . . , I), we have available units which occur naturally in pairs (such as the two channels of a microarray), and we can have 18 pairs of units available for the experiment. Assuming equal interest in each treatment, and hence equal replication, each treatment must occur four times, and it seems reasonable to assume that a particular treatment should therefore occur with four different treatments in the four blocks in which it appears. There is no BIB design with 18 blocks of two units each for nine treatments – the BIB design for nine treamtents in blocks of two units requires eight replicates and 36 blocks. But there is a partially balanced design such that the variance of an estimated difference between two treatments takes one of two values depending only on whether the two treatments appear in a block together. Writing the blocks down in treatment order, the block structure is

(AB) (AC) (AD) (AG) (BC) (BE) (BH) (CF) (CI)

(DE) (DF) (DG) (EF) (EH) (FI) (GH) (GI) (HI).

The pattern in this structure may not be obvious, but becomes clearer by re-arranging the order of the blocks. The treatments are effectively split into three groups of three, (A, B, C), (D, E, F) and (G, H, I), with each treatment occuring in a block with the other members of their group, and with the members of the other groups in the same position within the group (i.e. A with D and G, B with E and H, C with F and I). This structure can equally be thought of as being based on the three groups defined by these triples, as is suggested below:

(AB) (AC) (BC) (DE) (DF) (EF) (GH) (GI) (HI)

(AD) (AG) (DG) (BE) (BH) (EH) (CF) (CI) (FI).

This symmetry is the mathematical pattern which is the basis of this design. The design gives just two different variances (different by a factor $4/5$) for the differences between two estimated treatment effects, and is the only partially balanced design with just two variances of treatment difference for this particular block structure.

Another possible design for this problem has an obvious cyclical structure, and has been popularised in the design literature for two-channel microarrays as an 'interwoven loop' design (Wit, Nobile and Khanin, 2005; Kerr and Churchill, 2001). The design consists of two 'loops' of nine blocks connecting all nine treatments, and the motivation for using

Table 7.12. *Variance properties of the 16 alternative designs for the two-channel microarray experiment allocating four replicates of nine treatments to 18 blocks of size 2*

Design	Average variance	Number of distinct variances	Range of variance
(1)	0.978	9	0.862–1.128
(2)	0.982	4	0.865–1.093
(3)	0.988	6	0.857–1.103
(4)	0.996	9	0.842–1.155
(5)	1.000	2	0.889–1.111
(6)	1.001	21	0.835–1.173
(7)	1.004	19	0.835–1.209
(8)	1.006	11	0.833–1.167
(9)	1.014	21	0.837–1.239
(10)	1.021	8	0.800–1.257
(11)	1.023	20	0.835–1.241
(12)	1.027	17	0.800–1.250
(13)	1.033	6	0.844–1.244
(14)	1.049	4	0.850–1.242
(15)	1.052	10	0.835–1.314
(16)	1.096	9	0.800–1.400

these designs in microarray studies is from treatment sets comprising samples collected at a sequence of equidistant time points, with the 'loop' connecting the end of the time series back to the beginning.

This interwoven loop design (i) has steps of length 1 and 2, though it can be relabelled to have steps of length 1 and 4. A further composite design (ii) combines a simple loop design (comparing all treatments with their immediate neighbours in the ordered list) with the comparisons amongst the three treatments from three sets of three treatments (ADG, BEH, CFI) – essentially an interwoven loop design with steps 1 and 3, but for a step length of 3 the 'loop' is not connected:

(i) (AB), (BC), (CD),(DE), (EF), (FG), (GH), (HI), (IA),
    (AC), (CE), (EG), (GI), (IB), (BD), (DF), (FH), (HA), and
(ii) (AB), (BC), (CD), (DE), (EF), (FG), (GH), (HI), (IA),
    (AD), (DG), (GA), (BE), (EH), (HB), (CF), (FI), (IC).

Further research reveals that there are 13 further distinct designs for this problem. A reasonable criterion for comparing these designs is the average variance of a difference between two treatments, the average being over all possible 36 treatment pairs. The average variances for each design, together with the number of different variances and the range of these variances, are listed in Table 7.12 in order of increasing variance. The partially balanced design is (5), the interwoven loop design is (2), and the composite design is (14).

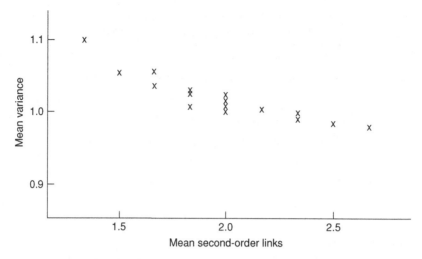

Figure 7.13 Relationship between mean variance and mean second-order links for the 16 possible designs for the two-channel microarray experiment, allocating four replicates of nine treatments to 18 blocks of size 2.

It can be seen from Table 7.12 that, in the sense of average variance, neither the partially balanced design nor the interwoven loop design nor the composite design prove the best, though all three provide relatively rather more homogeneous sets of variances than some of the other alternatives. Also, there are designs (numbers (10), (12) and (16)) which provide precise comparisons for particular treatment comparisons when a factorial treatment structure is assumed, inevitably at a price of less precise comparisons for others. Further investigation of the different designs shows that the most important characteristic in determining the average variance is the numbers of second-order links for those treatment pairs which do not occur together in a block. The relationship between average variance and mean second order links, s_{jk}, for those (j, k) combinations such that $p_{jk} = 0$ (using the notation suggested earlier in this section) is shown in Figure 7.13. There will be further consideration of this design situation for the particular case when the nine treatments are a 3×3 factorial set (in Chapter 15).

The principles considered here for designs in blocks of size 2 are particularly important for constructing the complex designs that are needed in microarray experimentation, with interwoven loop designs providing a particularly useful component where treatments can be considered to form an ordered set, and where the number of treatments is prime (so that loops with a range of different step lengths can be considered).

We now return to the third example from the beginning of the chapter, where we had 52 samples, collected as four biological replicates (plants) at each of thirteen time points, to compare in a two-channel microarray experiment (hence in blocks of size 2). Assuming that the experimenter could afford 104 arrays, we could construct a design with each of the 52 samples occuring on four separate microarrays. While the primary interest in this experiment will be in the comparison of samples from the 13 time points, we will also need to make comparisons among the biological replicates at each time point. A large number of designs are possible, but following the principles in the example above, we might first construct four loops connecting samples from adjacent time points (a step length of 1 – and looping back from time point 13 to time point 1) for different biological replicates. This would use half of the 104 arrays. Further paired comparisons could then link between time points at steps of 2 or 3 or 4 (all allow a single loop to connect through all thirteen samples because thirteen is a prime number), ensuring that these additional loops linked the different sets of biological replicates kept separate in the first part of the design.

7.6 Experimental designs for large-scale variety trials

Many large-scale variety trials are designed and completed every year. These are often organised in national and international research programmes, but they occur also within the research programmes of research institutes and commercial companies. They occur for all kinds of crops from wheat or rice to oil palms or apple trees, and including cauliflowers, maize, beans, carrots, chickpeas, tomatoes, cherries, plums, bananas, coconuts, pineapples and squashes. The capacity for searching for improved varieties of almost all crops grown commercially, and the consequent need to compare and assess a wide range of potential new best varieties, and to compare them with the current standard varieties seems to be unlimited.

Many large-scale variety trials are conducted on multiple sites, and some aspects of these trials are discussed further in Chapters 19 and 20. The trials will often involve large numbers of varieties, typically between 30 and 200 per trial. Most trials require quite a large area of land, or even, with for example oil palms, a very large area of land. In many cases the land area will have been managed uniformly for a substantial length of time prior to the experiment, but there may still be quite large variation of plant yields over the area, and a major component of the design for such trials will involve the control of this variation and the adjustment of yield comparisons to allow for the wide variation in the underlying soil quality.

The optimum form of design to adjust for the underlying variation of yield across the large area required for a large-scale variety trial is therefore a very important topic for a book on experimental design. There is, now, a classical design solution which will usually provide the most appropriate design for a large-scale variety trial. However, there are other methods of design and adjustment which some experimenters would argue can provide a comparable methodology, and we shall examine these methods also.

7.6.1 Alpha designs for large-scale variety trials

The class of alpha designs was developed specifically for large-scale variety trials by Patterson and Williams (1976). The designs are incomplete block designs where:

(i) the block size is small, usually between five and ten;
(ii) the number of varieties is large, usually between 30 and 200, and also clearly greater than the square of the block size; and
(iii) there are usually three or four complete replicates of the set of varieties.

Designs outside the 'usual' limits quoted are possible but not commonly needed.

For a design to be efficient it will be necessary that no pair of varieties occurs together more than once in a block across the whole experiment, and also that the pattern of the set

Table 7.13. *Initial sets of ten varieties for construction of the first replicate of an alpha design for 50 varieties in blocks of size 5*

Set 1	1	2	3	4	5	6	7	8	9	10
Set 2	11	12	13	14	15	16	17	18	19	20
Set 3	21	22	23	24	25	26	27	28	29	30
Set 4	31	32	33	34	35	36	37	38	39	40
Set 5	41	42	43	44	45	46	47	48	49	50

Table 7.14. *Shifted sets of ten varieties for construction of the second replicate of an alpha design for 50 varieties in blocks of size 5*

Set 1	1	2	3	4	5	6	7	8	9	10
Set 2	20	11	12	13	14	15	16	17	18	19
Set 3	29	30	21	22	23	24	25	26	27	28
Set 4	37	38	39	40	31	32	33	34	35	36
Set 5	44	45	46	47	48	49	50	41	42	43

of pairwise occurrences of varieties in blocks produces the lowest possible average variance of pairwise variety diferences. The method of construction of alpha designs ensures that variety pairs do not occur more than once, but the second part of the efficiency requirement is achieved through a computer search of the set of possible alpha designs to select the most efficient.

Example 7.15   We shall illustrate the construction of an alpha design for an experiment to compare 50 varieties using blocks of five, with three replicates.

The 50 varieties are first divided into five sets of ten varieties each, as shown in Table 7.13. Each block is then constructed to include one variety from each set. So, for the first replicate the ten blocks are simply defined by the ten columns of Table 7.13: $(1, 11, 21, 31, 41)$, $(2, 12, 22, 32, 42), \ldots, (10, 20, 30, 40, 50)$.

For the second and third replicates we shift the sets cyclically relative to each other, choosing the shifts carefully to ensure that the columns of the shifted sets produce further blocks such that no pair of varieties occurring together in a block are repeated in a second block. The shift mechanism can be defined by a set of generators, one for each subsequent replicate. For the second replicate, shown in Table 7.14, the new blocks (columns) have been obtained by applying the generator $(0, 1, 2, 4, 7)$. The first set is not shifted (this will always be the case), the second set is shifted one column to the right, the third set is shifted two columns to the right, the fourth set is shifted four columns to the right, and the fifth set is shifted seven columns to the right. The cyclical nature of the shift means that variety numbers shifted beyond column ten fill the empty columns at the start of each set.

The blocks of the second replicate are then defined by the ten columns of Table 7.14: $(1, 20, 29, 37, 44), (2, 11, 30, 38, 45), \ldots, (10, 19, 28, 36, 43)$, with the cyclical shift pattern

Table 7.15. *Shifted sets of ten varieties for construction of the third replicate of an alpha design for 50 varieties in blocks of size 5*

Set 1	1	2	3	4	5	6	7	8	9	10
Set 2	19	20	11	12	13	14	15	16	17	18
Set 3	26	27	28	29	30	21	22	23	24	25
Set 4	33	34	35	36	37	38	39	40	31	32
Set 5	42	43	44	45	46	47	48	49	50	41

and choice of generator ensuring that no pair of varieties occurring in the first replicate can be repeated in this second replicate.

For the third replicate we then have to find another shift generator which places the five sets in different relative positions from those in the first two replicates. This can be achieved with the generator $(0, 2, 5, 8, 9)$, which provides the shifted positions shown in Table 7.15.

The blocks of the third replicate are again defined by the columns of Table 7.15: $(1, 19, 26, 33, 42), \ldots, (19, 18, 25, 32, 41)$, and we can verify visually that no pair of varieties occues together in a block more than once. The complete design with 30 blocks is shown in Table 7.16.

The design we have constructed is satisfactorily efficient because it avoids any repetition of pairs of varieties together in a block. However, we can only determine exactly how efficient it is by reference to a computer program such as Alphagen.

The general method of constructing alpha designs involves the following steps:

(i) identify the number of varieties to be compared, the sensible block size (ofen constrained by the physical arrangement of plots), and the number of replicates;
(ii) identify the sets of varieties from which the blocks in each replicate will be formed by taking one variety from each set for each block; and
(iii) choose a shift generator for each replicate.

These generators must ensure that all pairs of sets are in different relative positions in each replicate. Avoiding any repeat of the relative positions of sets will ensure that the design is relatively good. To get the best design (with the smallest possible average variance) requires an optimised search as is provided by Alphagen or an equivalent program.

There are various aspects of the construction of alpha designs which deserve some comment.

First, alpha designs are resolvable, i.e. the blocks can be grouped together into superblocks which each contain a complete replicate of the set of varieties. This increased level of structure is generally preferred by experimenters both for ease of management, and because it allows observation of each of the varieties within a defined subset of the blocks.

Second, the design is strongly reliant on the cyclic structure of the shift generators for ensuring that all varieties occur in each replicate superblock, for avoiding repetition of pairs of varieties in a block and for providing a good mesh of within-block links.

Third, if the block size is not a factor of the total number of varieties, an alpha design can be constructed with block sizes differing by 1, by dividing the varieties into sets such that

Table 7.16. *A complete alpha design for three replicates of 50 varieties in blocks of size 5*

Replicate 1					
Block 1	1	11	21	31	41
Block 2	2	12	22	32	42
Block 3	3	13	23	33	43
Block 4	4	14	24	34	44
Block 5	5	15	25	35	45
Block 6	6	16	26	36	46
Block 7	7	17	27	37	47
Block 8	8	18	28	38	48
Block 9	9	19	29	39	49
Block 10	10	20	30	40	50
Replicate 2					
Block 11	1	20	29	37	44
Block 12	2	11	30	38	45
Block 13	3	12	21	39	46
Block 14	4	13	22	40	47
Block 15	5	14	23	31	48
Block 16	6	15	24	32	49
Block 17	7	16	25	33	50
Block 18	8	17	26	34	41
Block 19	9	18	27	35	42
Block 10	10	19	28	36	43
Replicate 3					
Block 21	1	19	26	33	42
Block 22	2	20	27	34	43
Block 23	3	11	28	35	44
Block 24	4	12	29	36	45
Block 25	5	13	30	37	46
Block 26	6	14	21	38	47
Block 27	7	15	22	39	48
Block 28	8	16	23	40	49
Block 29	9	17	24	31	50
Block 30	10	18	25	32	41

the number of sets is the next integer greater than (variety number)/(block size). The final set will be short of varieties and the deficiency is made up by dummy variety numbers. Each block may then include at most one dummy variety, dummy varieties do not require plots, and each block including a dummy variety will actually contain one fewer plot (variety) than the maximum block size. For example, suppose that in the previous example there had been only 48 varieties. Then variety numbers 49 and 50 are dummies, and these varieties do not have plots. In replicate 1 blocks 9 and 10 will have only four plots containing the sets of varieties (9, 19, 29, 39) and (10, 20, 30, 40), respectively. Similarly, in replicate 2 blocks 16 and 17 will include only four plots, and in replicate 3 the blocks with only four plots will be 28 and 29.

Fourth, the sets of generators for the different replicates determine the particular patterns of which variety pairs occur together in a block, and the combination of all such patterns determines the set of variances for all pairwise variety differences. Typically the variation within the set of variances is not very large, and the average variance will be fairly consistent across the various sets of generators. However, there can be some sets of generators which give both more variation of the variances, and also rather larger average variances. The only method of determining which sets of generators are best, or relatively good is to actually calculate the set of variances and their average, and this is what a program such as Alphagen does efficiently, identifying the set of generators which produces the best (most efficient) design.

If you have access to such a program, then obviously you will use it to identify the best design. If you do not have access, then you can calculate the set of variances for one or more designs produced by different sets of generators, using a standard analysis of variance program, and choose the best design from those examined. In general, most designs will produce an average variance within a few percentage points of the best, and there is no great benefit from trying to find the absolute best if this is at all difficult. Most alpha designs will do a very satisfactory job, and it is now quite clear that alpha designs with three or four replicates should normally be used in large-scale variety trials.

7.6.2 Other methods of adjusting for field variation

The other commonly used approach to allowing for the variation of yields across the large area required for a large-scale variety trial has been to attempt to model the underlying variation through the use of control varieties planted in a regular pattern. The control plots use a well-established standard variety, planted typically in every third position in every third row of plots. This requires that 1/9 of the plots are used for the control variety. Each new variety plot will be within a square of eight plots surrounding a control plot. An illustration of two blocks of a variety trial with controls on a 3×3 grid is shown in Figure 7.14, in which the control variety plots are indicated by letters (A, B, ... , M) and the varieties by numbers (1–48).

There is a range of possible adjustment methods which can be used to attempt to produce comparisons between the varieties by correcting for the natural yield variation across the experimental area. The three main forms of adjustment are to adjust by the nearest control plot yield, to adjust by the average of control yield for a rectangle of control plots surrounding each variety plot, and to fit a quadratic surface to the yields for a 3×3 array of control plot yields and adjust by the predicted control yield for each variety plot.

(*a*) The most simple method of adjustment is to use the yield for the control variety in the middle of each set of eight varieties, and to calculate for each variety the difference between the observed yield and the yield of the closest control plot. Thus in the first block in Figure 7.14 variety 1 would be measured relative to control D, variety 2 relative to control E, variety 5 relative to control F and variety 48 relative to control C.

In general this method of adjustment is not likely to be very precise, because the variability of a single yield will be quite large, and hence the differences between the variety yield and the control yield will also be very variable, so that benefits from adjusting

36	41	17	19	4	25	48	3	39
11	A	34	8	B	47	22	C	13
28	44	7	16	29	32	18	26	31
35	10	21	24	46	6	30	14	23
15	D	40	2	E	37	33	F	9
42	1	27	12	20	43	5	38	45
4	36	20	28	40	14	41	24	18
32	G	13	6	H	31	7	I	8
48	30	15	25	46	1	33	21	47
12	34	3	17	35	43	5	27	38
10	K	29	9	L	26	11	M	37
39	23	42	45	19	2	44	16	22

Figure 7.14 Two blocks of a 48-variety trial with control plots (shown as A, B, . . . , M) in a regular 3 × 3 grid.

to allow for yield variation over the field will be offset by the increased variability of the results.

(*b*) A more precise adjustment will often be achieved by using the average of several adjacent controls. A rectangular array of the four controls within which the variety occurs will provide the basis for a more precise adjustment. In the first block in Figure 7.14, variety 5 would be adjusted by the average of controls E, F, H and I; the other variety yields within the same four controls, 43 in block 1, and 14 and 41 in block 2 would be adjusted by the same average. However, variety 2 in block 1 is located directly between controls D and E and cannot be sensibly adjusted by an average of four controls.

To adjust the yield for variety 2 in block 1 we would use either the average of two controls, D and E, or the average of six controls, A, B, D, E, G and H. The choice between the two-control and the six-control adjustments is, again, a balance between the variability of the adjustment and the bias from using controls further away from the variety being adjusted. The two-control adjustment will introduce more variability, but will be less biased because D and E are very close to variety 2, whereas the six-control adjustment will be less variable but possibly more biased, if the yield variation in that area of the field is changing over small distances. There is no correct answer, just a range of alternatives of which some will sometimes be more efficient than others.

(*c*) The most complicated method of adjustment is to use an array of nine controls, such as A, B, C, D, E, F, G, H and I for variety 2 in block 1 of Figure 7.14, and to fit a second-degree polynomial to the 3×3 array of control yields, adjusting the variety 2 yield by the fitted control yield predicted at the location of the variety 2. This, though more complicated, is easily programmed and probably gives the most accurate and least variable adjustment.

There are further variations of the general method of adjusting yields for the natural yield variation across a large experimental area. Instead of using a 3×3 grid for the spacing between controls it would be possible to use wider spacing between controls. This would, of course, use less of the experimental area for controls, which might seem beneficial, but the benefit would be balanced by a less accurate adjustment process, and probably the 3×3 spacing will be the best compromise. It is unlikely that a closer spacing would be beneficial.

It is, of course, also possible to combine the ideas of alpha designs and a 3×3 grid of controls by using blocks of eight varieties with a control in the centre of each block. Then both adjustment and fitting block effects can be used to improve the precision of comparisons between the varieties being tested, using the two analysis methods either as alternatives, or in combination. Blocks of eight will often be a sensible size and this combined design may have some benefits. While we are not aware of experiments being designed to explicitly use these two methods of adjusting for spatial variation, the use of alpha designs including multiple replicates of standard (control) varieties per replicate superblock, has been commonly used by one of us in glasshouse based experiments screening breeding lines of a number of crop species for resistance to pests and diseases and nutrient responses. Such designs can be easily constructed using Alphagen or similar programs.

Exercises

7.1 Devise experimental designs for comparing seven treatments when all comparisons are of equal interest, given
   (*a*) seven blocks of six units each,
   (*b*) seven blocks of five units each,
   (*c*) seven blocks of four units each,
   (*d*) seven blocks of ten units each,
   (*e*) five blocks of six units each,
   (*f*) four blocks of six units, three blocks of five units and three blocks of four units.
   Explain briefly the reasons for your choice.

7.2 For an experiment to compare seven treatments 56 units are available. The 56 units occur in seven natural blocks containing eight plots each. Three experimental designs are suggested:
   (i) Discard one plot from each block and use a randomised complete block design.
   (ii) Split each block into two blocks of four plots each and use a balanced incomplete block design for seven treatments in 14 blocks of four plots.
   (iii) Use an extended block design with seven blocks and each treatment duplicated in one block.
   Compare the standard errors of a comparison of two treatment effects for each design. State and justify your preference order of the three designs.

7.3 An experimenter wishes to compare six treatments in blocks of three or four plots. For (*a*) three plots per block and (*b*) four plots per block:

(i) write out the unreduced BIB design;

(ii) find a reduced BIB design if one exists;

(iii) devise sensible unbalanced designs for six blocks and for eight blocks.

7.4   You are required to design an experiment to compare four treatments O, A, B, C.

The experimental material available consists of 20 units in two blocks of four units, two blocks of three units and three blocks of two units.

Use a general linear model computer package to obtain the variance–covariance matrix for various trial designs and thus find appropriate designs to satisfy in turn the following criteria:

(i) variances of treatment differences to be as nearly equal as possible;

(ii) variances of comparisons of O with other treatments to be approximately two-thirds those of comparisons between other treatments.

Comment on the implications of your results.

7.5   Four treatments are compared in four blocks of variable size, the design being given below. Obtain estimates of the treatment effects for O, A and B and standard errors of the estimated differences between effects.

block I	O	A	B	C	(four plots),
block II	O	A	B		(three plots),
block III	O	A	C		(three plots),
block IV	O	B	C		(three plots).

7.6   A horticulturist wishes to test which of seven new varieties of aubergine plant produces the best crop. He has a maximum of nine greenhouses available in which to conduct trials; each greenhouse is to be allocated aubergine plants of three different varieties. Construct a design for this experiment, assuming that the horticulturist can obtain as many plants of each variety as he needs. What are the advantages and disadvantages of the design? Calculate the efficiencies of the design relative to an orthogonal block design.

Suppose that before the experiment is started another new variety of aubergine plant becomes available. Redesign the experiment.

7.7   Construct the most efficient designs you can devise to compare six treatments in blocks of four units each, with no treatment occurring twice in a block:

(*a*) when all treatment comparisons are equally important, using nine blocks;

(*b*) when treatment A is a control for which all comparisons should be as precise as possible, using ten blocks.

Derive variances of the least squares estimates of treatment differences, $A - B$ and $B - C$ in (*b*) and compare these variances with the corresponding values for (*a*).

7.8   An experiment was performed to compare five drugs. Three of the drugs, V_1, V_2 and V_3 had an organic base and the remaining two, W_1 and W_2, had an inorganic base. It was found possible to test any subject with three of the drugs and the following design was used (with replication)

subject	1	2	3	4	5
	V_1	V_1	V_1	V_2	V_3
drugs	V_2	V_2	W_1	W_1	W_1
	V_3	V_3	W_2	W_2	W_2

Write down the least squares equations under the usual additive model and show how to estimate the effects of the drugs.

Consider the variances of the estimates of the differences between any two drugs. Show that these are in the ratio

$$9 : 8 : 11$$

for, respectively,

(*a*) comparison between two organic based drugs,

(*b*) comparison between two inorganic based drugs, and

(*c*) comparison between an organic and an inorganic based drug.

8

Multiple blocking systems and cross-over designs

8.0 Preliminary examples

(*a*) An experiment to examine the pattern of variation over time of a particular chemical constituent of blood involved sampling the blood of nine chickens on 25 weekly occasions. The principal interest is in the variation of the chemical over the 25 times, the nine chickens being included to provide replication. The chemical analysis is complex and long and a set of at most ten blood samples can be analysed concurrently. It is known that there may be substantial differences in the results of the chemical analysis between different sets of samples. How should the 225 samples (25 times for nine chickens) be allocated to sets of ten (or fewer) so that comparisons between the 25 times are made as precise as possible?

(*b*) In an experiment to compare diets for cows the experimenter has five diet treatments that he wishes to compare. The diets have to be fed to fistulated (surgically prepared) cows so that the performance of the cows can be monitored throughout the application of each diet. Nine such cows are available for sufficient time that four periods of observation can be used for each cow, providing a set of $9 \times 4 = 36$ observations. Concerned that there will be differences between cows, so that cows have to be treated as a blocking structure, the experimenter had already decided before consulting a statistician that he could only test four of the diet treatments, each cow receiving each diet once. Concerned also that the order of application of diets was important, he also wants to block for this, and suggests to the statistician that he will therefore probably have to discard one cow so that each of his remaining diet treatments can be tested twice in each time period. How can the statistician produce a better design?

8.1 Latin square designs and Latin rectangles

We have already mentioned in Chapter 7 the possibility that, in considering the units available for the proposed experiment, there may be more than one apparently reasonable blocking structure. In this chapter, we consider the problems of trying to accommodate two or more blocking systems in a single experiment.

A simple and well-known example is in the assessment of the wearing performance of car tyres. Different brands of tyre can be fitted in each of four wheel positions for each of several cars. There may be differences in performance between the four positions which are consistent for all cars. There will certainly be overall performance differences between cars. To compare four brands of tyres (A, B, C, D) using four test cars, we would like to allocate tyres to positions for each car so that each brand is tested on each car and also in each position. Can this be achieved?

Car

		1	2	3	4
Position	1	A	B	C	D
	2	B	D	A	C
	3	C	A	D	B
	4	D	C	B	A

Figure 8.1 Experimental plan for four brands of tyre with each brand on each car and in each position.

A solution is shown in Figure 8.1. Construction of such a design by trial and error leads rapidly to a solution, and there are in fact many solutions. A design with t^2 units arranged in a crossed double-blocking classification system with t blocks in each system, and with t treatments each occurring once in each block of each blocking system (so that there are t replicates of each treatment) is called a *Latin square* design, the name reflecting the common use of Latin letters, A, B, C, ..., to represent the treatments.

The model and corresponding analysis of variance for the Latin square design are simple extensions of those for the randomised complete block design. The two crossed blocking systems in the design are traditionally referred to as rows and columns, and the model for the response of the unit in row i and column j is written

$$y_{ij} = \mu + r_i + c_j + t_{k(ij)} + \varepsilon_{ij}. \tag{8.1}$$

There is one unusual feature of this model compared with those used previously. The responses, y_{ij}, only need to be classified by two of the three classifications, rows, columns and treatments. Each treatment occurs in each row, each treatment occurs in each column and each column 'occurs in' each row. But only two classifications are required to classify, uniquely, each observation. The treatment suffix k is completely defined by the row and column suffixes i and j, and this dependence is represented by the suffix notation $k(ij)$. Any two of the three classifications could be used to define the set of responses; the model (8.1) is that which is conventionally used. All three sets of effects may be estimated orthogonally, using the obvious restrictions:

$$\sum_i r_i = 0, \sum_j c_j = 0, \sum_k t_k = 0.$$

The analysis of variance structure has the form shown in Table 8.1. Methods for the random allocation of treatments in a Latin square design will be discussed in the next chapter.

Although the Latin square design is a neat mathematical solution to the problem of utilising two crossed blocking factors, it is extremely restrictive. The number of replicates of each treatment must be equal to the number of treatments. Examination of the analysis of variance shows that the df for error are $(t-1)(t-2)$, providing only 2 df when $t = 3$, and only 6 df when $t = 4$. In neither case would we expect to get an adequate estimate of σ^2. When using a Latin square design for three or four treatments, it is usually necessary, therefore, to include more than one square in the experiment.

If multiple Latin squares are to be used, then it is important to distinguish two different forms of design. The difference is based on whether the row or the column differences might

Table 8.1. *Analysis of variance for the Latin square design*

Source of variation	SS	df
Rows	$\sum_i Y_{i.}^2/t - Y_{..}^2/t^2$	$t - 1$
Columns	$\sum_j Y_{.j}^2/t - Y_{..}^2/t^2$	$t - 1$
Treatments	$\sum_k Y_k^2/t - Y_{..}^2/t^2$	$t - 1$
Error	By subtraction	$(t - 1)(t - 2)$
Total	$\sum_{ij} y_{ij}^2 - Y_{..}^2/t^2$	$t^2 - 1$

be expected to be consistent for the different squares. Examples of situations where one of the blocking systems should be consistent across squares are:

(i) The experimental unit is a leaf and the blocking systems are (*a*) plants and (*b*) leaf position. If two groups of plants are used for the two squares it would seem reasonable to assume that the differences between leaf position should be similar for plants in both groups.

(ii) In the car tyre example, if two groups of four cars each are used to form two squares with the four positions forming the second blocking system for each group, then, if the differences between position are consistent enough to define a blocking system for one group, they should be consistent over the two groups.

This approach could also be used for the cow diet problem, as 'simplified' by the experimenter, for four diets allocated to eight cows each over four periods, with the periods considered to be consistent across cows, though we might want to do more about the order in which the diets are allocated to periods within cows (see Section 8.5).

The structure of these situations is illustrated in Figure 8.2(a).

When one of the blocking systems is common to the two or more squares, then the appropriate model for the yields is described simply in terms of rows and columns

$$y_{ij} = \mu + r_i + c_j + t_{k(ij)} + \varepsilon_{ij}.$$

Arbitrarily, we assume for this model that row effects are consistent over several sets of columns so that, in the model, i takes values 1 to t and j takes values 1 to nt. The model is formally identical to that written down for a single Latin square. Essentially this form of multiple Latin square design has the same philosophy as a single Latin square design. Because the row effects are assumed to be consistent over columns, a less stringent requirement is appropriate for treatment occurrences in rows than would be required for separate Latin squares, namely that each treatment appears n times in each row, but not necessarily in groups of t consecutive columns. The design may be thought of as a Latin rectangle. A typical Latin rectangle is shown in Figure 8.3. Note how treatment A occurs in row 2 twice within the first four columns, unlike the arrangement seen in Figure 8.2. The randomisation procedure will be discussed in the next chapter.

Table 8.2. *Analysis of variance for Latin rectangle design*

Source	SS	df
Rows	$\sum_i Y_{i.}^2/(nt) - Y_{..}^2/(nt^2)$	$t-1$
Columns	$\sum_j Y_{.j}^2/t - Y_{..}^2/(nt^2)$	$nt-1$
Treatments	$\sum_k T_k^2/(nt) - Y_{..}^2/(nt^2)$	$t-1$
Error	By subtraction	$(nt-2)(t-1)$
Total	$\sum_{ij} y_{ij}^2 - Y_{..}^2/(nt^2)$	nt^2-1

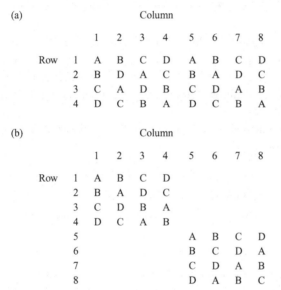

(a)

		Column							
		1	2	3	4	5	6	7	8
Row	1	A	B	C	D	A	B	C	D
	2	B	D	A	C	B	A	D	C
	3	C	A	D	B	C	D	A	B
	4	D	C	B	A	D	C	B	A

(b)

		Column							
		1	2	3	4	5	6	7	8
Row	1	A	B	C	D				
	2	B	A	D	C				
	3	C	D	B	A				
	4	D	C	A	B				
	5					A	B	C	D
	6					B	C	D	A
	7					C	D	A	B
	8					D	A	B	C

Figure 8.2 Multiple Latin squares: (a) with common row effects, (b) completely separate squares.

		1	2	3	4	5	6	7	8
Row	1	A	B	C	B	D	A	D	C
	2	B	D	A	A	B	C	C	D
	3	C	C	B	D	A	B	B	A
	4	D	A	D	C	C	D	A	B

Figure 8.3 Latin rectangle design.

The analysis for the Latin rectangle follows, as usual, directly from the model (see Table 8.2).

The alternative situation, demonstrated in Figure 8.2(b), where neither the rows nor the columns in the different squares have any relation to each other, might occur in an agricultural crop experiment where the geographical arrangement of plots could directly resemble

Table 8.3. *Analysis of variance for two separate Latin squares*

Source	SS	df
Squares	$\sum_h Y_{h..}^2/t^2 - Y_{...}^2/(nt^2)$	$n-1$
Rows in squares	$\sum_{hi} Y_{hi.}^2/t - \sum_h Y_{h..}^2/t^2$	$n(t-1)$
Columns in squares	$\sum_{hj} Y_{h.j}^2/t - \sum_h Y_{h..}^2/t^2$	$n(t-1)$
Treatments	$\sum_k T_k^2/(nt) - Y_{...}^2/(nt^2)$	$t-1$
Error	By subtraction	$(nt-n-1)(t-1)$
Total	$\sum_{hij} y_{hij}^2 - Y_{...}^2/(nt^2)$	nt^2-1

Figure 8.2(b). Another example could be found in psychological experiments where different individuals (one blocking system) are exposed to different treatments on different occasions (the second blocking system). Both the sets of individuals and the sets of occasions will usually be different in the different squares so that neither set of block differences would be consistent between squares.

The model for the latter situation, where row and column effects are particular to the square in which they occur, is

$$y_{hij} = \mu + s_h + r_{hi} + c_{hj} + t_{k(hij)} + \varepsilon_{hij},$$

where suffixes *h, i* and *j* pertain to square, row and column, respectively. The structure of the corresponding analysis of variance for *n* separate Latin squares, each $t \times t$, for *t* treatments is given in Table 8.3.

8.2 Multiple orthogonal classifications and sequences of experiments

At various points in the development of the theory of experimental design, the mathematical ideas involved in the construction of useful designs offer a temptation to divert from the path of usefulness. Extensions of the ideas of constructing Latin squares are one such point and, with the excuse that some of the extensions are occasionally useful, we now succumb to temptation and explore, briefly, the problems of superimposing multiple Latin squares.

For the 3×3 Latin square shown in Figure 8.4(a), a second Latin square using symbols (α, β, γ) can be superimposed as shown in Figure 8.4(b), so that not only does each Greek letter occur in each row and in each column, but also each Greek letter occurs once with each Latin letter. There are four orthogonal classifications: rows, columns, Latin letters and Greek letters. The design of Figure 8.4(b) is called a *Graeco-Latin square*, for obvious reasons. By considering the df in the analysis of variance we can recognise that each of the four orthogonal classifications has 2 df and, since the total df for the nine observations is 8, we can deduce that no further orthogonal classifications may be added. Note that there are also no df remaining for the residual term. The 3×3 Graeco-Latin square is an example of a *completely orthogonal square*.

(a)   Column
                        1   2   3

Row   1   A   A   C
        2   B   C   A
        3   C   A   B

(b)   Column
                        1   2   3

Row   1   Aα Aβ Aγ
        2   Bγ Cα Aβ
        3   Cβ Aγ Bα

Figure 8.4 (a) Latin square, (b) Graeco-Latin square.

(a)   Aα Bβ Cγ Dδ (b)   A   B   C   D
      Bγ Aδ Dα Cβ B   C   D   A
      Cδ Dγ Aβ Bα C   D   A   B
      Dβ Cα Bδ Aγ D   A   B   C

Figure 8.5 Two Latin squares, of which (a) allows a Graeco-Latin square
(as shown) and (b) cannot be extended.

More generally, we might expect to be able to superimpose sets of multiple Latin squares for any size of square, t. The total df, $(t^2 - 1)$, could potentially be split into $(t + 1)$ sets, each with $(t - 1)$ df, corresponding to $(t + 1)$ mutually orthogonal classifications: these classifications would be rows, columns and $(t - 1)$ superimposed Latin squares. However, this general expectation would be only partially correct. For some Latin squares, it is impossible to construct a Graeco-Latin square by superimposing a second Latin square. Two 4×4 Latin squares are shown in Figure 8.5(a) and (b). The first has a second Latin square imposed to achieve a Graeco-Latin square; the second does not allow a Graeco-Latin square to be constructed. The criteria governing the existence of Graeco-Latin (or completely orthogonal) squares are related to the algebraic theory of Galois fields. It can be shown quite simply that, for any group of elements satisfying the conditions of a Galois field, there is a corresponding completely orthogonal square (with $(t - 1)$ superimposed Latin squares), and hence a Graeco-Latin square. From a practical viewpoint, the important consequences are that Graeco-Latin and completely orthogonal squares may be constructed for any Latin square of size 3, 5 or 7 (and of course for many larger sizes). Graeco-Latin squares exist for some Latin squares of size 4 or 8. But there are no Graeco-Latin squares of size 6. There is much literature on the mathematical properties of completely orthogonal squares but further discussion of this topic is not appropriate for this book.

Having succumbed to the temptation to consider this mathematical diversion, we return to the practical considerations associated with designing real experiments, and consider whether there any uses of Graeco-Latin squares for the practical experimenter. It is clearly unrealistic in general to expect to be able to use three blocking criteria represented by rows and columns and Greek letters because of the requirements that the experimental units would have to group into equal size 'blocks' for each of the three blocking criteria, and that for each pair of blocking criteria all possible combinations of the three blocking systems would have to be represented exactly once. Such an occurrence in a set of experimental units could happen only by an extraordinary chance (which the disciple of significance would clearly reject!) or by gross premeditation. However, Graeco-Latin squares may be useful when experimental units are used repeatedly for experiments, though even here gross premeditation is needed to control the number of treatment levels to be the same for each repeat use of the experimental units (and equal to the number of rows and columns) with replication also equal to the number of treatment levels. Consider a set of 64 experimental units used in an 8×8 Latin square design for an experiment, such that at the end of the experiment the units could be used for a further experiment, but with the additional requirement that planning of the succeeding experiment should make allowance for the possible long-term effects of the treatments in the first experiment. Such a situation can arise with experiments on fruit trees or other perennial crops whose useful experimental life may last 20 years, compared with a duration of three or four years for a particular experiment. Other examples may be found in animal experiments.

If the initial Latin square is of the type that allows a subsequent Graeco-Latin square, then the set of treatments of the first experiment may be regarded as a third blocking classification for the second experiment, and all effects of the original two blocking factors and of the treatments for the first experiment will be orthogonal to comparisons of the treatments in the second experiment. An implication of the use of this form of sequence of experiments is that initial experiments must allow for the possibility of later experiments. In particular, the choice of Latin square or of multiple Latin squares could be modified to include only those Latin squares for which Graeco-Latin squares exist. Where it is likely that several experiments will be performed using the same experimental units, then, even when the initial design is a randomised block design rather than a Latin square, the choice of initial design may have consequences for the design of subsequent experiments, and it is important for the experimenter and statistician to be clear about such possibilities.

8.3 Row-and-column designs with more treatments than replicates

We have already identified that Latin square designs are extremely restrictive, with the number of replicates per treatment required to be equal to the number of treatments, t. Where the number of treatments is small this constraint has implications for the residual df and estimation of the underlying unit-to-unit variation, σ^2, but this can be overcome by using multiple squares, possibly combined into a Latin rectangle. But the constraint of needing t^2 or nt^2 units, suitably blocked, will frequently involve either not using available units or including units which are not entirely suitable. Thus, in animal nutrition experiments, where the two blocking criteria of genetic similarity (litters = blocks) and initial size (weight class = block) provide an ideal Latin square structure, the experimenter may have to omit suitable animals from larger litters, or include undersized and atypical animals from smaller litters.

When the number of treatments is large, then the constraints imposed by the Latin square design will often require the inclusion of more replicates than is possible or appropriate. We have already seen in Chapter 7 how we can effectively allocate treatments to blocks where the block size is smaller than the number of treatments, and a related approach can be applied to this double-blocking scenario, though not with quite the same simplicity. This leads to the general idea of *non-orthogonal row and column* designs (Section 8.3.1).

An alternative approach can be taken where both the resources and treatments have some additional structure, based on the idea of superimposing multiple Latin squares, as introduced in Section 8.2. However, in the simplest case, rather than applying successive sets of treatments to the same units, t replicates of st treatments are allocated within an array of st by t experimental units. The set of treatments needs to naturally fall into s sets of t treatments, with the resources naturally falling into groups of s units within each of the t^2 combinations of t rows and t columns. This leads to the idea of *semi-Latin squares*, and the special case of *Trojan squares*, where s mutually orthogonal Latin squares are used to allocate the s sets of t treatments (Section 8.3.2).

8.3.1 Non-orthogonal row and column designs

The ideas of general block designs do not apply with quite the same simplicity to double-blocking systems as to the single-blocking system discussed in Chapter 7. Nevertheless, the range of useful designs is very much wider than is generally realised, and the possibility of matching a suitable design to the natural structure of the experimental units is much greater than the overfrequent use of the restrictive Latin square design would suggest.

Consider first the case where the number of units in each block is the same within a block system but different in the two blocking systems, with each combination of the two blocking systems occurring just once. The resultant structure of experimental units can be represented as a rectangular array, and designs imposed on such a structure are described generally as row-and-column designs. The first group of designs to be considered is for those row-and-column structures where the number of columns (or rows) is equal to the number of treatments t or some multiple of t.

If the number of columns equals t, then obviously treatments can be arranged orthogonally to rows, with each treatment appearing exactly once in each row. Since columns and rows are orthogonal (because of the rectangular form of the row-and-column array), the design structure is determined by the pattern of allocation of treatments to columns. This is the same problem as the allocation of treatments to blocks for a single-blocking factor when the block size is not equal to the number of treatments. The subsequent arrangement of treatments within columns so that each treatment occurs in each row is trivial. The design for treatments within columns may be constructed exactly as in Section 7.2 (with treatments allocated so that all treatment pairs occur equally in columns if possible, following a balanced incomplete block design), and the treatments are then reordered within columns to achieve the orthogonality with respect to rows.

This approach is illustrated in Figure 8.6. The balanced design for seven treatments in seven blocks of three units is shown in Figure 8.6(a), the treatments being written down in the natural order of construction. The rearrangement of letters within each column so that each letter appears once in each row is achieved by working through the columns sequentially. There are some arbitrary choices and it is possible to make these in such a way as to fail to achieve a solution. However, with reasonable forethought a solution can always be found. In column 2, A is put in row 2, and in column 3, A is put in row 3. Since each of the other letters has occurred only once in the first three columns, their row positions are chosen arbitrarily. In column 4, given the previous arbitrary decisions, B must go in row 1, and we now have a choice between (D in row 2, F in row 3) or (D in row 3, F in row 2). We defer the choice,

(a)

				Block			
	I	II	III	IV	V	VI	VII
	A	A	A	B	B	C	C
	B	D	F	D	E	D	E
	C	E	G	F	G	G	F

(b)

				Column			
	1	2	3	4	5	6	7
Row 1	A	D	F	B	G	C	E
2	B	A	G	F	E	D	C
3	C	E	A	D	B	G	F

Figure 8.6 Experimental plan for seven treatments: (a) in seven blocks of three units, (b) in seven columns × three rows.

(a)

					Column	
		1	2	3	4	5	6
Row	1	A	B	C	D	E	F
	2	B	D	A	F	C	E
	3	C	F	E	B	A	D
	4	D	A	B	E	F	C
	5	E	C	F	A	D	B

(b)

					Column	
		1	2	3	4	5	6
Row	1	A	B	C	D	E	F
	2	B	F	A	E	D	C
	3	C	D	E	B	F	A
	4	D	A	B	F	C	E
	5	E	C	F	A	B	D
	6	F	E	D	C	A	B
	7	A	B	C	D	E	F

Figure 8.7 Experimental plans for six treatments: (a) in five rows × six columns, (b) in seven rows × six columns.

proceeding to column 5, where there is no choice; B must go in row 3, G consequently in row 1 and E in row 2. Now, in column 6, G must go in row 3, and D consequently in row 2, with C in row 1. This determines the choice in column 4; D must go in row 3 and F in row 2. The disposition in column 7 is now inevitable. The resulting design is shown in Figure 8.6(b).

Now consider again the designs for six treatments developed in Section 7.2. The designs in six columns of five units or six columns of seven units can be simply adapted to row and column designs as shown in Figure 8.7. These designs are examples of two particularly useful classes of designs, being Latin squares with either a row missing or an additional row. Any Latin square with a row missing or an extra row will have treatments balanced with respect to columns, and such designs are clearly optimal for t treatments in $t \times (t-1)$ or $t \times (t+1)$ row and column designs. Without considering any other designs with treatments balanced with respect to columns, the set of useful designs immediately increases to three times the number of Latin square designs.

Row-and-column designs in which the number of columns is equal to the number of treatments, and in which the treatments are balanced for their occurrence in columns, as is the case for the examples shown in Figures 8.6 and 8.7, are called Youden squares, because

		\multicolumn{6}{c}{Column}
		1	2	3	4	5	6
Row	1	A	F	E	D	C	B
	2	B	A	F	E	D	C
	3	C	B	A	F	E	D
	4	D	C	B	A	F	E

Figure 8.8 Experimental plan for six treatments in four rows × six columns.

(a)

\multicolumn{10}{c}{Block}
I	II	III	IV	V	VI	VII	VIII	IX	X
A	A	A	A	A	B	B	B	C	C
B	B	C	D	E	C	D	E	D	D
C	D	E	F	F	F	E	F	E	F

(b)

	\multicolumn{10}{c}{Column}
	1	2	3	4	5	6	7	8	9	10
Row 1	A	D	C	A	F	B	E	E	C	F
2	B	A	E	D	A	C	B	F	D	C
3	C	B	A	F	E	F	D	B	E	D

Figure 8.9 Experimental plan for six treatments: (a) in ten blocks of three units, (b) in three rows × ten columns.

they were systematically developed by the statistician W. J. Youden. Of course they are not squares, because the number of rows is not equal to the number of columns, but the name has stuck.

Unbalanced block-treatment designs can also be easily converted into row-and-column designs when the number of blocks is equal to the number of treatments. For example, the design in Example 7.3 with six treatments in six blocks of four units each, which was argued to be the most efficient available within the restrictions of the experimental units, can be rearranged into a four-row × six-column array, with each row containing a complete replicate of the six treatments, as shown in Figure 8.8.

The structure of the row-and-column designs considered so far has always included two pairs of orthogonal sets of effects – in each instance rows and columns have been orthogonal, and so have treatments and rows. But if we try to adapt other block-treatment designs of the balanced or nearly balanced types developed in Chapter 7, then it will rarely be possible to arrange treatments and rows to be orthogonal. Consider the design from Example 7.4 for six treatments in ten blocks of three units. If we attempt to transfer this to a three-row × ten-column design, then treatments can obviously be balanced with respect to columns. But with three rows of ten units each, it is clear that the nearest we can come to balance is to regard each row as containing two units per treatment, with two treatments omitted once in each row, and with each treatment omitted once overall. If we start with the block-treatment design shown in Figure 8.9(a) (as developed in Example 7.4), we can then construct the row-and-column design shown in Figure 8.9(b), though the choice of allocation combinations for treatments to rows within each column is considerable.

Table 8.4. *Relationships between rows and columns and treatments*

Design	Row and treatment	Column and treatment
Figure 8.6 7R × 3C	Orthogonal	Balanced
Figure 8.7 5R × 6C	Orthogonal	Balanced
Figure 8.7 7R × 6C	Orthogonal	Balanced
Youden squares generally	Orthogonal	Balanced
Figure 8.8 4R × 6C	Orthogonal	Not balanced
Figure 8.9 3R × 10C	Balanced	Not balanced

The different possible relationships between the three classifications (rows, columns and treatments) have been used by many authors, notably Pearce (1963, 1975), to classify row-and-column designs. Thus far we have only considered designs for which rows and columns are orthogonal, with a range of possibilities for the other relationships, as summarised in Table 8.4.

The overall relationship for treatment comparisons in the designs listed in Table 8.4 is always the 'worse' of the two relationships with rows and columns. Thus, except for the last two, all the listed designs have balanced treatment comparisons. However, it is theoretically possible to have a design with overall balanced treatment comparisons when neither the row–treatment nor column–treatment comparisons are balanced. The crucial result was presented by Pearce (1963). If there are r rows and c columns with

$$n_{1ik} \text{ observations on treatment } k \text{ in row } i,$$

and

$$n_{2\,jk} \text{ observations on treatment } k \text{ in column } j$$

then, after elimination of row-and-column effect estimators, the least squares equations for treatment effects are

$$\hat{t}_k \left(N_{.k} - \sum_i n_{1ik}^2/c - \sum_j n_{2\,jk}^2/r \right) - \sum_{k' \neq k} t_{k'} \left(\sum_i n_{1ik} n_{1ik'}/c + \sum_j n_{2\,jk'} n_{2\,jk'}/r \right)$$

$$= T_k - \sum_i n_{1\,ik} Y_{i.}/c - \sum_j n_{2jk} Y_{.j}/r.$$

Hence, if the weighted sum of the concurrences for rows and for columns

$$\sum_i n_{1\,ik} n_{1\,ik'}/c + \sum_j n_{2\,jk} n_{2\,jk'}/r$$

is invariant over all pairs (k, k'), then the treatments are balanced. Clearly, this could be achieved without necessarily having balance with respect to either rows or columns, for which the requirements would be the invariance over (k, k') of either

$$\sum_i n_{1\,ik} n_{1\,ik'}/c$$

Column

		1	2	3	4	5	6
Row	1	G	C	H	E	F	D
	2	A	G	E	B	H	F
	3	C	B	D	G	A	H
	4	E	F	A	D	C	B

Figure 8.10 Experimental plan for eight treatments in four rows × six columns.

or

$$\sum_j n_{2jk} n_{2jk'} / r$$

respectively. As with the designs for a single-blocking criterion, this Pearce criterion provides an ideal at which the designer may aim even when the ideal is not achievable.

In constructing general row-and-column designs the allocation of treatments to rows and to columns should initially be considered separately. For each blocking system the allocation should be such as to achieve orthogonality if that is possible, balance if orthogonality is not possible, and if balance is not possible the joint occurrences in rows (or columns) should be made as nearly equal for all treatment pairs as is possible, just as for the allocation to blocks in a single-blocking system. Where balance is not possible for either blocking system, some manipulation of the two designs is needed. One approach is to construct the best possible designs for each, and then to relabel the treatments in each allocation to match high pairwise occurrences in one blocking system with low pairwise occurrence in the other. Other possible approaches include finding the best design for the blocking system that is closest to achieving balance (equal pairwise occurrence) and then allocating treatments in the other blocking system to compensate for low pairwise occurrences in the first blocking system (in terms of the Pearce criterion), or finding the best design for the blocking system that is furthest from achieving balance and again allocating treatments in the other blocking system so as to get closest to the Pearce criterion. When the row-and-column allocations are complete the joint allocation is constructed to be compatible with the separate allocations. This last step is relatively trivial except for incomplete rectangular arrays, but may require some trial and error investigations for such arrays. Using these several guidelines we can deal with a variety of situations.

Figure 8.10 shows a solution that is certainly nearly optimal for comparing eight treatments in a 4 × 6 array. The obvious solution for the rows is for a different pair of treatments to be omitted from each row, with each of the eight treatments being omitted once across the four rows. This means that the pairs of omitted treatments occur together three times in rows, and all other pairs occur together twice. Then in columns these omitted treatment pairs should occur together less often than the other pairs, and by constructing blocks of size 4 based on taking one treatment from each omitted pair we can achieve a design such that pairs of treatments occur together in either rows of columns either three or four times, with weighted sums of occurences of 0.5, 0.57 or 0.83 for different treatent pairs. Of course, this might be particularly sensible if the treatments formed a 2 × 4 factorial set.

A slightly more complicated design is for comparing nine treatments in a 5 × 7 array. The first point to note is that not all nine treatments can be equally replicated, with the obvious solution being that one treatment has only three replicates, in this case treatment I. Then

```
                        Column
              1   2   3   4   5   6   7
     Row  1   I   B   A   G   C   D   H
          2   A   F   G   E   B   C   D
          3   G   I   B   H   F   E   A
          4   C   H   I   D   E   G   F
          5   E   D   F   A   H   B   C
```

Figure 8.11 Experimental plan for nine treatments in five rows × seven columns.

```
                          Column
              1   2   3   4   5   6   7   8
     Row  1   ×   ×   ×   ×   ×           ×   ×
          2   ×   ×   ×           ×   ×   ×
          3   ×       ×   ×   ×   ×           ×
          4   ×   ×   ×   ×   ×   ×
          5   ×   ×   ×   ×           ×   ×   ×
          6   ×   ×       ×           ×
```

Figure 8.12 Experimental row-and-column structure of 36 units.

treatment I can only occur in three rows, so has 18 comparisons with other treatments in rows, most evenly achieved by occurring with two treatments (G, H) three times and each of the other treatments twice. Each other treatment appears in four of the five rows, with three pairs of the treatments that only occur with I twice having to appear together in all four rows in which they appear – in this case the pairs AB, CD and EF. Having found the best allocation of treatments to rows, we then allocate treatments to columns, ideally so that treatment I occurs more often with treatments A–F than with G and H, and so that the pairs occurring together in four rows appear less often together in columns. A reasonably good solution has most pairs occurring together twice in columns, with some occuring either three times or once. A possible solution is shown in Figure 8.11.

Finally we consider comparing six treatments in an incomplete 6 × 8 array, in which only 36 of the possible 48 combinations occur. Such a scenario might occur in orchard trials where some trees have died, or in field trials associated with weed control where the initial weed presence is rather patchy. We discuss the derivation of a solution to this example in more detail to illustrate the general approach.

Example 8.1 The structure of the 36 available units is shown in Figure 8.12. We first consider rows and columns separately. The six rows have seven, six, six, six, seven and four units, respectively; the eight columns have six, five, five, five, four, four, four and three units.

An obvious allocation of treatments to rows is to have all six treatments (A, B, C, D, E and F) in each of the first five rows; there can be only four treatments in the final row and the two treatments omitted from that row must then appear as the extra treatments in rows 1 and 5. If the treatments in row 6 are A, B, C, D, then the set of weighted concurrences for the 15 treatment pairs in rows is shown in Table 8.5(a). The variances of treatment comparisons that would be achieved if columns were ignored are given in Table 8.5(b).

The allocation to columns is less inevitable. All six treatments (1, 2, 3, 4, 5, 6) will appear in column 1, and columns 2, 3 and 4 will have a different treatment omitted from each (treatments 4, 5 and 6); there is then a choice of which three treatments to put in column 8,

Table 8.5. *Weighted concurrences and variances of treatment differences, allowing for rows: (a) weighted concurrences for rows; (b) variances of treatment differences allowing for rows only*

(a)

	B	C	D	E	F
A	1.04	1.04	1.04	0.93	0.93
B		1.04	1.04	0.93	0.93
C			1.04	0.93	0.93
D				0.93	0.93
E					1.07

(b)

	B	C	D	E	F
A	0.333	0.333	0.333	0.345	0.345
B		0.333	0.333	0.345	0.345
C			0.333	0.345	0.345
D				0.345	0.345
E					0.342

and then the various ways of completing columns 5–7. The choice for column 8 is in terms of how many of the treatments omitted in columns 2–4 to include. An obvious approach is to include all three, with the obvious alternative to exclude all three. The former leads to the most even distribution of weighted concurrences, with two of these three treatments occurring in each of columns 5, 6 and 7, together with two of the other three treatments to produce a column allocation as follows:

1st column 1, 2, 3, 4, 5, 6;
2nd column 1, 2, 3, 4, 5;
3rd column 1, 2, 3, 4, 6;
4th column 1, 2, 3, 5, 6;
5th column 2, 3, 4, 5;
6th column 1, 2, 5, 6;
7th column 1, 3, 4, 6;
8th column 4, 5, 6.

The weighted concurrences and variances of treatment differences, assuming that the columns provide the only blocking systems for the treatments, are shown in Table 8.6(a) and (b).

Note that we have intentionally used different sets of symbols for the treatments when considering the row-and-column allocations. The final step to complete the total design is the identification of treatments (A, B, C, D, E, F) with treatments (1, 2, 3, 4, 5, 6), and this is simply a question of examining the two sets of weighted concurrences and matching them

Table 8.6. *Weighted concurrences and variance of differences, allowing for columns: (a) weighted concurrences for columns; (b) variations of treatment differences allowing for columns only*

(a)

	2	3	4	5	6
1	1.02	1.02	0.82	0.82	1.07
2		1.02	0.82	1.07	0.82
3			1.07	0.82	0.82
4				0.95	0.95
5					0.95

(b)

	2	3	4	5	6
1	0.348	0.348	0.365	0.365	0.350
2		0.348	0.365	0.330	0.365
1			0.350	0.365	0.365
4				0.361	0.361
5					0.361

		1	2	3	4	5	6	7	8
					Column				
Row	1	C	E	B	F	A		E	D
	2	E	A	D		F	C	B	
	3	F		E	A	D	B		C
	4	D	C	A	B	E	F		
	5	B	D	F	E		A	C	F
	6	A	B		C		D		

Figure 8.13 Experimental plan for six treatments in 36 units.

to produce as even a set of totals as possible. This is achieved by equating A with 1, B with 2, E with 3, C with 4, D with 5 and F with 6, eventually producing the design shown in Figure 8.13 after a trial and error allocation within these row-and-column constraints. The actual precision of the comparison of pairs of treatments can be obtained by using a general analysis package, and the variances of treatment pairs so obtained are shown in Table 8.7(b), with the sums of the two sets of weighted concurrences shown in Table 8.7(a).

It is clear from these variances that the designs achieve a very even level of precision of comparison with a maximum deviation from the average variance of less than 5%. This design may not be the most efficient that can be achieved, but it will be very close to the optimal design, and this has been achieved by simple application of the principles of good block-treatment design. It should therefore be clear that efficient designs for quite general row-and-column designs may be constructed by simple methods.

Table 8.7. *Sums of weighted concurrences and variances: (a) overall sums of weighted concurrences; (b) variances of treatment differences allowing for rows and columns*

(a)

	B	C	D	E	E
A	2.06	1.86	1.86	1.95	2.00
B		2.11	1.86	1.95	1.75
C			1.99	1.75	1.88
D				2.00	1.88
E					1.89

(b)

	B	C	D	E	F
A	0.350	0.368	0.368	0.366	0.359
B		0.370	0.367	0.360	0.377
C			0.362	0.380	0.373
D				0.369	0.373
E					0.378

8.3.2 Semi-Latin squares and Trojan squares

A further approach to combining row-and-column designs, initially based on sets of orthogonal Latin squares, can be useful where the rows (or columns) of an array of experimental units can be divided into equal sized sets of adjacent rows (or columns) such that the experimental units in one column (or row) within a set of adjacent rows (or columns) are likely to be more similar in response than those in neighbouring sets of experimental units. It is then clear that the experimental units should be considered to be blocked by column (or row), and by sets of adjacent rows (or columns) (termed a main row or main column), with each combination of column (or row) and main row (or column) forming a further nested blocking structure. Where the numbers of main rows (or columns) and columns (or rows) are equal, and if the treatments can be divided into a number of distinct sets each containing the same number of treatments as main rows and columns, then an appropriate design can be obtained by superimposing multiple Latin squares, with the treatments in each row–column combination in each square occupying the expermiental units in each main row–column combination in the new design. With any choice of Latin squares to be superimposed in this way, this design ensures that each treatment occurs exactly once in each column and exactly once in each main row, and such designs are referred to as semi-Latin squares and were identified by Yates (1935). An example design for eight treatments in an 8 × 4 array of experimental units is shown in Figure 8.14. Treatments A–D are arranged in one Latin square and treatments E–H in a second, with columns 1 and 2 forming main column 1, 3 and 4 forming main column 2, and so on.

		1	2	3	4	5	6	7	8
Row	1	A	H	G	B	C	F	E	D
	2	B	G	E	A	H	D	C	F
	3	F	C	D	H	E	A	G	B
	4	D	E	F	C	B	G	A	H

Figure 8.14 Semi-Latin square design.

		1	2	3	4	5	6	7	8
Row	1	A	E	F	B	C	G	H	D
	2	B	H	G	A	F	D	C	E
	3	F	C	D	E	H	A	G	B
	4	D	G	H	C	B	E	A	F

Figure 8.15 Trojan square design.

Considering the pairs of treatments that occur together in the main row-and-column blocks of the above design, however, identifies a possible disadvantage of such designs – treatments B and G always occur together, as do treatments C and F, so that these pairs of treatments are always going to be more precisely compared than will treatments B and C with the other treatments in the second Latin square. That the treatments in a particular Latin square never appear together in a main row-and-column block suggests that the allocation of treatments to Latin squares needs to be done carefully to ensure good comparisons between treatment pairs of interest, and possibly that such designs can be most beneficially used where the treatments form a suitable set of factorial combinations (see Chapter 13).

The first problem, relating to the uneven pairwise occurrences of treatments from the different Latin squares in main row-and-column blocks, can be removed by requiring that the superimposed Latin squares are mutually orthogonal (as was required for the Graeco-Latin squares identified earlier). Imposing this constraint then means that each pair of treatments, one from each Latin square, will occur exactly once in a main row-and-column block. Such designs have been termed Trojan squares (though they are no longer squares as the numbers of rows and columns are no longer equal), a description first introduced by Darby and Gilbert (1958), and an example for eight treatments in an 8×4 array of experimental units is shown in Figure 8.15. The allocation of treatments to the two mutually orthogonal Latin squares is as specified for Figure 8.14.

The class of Trojan squares can be extended for as many mutually orthogonal squares as are available for a particular size of Latin square, but, as with the other fully orthogonal row-and-column designs introduced earlier, it should be obvious that these designs are rather restrictive on the numbers of replicates and treatments that can be included. In particular the number of treatments must be an exact multiple of the number of replicates, main rows and columns. Again, as with simple Latin squares, incomplete and extended Trojan squares can be easily generated by removing or adding one main row or column (Edmondson, 1998), without overcomplicating the subsequent analysis. More general designs constructed around these additional blocking constraints are certainly possible, but will introduce further non-orthogonality into the treatment comparisons.

Complete and incomplete Trojan squares have been reasonably widely used in crop research, particularly for protected crops grown in glasshouse environments, where the additional blocking constraints do relate directly to the physical structures in which the experiments are arranged. Of course, where the grouping of rows into main rows is rather arbitrary, such designs are not really appropriate, though the analysis can be adapted to merge the information provided by comparisons between and within the main row-and-column blocks.

8.4 Three-dimensional designs

Though fairly rare, experimental situations do occasionally occur where it is necessary to block in three orthogonal directions: One example is in laboratory experiments using incubators with variability between shelves, across the width of each shelf, and from front to back of the incubator. Another example, in which one author (AM) has considerable experience, is in designing trials to assess the impacts of different agronomic factors on the production of mushrooms in growing tunnels, where experimental units (trays) are stacked up to four high with multiple stacks across the width and along the length of the growing tunnel.

A particular example considered the allocation of 64 treatment combinations (consisting of all combinations of six two-level factors) into a set of 128 trays (experimental units) arranged in a 4 (columns) by 8 (rows) by 4 (layers) array. The design allows two replicates of each treatment combination. Ideally, the full complex three-dimensional blocking structure should be retained, with combinations of treatment factors arranged such that high-order interactions are confounded (see Chapter 15) with the different blocking factors, so that full information is available about the main factor effects and most low-order interactions can also be estimated. In practice, the structure was used to divide the units into 16 blocks of 8 units each, the blocks comprised positions (units) in which a similar response would be expected if the same treatment combination was applied (Figure 8.16(a)). The 16 blocks took full account of the three orthogonal blocking factors (four layers by two positions across the tunnel (outside stacks or inside stacks) by two positions along the tunnel (left hand end or right hand end), with different high-order interactions confounded with each blocking factor. The allocation of treatment combinations (expressed as levels 0 and 1 for each of the six factors) to experimental units is shown in Figure 8.16(b), obtained by confounding different carefully selected high-order interactions with each of the three blocking factors. Within each of the 16 blocks, each of the 4 layers, each of the positions along (columns 1–4 or 5–8) and across the tunnel (rows 1 and 4, or 2 and 3), and each of the combinations of pairs of blocking factors (for example consider the set of treatment combinations in columns 1–4 of the top layer across all four rows), it is relatively easy to check that each main effect is estimable (equal numbers of 0s and 1s). Various different two- and three-factor interactions are also estimable within the different blocking strata (layers, positions across and along the tunnel), so that it is possible to estimate all main effects and two- and three-factor interactions within blocks using this design.

With a simpler treatment structure it would be possible to allow for the three-dimensional blocking structure in a more precise way, for example considering each of the four stacks across the tunnel or each position (column) along the tunnel as a separate level of a blocking factor, though it is also important to note the constraints (even greater than for designs including a pair of orthogonal blocking structures) on treatment allocations that must be imposed by considering such three-dimensional blocking structures.

(a)

Top layer

Column	1	2	3	4	5	6	7	8
Row 1	A	A	A	A	B	B	B	B
Row 2	C	C	C	C	D	D	D	D
Row 3	C	C	C	C	D	D	D	D
Row 4	A	A	A	A	B	B	B	B

Second layer

Column	1	2	3	4	5	6	7	8
Row 1	E	E	E	E	F	F	F	F
Row 2	G	G	G	G	H	H	H	H
Row 3	G	G	G	G	H	H	H	H
Row 4	E	E	E	E	F	F	F	F

Third layer

Column	1	2	3	4	5	6	7	8
Row 1	I	I	I	I	J	J	J	J
Row 2	K	K	K	K	L	L	L	L
Row 3	K	K	K	K	L	L	L	L
Row 4	I	I	I	I	J	J	J	J

Bottom layer

Column	1	2	3	4	5	6	7	8
Row 1	M	M	M	M	N	N	N	N
Row 2	O	O	O	O	P	P	P	P
Row 3	O	O	O	O	P	P	P	P
Row 4	M	M	M	M	N	N	N	N

(b)

Top layer

Column	1	2	3	4	5	6	7	8
Row 1	100000	010001	111010	000110	110110	111011	000111	100001
Row 2	011011	000010	110111	101110	111001	011111	000101	001000
Row 3	111000	010100	001101	100001	101110	110100	100011	010010
Row 4	101101	110111	011100	001011	101100	001010	011101	010000

Second layer

Column	1	2	3	4	5	6	7	8
Row 1	100101	011111	001001	010000	111100	101011	110001	001101
Row 2	100101	000011	001110	110010	001000	010001	101011	111101
Row 3	101000	010100	011001	111111	000111	100100	110010	011110
Row 4	101010	110011	111100	000110	100110	010111	011010	000000

Third layer

Column	1	2	3	4	5	6	7	8
Row 1	110000	111111	010011	101001	000001	101010	100111	010110
Row 2	001110	101101	000001	100010	110011	011000	111110	010101
Row 3	011000	110100	111011	010111	000010	101001	001111	100100
Row 4	000101	100110	001010	011100	111101	001100	110000	011011

Bottom layer

Column	1	2	3	4	5	6	7	8
Row 1	111000	100010	001001	110101	010110	100011	110101	011001
Row 2	001100	111001	110110	010101	111110	011101	000100	100111
Row 3	101111	100000	011010	000011	010010	101000	001010	110001
Row 4	000100	011110	010011	101111	101100	000000	111010	001111

Figure 8.16 Design with 16 blocks created from three orthogonal blocking factors: (a) allocation of units to the blocks; (b) allocation of treatment combinations to units.

8.5 The practical choice of row-and-column design

The general philosophy of choosing a design for a particular number of treatments and for a particular set of units using two blocking systems is similar to that for a single blocking system. There are some classes of standard designs (for example Latin squares, Youden squares, Graeco-Latin squares) with particularly desirable properties which should be used if they happen to coincide with the requirements of the particular problem. Where there is no such standard design for a problem, then the choice lies between modifying the problem (slightly) to fit a standard design (the Procrustean philosophy) or producing a tailor-made solution to fit the peculiar features of the problem. As for the designs with a single blocking system, usually the production of a tailor-made solution will provide a good comparison of the treatments of interest without much loss of orthogonality, and modern analysis algorithms are easily able to produce an appropriate analysis. As with the single blocking system the design of experiments for comparing large numbers of treatments requires special consideration and is discussed later in this section.

A fully orthogonal design requires single or multiple Latin squares. As with the randomised complete block design for a single blocking system, a Latin square design is ideal if appropriate. To a greater extent than for the randomised block design it is rare for the conditions to be exactly appropriate, and many Latin square designs used in practice represent a major compromise of one of the ideal requirements, either through modification of the desired number of treatments or through accepting either more or less replication than is sensible. Although the possibility of single or multiple Latin square designs should always be considered, they should be used only when the circumstances happen to fit the design exactly or so nearly that the amendment to the experimental design problem to persuade it to fit a Latin square solution is slight. The extension to a semi-Latin square or Trojan square provides an orthogonal design alternative for larger numbers of treatments though with some structured variability in the precision of different treatment comparisons.

Similarly the conditions for designs such that rows and columns are orthogonal, treatments and rows are orthogonal and treatments are balanced against columns are very restrictive, and the probability of finding an appropriate standard balanced design is small. Therefore to use a double blocking system efficiently it is frequently necessary for the experimenter and statistician to construct a design which is as nearly balanced as possible for each particular situation.

The two examples at the beginning of the chapter illustrate the two possible approaches clearly. In the first there are 25 treatments (= times) for each of nine chickens (= blocks), and the 225 observations must be grouped into sets of ten or less, each set being a block of the second blocking system. This is an example of a problem for which there is a standard design which is very close to the original specifications. Using sets of ten observations obviously cannot provide a suitable design since the twenty-third set will contain only five observations. However, if we consider using nine observations per set there is a clear hint of a satisfactory pattern. Using 25 blocks of nine observations with a second blocking factor having nine levels (chickens) provides a design problem in 25 rows and nine columns for 25 treatments. Clearly the 25 treatments can be arranged orthogonally to columns, each column containing a complete set of the 25 treatments. A treatment occurring nine times must occur in nine different rows with a total of $9 \times 8 = 72$ comparisons with other

		Chicken								
		A	B	C	D	E	F	G	H	I
Set	1	1	2	3	4	5	6	7	8	9
	2	2	4	9	10	24	17	15	22	12
	3	3	24	8	23	18	21	13	4	10
	4	4	22	25	8	20	12	11	3	19
	5	5	15	17	18	8	11	2	13	20
	6	6	8	12	13	1	14	24	25	15
	7	7	16	5	22	3	10	25	15	13
	8	8	10	11	16	6	22	23	1	17
	9	9	13	20	5	12	23	1	21	22
	10	10	19	14	12	16	2	8	5	21
	11	11	18	19	24	10	1	5	9	25
	12	12	6	10	25	7	18	20	2	23
	13	13	11	4	9	23	25	14	16	2
	14	14	3	7	17	11	5	12	23	24
	15	15	20	21	1	14	7	10	11	4
	16	16	17	18	7	13	19	4	12	1
	17	17	14	13	20	19	3	9	10	6
	18	18	9	22	14	25	8	21	17	7
	19	19	7	23	15	9	20	16	24	8
	20	20	5	16	6	4	24	22	14	18
	21	21	12	6	11	15	9	3	18	16
	22	22	21	24	19	2	13	6	7	11
	23	23	1	2	3	22	15	18	19	14
	24	24	25	1	2	21	16	17	20	3
	25	25	23	15	21	17	4	19	6	5

Figure 8.17 Experimental plan for 25 treatments in 25 rows $\times$ 9 columns.

treatments in those nine rows. With 24 other treatments this would suggest a replication of each treatment pair of $\lambda = 72/24 = 3$. The integral λ value is sufficient encouragement for the statistician to consult reference books with a strong expectation of finding a balanced design. With the large numbers involved it is plainly easier to refer to an index of designs (for example Cochran and Cox (1957)) rather than construct the design of Figure 8.17 directly. The detailed design may alternatively be obtained from programs for computer-aided design, but the initial investigation (finding $\lambda = 3$) is still required to provide encouragement that such a program will be able to construct the required design.

The second problem, with five treatments to be compared within a double block structure of 4 rows $\times$ 9 columns, offers several possibilities. The experimenter's presumption that it would be necessary to reduce the treatments to four and the number of cows to eight was based on knowledge of the Latin square design, using a pair of Latin squares to accommodate the eight cows. With a variance of treatment differences of $2\sigma^2/8$, this is a reasonable but pessimistic starting point – certainly a waste of the available resources, but also likely to compromise the initial aims of the experiment. If it is decided to accept the reduction to four treatments, then hopefully the statistician and experimenter will decide to include the ninth cow giving nine replicates per treatment with each row (= period) having a different one of the four treatments repeated a third time. The variance of treatment differences for this design (treatments balanced over rows, orthogonal to columns), shown in Figure 8.18(a), is

(a)

		1	2	3	4	Cow 5	6	7	8	9
Period	1	A	B	C	A	D	B	C	D	A
	2	B	D	B	C	A	D	A	C	B
	3	C	A	D	B	C	A	D	B	C
	4	D	C	A	D	B	C	B	A	D

(b)

		1	2	3	4	Cow 5	6	7	8	9
Period	1	A	B	C	D	E	A	B	C	D
	2	B	C	D	E	A	B	C	D	E
	3	C	D	E	A	B	C	D	E	A
	4	D	E	A	B	C	D	E	A	B

Figure 8.18 Experimental plans for a 4 row × 9 column structure for: (a) four treatments, (b) five treatments.

$9\sigma^2/40$. If a design is sought for five treatments, then with four rows (periods) the advantages of Latin squares with a row omitted should be remembered. Such a design for five treatments arranged for the first five cows and the four periods is highly efficient. The same logical approach for the remaining four cows times four periods would suggest using a Latin square omitting one row and one column. An example total design is illustrated in Figure 8.18(b), and the variances of treatment differences are $0.284\sigma^2$ for comparisons including treatment D (which occurs 8 times, while all other treatments only occur 7 times) and $0.313\sigma^2$ for other comparisons. The two designs in Figure 8.18 are optimal for four and for five treatments, and the choice of design reduces to deciding whether the benefit of including the fifth treatment is sufficiently advantageous to outweigh the increase of the variances of treatment difference from $0.225\sigma^2$ to $0.284\sigma^2$ and $0.313\sigma^2$.

For design problems with many treatments it may often be appropriate to use several rectangular arrays, within each of which there are row and column effects. An advantage of using two blocking systems is that precision of comparison between two treatments depends on the joint occurrences of the two treatments either in a row or in a column. Thus, pairs of treatments which do not occur together in any row can usually be arranged to occur together in a column, and vice versa. This is a crucial implication of the Pearce weighted joint occurrence result derived in the previous section. As usual there are standard designs, the major design pattern being the lattice square based on writing the treatments in several square arrays with the rows and columns of each square defining the blocks of each blocking system. Also as usual these designs in their complete, balanced, form are very restrictive; the number of treatments must be the square of an integer, k, and the number of replicates is $(k + 1)$ or $(k + 1)/2$. The design for $k = 7$, 49 treatments with four replicate squares, is shown in Figure 8.19.

Often the size of the design requires that the average number of joint occurrences will be less than 1 so that ideally each treatment pairs should occur together once or not at all. A simple approach to constructing reasonably efficient designs is to write out rectangular arrays for each replicate using the rows and columns for each array as the blocks of the two blocking systems. If we try this and attempt to construct more than one replicate, then we rapidly discover that it is not possible to avoid replication of treatment pairs, unless the

Replicate I   Replicate II

1	2	3	4	5	6	7		1	38	26	14	44	32	20
8	9	10	11	12	13	14		21	2	39	27	8	45	33
15	16	17	18	19	20	21		34	15	3	40	28	9	46
22	23	24	25	26	27	28		47	35	16	4	41	22	10
29	30	31	32	33	34	35		11	48	29	17	5	42	23
36	37	38	39	40	41	42		24	12	49	30	18	6	36
43	44	45	46	47	48	49		37	25	13	43	31	19	7

Replicate III   Replicate IV

1	19	30	48	10	28	39		1	42	27	12	46	31	16
40	2	20	31	49	11	22		17	2	36	28	13	47	32
23	41	3	21	32	43	12		33	18	3	37	22	14	48
13	24	42	4	15	33	44		49	34	19	4	38	23	8
45	14	25	36	5	16	34		9	43	35	20	5	39	24
35	46	8	26	37	6	17		25	10	44	29	21	6	40
18	29	47	9	27	38	7		41	26	11	45	30	15	7

Figure 8.19 Lattice square design for 49 treatments in four arrays each of
7 rows × 7 columns.

row-and-column arrays are square. For example, consider an 8 row × 10 column array for 80 treatments. In each row of a second 8 × 10 array at least two pairs of treatments must occur which also occurred together in some row of the first array, simply because there are only eight rows of the original array. The reason for the advantages of the lattice square should now be clear. Nevertheless, nearly square arrays of suitably small size can be constructed when many treatments are to be compared.

Another alternative approach to the construction of these designs is based on the alpha designs introduced in Chapter 7. A further extension of these considers each replicate block to be a row-and-column array, so that comparisons within both rows and columns contribute to the overall efficiency. The currently used search algorithm (AlphaGen) first constructs a good design based on a single blocking system, and then modifies the design to improve the efficiency of comparisons within columns, resulting in the best design allowing for comparisons in both rows and columns.

8.6 Cross-over designs – time as a blocking factor

One particular form of row-and-column design has time as one blocking factor with the different treatments applied in sequence to each experimental unit, the treatment sequences being different for different units. This practice is particularly common in medical, psychological or agricultural animal experiments (and might be considered appropriate for the second example introduced at the start of this chapter, concerned with the allocation of treatments to cows over four time periods), and requires special consideration. The experiment will include a number of patients, subjects or animals who each receive different treatments in each of a number of successive periods. Essentially the experimental unit is redefined to be an observation for an individual subject in a short period of time. There are

							Subject					
		1	2	3	4	5	6	7	8	9	10	11	12
Period	1	A	B	B	A	A	B	A	B	B	B	A	A
	2	B	A	A	B	B	A	B	A	A	A	B	B

Figure 8.20 Multiple Latin square, or cross-over design for two treatments.

assumed to be consistent differences between periods in addition to those between subjects. In the simplest case with two treatments, A and B, subjects either receive treatment A in the first period followed by treatment B in the second period, or the reverse ordering. By having equal numbers of subjects for the two orderings, the effects of the order of treatments are eliminated from the comparison between the treatments A and B. The resulting design will be a multiple 2×2 Latin square, known as a cross-over design and illustrated in Figure 8.20.

Before considering some of the technical problems of design and analysis involved in the cross-over design, the general advantages and disadvantages should be examined. The general reason for wishing to use cross-over designs is the anticipated high level of variability between patients, subjects or animals. In many experiments, it is found that the variance of observations for different patients or animals may be more than ten times greater than the variation between observations at different times for the same patient or animal. This suggests that enormous gains in precision can be achieved by observing more than one treatment on each patient or animal. A further advantage is that fewer patients or animals are required. This improvement in statistical precision might seem to provide an overwhelming argument for the use of cross-over designs. However, there are other statistical considerations than precision in the design of experiments, and, in particular, the questions of validity and of the population for which the experimental results are relevant must be considered.

The difficulty with the cross-over design is that the conclusions are appropriate to units similar to those included in the experiment, i.e. to subjects to which treatments are given for a short time period in the context of a sequence of different treatments. We then have to ask if the observed difference between two treatments would be expected to be the same if the same treatment is applied consistently to each subject to which it is allocated. This is a problem of the interpretation of the results from an experiment to subsequent use, and it is a problem which must be considered carefully for all experiments. It is particularly acute in the case of cross-over designs because the experiment is so different from subsequent use. After all, no farmer is going to continually swap the diets for his cattle!

The decision on whether to use cross-over designs is therefore a balance between a hoped for gain in precision, which may be considerable, and a potential loss of relevance. If it is decided that the benefits of the cross-over design outweigh the deficiencies, then the range of available designs can be considered.

Only rarely would more than four periods be used in a cross-over design, and two- or three-period designs predominate. It is useful to discuss cross-over designs in two contexts. First, we consider briefly designs in which the separate periods are considered simply as a second set of blocks. For these designs we assume that the ordering of time is not relevant, and that the effect of a treatment applied to a subject in one period has no residual effect on the response of the subject in the following period. The second context for considering cross-over designs is when there are assumed to be residual, or carry-over, effects of the treatments applied in the previous periods. The residual effect complicates the models for the experimental data and the consequent estimation of treatment effects, and we consider such models in Section 8.7.

For this first situation, with no residual effects, the model for the result from the ith subject in the jth time period, during which the subject receives treatment k, is

$$y_{ij} = \mu + s_i + p_j + t_{k(ij)} + \varepsilon_{ij},$$

where s_i is the subject effect, p_j is the period effect and $t_{k(ij)}$ is the treatment effect.

This is exactly the same form of model that was required for the Latin square and Latin rectangle designs, and for the incomplete row-and-column designs of Section 8.3. Essentially, we assume that there may be consistent differences between periods but assume also that, after allowing for these differences, the results are as they would be if the periods occurred in a quite different order.

If those simple model assumptions are felt to be appropriate, then the design problem does not differ from the general row-and-column design situation considered earlier in this chapter. The number of periods will not usually be large and the complete Latin square will not often provide solutions to practical problems, except that multiple 4×4 Latin squares are used quite frequently and effectively for the comparison of four diets or treatment regimes. The most common situation in dietary trials is probably that with between six and eight treatments using three or four periods, and a number of subjects sufficient to give about four replications per treatment. The general principles for general row-and-column designs developed during Section 8.4 will enable any particular experimental situation to be solved. In seeking to achieve near-balance, it is usually advisable to first consider the allocation of treatments to subjects and then to order the treatments within each subject, to achieve improved balance from the distribution of treatments between periods.

Example 8.2   At this point we consider again the example on pain relief discussed in Chapters 4 and 6. In this experiment patients were given drugs on request. Three drugs were compared in the experiment and each patient received two different drugs. The allocation of drugs to patients at the first request was random within overall restrictions of approximate equality of drug replication. The allocation of the drug at the second request, the second drug being not the same as the first drug, was similarly random. The full results of the trial are shown in Table 8.8.

In Chapter 6 the results from the first period only were analysed. The analysis of variance for a model including patient, period and drug effects, is as shown in Table 8.9.

The treatment effect estimates are

$$
\begin{array}{lll}
T_1 - T_2 & +3.42 & \text{SE } 1.05, \\
T_1 - T_3 & +2.04 & \text{SE } 0.99, \\
T_2 - T_3 & -1.38 & \text{SE } 0.96.
\end{array}
$$

The advantage of drug 1 is clear, though it is interesting to note that the results both in terms of the size of the effects and the precision of estimates are little changed from the previous analysis of the first administration data only.

Table 8.8. *Hours of relief from pain for each patient after each drug application*

Period	Drug	Patients							
		1	2	3	4	5	6	7	8
1	T_1	2	6	4	13	5	8	4	
2	T_2	10	8	4	0	5	12	4	
1	T_2	2	0	3	3	0			
2	T_1	8	8	14	11	6			
1	T_1	6	7	6	8	12	4	4	
2	T_3	6	3	0	11	13	13	14	
1	T_3	6	4	4	0	1	8	2	8
2	T_1	14	4	13	9	6	12	6	12
1	T_3	12	1	5	2	1	4	6	5
2	T_2	11	7	12	3	7	5	6	3
1	T_2	0	8	1	4	2	2	1	3
2	T_3	8	7	10	3	12	0	12	5

Table 8.9. *Analysis of variance for pain relief trial*

	SS	df	MS
Patients (ignoring drugs)	607	42	
Periods (ignoring drugs)	262	1	
Drugs (allowing for patients and periods)	116	2	58.0
Residual	432	40	10.8
Total	1417	85	

8.7 Cross-over designs for residual or interaction effects

Returning to the more complex problem where residual or carry-over effects are anticipated, we first consider designs in which all possible combinations and orders are included. Such a design will be appropriate if either:

(i) the number of treatments is so small and the replication is so large that there really is no case for not acquiring the more complete information provided by using all possible combinations, or

(ii) the possible interactions between treatments and orders, and between treatments in different periods, are sufficiently important, or in doubt, that it is necessary to examine all combinations in order to understand the biological situation.

In medical experiments, there has been considerable investigation of the two-treatment, two-period cross-over, notably by Hills and Armitage. If more than two treatments are to be compared, then the subexperiment for each pair of treatments can be analysed separately. The form of analysis given here is based on that described by Hills and Armitage (1979).

Consider the two-treatment, two-period cross-over design with n_1 patients receiving treatment 1 in period 1 followed by treatment 2 in period 2, and n_2 patients receiving treatment 2 in period 1 followed by treatment 1 in period 2. Instead of approaching an analysis through the analysis of variance, the sums and differences of observations for each patient are used to obtain a more simply interpretable analysis. Consider the difference between the observations for the first and second periods for each patient. If it is assumed that there are no residual effects, so that the treatment effects are the same in both periods, then for patients receiving treatment 1 first

$$d_{1i} = (t_1 - t_2) + (p_1 - p_2) + \varepsilon_{1i},$$

where t_1, t_2 are the treatment effects, p_1, p_2 the period effects, and ε_{1i} represent the error variation. Similarly,

$$d_{2i} = (t_2 - t_1) + (p_1 - p_2) + \varepsilon_{2i}.$$

The two sample means $\bar{d}_1, \bar{d}_2$ provide estimates of $(t_1 - t_2) + (p_1 - p_2)$ and $(t_2 - t_1) + (p_1 - p_2)$, respectively. Variances of the sample means are obtained from the sample variances of differences. The difference $\bar{d}_1 - \bar{d}_2$ with variance $s_1^2/n_1 + s_2^2/n_2$ provides an estimate of $2(t_1 - t_2)$, the period effect being eliminated in the subtraction.

Now suppose that there are residual effects of the first period treatment on the second period observation, or equivalently that the difference between treatment effects is modified in the second period. The analysis of differences cannot now provide the required information about the simple treatment effect, $t_1 - t_2$. Suppose that in the second period the treatment effects are altered from t_1, t_2 to t_1', t_2'. Then

$$d_{1i} = (t_1 - t_2') + (p_1 - p_2) + \varepsilon_{1i},$$
$$d_{2i} = (t_2 - t_1') + (p_1 - p_2) + \varepsilon_{2i}$$

and

$$\begin{aligned}
\bar{d}_1 - \bar{d}_2 &= (t_1 - t_2') - (t_2 - t_1') + \bar{\varepsilon}_1 - \bar{\varepsilon}_2 \\
&= (t_1 - t_2) + (t_1' - t_2') + \bar{\varepsilon}_1 - \bar{\varepsilon}_2.
\end{aligned}$$

Thus the difference between the means of the within-patient differences estimates the sum of the first period treatment effect and the second period treatment effect. The analysis of differences cannot yield any information about the difference between $(t_1 - t_2)$ and $(t_1' - t_2')$. The only source of information about this interaction difference is contained in comparisons between the sums of observations for the different patients.

The model for the sum of observations for each patient includes the sum of the two period effects, and the sum of the treatment effects applied in the two periods:

$$s_{1i} = 2\mu + (t_1 + t_2') + (p_1 + p_2) + \eta_{1i},$$
$$s_{2i} = 2\mu + (t_2 + t_1') + (p_1 + p_2) + \eta_{2i},$$

where the η error terms represent the between-patient variation and, if the reasons for using cross-over designs to improve precision are valid, $\text{Var}(\eta)$ will be much larger than $\text{Var}(\varepsilon)$.

The difference between the sample means of the sums provides information about the difference between $(t_1 - t_2)$ and $(t'_1 - t'_2)$:

$$\bar{s}_1 - \bar{s}_2 = (t_1 + t'_2) - (t_2 + t'_1) + \bar{\eta}_1 - \bar{\eta}_2$$
$$= (t_1 - t_2) - (t'_1 - t'_2) + \bar{\eta}_1 - \bar{\eta}_2.$$

Hence, calculating $\bar{s}_1 - \bar{s}_2$, and the variance of $\bar{s}_1 - \bar{s}_2$ obtainable directly from the sample variances for s_1 and for s_2, we may test the difference between $(t_1 - t_2)$ and $(t'_1 - t'_2)$,

$$(\bar{s}_1 - \bar{s}_2)/\{\mathrm{Var}(\bar{s}_1 - \bar{s}_2)\}^{1/2}.$$

If the test shows negligible evidence of a difference, then the estimate of $(t_1 - t_2)$ from $\bar{d}_1 - \bar{d}_2$ can be used.

The problems with this apparently simple approach are two-fold. Whereas the estimation of $(t_1 - t_2)$, assuming no interaction or residual effects, is in terms of within-patient variation (differences) and is likely to be relatively precise, the test of the existence of interaction or residual effects is in terms of between-patient variation (sums), and is likely to be relatively imprecise. But the interaction must be tested before the estimate of the treatment difference can be interpreted. It is quite possible that the lack of precision for the test of interaction would lead to a result that a difference between $(t_1 - t_2)$ and $(t'_1 - t'_2)$ bigger than the average of the two values was not significantly different from zero.

The second problem arises if it is decided that the interaction or residual effects are non-negligible. We then have a choice of estimating treatment differences from within-patient differences which are precise but which provide an estimate of $(t_1 - t_2) + (t'_1 - t'_2)$, or of using only the information from the first period, which will be imprecise (because it relies on between-patient information), but which provides an estimate of $(t_1 - t_2)$. As always in statistics, then, we must choose between greater precision with a possibly less relevant estimate, or lesser precision with a more relevant estimate. And, of course, we have returned to the philosophical problem which is discussed at the start of Section 8.5.

Example 8.3 (continued) The methods for the two-treatment, two-period cross-over are applied to each pair of drugs in turn.

Drugs 1 and 2

T_1 before T_2, $n = 7$	difference (period 1 − period 2)	sum
mean	−0.14	12.14
variance	42.14	17.48
T_2 before T_1, $n = 5$		
mean	−7.80	11.00
variance	4.70	20.00

To test whether the differences between the drugs are consistent between the two periods we use the difference between the two sum means:

$$(12.14 - 11.00)/(17.48/7 + 20.00/5)^{1/2} = 0.45.$$

There is clearly no evidence of an interaction. We therefore estimate the difference between the effects of the drugs from the mean difference:

$$\hat{t}_1 - \hat{t}_2 = \tfrac{1}{2}\{-0.14 - (-7.80)\} = 3.83,$$

and the SE of this estimate is

$$0.5(42.14/7 + 4.20/5)^{1/2} = 1.31.$$

Drugs 1 and 3

T_1 before T_3, $n = 7$	difference	sum
mean	−3.00	15.29
variance	30.00	40.57
T_3 before T_1, $n = 8$		
mean	−5.38	13.62
variance	9.70	37.41

To test for an interaction we again use the difference between the two sum means:

$$(15.29 − 13.62)/(40.57/7 + 37.41/8)^{1/2} = 0.52.$$

Again there is negligible evidence for an interaction:

$$\hat{t}_1 − \hat{t}_3 = \tfrac{1}{2}\{−3.00 − (−5.38)\} = 1.19,$$

with SE

$$0.5(30.00/7 + 9.70/8)^{1/2} = 1.17.$$

Drugs 2 and 3

T_2 before T_3	difference	sum
mean	−4.50	9.75
variance	30.57	18.79
T_3 before T_2		
mean	−2.25	11.25
variance	12.50	35.36

To test for an interaction we again use the difference between the two sum means:

$$(11.25 − 9.75)/(18.79/8 + 35.36/8)^{1/2} = 0.58.$$

The evidence for an interaction is again negligible:

$$\hat{t}_2 − \hat{t}_3 = \tfrac{1}{2}\{−4.50 − (−2.25)\} = −1.12,$$

with SE

$$0.5(30.57/8 + 12.50/8)^{1/2} = 1.16.$$

The three sets of data provide a negligible level of evidence for any interaction with re-markable consistency. They also provide three reasonably consistent estimates of differences between the drug effects

$$\hat{t}_1 − \hat{t}_2 = 3.83 \quad \text{SE } 1.31,$$
$$\hat{t}_1 − \hat{t}_3 = 1.19 \quad \text{SE } 1.17,$$
$$\hat{t}_2 − \hat{t}_3 = −1.12 \quad \text{SE } 1.16.$$

Since these estimates are independent they are inevitably not exactly consistent, the dif-ference between t_1 and t_2 being greater than the sum of the differences for $(t_1 − t_3)$ and

(a) Subject

		1	2	3	4
Period	1	A	B	C	D
	2	D	C	A	B
	3	B	A	D	C
	4	C	D	B	A

(b) Subject

		1	2	3	4
Period	1	A	B	C	D
	2	D	C	A	B
	3	C	D	B	A
	4	B	A	D	C

Figure 8.21 Latin square designs (a) not suitable, (b) suitable for residual effects.

$(t_3 - t_2)$. A set of consistent estimates can be calculated using simple least squares theory. If the variance estimates are pooled to estimate an assumed common variance, then the estimates obtained from the combined estimation procedure will in fact be those obtained from the within-patient analysis in the previous section:

$$\hat{t}_1 - \hat{t}_2 = 3.42,$$
$$\hat{t}_1 - \hat{t}_3 = 2.04,$$
$$\hat{t}_2 - \hat{t}_3 = -1.38.$$

The example, which has been discussed at some length, shows a relatively small level of between-patient variation, and the benefits and the dilemmas of the use of cross-over designs do not materially affect the conclusions.

8.7.1 More than two periods

With only two periods it is not possible to escape from the problem that the treatment difference may differ between periods. If more than two periods are available, then we can obtain information about the treatment effects even if these effects vary between periods, provided that we assume a particular form for that interaction variation. Instead of assuming simply that the treatment effects differ between periods we assume that each observation is influenced, not only by the treatment applied in the current period, but also by the treatment applied to the same subject in the previous period, i.e. we assume residual treatment effects. With only two periods the model assuming residual effects is the same as that assuming that the treatment effects change. With more than two periods the residual effect model is simpler than assuming different treatment effects in each period.

To allow separation of treatment and residual effects using row-and-column designs, with periods as rows, the choice of appropriate row-and-column designs must be made with care. We remarked earlier that, if residual effects were of no importance, then any of the row-and-column designs considered previously were appropriate. This is not true if residual effects are important. Consider the two Latin square designs shown in Figure 8.21 for the comparison of four diets in four periods using four subjects. In (a), treatment A is always preceded by treatment C, B by D, C by B and D by A, so that the estimates of the treatment effect of A and of the residual effect of C will be indistinguishable and the other treatment and residual effects will be similarly linked. In contrast, design (b) has each treatment preceded by each other treatment exactly once. Obviously, design (b) will provide information about residual treatment effects in addition to direct treatment effects, while design (a) will not. However, the investigation of residual effects requires a more complex analysis than for the row-and-column designs considered earlier, because the observations in the first period are not subject to any residual effects of a treatment in the previous period, and consequently residual effects are not orthogonal to subject effects or treatment effects.

Consider the model and analysis for the design in Figure 8.21(b):

$$y_{ij} = \mu + s_i + p_j + t_{k(ij)} + r\,t_{l(ij)} + \varepsilon_{ij},$$

where $s_i, p_j, t_k, r\,t_l$ represent subject effects, period effects, direct treatment effects and residual treatment effects, respectively. The $rt_{l(ij)}$ term is included in the model only for $j \neq 1$, i.e. for periods 2–4. The least squares equations for the four sets of effects are not fully orthogonal, as is demonstrated below (using the usual form of constraints):

$$4\hat{\mu} + 4\hat{s}_1 - r\hat{t}_B = Y_{1.},$$
$$4\hat{\mu} + 4\hat{s}_2 - r\hat{t}_A = Y_{2.},$$
$$4\hat{\mu} + 4\hat{s}_3 - r\hat{t}_D = Y_{3.},$$
$$4\hat{\mu} + 4\hat{s}_4 - r\hat{t}_C = Y_{4.},$$
$$4\hat{\mu} + 4\hat{p}_j = Y_{.j}, j = 1, 2, 3, 4,$$
$$4\hat{\mu} + 4\hat{t}_A - r\hat{t}_A = T_A,$$
$$4\hat{\mu} + 4\hat{t}_B - r\hat{t}_B = T_B,$$
$$4\hat{\mu} + 4\hat{t}_C - r\hat{t}_C = T_C,$$
$$4\hat{\mu} + 4\hat{t}_C - r\hat{t}_C = T_C,$$
$$3\hat{\mu} + 3r\hat{t}_A - \hat{s}_2 - \hat{t}_A = R_A,$$
$$3\hat{\mu} + 3r\hat{t}_B - \hat{s}_1 - \hat{t}_B = R_B,$$
$$3\hat{\mu} + 3r\hat{t}_C - \hat{s}_4 - \hat{t}_C = R_C,$$
$$3\hat{\mu} + 3r\hat{t}_D - \hat{s}_3 - \hat{t}_D = R_D.$$

The non-orthogonality is not complex, the least squares equations being linked in groups of three, for example the three equations relating to $Y_{1.}$, T_B and R_B all involve the term $(\hat{s}_1, \hat{t}_B, r\hat{t}_B)$, but it does mean that the testing of significance for direct treatment effects, residual treatment effects and even subject effects (if it is considered useful to test them) requires different orders of fitting terms in the analysis of variance. To test the significance of residual effects, the analysis of variance is calculated in the form shown in Table 8.10. To test treatment effects, allowing for residual effects, a second analysis of variance is calculated as shown in Table 8.11. And, should it be important to assess the significance of between-subject variation, then a third ordering would be used, with subject effects fitted last.

In practice, the level of non-orthogonality for designs with four or more periods is rarely such as to cause any substantial change in the pattern of SS in the different orders of fitting, and the interpretation is not usually difficult. However, the different relevant orders of fitting should be examined to provide properly stringent tests.

Table 8.10. *Analysis of variance for testing residual effects*

Source	df
Periods	3
Subjects (ignoring residuals)	3
Treatments (ignoring residuals)	3
Residuals (eliminating S and T)	3
Error	3
Total	15

Table 8.11. *Analysis of variance for testing treatment effects, adjusting for residual effects*

Source	df
Periods	3
Subjects (ignoring residuals)	3
Residuals (eliminating S, ignoring T)	3
Treatments (eliminating R)	3
Error	3
Total	15

Designs allowing investigation of, and adjustment for, residual effects are subject to the usual considerations for row-and-column designs with additional restrictions of the type illustrated in the earlier discussion of the two 4×4 Latin square designs (Figure 8.21). If the number of treatments is equal to the number of periods, then obviously a Latin square design is appropriate, and all that is usually necessary is the avoidance of Latin squares which have each treatment always preceded by the same other treatment. The same philosophy applies when the number of treatments is one greater than the number of levels, when a Latin square omitting the last row should be used with the same restrictions. If the number of treatments is greater than the number of periods by more than 1, then the usual approach to selecting row-and-column designs will tend to produce designs with different combinations of treatments occuring in each column (so different combinations of treatments for each subject), and it should not then be difficult to impose the additional requirement that the treatments preceding a particular treatment should be as evenly distributed as possible over the set of treatments.

Exercises

8.1 Construct designs for comparing seven treatments in row-and-column designs with the following structure:
  (i) 49 units in seven rows × seven columns;
  (ii) 35 units in seven rows × five columns;

		Column							
		1	2	3	4	5	6	7	8
Row	1	×	×	×	×	×	×	×	×
	2	×	×		×	×	×	×	
	3	×	×	×			×	×	×
	4	×	×	×	×	×		×	×
	5	×		×	×	×	×		
	6	×	×		×		×	×	×
	7	×					×	×	×

Figure 8.22 Occurrence of a set of 42 units in seven rows and eight columns.

  (iii) 70 units in seven rows × ten columns;
  (iv) 48 units in six rows × eight columns;
   (v) 42 units in an incomplete seven rows × eight columns as shown in Figure 8.22.

8.2 The 16 students on an MSc course in statistics discover that they form an exceptional body. Four are English, four are Scottish, four are Irish and four are Welsh. From their undergraduate careers, there are four graduates each from the four universities Cambridge, Oxford, Reading and Edinburgh and no two of the same nationality went to the same university. In their choice of option courses in the MSc, four took three pure options (PPP), four took three applied options (AAA), four took two pure and one applied (PPA) and four took one pure and two applied (PAA), and no two with the same choice of options are graduates of the same university, or are of the same nationality. Four students intend to enter the Civil Service, four to remain in academic life, four to train to be actuaries and four to become industrial statisticians, and no two of the same intended profession took the same options or graduated at the same university or are of the same nationality.

Three of the intending academic statisticians are the English Edinburgh graduate who took AAA, the Scottish Oxford graduate who took PPP and the Irish Cambridge graduate who took PAA. The Scottish Edinburgh graduate intends entering the Civil Service, the English Cambridge graduate plans to be an industrial statistician, and the Scottish Reading graduate took AAA. What can you say about the Welshman who went to Oxford?

8.3 A design is required to compare four dietary treatments for cows, using seven cows for three periods of observation each in a 3-row × 7-column structure. It is believed that residual effects are not important. One of the treatments is a control. An initial proposal is to modify the Youden square design given below by allocating the control (C) to A, treatment T_1 to B and D, treatment T_2 to C and F and treatment T_3 to E and G:

$$
\begin{array}{ccccccc}
A & D & F & B & G & C & E \\
B & A & G & F & E & D & C \\
C & E & A & D & B & G & F.
\end{array}
$$

How much can you improve the design in the sense of achieving more even precision and smaller variances, for treatment comparisons

  (i) using three replications for C and six replications for T_1, T_2 and T_3,
 (ii) using six replications for C and five replications for T_1, T_2 and T_3?

Table 8.12. *Yields for Exercise 8.5: (a) total yields; (b) RSS*

(a)

Period	1	2	3	4	5					
	5985	6153	6212	5932	5822	Total = 30 104				
Animal	1	2	3	4	5	6	7	8	9	10
	2791	3093	3240	2889	3032	3049	3117	2688	2988	3217
Diet	C	D	E	F	G					
	5483	6097	6124	6217	6183					
Observation following diet (residual total)	C	D	E	F	G					
	5035	4808	4828	4830	4617					

(b)

Fitting	RSS	df
General mean	118 298	49
+ period effects	108 025	45
+ animal effects	52 461	36
+ residual diet effects	48 569	32
+ direct diet effects	13 186	28
omit residual diet effects	15 472	32

		Cow			
		1	2	3	4
Day	1	A	B	C	D
	2	B	D	A	C
	3	C	A	D	B
	4	D	C	B	A

Figure 8.23 Cross-over design for four milkmaids milking four cows on four days.

8.4 An experiment to investigate variation between milkmaids was designed as a cross-over design as in Figure 8.23. The model assumed is

$$y_{ij} = \mu + d_i + c_j + m_k + r_l + e_{ij},$$

where d_i = day effect, c_j = cow effect, m_k = milkmaid effect, and r_l = residual effect of previous maid. Using a least squares estimation, show that

$$40(\hat{m}_A - \hat{m}_B) = 11(M_A - M_B) + 4(R_A - R_B) + (C_4 - C_3),$$

where capital letters denote totals, i.e. M_A = total for maid A, and derive the variance of

$$(\hat{m}_A - \hat{m}_B).$$

8.5 An experiment to compare five diets for dairy cattle was designed as a cross-over design using two 5 × 5 Latin squares with five periods for each of ten animals, allowing residual effects of diets to be estimated. The five diets were a control diet (C) and a 2 × 2 factorial set of two

Table 8.13. *Totalpip measurements for Exercise 8.6*

	Group I					Group II			
	Drug A		Drug B			Drug B		Drug A	
Weeks	2	4	6	8	Weeks	2	4	6	8
Patient					Patient				
1	553	569	568	568	15	652	649	640	613
2	650	632	617	603	16	552	551	533	521
3	639	641		657	17	731	711	676	705
4	605	594	591	582	18	598	621	605	605
5	552	532	545		19	566	574	588	
6	663	627	616	618	20	600	600	605	603
7	702	722	714	727	21	725		705	753
8	655	644	637	644	22	582	589		
9		567	578	580	23	540	539	540	557
10	622	631	618	653	24	676	679		689
11	616	607	614	600	25	682		682	
12	681	653	646						
13	555	553	570	547					
14	595	600	590	598					

Gaps in the table occur where patients were not available for observation at the relevant time.

			Animal								
		1	2	3	4	5	6	7	8	9	10
Period	1	F	C	E	G	D	C	G	E	F	D
	2	G	D	F	C	E	E	D	G	C	F
	3	D	F	C	E	G	D	C	F	G	E
	4	E	G	D	F	C	G	F	D	E	C
	5	C	E	G	D	F	F	E	C	D	G

Figure 8.24 Cross-over design for five diets applied to ten animals over five periods.

different additives each on two levels, a_1b_1 (D), a_1b_2 (E), a_2b_1 (F), a_2b_2 (G). The design is displayed in Figure 8.24.

Show that the least squares estimate of the direct effect of diet C (d_C) is given by

$$240\hat{\mu} + 180\hat{d}_C = 19D_C + 5R_C + A_1 + A_8,$$

where D_C, R_C, A_i are totals for observations for diet C, for observations following diet C and for observations for animal i, respectively. Total yields (litres of milk over four weeks) are listed in Table 8.12(a) together with the residual SS obtained from fitting a sequence of models in Table 8.12(b). Construct the analysis of variance and write a full report on the results, quoting numerical values and their standard errors for the important effects.

8.6　A clinical trial was conducted to compare the effects of two drugs A and B in alleviating the painful effects of rheumatism. Patients who consented to participate in the trial were allocated at random to one of two groups I and II. Before entering the trial, the amount of swelling present in each patient's finger joints was measured by taking the total circumference of the knuckles of all ten digits, called *totalpip*. Group I was then given drug A for four weeks followed by drug

B for a further four weeks, while group II received the two drugs in the reverse order. Patients knew that their treatment was being changed after four weeks, but neither they nor the doctors running the trial knew which group they were in. The totalpip measure was taken from each patient at fortnightly intervals and the results were as in Table 8.13. Analyse the data in any way you consider appropriate and write a short report on your findings.

9

Multiple levels of information

9.0 Preliminary examples

(*a*) In an experiment to investigate the effect of training on human–computer–human inter-actions, six subjects were randomly allocated to each of four training programmes. Subjects were then paired into 12 blocks using two replicates of an unreduced balanced incomplete block design. Each pair carried out a conversation through a computer 'chat' program. In addition to several response variables measured on each subject individually, each pair was given a score by an independent observer, for the success of their interaction.

We have only a single response representing each block. Can we use this information and, if so, how? If we can, do the block totals contain useful information about the effects of treatments on other responses? Does this affect how we should design the experiment? In particular, for this response, should we have used some blocks with both subjects getting the same treatment?

(*b*) Eight feeds are to be compared for their effects on the growth of young chickens. The experiment will be carried out using 32 cages, arranged in four brooders, with each brooder having four tiers of two cages. Should the experiment be designed to ensure that each treatment appears once in each brooder and once in each tier, or should we consider the brooder×tier combinations as blocks of size 2 and choose a good design for this setup? Can we do both simultaneously?

9.1 Identifying multiple levels in data

In Section 7.3 we considered the analysis for general block–treatment designs. However, in that analysis only the information about treatments from comparisons within blocks was considered. The differences between block totals also contain information about treatment differences. Consider the two incomplete block designs shown in Figure 9.1:

(a) $t = 6, k = 5, b = 6, r = 5, \lambda = 4$ (unreduced), and
(b) $t = 10, k = 4, b = 15, r = 6, \lambda = 2$ (1/14 of unreduced).

The analysis used so far is based on the idea that if, in block I, the observation for treatment A is higher than the other observations in that block, then we tend to believe that treatment A is a higher-yielding treatment. We should, however, wish to examine whether treatment A is also relatively high yielding within each of the other blocks in which it occurs before we could rely on the supremacy of A. This is the within-block form of comparison which is made more precise through the use of small blocks. Now clearly if in design (a) the

Block

(a)	I	II	III	IV	V	VI
	A	A	A	A	A	B
	B	B	B	B	C	C
	C	C	C	D	D	D
	D	D	E	E	E	E
	E	F	F	F	F	F

Block

(b)	I	II	III	IV	V	VI	VII	VIII
	A	A	A	A	A	A	B	B
	B	B	C	D	E	H	C	D
	C	E	G	F	G	I	I	G
	D	F	H	I	J	J	J	J

	IX	X	XI	XII	XIII	XIV	XV
	B	B	C	C	C	D	D
	E	F	D	E	F	E	F
	H	G	E	F	G	G	H
	I	H	H	J	I	I	J

Figure 9.1 Incomplete block designs: (a) $t = 6, k = 5, b = 6, r = 5, \lambda = 4$, (b) $t = 10, k = 4, b = 15, r = 6, \lambda = 2$.

total of the observations from block I is markedly larger than the other five block totals we should interpret this as evidence that treatment F gives lower values, though since there is no replication (of blocks with a particular treatment missing) the evidence is slight and no measure of precision is possible. In the second design the potential for evidence is greater. If blocks II, IV, X, XII, XIII and XV gave the six highest block totals, then it would be difficult to avoid the suspicion that F yields more than the other treatments. Note that the differences between the individual values making up each block total are irrelevant to this argument. The essential idea (which we shall develop further in Chapter 10) is that the selection of six blocks out of 16 can be made in $(16!)/(6!10!) = 8008$ ways and that the occurrence of one treatment in all six blocks giving the highest totals is a rather extreme level of coincidence (or significance).

The information from variation between block totals is not limited to arguments about coincidence. By examining the set of 16 block totals and relating the differences between these totals to the sets of treatments occurring in each block we can obtain estimates of differences between treatments. These estimates may not be very precise because blocks are used deliberately to make the variation between units in each block small, and consequently the sampling variation between blocks will tend to be large. Nevertheless, the estimates of treatment differences from interblock variation do provide information additional to that obtained from the variation within blocks. Somewhat confusingly, in contrast to the interblock information, the within-block information is often referred to as 'intrablock' information, a term which will be familiar to classical scholars but not to other readers.

9.2 The use of multiple levels of information

The possibility of obtaining information about treatment effects from two levels of analysis leads to various questions:

(i) How is the interblock information obtained?
(ii) How should the within-block and interblock information about treatment differences be combined?
(iii) When is it worthwhile to obtain the interblock information?

We consider these questions using the class of BIB designs as the basis for illustration. However, we start by reconsidering the analysis of the within-block information. The principles of the analysis are those of Section 7.3 but because of the particular pattern of BIB designs there are certain simplifying features of the analysis.

9.2.1 Within-block analysis

The model for the observation for treatment j in block i is

$$y_{ij} = \mu + b_i + t_j + \varepsilon_{ij}.$$

The least squares equations are

$$bk\hat{\mu} = Y_{..}, \tag{9.1}$$

$$k\hat{\mu} + k\hat{b}_i + \sum_{j(i)} \hat{t}_j = Y_{i.}, \tag{9.2}$$

$$r\hat{\mu} + \sum_{i(j)} \hat{b}_i + r\hat{t}_j = Y_{.j}, \tag{9.3}$$

where $\sum_{j(i)}$ represents the sum for all treatments (j) which occur in block i, and, similarly, $\sum_{i(j)}$ represents the sum for all blocks (i) in which treatment j occurs. Using (9.2) to eliminate the $\hat{b}_i$ from (9.3) we obtain

$$r\hat{t}_j - r(1/k)\hat{t}_j - \lambda(1/k)\sum_{j' \neq j} \hat{t}_{j'} = Y_{.j} - \sum_{i(j)} (1/k)Y_{i.}, \tag{9.4}$$

since each treatment other than j occurs λ times in the set of blocks, $i(j)$, which include treatment j. Since the treatment effects are, as usual, constrained to sum to zero,

$$t_j = -\sum_{j' \neq j} t_{j'}.$$

Hence (9.4) can be written

$$\hat{t}_j(r - r/k + \lambda/k) = Y_{.j} - (1/k)\sum_{i(j)} Y_{i.},$$

whence

$$\hat{t}_j = kQ_j/(\lambda t),$$

Table 9.1. *Values of E for k = 2, 3, 4, 6*
and 8 and t = 4, 6, 8, 12, 16, 24 and 32

			k		
t	2	3	4	6	8
4	0.67	0.88			
6	0.60	0.80	0.90		
8	0.57	0.76	0.86	0.95	
12	0.55	0.73	0.82	0.91	0.96
16	0.53	0.71	0.80	0.89	0.93
24	0.52	0.70	0.78	0.87	0.91
32	0.52	0.69	0.77	0.86	0.90

using the relationship $\lambda(t - 1) = r(k - 1)$, where $Q_j = Y_{.j} - (1/k) \sum_{i(j)} Y_{i.}$ may be thought of as the treatment total effect relative to the block mean yields for those blocks in which the treatment occurs.

The variance of $\hat{t}_j$ and of the estimate of the difference between two treatment effects, $\hat{t}_j - \hat{t}_{j'}$, may be obtained by consideration of the variances and covariances of $Y_{.j}$, $\sum_{i(j)} Y_{i.}$ and hence Q_j:

$$\text{Var}(Q_j) = r(k - 1)\sigma^2/k,$$
$$\text{Cov}(Q_j, Q_{j'}) = -\lambda\sigma^2/k,$$
$$\text{Var}(\hat{t}_j) = \{(t - 1)/t\}k\sigma^2/(\lambda t),$$
$$\text{Var}(\hat{t}_j - \hat{t}_{j'}) = 2k\sigma^2/(\lambda t).$$

In assessing incomplete block designs, the efficiency of the design is measured by comparing the variance of the BIB design with the variance for a complete block design or unblocked design, with the same replication per treatment and assuming, which is most unrealistic, the same σ^2. From the variance for $\hat{t}_j - \hat{t}_{j'}$ the efficiency of the BIB design is

$$E = (2\sigma^2/r)/(2k\sigma^2/\lambda t) = \lambda t/rk,$$

which can be rewritten as $(1 - 1/k)/(1 - 1/t)$ whence it is clearly less than 1 since $k < t$. Note that this efficiency depends critically on k as is shown in numerical values of E given in Table 9.1.

To estimate σ^2, the RSS is derived in a simple algebraic form as follows:

$$\text{RSS} = \sum_{ij} y_{ij}^2 - \hat{\mu}Y_{..} - \sum_i \hat{b}_i Y_{i.} - \sum_j \hat{t}_j Y_{.j}.$$

Substituting for $\hat{\mu}$ and $\hat{b}_i$ from (9.1) and (9.2) and simplifying gives

$$\text{RSS} = \sum_{ij} y_{ij}^2 - (1/k) \sum_i Y_{i.}^2 - \sum_j \hat{t}_j Q_j.$$

This may be recognised as defining the analysis of variance with the last two items being SS for blocks (ignoring treatments) and treatments (allowing for blocks).

9.2.2 *Interblock analysis*

The model for block totals may be derived from the original model as

$$Y_{i.} = k\mu + \sum_{j(i)} t_j + kb_i + \sum_{j(i)} \varepsilon_{ij}.$$

Now the b_i represent, with the ε_{ij}, the variation in $Y_{i.}$ additional to that of the treatment and overall mean effects, and we define

$$\eta_i = kb_i + \sum_{j(i)} \varepsilon_{ij}$$

with variance

$$\mathrm{Var}(\eta_i) = k^2\sigma_b^2 + k\sigma^2 = k\sigma_1^2,$$

where σ_b^2 represents the variance of the population from which the experimental b_is are a sample.

The least squares equations for μ and t_j are

$$bk\hat{\mu} \qquad\qquad\qquad = Y_{..},$$
$$rk\hat{\mu} + r\hat{t}_j + \lambda \sum_{j' \neq j} t_{j'} = \sum_{i(j)} Y_{i.}.$$

Hence,

$$\hat{t}_j = \left\{ \sum_{i(j)} Y_{i.} - (r/b)Y_{..} \right\} \Big/ (r - \lambda)$$

and

$$\hat{t}_j - \hat{t}_{j'} = \left(\sum_{i(j)} Y_{i.} - \sum_{i(j')} Y_{i.} \right) \Big/ (r - \lambda),$$

whence

$$\mathrm{Var}(\hat{t}_j - \hat{t}_{j'}) = \{1/(r - \lambda)^2\} k\sigma_1^2(2r - 2\lambda) = 2k\sigma_1^2/(r - \lambda).$$

We can estimate σ_1^2 from the residual MS of the interblock analysis which is simply derived from the interblock RSS:

$$\text{Interblock RSS} = \sum_i Y_{i.}^2 - k\hat{\mu}Y_{..} - \sum_j t_j \left(\sum_{i(j)} Y_{i.} \right)$$

$$= \sum Y_{i.}^2 - Y_{..}^2/b - 1/(r - \lambda) \left\{ \left(\sum_{i(j)} Y_{i.} \right)^2 - rkY_{..}^2/b \right\}.$$

Then $k\sigma_1^2$ can be estimated from the interblock residual mean square, by dividing the interblock RSS by the residual df, $b - t$. If the total SS for the interblock analysis is made compatible with the block SS in the intrablock analysis by dividing all the terms in the expression for RSS by k, then σ_1^2 is estimated by this scaled interblock RSS divided by the residual df $(b - t)$.

9.2.3 Combination of information

The within-block and interblock analyses provide two independent sets of estimates of treatment effects. There must, inevitably, be a combined estimate, calculated from the two separate estimates of each treatment effect, which will be the best estimate. Since both sets of estimates are unbiased by the general linear model theory, any linear combination of the estimates will also be unbiased and the linear combination with the smallest variance will be the best linear unbiased estimator. This minimum variance combined estimator can be shown to require weighting each estimate inversely by its variance. Thus the best combined estimate is

$$\frac{(kQ_j/\lambda t)/(k\sigma^2/\lambda t) + \left[\left\{\sum_{i(j)} Y_{i.} - (r/b)Y_{..}/(r - \lambda)\right\} \Big/ \left\{k\sigma_1^2/(r - \lambda)\right\}\right]}{1/(k\sigma^2/\lambda t) + 1/\left\{k\sigma_1^2/(r - \lambda)\right\}}, \tag{9.5}$$

which would be somewhat simplified for calculation. The variance of this combined estimate is

$$\left[1/(k\sigma^2/\lambda t) + 1/\left\{k\sigma_1^2/(r - \lambda)\right\}\right]^{-1}. \tag{9.6}$$

There is one further level of complexity which must usually be added to this combined use of within-block and interblock analyses. In the formula for the combined estimate, values of σ^2 and σ_1^2 are required to calculate the actual weights. In almost all experiments, these will have to be replaced by estimates. The estimates for σ^2 and σ_1^2 may be obtained from the residual mean squares for the two analyses but the df for the interblock residual may often be rather few with the result that the second weight is poorly determined and the estimated SE of the combined estimate may also be poorly estimated. It is possible to obtain a better estimate of σ_1^2 and this is achieved by considering two alternative ways of constructing the analysis of variance. The two forms of analysis of variance are:

(1)	(2)
block SS (ignoring treatments)	treatment SS (ignoring blocks)
treatment SS (allowing for blocks)	block SS (allowing for treatments)
RSS	RSS

The RSS are, of course, identical, and therefore

$$\text{block SS (ign. T)} + \text{treatment SS (all. B)}$$
$$= \text{treatment SS (ign. B)} + \text{block SS (all. T)}.$$

By evaluating the expected values of the first three SSs it can be shown that

$$\text{E (block SS (allowing for treatments))}$$
$$= (kb - t)\sigma_1^2/k + (t - k)\sigma^2/k.$$

Hence, estimating σ^2 from the within-block residual MS, an estimate of σ_1^2 can be obtained from the block SS (allowing for treatments) based on $(b - 1)$ df.

Table 9.2. *Analysis procedure for balanced incomplete block design*

SS	df	SS	df
(5) Interblock treatment SS	$t - 1$		
(6) Interblock error SS	$b - t$		
(2) Block SS (ignoring treatments)	$b - 1$	(7) Treatment SS (ignoring blocks)	$t - 1$
(3) Treatment SS (allowing for blocks)	$t - 1$	(8) Blocks SS (allowing for treatments)	$b - 1$
(4) Residual	$bk - t - b + 1$	(4) Residual	$bk - t - b + 1$
(1) Total	$bk - 1$		

The complete procedure for the possible analysis of variance structure for a BIB design is shown in Table 9.2. The SSs are calculated in the order indicated by the numerical labelling (1)–(8).

The logic of the calculation and interpretation is as follows.

(A) The principal form of information about treatment effects is obtained by eliminating the effects of blocks and estimating the treatment differences from the within-block comparisons, the precision of these estimates being determined by the residual MS (stages (1), (2), (3) and (4)).

(B) There is additional information on treatments from the differences between block totals and a second set of independent estimates of the treatment effects may be obtained from the interblock analysis with precision determined by the interblock error MS (stages (5) and (6)).

(C) If the precision (df) of the SE for the interblock estimates is poor, then the second division of the fitting SS for the initial analysis may be used to get better estimates of the SE of interblock estimates and of the consequent combined estimates (stages (7) and (8)).

Example 9.1   To illustrate the combination of estimates we reconsider the data of Example 7.10. First we must obtain the interblock estimates from the block (cage) totals (see Table 9.3; species numbering as in Example 7.10). The two sets of estimates are shown in Table 9.4. Note that the two sets of estimates are similar only in a fairly broad sense. Also note that the choice of treatment 6 (Romanesco) as an origin for the set of differences is arbitrary and does not affect the results.

The SEs of the two sets of estimates of differences are calculated from the variance formulae, using the estimate of σ_1^2 based on the block SS allowing for treatments, and are

$$\text{within-block, } 1.95;$$

$$\text{interblock, } 9.30.$$

It is immediately clear that the between-block estimates are much less precise than the within-block estimates. The combined estimates, weighted by the reciprocals of the variances, are shown in Table 9.5. Clearly in this case, as in many others, the use of interblock information adds very little.

Table 9.3. *Cage totals and treatments present in experiment on cabbage root flies*

Cage	Species present
38.7	1, 2, 3
47.6	1, 2, 4
34.1	1, 3, 5
44.9	1, 4, 6
36.6	1, 5, 6
37.5	2, 3, 6
48.8	2, 4, 5
20.8	2, 5, 6
25.2	3, 4, 5
10.2	3, 4, 6

Table 9.4. *Within-block and interblock estimates for cabbage root fly experiment*

Treatment difference	Within-block	Interblock
1 − 6	26.2	17.3
2 − 6	8.7	14.5
3 − 6	6.1	−1.4
4 − 6	1.8	8.9
5 − 6	2.9	6.2

Table 9.5. *Combined estimates for cabbage root fly experiment*

Treatment difference	Combined estimate
1 − 6	25.8
2 − 6	8.9
3 − 6	5.8
4 − 6	2.1
5 − 6	3.0

SE = 1.91

In row-and-column designs, we can also obtain information about treatments from comparisons between row totals or between column totals. For many row-and-column designs there are insufficient rows (or columns) to provide useful information about treatment differences in this way. However, in some cases the information can be useful.

Example 9.2 We return to the pain relief trial of Example 8.2. Here there are 43 patients (columns) to provide plentiful information.

Table 9.6. *Interpatient analysis of variance for pain relief trial*

	SS	df	MS
Between drugs	61	2	30.5
Residual	546	40	13.6
Total	607	42	

The interpatient analysis gives the results shown in Table 9.6. The interpatient estimates of differences of drug effects are:

$$T_1 - T_2 \quad +3.90 \quad \text{SE } 1.33,$$
$$T_1 - T_3 \quad +1.17 \quad \text{SE } 1.41,$$
$$T_2 - T_3 \quad -2.73 \quad \text{SE } 1.43.$$

The pattern is the same as for the results from the patient × period analysis. The SEs are not very much larger than those for the within-patient analysis reflecting the possibly surprising fact that the between-patient variation is not much larger than the within-patient residual variation. If the results of the two analyses are combined using reciprocal weighting by variances, the combined estimates and their SEs are:

$$T_1 - T_2 \quad +3.60 \quad \text{SE } 0.82,$$
$$T_1 - T_3 \quad +1.75 \quad \text{SE } 0.81,$$
$$T_2 - T_3 \quad -1.80 \quad \text{SE } 0.79.$$

Note that because the relative sizes of the SEs vary for the three differences the three combined estimates of differences are not exactly consistent. This could be corrected by a least squares analysis.

Is it always necessary or useful to calculate the full analysis of variance with both sets of estimates? In practice often only the within-block analysis is calculated and it is proper to consider when we should additionally attempt to use the interblock information.

The principle of using incomplete blocks stems from a recognition that σ^2 should be reduced by using smaller blocks. As the selection of blocks becomes more successful the ratio σ^2/σ_1^2 will become smaller and consequently the contribution of the interblock information will become less. Thus the more successful we are in making the within-block analysis efficient the less benefit will accrue from using the interblock information.

Another way of looking at this situation is to consider the use of an incomplete block design as splitting information about treatments into two components. By the allocation of treatments to blocks we design to maximise the proportion of the information in the within-block analysis. By choosing the blocks efficiently the precision of that within-block proportion of information will be made as much greater than the precision of the inter-block information as is possible.

It therefore follows that we should expect to use the interblock information when the within-block proportion of information is not very large, or the gain from using small blocks is not very substantial. The within-block proportion of information is measured by E and is dependent on the ratio k/t. Even when E is not large (say less than 0.8) it will be appropriate to use the interblock information only when the ratio σ_1^2/σ^2 is not large. Overall the relative information in the two parts of the analysis is in the ratio

within-block $\quad k\sigma_1^2/(r-\lambda) : k\sigma^2/\lambda t \quad$ interblock

which can be rewritten

$$\sigma_1^2 E : \sigma^2(1-E).$$

A further consideration is that the use of the combined estimate involves SEs based on the combination of two estimates of variances and consequently the inferences about the combined estimates cannot be simply based on the t-distribution.

Overall it seems reasonable to recommend that the conclusions from BIB designs be based on the within-block analysis only, except where E is less than 0.8 and σ_1^2/σ^2 is less than 5, noting that these conditions imply that at least 4% of the total information is contained in the interblock analysis. In practice this means that the interblock analysis would be used if small block sizes of two, three or four units are used, with the blocking resulting in considerably less reduction in σ^2 than would be hoped. This suggests that the usual practical procedure of ignoring interblock information is not unreasonable.

For designs other than BIB designs, the recovery of inter-block information can be both more useful, especially for some complex treatment structures described in Part III, and more difficult. Although the steps in the analysis described in Table 9.2 can be followed, with unbalanced block designs the amount of interblock information will be different for different treatment comparisons and it is often the case that there is no interblock information on several comparisons. This means that the df in steps (5) – (8) of that analysis must be modified. In other cases, the interblock error SS can have 0 df and so cannot be estimated. In this case, although interblock information exists, there is no good general way to make use of it to obtain a useful combined analysis. The next two sections deal with more automatic methods of analysis which rely on the assumption of normally distributed errors.

9.3 Random effects and mixed models

In Chapter 7, the model for a general block design was given, with

$$y_{ijk} = \mu + b_i + t_j + \epsilon_{ijk} \tag{9.7}$$

representing the response from the kth observation on treatment j in block i. Initially b_i was treated like any other parameter for the purposes of least squares estimation. Doing so defines the within-block analysis. In the interblock analysis described in Section 9.2, we have implicitly treated the b_i as if they are random variables, i.e. they are sampled from some population. We now consider making distributional assumptions about these *random effects*, usually that they are independent and identically normally distributed.

First we should consider the practical question of whether it is appropriate to treat the block effects as if they are a random sample from a normally distributed population. The most obvious justification for such an assumption is that the blocks are in effect a sample from a population of possible blocks. In an animal experiment, the litters chosen for the experiment and used as blocks in the design might have been a random sample from a large population of litters. Much more often, no formal random sampling will have taken place, but the litters used will have arisen essentially by chance in a way that can reasonably be modelled as being random.

Slightly more problematic is the situation which arises commonly in industrial and laboratory experiments in which the blocks represent days, or shifts, on which the experiment is run. A series of consecutive days cannot be considered as a random, or even representative, sample of future days of production. However, if the process is well enough understood that the experimenters know that there is no additional variation over long periods of time beyond that between days, then it might be acceptable to model the block effects as coming from a normal distribution.

In other situations, such as with spatial contiguous blocks in an agricultural field experiment, or with blocks defined by age groups in a clinical trial, there is no sense in which the blocks can be considered as a sample from a population. However, there is another possible justification for assuming normally distributed random effects. In developing the interblock analysis, initially in Section 9.1 and then in detail in Section 9.2, we appealed to the probability of all blocks with a particular treatment giving greater responses than all other blocks. The argument was based on how unlikely a particular pattern of responses was under randomisation. This will be discussed in more detail in Chapter 10, but if the sets of treatments within blocks $1, \ldots, b$ are randomly allocated to the b physical blocks, then the particular b_i appearing in model (9.7) is a random outcome. Using arguments similar to those in Chapter 10, we can show that these have expectation zero and constant variance. It can also be shown that asymptotically, as the number of blocks becomes large, they become uncorrelated and, by the Central Limit Theorem, they converge to a normal distribution.

Under the assumption of normality, model (9.7) becomes a *linear mixed model*, which, in order to distinguish it from the usual model, we will write as

$$y_{ijk} = \mu + t_j + \delta_i + \epsilon_{ijk}, \tag{9.8}$$

where $\delta_i \sim N(0, \sigma_b^2)$, $\epsilon_{ijk} \sim N(0, \sigma^2)$ and all random variables are independent. The existence of both fixed effects (in addition to the overall mean μ), i.e. the terms t_j, and random effects (in addition to the errors ϵ_{ijk}), i.e. the terms δ_i, is what marks it out as a mixed model. If only one of these two terms were in the model, it would be a fixed effects model or a random effects model respectively. Note also that, by convention, we write all fixed effects first and then all random effects, so that the block effects appear in a different place in (9.8) from where they did in (9.7).

The analysis of data from all but the simplest designs using mixed models has become increasingly popular over the last few decades, to the extent that they are now often used routinely. The preceding discussion should give warning that this might not be very wise. The justifications for assuming that the block effects come from a normal distribution are not convincing and at the very least should be treated with caution. In particular, the analyses described in this section sometimes appear to give much more information than the simpler analyses described in the previous section. It should be noted that this gain in information comes at the cost of the additional assumptions and is heavily dependent on them. On the other hand, as long as no major departures from normality occur and the number of blocks is large, mixed models give a simple, unified, approach to analysing data from even the most complex designs.

9.4 Analysis of multiple level data using REML

Given the distributional assumptions of model (9.8), it might seem natural to estimate the parameters by the method of maximum likelihood. Although this is straightforward using widely available numerical optimisation routines, it is not usually recommended. One reason is that, if it is applied to simple orthogonal designs, such as completely randomised or randomised complete block designs, although we get the usual estimates of treatment comparisons, we do not get the usual estimates of σ^2. In fact, rather than $s^2 = S_r/(n - p)$, as in Chapter 4, the maximum likelihood estimator of σ^2 is $\hat{\sigma}^2 = S_r/n$. Since s^2 is an unbiased estimator of σ^2, it is clear that $\hat{\sigma}^2$ is negatively biased, i.e. it will tend to underestimate the variance, and therefore inferences based on maximum likelihood will tend to be overoptimistic. Note that, of course, the maximum likelihood estimator is asymptotically unbiased, although in many experiments $n - p$ will not be large enough for that to reassure us.

Although unbiasedness is the simplest justification for preferring to use s^2 as our estimator of σ^2, a deeper reason arises in the the proof of the theorem in Appendix 4.A6. There, we used a transformation of the model, $\mathbf{z} = \mathbf{H}\boldsymbol{\theta} + \boldsymbol{\eta}$, to show that S_r could be written as the sum of $n - p$ independent squared $N(0, \sigma^2)$ variables, where

$$\mathbf{H} = \mathbf{LA} = \begin{bmatrix} \mathbf{T} \\ \mathbf{0} \end{bmatrix}$$

and $\mathbf{z} = \mathbf{Ly}$. This transformation separates the information in the data which contributes towards estimating the fixed parameters $\boldsymbol{\theta}$ from the residual information which contains information only about σ^2. If we assume normality and maximise the likelihood of this residual part of the data, we obtain the REML estimator of σ^2, which turns out to be s^2. Maximising the other part of the likelihood leads to the usual least squares estimators of the fixed effects parameters.

This idea can be extended to models with more than one random effect. An orthogonal matrix $\mathbf{L}$ is chosen so that $\mathbf{z} = \mathbf{Ly}$ has the information which contributes to estimating the fixed effects separated from that which estimates the random effects. The likelihood of the residual part of the data is maximised to obtain the REML estimates of σ_b^2 and σ^2 (or more generally all variance components corresponding to random effects). In the case of orthogonal designs, the REML estimates are identical to the least squares estimates. For general designs, they can be shown to be much less biased than maximum likelihood estimators and generally to have good properties, given the assumption of normality.

REML only gives estimates of the random effects and their variance components. The remainder of the likelihood is conditioned on these and so cannot be maximised directly. However, maximising this conditional likelihood, given values of the variance components, can be shown to give identical estimates to generalised least squares estimation, which for a BIB design is the combined estimate (9.5). Hence in practice, as with the method described in Section 9.2, we usually plug in the REML estimates of the variance components and obtain the empirical generalised least squares (GLS) estimates at these values.

REML-GLS analysis has the great benefit that, once the design structure has been identified, any design can be analysed using standard statistical software. The easy availability of such a general method of analysis has removed most of the remaining doubts experimenters had about using non-orthogonal designs and allows them to concentrate on choosing efficient designs. While the overall effect of this is undoubtedly beneficial, there are dangers in relying thoughtlessly on REML-GLS analysis in all circumstances.

The first point to note is that, although the estimators of the treatment effects are unbiased and the estimators of the variance components are almost unbiased, the same is not true for the estimated SEs of the treatment effects. These are obtained by plugging in the REML estimates of the variance components, for example into (9.6). However, the variance (9.6) takes no account of the uncertainty in estimating the variance components and so is an underestimate. Hence, although we rejected maximum likelihood because it underestimated the SEs, our preferred method REML-GLS also underestimates them, albeit to a lesser extent.

A much more serious problem arises when the information on some treatment comparisons of interest appears entirely or mostly in the interblock analysis, but the number of blocks is not very large. We will see many examples of designs like this in Part III. If there are not many blocks, it might be difficult, or even impossible, to estimate the interblock variance component σ_b^2. In such circumstances its REML estimator will usually turn out to be zero. Then, if we use this to estimate the SEs of the relevant treatment comparisons, the estimates will turn out to be those from the analysis which ignores blocks. In other words, if we have no information to estimate a particular SE, REML-GLS analysis reports an estimate which is the smallest it could possibly be. All packages we have seen simply report this value with no warning that something might be wrong.

In practice, we would recommend always looking at the within-block analysis in conjunction with the REML-GLS analysis. Any major differences between them should be taken as an indication that we might be relying on unreasonable assumptions in using the REML-GLS analysis. If some effect is not estimable from the within-block analysis, its estimate in the REML-GLS analysis should usually have a considerably larger SE than other treatment comparisons. If it does not, this can be taken as an indication that the REML-GLS analysis is not reliable with respect to that comparison. In such circumstances, the only reasonable approach, beyond sticking to the within-block analysis, is to carry out a fully Bayesian analysis, but this is beyond the scope of this book.

A final point to consider is what are the design implications of the above discussion? We already noted in Section 9.2 that the experiment should be designed to make the within-block analysis as informative as possible. If, however, practical restrictions mean that some treatment comparison of interest can only be estimated at the interblock level, then we must ensure that there are enough blocks to make this informative. This can be difficult to achieve in practice, because the same restrictions which force us to estimate this effect at the interblock level often make it difficult for us to have many blocks. We will return to this issue in Chapter 18. Sometimes, we might have to simply accept that we will not obtain much information on this effect in the current experiment. If we limit our ambitions to looking for an indication of a very large effect, we can still obtain useful information.

9.5 Multiple blocking systems

So far in this chapter, we have discussed only designs with a single set of blocks. However, the ideas apply equally well to row-and-column designs and more complicated structures with multiple blocking systems. Information on some treatment effects might appear at several

Table 9.7. *Trojan design for the chick experiment*

Tier	Brooder							
	I		II		III		IV	
I	A	E	B	F	C	G	D	H
II	C	F	D	E	A	H	B	G
III	D	G	C	H	B	E	A	F
IV	B	H	A	G	D	F	C	E

levels, so that we can combine the usual estimates, which are from between-row-and-column combinations, with interrow estimates and intercolumn estimates. Again, subject to there being no problems with inestimable variance components, mixed models provide a simple way of obtaining a combined analysis.

For a simple row-and-column design, with just one experimental unit in each row-and-column combination, a suitable mixed model will be

$$y_{ij(k)} = \mu + t_k + \gamma_i + \delta_j + \epsilon_{ij(k)},$$

where $\gamma_i \sim N(0, \sigma_r^2)$ represents the random row effect, $\delta_j \sim N(0, \sigma_c^2)$ represents the random column effect, $\epsilon_{ij(k)} \sim N(0, \sigma^2)$ is the error in the unit defined by row i and column j and all random variables are independent.

The more blocking systems there are, the less likely it is that we will be able to have orthogonal, or near-orthogonal, blocking for all of them and, therefore, the more likely it is that the REML-GLS analysis will estimate some variance components to be zero. As always, caution is needed in interpreting the results of such analyses. If there is any doubt, then it is safer to rely on only the analysis from between row–column combinations.

There might be some situations in which we expect one of the blocking systems to have very little effect where, for example, it is included mainly for the purposes of management of the experiment. In such cases, it might be more important to ensure that the design is efficient with respect to the other blocking factor and to find the best possible design subject to this constraint. As always, however, this is a minor issue and seems less important than getting a good design overall. We also need to recognise that the recovery of the interblock information is likely to be overoptimistic.

Finally, we reconsider design questions in complex block structures, such as the semi-Latin squares described in Chapter 8. We can do this in the context of motivating example (b) from Section 9.0. In this experiment, there are eight treatments, each replicated four times. The experimental units are 32 cages, arranged in four rows by four columns, with two cages in each row-and-column combination.

From Chapter 8, a sensible design might be a Trojan square, as in Table 9.7. This design has a complete replicate in each row (tier) and a complete replicate in each column (brooder) and, subject to these restrictions, is clearly the best design available. However, from the discussion in this chapter, this seems like the wrong way to approach the problem. If we call row-and-column combinations 'blocks', most of the information for comparing treatments comes from the within-block analysis. Therefore a better approach might be to find a good design in blocks of size 2 and then to try to arrange the blocks in rows and columns to minimize the lack of orthogonality.

It turns out that, for this experiment, if we assume that all pairwise comparisons are of equal interest, then the block design implied in Table 9.7 is still the best. However, if the eight treatments are combinations of levels of three factors with two levels each or a two-level and a four-level factor, then better designs can be found with respect to the blocks of size 2. In general, it is not known when the optimal block design is also the optimal generalised row-and-column design in such structures. This is a topic which requires further research.

In practice, the best advice that can be given is to consider designs built in each way and to study their properties in detail. Often the best generalised row-and-column design will be optimal, or very close to optimal for the implied block structure. In cases where it is not, experimenters have to consider carefully whether they expect substantial differences between blocks beyond that implied by the differences between rows and columns. Very often the answer will be yes: why should we expect row and column effects to be additive? In such cases, we should almost always stick to the best available block design and then try to arrange it as well as possible in rows and columns. This, however, goes against the advice given in much of the statistical literature.

10

Randomisation

10.1 What is the population?

The fundamental results on which most inferences for data from experiments are based were described and proved in Chapter 4. All these results are based on the assumption that the experimental units are a sample from a normally distributed population. This assumption is made for many other sampling situations, and is frequently accepted as being a reasonable approximation. Transformations of the observed variable may be necessary to ensure that the assumptions of normality and homogeneity of variance are satisfied. However, for most analyses of sample data from observational studies, it is possible to envisage a population of which the sample may be realistically accepted as representative. In designed experiments, the very careful control which is exercised often makes it difficult to identify a population for which the sample is relevant.

In field crop experiments, small plots are marked out and treated with great care so that the plots may be regarded as homogeneous, and the treatments may be compared as precisely as possible. In some animal nutrition experiments, animals are surgically treated so that more detailed observations can be made. In psychological experiments, subjects are subjected to stresses in controlled situations, and are required to complete specified tasks, often in very limited times. In industrial experiments, machinery is operated under each set of conditions for a short period of time. Often the machinery will be of the same general type as, but not identical to, those used in production.

Two questions should be asked about the results from small-scale experiments. There is the general question of whether the conclusions from such rigidly controlled experiments have any validity for future practical situations. This is a difficult question, which extends beyond statistics and the scope of this book, though some examples where the relevance of experimentation seems particularly dubious are discussed. The importance of including a wide range of factors in an experiment to provide greater validity for conclusions has already been discussed in Chapter 3, and will recur in Part III.

The second question is a rather more technical one. When the experimental units have been carefully selected and monitored, is there any sense in which they are a sample from a population? And if the assumption of the existence of a population from which the sample is drawn is untenable, how does this affect the inferences from the data? A solution to this difficulty is the device of randomisation, and in this chapter we discuss the full statistical purpose and theory of randomisation, which is a much more important idea than that suggested in Chapter 2 of simply 'giving each treatment the same chance of allocation to each unit'.

Before developing the formal ideas of randomisation, consider possible inferences from a sample set of data. Suppose eight students, Alan, Carol, Deborah, James, Mariam, Nigel, Sally and Thomas, take a test and get the following marks:

Alan	14
Carol	15
Deborah	15
James	13
Mariam	16
Nigel	12
Sally	19
Thomas	10.

Someone looking at their marks might notice that all the four female students had scored higher marks than all four male students. A possible inference, without implying any particular further population of students since these were the only students taking the course leading to the test, might be that females are obviously better than males (at least at this test). The underlying argument is that it is surely improbable that, if there is really no male–female difference, the four highest scores should be achieved by females. It is essentially an extension of the idea of 'coincidence' and is, in fact, plausible since the probability of a prespecified group of four getting the four highest scores is 1 in 70. Since the same level of surprise could have been engendered by the four males getting the highest scores, then the probability of the four highest scores being achieved by four individuals of the same gender is 1 in 35. The logic behind the subsequent inference that gender must be related to genius in that test is based only on the eight scores actually achieved (10, 12, 13, 14, 15, 15, 16 and 19) and implies that a priori any pairing of these eight scores and the eight students (A, C, D, J, M, N, S and T) was equally likely. Any less extreme pattern such as

A	16	C	15
J	13	D	15
N	12	M	14
T	10	S	19

would be less convincing evidence for a relationship between gender and ability. However, the crucial idea is that arguments about inferences can be based only on the eight results actually obtained, and that the arguments depend only on considering the actual pairings of students and marks in contrast to all the other possible pairings, and then contending that the result for the actual pairing is (surprisingly) extreme.

10.2 Random treatment allocation

The device of randomisation is intended to provide a valid basis for the general form of argument illustrated by the student results just discussed. The weakness of the argument that gender and ability are related is that it is a purely post hoc recognition of pattern in the grouping of high and low scores. Alternative groupings could be noticed: four students are older; four are following a different course; four have red hair; or four live in London. And each of these divisions of the eight students into two groups of four could conceivably be used to 'explain' the results. In other words, there are so many coincidences that could have happened, we should not be surprised that one did.

The advantage of random allocation of treatments in a controlled experiment is that for the division of interest the occurrence of any particular allocation of units to different treatments is exactly as probable as any other allocation. Suppose we are doing a psychological experiment and eight students are randomly split into two groups, one group being given special training and the other no special training, the students being otherwise apparently similar. Andrew, John, Neil and Tai-hua have had no specialised training, while those randomly selected for special training are Charles, David, Mohammed and Sanjiv. Now, when the scores obtained are:

Andrew	14
Charles	15
David	15
John	13
Mohammed	16
Neil	12
Sanjiv	19
Tai-hua	10

the argument that special training is beneficial (C 15, D 15, M 16, S 19) compared with no training (A 14, J 13, N 12, T 10) is much stronger because the alternative to that argument is that the special training has no effect (null hypothesis) and the random allocation of students to the special training group happened to pick out the four students who were going to score the four highest results. The deliberate random allocation for the one potentially important question provides a stronger basis for the use of the coincidence argument.

The procedure of random allocation of treatments to units within a set of restrictions imposed by the design structure is illustrated by the following three designs:

(i) A completely randomised design with two treatments and eight units, each treatment being allocated to four units. Since each treatment is to be applied to the same number of units, each unit is equally likely to be allocated to treatment A or to treatment B. One correct procedure is, for each unit in turn, to allocate treatment A with a probability of $(4 - a)/(8 - a - b)$, where a is the number of units already allocated to A and b the number of units already allocated to B. At any point in the allocation procedure, if either A or B has been allocated to four units, then all the remaining units are inevitably allocated to the other treatment.

(ii) A randomised block design with three treatments and nine units divided into three blocks of three units each. Here, the randomisation procedure is restricted, since each treatment must appear once in each block. This, in fact, makes the procedure easier, since in each block of three units one of the three treatments, A, B or C, can be selected randomly for the first unit in the block, one of the other two treatments is then selected randomly for the second unit in the block, and the remaining treatment is then allocated to the third unit.

(iii) A completely randomised design with two treatments and 20 units available for testing. One treatment is to be applied to four units and the other treatment, which is a control, to the remaining 16 units. A simple procedure for this case is to select the four units for the first treatment. To do this, we randomly select one unit from the 20, then one from the remaining 19, then one from the remaining 18 and finally one from the remaining 17. These four units are allocated to the new treatment and the remaining 16 to the control.

In each case randomisation can be achieved either by selecting a treatment for a particular unit, or by selecting a unit for a particular treatment. For each example only one method has been described, not always the simpler one. Whichever method is used, the logic of the method must be checked to ensure that, for each unit, the probability of allocation to each treatment is constant for that treatment. For example, if in (iii) we tried to choose a treatment for the first unit, then one for the second and so on, we must not make the probabilities of allocation to the first units equal for treatments A and B; in that case the last four units would be virtually certain to be allocated to B because only four are to be allocated to A. Instead, the allocation must be such that the first unit has a 4/5 probability of getting B, and 1/5 of A, and for each subsequent unit the probabilities will depend on the numbers already allocated to A and B. A more general way to implement any randomisation is to write down the unrandomised designs with experimental units labelled for each blocking system, as was done for the responses in Chapter 7 when discussing the modelling of data. Then randomly rearrange each set of labels, for example the block labels $1, \ldots, b$ and the unit labels $1, \ldots, n_b$ within each block. This can be cumbersome, but is a fairly clear way of obtaining a valid randomisation.

In clinical trials, the random allocation of treatments to patients is particularly important, but is made difficult by further practical considerations. Essentially none of the clinicians involved in the conduct of the trial should know or be able to deduce the identity of the treatment applied to any particular patient. This is important because such knowledge could not be assumed to have no influence on the general treatment of the patient. The double negative is deliberate. It is the lack of assurance that there would be no effect of knowledge on the treatment of the patient that would leave the conclusions of the trial open to doubt.

The requirement that the trial be conducted 'blind' so that those concerned with the routine administration of the trial cannot know the identity of the treatment for any patient has two implications. First, and obviously, the clinician does not have access to the information about treatment allocation. Second, the clinician should not be able to deduce the identity of the treatments applied to particular patients. Ideally the clinician should not even be able to deduce that the treatments for two patients were different. However, it is important that the clinician be involved in the decisions about the principles used to control experimental variability, and he or she will subsequently be able to recognise the characteristics on which the principles are based. Therefore for small numbers of treatments, 'blocks' should contain more than a single replication of each treatment. If the number of treatments is large, each block will contain a subset of treatments, the principle, but not the detail of the subset arrangements, being determined in advance. These design problems will be considered further in Chapter 11.

10.3 Randomisation tests

The device of randomisation creates a population of experiments that could have been performed, though only one experiment has actually occurred. The identification of the

Table 10.1. *Numbers of accidents at two road sites*

	Improved site				Unchanged site			
	A	B	C	D	E	F	G	H
Accidents before	3	5	4	11	7	1	5	8
Accidents after	0	2	0	5	8	3	2	6
Difference	−3	−3	−4	−6	+1	+2	−3	−2

population allows assessment of the value of a test statistic against the background of the distribution of values of that test statistic for all other members of the population of possible experiments, assuming that the null hypothesis of no treatment effect is true. The idea of a randomisation test is developed initially through examples for the three cases for which randomisation procedures were defined in the previous section.

Example 10.1 An experiment is conducted in a city to see whether road improvements affect the incidence of accidents. Eight accident black spots, of a similar nature and each capable of improvement, are selected, and of these eight four are chosen at random and are actually improved, while the other four remain unchanged. The numbers of accidents for (*a*) the year before the date of improvement and (*b*) the year after the improvement are recorded in Table 10.1. Various criteria to measure the effect (after/before) could be used. The simple difference (after − before) ignores possible scale differences between the sites, but demonstrates the principles of the test clearly. Consider the hypothesis that improvement has no effect on accidents. If the hypothesis is true we should not expect to find a clear-cut difference between the four differences for 'improved' sites and those for the 'unchanged' sites. Further, if improvement has no effect, then the accidents observed at sites A–H would have occurred at those sites regardless of which four sites had been, randomly, selected for improvement. If the total of the four differences for each possible set of four sites out of the eight is calculated, it is possible to assess whether the total which was actually observed for the four randomly selected sites is significantly extreme. We can make this assessment rather tediously by calculating for each set of four sites the total of the four observed differences as follows:

A	B	C	D	−3	−3	−4	−6 = −16
A	B	C	E	−3	−3	−4	+1 = −9
A	B	C	F	−3	−3	−4	+2 = −8
A	B	C	G	−3	−3	−4	−3 = −13
⋮	⋮	⋮	⋮	⋮	⋮	⋮	⋮ ⋮
E	F	G	H	+1	+2	−3	−2 = −2.

The extremeness of the observed result is assessed by counting how many of the possible ($70 = 8!/(4!4!)$) possible allocations would have given a result as extreme as that actually observed. By inspection, we can see that no sets of four sites could have given a total of *more*

Table 10.2. *Data from randomised block design*

Treatment	Block I	Block II	Block III	Treatment total
A	14(a)	11(d)	14(g)	39
B	10(b)	12(e)	10(h)	32
C	15(c)	17(f)	13(i)	45
Block total	39	40	37	

Table 10.3. *Analysis of variance for randomised block design*

	SS	df	MS	F
Blocks	1.6	2		
Treatments	28.2	2	14.1	3.7
Error	15.1	4	3.8	
Total	44.9	8		

than -16 and that exactly three sets would give a total of -16, namely:

$$ABCD, ACDG \text{ and } BCDG.$$

Consequently, the significance level of the observed effect is $3/70$ or 0.04, and the effect would be classified as 'significant at the 5% level'. Note that we have implicitly used a one-tailed test, with the alternative hypothesis being that the improvements lower the accident rate, and the interpretation of the significance level must be made in that context.

Example 10.2 Three treatments, A, B and C, are compared in three randomised blocks of three units each (see Table 10.2). A standard analysis of variance is shown in Table 10.3.

For the randomisation test, all possible random allocations of the treatments A, B and C to the units a, b, c, d, e, f, g, h, i, subject to the blocking restriction, are equally likely. If for each possible allocation the analysis of variance and the F ratio are calculated, the significance of the observed value of 3.7 can be assessed. Since there are 3! possible allocations within each block, this could give $(3!)^3$, or 216, sets of calculations, which seems rather laborious. However, these 216 split into 36 groups of six identical results, because any permutation of the three treatments, using the same permutation in each block, would give the same set of treatment totals (with different treatment labels), and therefore the same F ratio. The calculations may be further reduced by noting that:

(i) the total SS and block SS will remain unaffected under the hypothesis that for each experimental unit the different treatments would produce identical results, given that the units remain in the same blocks.

Table 10.4. *All possible allocations under randomisation*

Units allocated to treatments			Treatment totals	T	Units allocated to treatments			Treatment totals	T
A	B	C			A	B	C		
adg	beh	cfi	39, 32, 45	4570	aeg	bfh	cdi	40, 37, 39	4490
adg	bei	cfh	39, 35, 42	4530	aeg	bfi	cdh	40, 40, 36	4496
adh	beg	cfi	35, 36, 45	4546	aeh	bfg	cdi	36, 41, 39	4504
adh	bei	cfg	35, 35, 46	4566	aeh	bfi	cdg	36, 40, 40	4496
adi	beg	cfh	38, 36, 42	4524	aei	bfg	cdh	39, 41, 36	4504
adi	beh	cfg	38, 32, 46	4584	aei	bfh	cdg	39, 37, 40	4490
adg	bfh	cei	39, 37, 40	4490	afg	bdh	cei	45, 31, 40	4586
adg	bfi	ceh	39, 40, 37	4490	afg	bdi	ceh	45, 34, 37	4550
adh	bfg	cei	35, 41, 40	4506	afh	bdg	cei	41, 35, 40	4506
adh	bfi	ceg	35, 40, 41	4506	afh	bdi	ceg	41, 34, 41	4518
adi	bfi	ceh	38, 41, 37	4494	afi	bdg	ceh	44, 35, 37	4530
adi	bfh	ceg	38, 37, 41	4494	afi	bdh	ceg	44, 31, 41	4578
aeg	bdh	cfi	40, 31, 45	4586	afg	beh	cdi	45, 32, 39	4570
aeg	bdi	cfh	40, 34, 42	4540	afg	bei	cdh	45, 35, 36	4546
aeh	bdg	cfi	36, 35, 45	4566	afh	beg	cdi	41, 36, 39	4505
aeh	bdi	cfg	36, 34, 46	4568	afh	bei	cdg	41, 35, 40	4506
aei	bdg	cfh	39, 35, 42	4530	afi	beg	cdh	44, 36, 36	4528
aei	bdh	cfg	39, 31, 46	4598	afi	beh	cdg	44, 32, 40	4560

(ii) The implication of (i) is that the sum (treatment SS + error SS) is not altered under permutation of treatment allocation and therefore larger treatment SS implies larger *F*.

(iii) To rank the 36 different allocations in order of increasing *F* ratio the allocations need only be ranked in order of increasing treatment SS, or equivalently in order of increasing values of $(T_A^2 + T_B^2 + T_C^2)$.

The 36 different allocations and the corresponding values of

$$T = \left(T_A^2 + T_B^2 + T_C^2\right)$$

that would have been observed if the unit yields are unaffected by treatment are listed in Table 10.4. By inspection, there are seven values of T as large as or larger than the observed value of 4570, and hence the significance level of the observed results against the null hypothesis of no treatment difference is $7/36 = 0.19$.

Example 10.3 Two treatments are compared, one of which is applied to four randomly selected mice, and the other to the 16 remaining mice. The survival times of the mice in days are as follows:

treatment A: 64, 35, 39, 21
treatment B: 7, 15, 32, 5, 43, 19, 26, 8, 6, 19, 45, 31, 33, 14, 21, 12.

To assess how strongly these data support the apparent supremacy of treatment A in prolonging survival, consider what proportion of the sets of four values selected from the 20

Table 10.5. *Randomisations giving a treatment total $\geq$ 159*

(1)	64	45	43	39	Total 191
(2)	.	.	.	35	187
(3)	.	.	.	33	185
$\vdots$	$\vdots$	$\vdots$	$\vdots$	$\vdots$	$\vdots$
(14)	64	45	43	8	160
(15)	.	.	.	7	159
(16)	64	45	39	35	183
(17)	.	.	.	33	181
$\vdots$	$\vdots$	$\vdots$	$\vdots$	$\vdots$	$\vdots$
(26)	64	45	39	14	162
(27)	..	..	..	12	160
(28)	64	45	35	33	171
(29)	$\vdots$	$\vdots$	$\vdots$	$\vdots$	$\vdots$
$\vdots$					
(104)	64	33	32	31	160
(105)	45	43	39	35	162
(106)	45	43	39	33	160
(107)	45	43	39	32	159

observed survival times would give at least as strong evidence of the superiority of treatment A. First, the criterion for the randomisation test must be chosen. Because there are only two treatments the simple criterion is the sum of the observed survival times for treatment A, which for the observed experiment is $64 + 35 + 39 + 21 = 159$. Next, consider how many values there are in the randomisation distribution, i.e. how many possible ways are there of selecting four mice from 20. This is

$$\frac{20!}{16!4!} = 4845.$$

Clearly to consider the set of 4845 possible experiments by hand calculation is not practical, though it would be trivial by computer. We can enumerate all those randomisations which give a treatment total greater than, or equal to, 159. The initial stages of such an enumeration are shown in Table 10.5. The significance level found through such an enumeration is 107/4845, or 2.2%.

In cases when the number of values in the randomisation distribution is large, such computations become time consuming, or perhaps even impossible. Another approach is to estimate the significance level by sampling from the large number of possible randomisations. Suppose that, for a sample of 500 randomisations out of the possible 4845 in the last example, exactly 14 of the sample experiments give a treatment A total of 159 or more, then the estimated significance level of the observed results would be 14/500 or 2.8%. Using the binomial sampling variance, the variance of this estimated significance level is $(0.028)(0.972)/500 = 0.000054$, the corresponding SE being 0.0075 or 0.75%. In practice, we should use more than 500 and it is common to use 10 000. For almost all experiments, this gives a precision which is perfectly adequate and most likely a better approximation than would be obtained from the F-distribution.

Note that, in this last example, the calculated significance level is for a one-sided test. To determine the significance level for a two-sided test, it is necessary to determine what difference between treatments A and B, with treatment A giving the lower average survival time, is equivalent to the observed difference. There is an element of subjective choice of criterion here, but a simple form of argument is as follows. For the observed ratio the treatment means are

$$159/4 = 39.75 \text{ for A}$$

and

$$336/16 = 21.00 \text{ for B},$$

giving a difference of 18.75 between the means. The total of the 20 survival times is 495. To obtain a difference of 18.75 between the two treatment means, with the mean for A being the lower of the two means, the total for treatment A, T_A, must satisfy

$$\frac{T_A}{4} + 18.75 = \frac{495 - T_A}{16},$$

whence $T_A = 39$. By enumerating or estimating the proportion of the randomisation distribution giving treatment A totals either greater than or equal to 159, or less than or equal to 39, a two-sided test significance level for this particular criterion can be obtained.

10.4 Randomisation theory of the analysis of experimental data

To demonstrate the implications of randomisation theory for the theory of the analysis of data from randomised experiments, we consider the example of the randomised complete block design. The idea of randomisation within a fixed set of experimental units requires a slightly different model from that developed in Chapter 2. There are assumed to be t units in each of b blocks, and the model for unit yields if all units are treated identically is

$$y_{ij} = \mu + b_i + \varepsilon_{ij}, \quad i = 1, 2, \ldots, b, \quad j = 1, 2, \ldots, t,$$

where b_i represents the average deviation for block i from the overall mean, and ε_{ij} represents the deviation for unit j within block i from the mean for that block. These assumptions imply

$$\sum_i b_i = 0$$

and, most importantly,

$$\sum_j \varepsilon_{ij} = 0, \quad i = 1, 2, \ldots, b.$$

The yield that would be obtained if treatment k were applied to unit j in block i would be

$$y_{ij} = \mu + b_i + t_k + \varepsilon_{ij}.$$

However, each treatment is allocated to only one unit in each block, and the allocation procedure is defined by a random variable δ_{ij}^k for which the properties are as follows:

$$\delta_{ij}^k = \begin{cases} 1 & \text{if treatment } k \text{ occurs on unit } j \text{ in block } i, \\ 0 & \text{otherwise,} \end{cases}$$

$$\text{prob}(\delta_{ij}^k = 1) = 1/t,$$

since there are t units in block i and treatment k occurs on just one,

$$\text{prob}(\delta_{ij'}^k = 1 | \delta_{ij}^k = 1) = 0,$$

since treatment k can only occur on one unit in block i,

$$\text{prob}(\delta_{ij}^{k'} = 1 | \delta_{ij}^k = 1) = 0,$$

since only one treatment can occur on unit j in block i and

$$\text{prob}(\delta_{ij'}^{k'} = 1 | \delta_{ij}^k = 1) = 1/(t-1),$$

since once treatment k has been allocated to unit j there remain $(t-1)$ units for treatment k'.

The model, considering the population of possible randomisations, for the yield of treatment k in block i, y_{ik}, is therefore

$$y_{ik} = \mu + \text{b}_i + \text{t}_k + \sum_j \delta_{ij}^k \varepsilon_{ij},$$

where the summation is over all units in block i. Using this model, we consider first the estimates of treatment effects and their variances, then the terms in the analysis of variance identity and finally the relationship between the two.

First, the treatment total for treatment k is

$$\text{T}_k = b\mu + b\text{t}_k + \sum_i \sum_j \delta_{ij}^k \varepsilon_{ij}.$$

Since $\text{E}(\delta_{ij}^k) = 1/t$ and $\sum_j \varepsilon_{ij} = 0$,

$$\text{E}(\text{T}_k) = b\mu + b\text{t}_k,$$

whence

$$\text{E}(\text{T}_k - \text{T}_{k'}) = b(\text{t}_k - \text{t}_{k'}).$$

Also

$$\text{T}_k - \text{E}(\text{T}_k) = \sum_i \sum_j \delta_{ij}^k \varepsilon_{ij},$$

and so

$$\text{E}[\{\text{T}_k - \text{E}(\text{T}_k)\}^2] = \text{E}\left\{ \left(\sum_i \sum_j \delta_{ij}^k \varepsilon_{ij} \right)^2 \right\}.$$

Expanding the double summation in terms of different sets of ε_{ij}s gives

$$\mathrm{E}\left\{\sum_i \sum_j (\delta_{ij}^k)^2 \varepsilon_{ij}^2 + \sum_i \sum_{j' \neq j} \delta_{ij}^k \delta_{ij'}^k \varepsilon_{ij} \varepsilon_{ij'} + \sum_{i' \neq i} \sum_j \sum_{j'} \delta_{ij}^k \delta_{i'j'}^k \varepsilon_{ij} \varepsilon_{i'j'}\right\}.$$

Since $\sum_j \varepsilon_{ij} = 0$ and $\delta_{ij}^k \delta_{i'j'}^k$ are independent (randomisation is independent in different blocks) the third term is zero. Also, from the properties of $\delta_{ij}^k \delta_{ij}^k$ the second term is zero. Finally, since the possible values of δ_{ij}^k are only 0 or 1, the expectation of $(\delta_{ij}^k)^2$ is $1/t$. Hence,

$$\mathrm{Var}(\mathrm{T}_k) = \mathrm{E}[\{\mathrm{T}_k - \mathrm{E}(\mathrm{T}_k)\}^2] = (1/t)\sum_{ij} \varepsilon_{ij}^2$$

and

$$\mathrm{Var}(y_{.k}) = \sum_i \sum_j \varepsilon_{ij}^2 / (b^2 t). \tag{10.1}$$

Similarly,

$$\mathrm{E}[\{\mathrm{T}_k - \mathrm{E}(\mathrm{T}_k)\}\{\mathrm{T}_{k'} - \mathrm{E}(\mathrm{T}_{k'})\}] = \mathrm{E}\left\{\left(\sum_{ij} \delta_{ij}^k \varepsilon_{ij}\right)\left(\sum_{ij} \delta_{ij}^{k'} \varepsilon_{ij}\right)\right\}.$$

Again expansion in terms of different combinations of ε_{ij}s gives

$$\mathrm{E}\left(\sum_i \sum_j \delta_{ij}^k \delta_{ij}^{k'} \varepsilon_{ij}^2 + \sum_i \sum_{j' \neq j} \delta_{ij}^k \delta_{ij'}^{k'} \varepsilon_{ij} \varepsilon_{ij'} + \sum_{i' \neq i} \sum_j \sum_{j'} \delta_{ij}^k \delta_{i'j'}^{k'} \varepsilon_{ij} \varepsilon_{i'j'}\right).$$

Again, the last term is zero because of the independence of δ_{ij}^k and $\delta_{i'j'}^{k'}$ and because $\sum_j \varepsilon_{ij} = 0$. The first term is zero because of the joint properties of δ_{ij}^k and $\delta_{ij}^{k'}$. In the second term, the expected value of $\delta_{ij}^k \delta_{ij'}^{k'}$ is $(1/t)\{1/(t-1)\}$ and

$$\sum_i \sum_{j' \neq j} \varepsilon_{ij} \varepsilon_{ij'} = \sum_i \sum_j \varepsilon_{ij}(-\varepsilon_{ij}) = -\sum_i \sum_j \varepsilon_{ij}^2.$$

Hence

$$\mathrm{E}[\{\mathrm{T}_k - \mathrm{E}(\mathrm{T}_k)\}][\{\mathrm{T}_{k'} - \mathrm{E}(\mathrm{T}_{k'})\}] = [-1/\{t(t-1)\}]\sum_i \sum_j \varepsilon_{ij}^2$$

and

$$\mathrm{E}[\{(\mathrm{T}_k - \mathrm{T}_{k'}) - \mathrm{E}(\mathrm{T}_k - \mathrm{T}_{k'})\}^2] = (2/t)\sum_i \sum_j \varepsilon_{ij}^2 + [2/\{t(t-1)\}]\sum_i \sum_j \varepsilon_{ij}^2$$

$$= \{2/(t-1)\}\sum_i \sum_j \varepsilon_{ij}^2,$$

so

$$\mathrm{Var}(y_{.k} - y_{.k'}) = [2/\{b^2(t-1)\}]\sum_i \sum_j \varepsilon_{ij}^2. \tag{10.2}$$

Notice at this point from (10.1) and (10.2) that the variance of $(y_{.k} - y_{.k'})$ is not twice the variance of $y_{.k}$ as it would be in the usual linear model analysis of Chapter 4.

Now consider the analysis of variance identity developed in Chapter 2:

$$\sum_{ik} (y_{ik} - y_{..})^2 = t \sum_i (y_{i.} - y_{..})^2 + b \sum_k (y_{.k} - y_{..})^2$$
$$+ \sum_i \sum_k (y_{ik} - y_{i.} - y_{.k} + y_{..})^2.$$

The first term on the right hand side of the identity involves only the block mean yields, $y_{i.}$, and the overall mean yield, $y_{..}$. These mean yields are unaffected by randomisation since each block total yield includes each treatment effect once, and each unit effect in that block once, and both these sets of effects sum to zero.

The second term on the right hand side of the identity is the treatment SS which can be written

$$\sum_k (y_{.k} - y_{..})^2 = \sum_k \left(T_k^2 / b^2 \right) - Y_{..}^2 / b^2 t.$$

Again, $Y_{..}$ is unaffected by randomisation. Also from the earlier results for $E(T_k)$ and $\text{Var}(T_k)$,

$$E\left(T_k^2\right) = \{b(\mu + t_k)\}^2 + (1/t) \sum_i \sum_j \varepsilon_{ij}^2$$
$$= b^2 \mu^2 + b^2 t_k^2 + 2b\mu t_k + (1/t) \sum_i \sum_j \varepsilon_{ij}^2,$$

since $\sum_k t_k = 0$. Hence,

$$E\left\{ \sum_k (y_{.k} - y_{..})^2 \right\} = t\mu^2 + \sum_k t_k^2 + (1/b)^2 \sum_i \sum_j \varepsilon_{ij}^2 - t\mu^2$$
$$= \sum_k t_k^2 + (1/b)^2 \sum_i \sum_j \varepsilon_{ij}^2.$$

Finally, consider again y_{ik}:

$$y_{ik} = \mu + b_i + t_k + \sum_j \delta_{ij}^k \varepsilon_{ij}$$
$$\Rightarrow E\left(y_{ik}^2\right) = \mu^2 + b_i^2 + t_k^2 + 2\mu b_i + 2\mu t_k + 2b_i t_k + (1/t) \sum_j \varepsilon_{ij}^2,$$

since all cross-products involving ε_{ij} are zero because $\sum_j \varepsilon_{ij} = 0$. Further, since $\sum_i b_i = 0$ and $\sum_k t_k = 0$,

$$E\left(\sum_{ik} y_{ik}^2 \right) = bt\mu^2 + t \sum_i b_i^2 + b \sum_k t_k^2 + \sum_i \sum_j \varepsilon_{ij}^2.$$

Table 10.6. *Analysis of variance for randomised complete block design under finite population model*

Source	df	E(SS)	E(MS)
Blocks	$b-1$	$t\sum_i b_i^2$	
Treatments	$t-1$	$\dfrac{1}{b}\sum_i\sum_j \varepsilon_{ij}^2 + b\sum_k t_k^2$	$1/\{b(t-1)\}\sum_i\sum_j \varepsilon_{ij}^2$ $+\{b/(t-1)\}\sum_k t_k^2$
Residual (by subtraction)	$(b-1)(t-1)$	$\{(b-1)/b\}\sum_i\sum_j \varepsilon_{ij}^2$	$1/\{b(t-1)\}\sum_i\sum_j \varepsilon_{ij}^2$
Total	$bt-1$	$t\sum_i b_i^2 + b\sum_k t_k^2 + \sum_i\sum_j \varepsilon_{ij}^2$	

Hence the total SS, the left hand side of the analysis of variance identity, has expected value

$$E\left\{\sum_{ik}(y_{ik}-y_{..})^2\right\} = bt\mu^2 + t\sum_i b_i^2 + b\sum_k t_k^2 + \sum_i\sum_j \varepsilon_{ij}^2 - bt\mu^2$$

$$= t\sum_i b_i^2 + b\sum_k t_k^2 + \sum_i\sum_j \varepsilon_{ij}^2.$$

The expected values for the analysis of variance structure are therefore as in Table 10.6. If we now write σ^2 for the quantity $\sum_i\sum_j \varepsilon_{ij}^2/\{b(t-1)\}$, which is the expected value of the residual MS, then the variances for differences between treatment means and for treatment means can be written

$$\text{Var}(y_{.k}-y_{.k'}) = 2\sigma^2/b$$

and

$$\text{Var}(y_{.k}) = (t-1)\sigma^2/(bt).$$

Hence the pattern of MS in the analysis of variance and the relation between the residual MS and the variance of a difference between two treatment means is exactly the same as in the normal distribution theory model. The variance for a single treatment mean is, as noted earlier, smaller by a factor $(t-1)/t$ than would be the case for the normal distribution theory model.

Here we have developed the finite population model under randomisation for the randomised complete block design. In fact, the theory can be extended in a similar way to incomplete block designs, complete row-and-column designs and many other structures. The theory follows in essentially the same way for any block structure which involves complete nesting and/or crossing of block factors which define blocks of equal sizes. The appropriate randomisation is that described in the previous section, i.e. independently, randomly rearrange the labels for each blocking factor within the levels of any other blocking factor within which it is nested. For other structures, such as block designs with unequal sized blocks, incomplete row-and-column structures, etc., there is generally no satisfactory generalisation of the randomisation theory, although a similar theory can be developed for particular special cases.

10.5 Practical implications of the two theories for the analysis of experimental data

The practical analysis of experimental data can usually proceed without detailed consideration of which of the two alternative theories of analysis is assumed. However, it is useful to summarise and compare the assumptions and implications of the two theories.

A: Infinite model

1. This model, for which the theory was developed in Chapter 4, assumes that the experimental units are a random sample from a population which, by the implication of the distributional assumptions for the unit errors, must be of infinite size.
2. The unit errors are assumed to be normally and independently distributed with a common variance σ^2.
3. The experimental units are assumed to be subject to fixed environmental (block) and treatment effects.
4. The inferences from the analysis of the data apply to the infinite population from which the experimental units are a sample.

B: Finite model

1. This model, for which the theory has been developed in the previous section, assumes a finite set of experimental units, all of which are used in the experiment; it further assumes that, within restrictions imposed by the blocking structure of the experiment, the treatments are allocated randomly to experimental units.
2. The unit errors are assumed to be fixed and the set of unit errors in a block is assumed to have a zero sum.
3. The inferences from the analysis of the data apply strictly to the finite set of units used in the experiment.

Least squares estimates For both models, the least squares estimates of block and treatment effects are appropriate. The analysis of variance corresponds to the subdivision of variation between the various sets of fitted estimates for the two models, and the relationship between the residual MS and the variance of a difference between two treatment means is identical for both models. The slightly different result for the variance of a single treatment mean, for the finite model, is a result of the non-independence of treatment means, which is a consequence of the random allocation of treatments to units. Essentially, treatment mean yields are not independent because, if in a particular block treatment A is allocated to unit j, then treatment B must be allocated to one of the remaining units in that block.

Inferences For the infinite model, with its normal distribution assumptions, the F test for the comparison of mean squares in the analysis of variance and the t test for the comparison of treatment means are correct.

For the finite model, it can be shown by considering the moments of the distribution of the ratio

$$\frac{\text{treatment MS}}{\text{treatment MS} + \text{error MS}}$$

that the null hypothesis distribution of the ratio of the treatment MS and error MS is approximately represented by the F-distribution on $(t-1)$ and $(b-1)(t-1)$ df, provided the ε_{ij} are homogeneous between blocks. Example 10.2 for data from three blocks of three treatments provides as much support for this general rule as any single example can, the randomisation test giving a significance level of 0.19, which agrees fairly closely with the significance of an F statistic of 3.7 on 2 and 4 df, which is 0.12. This, of course, is a very small experiment and the conclusions are usually much closer for larger experiments.

For the application of t tests, an analysis based on the finite model has to rely on the general robustness of the t test to departures from an underlying normal distribution. Similarly, no general method for obtaining confidence intervals of treatment comparisons is available. Methods have been developed for particular cases, but these are not as intuitive as the overall hypothesis test and can be awkward to implement.

In addition to the common requirement (3) of fixed environmental and treatment effects, both models assume additive and homogeneous error terms. Departure from these assumptions can frequently be corrected by using a transformation of the observed variable.

Choice between models It should be clear from the foregoing discussion that the practical analysis, inference and interpretation of experimental data are not materially affected by the choice of model. If the assumptions of the infinite model are felt to be reasonable, then there is no reason not to use that model. However, since those assumptions are often found to be difficult to credit, and the recognition that the assumptions are unrealistic may not come until the experiment is in progress, it is sensible always to assume that treatments should be randomly allocated, unless there are very good practical reasons for using a systematic allocation (some situations where systematic designs may be appropriate will be discussed in Chapters 16 and 18).

Not only does the device of randomisation define a procedure for allocating treatments to units, which would otherwise have to be devised to avoid a purely subjective allocation, it also provides an alternative basis for the analysis of experimental data. The real beauty of randomisation theory is that, in most practical applications, it comes for free. Even if we fully believe in the infinite population model, randomisation does no harm and, if we carefully define blocking factors to correspond to the fixed environmental factors in our infinite model, the analysis we carry out will be identical. Therefore any potential user of the results from the experiment who is sceptical about our model should still believe in the results of our analysis, since they are justified by randomisation. The only difference is in how broadly the conclusions can be assumed to apply. This, however, becomes a debate about whether the experimental units can truly be considered as a random sample from the population.

One other issue which causes controversy is the analysis when blocks are used which do not correspond to any expected environmental factors, but are included purely for the purposes of simplifying the management of the experiment. This occurs in clinical trials, where blocks of a few patients are used to ensure that the allocation remains balanced between treatments even if recruitment is stopped early, and in agricultural field experiments, where a block might correspond to the area of land which can be monitored by a single data collector. Some argue that block effects should not be included in the model, because they do not correspond to any expected fixed environmental effects in the infinite model. From the randomisation viewpoint, on the other hand, they must be included. We strongly advocate the latter view since, except in very small experiments, nothing is lost by including these effects in the analysis and they add considerably to the robustness of the conclusions by removing the need to fully believe the infinite model. In very small experiments, the loss of df for blocking might seriously harm the estimation of error and here more careful thought is needed. However, this thought must be done at the design stage, so that either more units or perhaps larger blocks are used.

In the remainder of this book, we shall usually discuss the analysis of data in the context of the infinite model because it is desirable to discuss models for experimental results in terms of a single form of model and the formal model is simpler to write down for the infinite model. This does not mean, however, that we really believe such a model. Rather, it is considered to be an infinite model whose analysis corresponds to that, under randomisation, of the finite model which we really believe. The randomisation ideas will never be far away from the development of design principles and it will be assumed throughout that treatments should be allocated randomly unless a non-random allocation is explicitly defined.

10.6 Practical randomisation

In this section we consider the randomisation procedures required for the design structures discussed in earlier chapters, particularly Chapters 7 and 8. The simple practical discussion of randomisation at the end of Chapter 2 must now be modified and extended to allow for the more complex design structures developed.

The first requirement of a randomisation procedure is to define precisely the restrictions within which the randomisation must be applied. Thus, if a completely randomised design is to be used, the numbers of observations for each treatment must be specified. The randomisation scheme used must then ensure that, for each unit, the probabilities that the different treatments are allocated to the unit *must be identical for each unit, regardless of the order of allocation*. The random allocation may be achieved either by allocating units to treatments or by allocating treatments to units. The former requires that the units be individually labelled and that the required number of units for treatment A be selected randomly and sequentially, all units not previously selected being equally likely to be selected at any particular stage of the selection. After all the necessary units for treatment A have been selected, those for B are selected, and thereafter for each other treatment in turn. The procedure of allocating treatments to units is more complex. If treatments A, ..., T are to have $n_A, n_B, \ldots, n_T$ units, respectively, then the allocation probabilities for the first unit are $n_A / \sum n, n_B / \sum n, \ldots, n_T / \sum n$. As units are allocated the probabilities change so that each treatment allocation probability is proportional to the number of units still required for that treatment.

If units are grouped in blocks for a single blocking system, then there are three possible components of randomisation. The grouping of treatments into sets for different blocks may involve a random element. The allocation of treatment sets to particular blocks should usually involve a random element. Finally, the actual allocation of treatments within a set to the units within a block should always involve randomisation. In the randomised complete

Block						
	I	A	B	C	D	(omitting E F)
	II	A	B	C	E	(omitting D F)
	III	A	B	D	F	(omitting C E)
	IV	A	C	E	F	(omitting B D)
	V	B	D	E	F	(omitting A C)
	VI	C	D	E	F	(omitting A B)

Figure 10.1 Design to compare six treatments in six blocks of four units.

block design all treatments occur in each block so that the first two components of the random allocation procedures have no meaning and only the third component is appropriate. In designs with incomplete blocks all three components should be considered.

Consider, for example, the design shown in Figure 10.1 for comparing six treatments in six blocks of four units each (Example 7.3). This design is not balanced but, in terms of attempting to achieve equality of treatment comparison, it is as close to balance as may be achieved within the restrictions imposed by the blocking system. Nevertheless, the inequality of comparisons means that a particular treatment, for example A, will be compared more precisely with those treatments with which it is jointly omitted from a block than with the other treatments. Thus, comparisons of A with B, A with C, B with D, C with E, D with F and E with F will be more precise than other comparisons. The experimenter must decide whether he wishes to choose which treatment pairs have this very slight advantage of precision, or whether the selection of the six (linked) pairs for omission should be determined randomly. This randomisation has no implications for the validity of the analysis, but avoids having to make a subjective decision. Having determined the six sets of four treatments each, these six sets should be allocated randomly to the six blocks. Finally, the four treatments in the set allocated to a particular block must be randomly allocated to the four units in that block. Thus the design has two components for which random allocation is required, and a third where the selection of the six treatment sets may be deliberately non-random.

The question of whether the allocation of treatments to treatment sets for blocks should be random or deliberate depends on the treatment structure. Consider the problem of example (*a*), Section 7.0 (and Example 7.10), in which eight treatments consisting of two heating levels × two lighting levels × two carbon dioxide levels are to be allocated to 12 units in two blocks of six units each. Two sets of six treatments must be selected, one for the first block and the other for the second block. The two sets will inevitably include four common treatments, and each set will also include two of the four remaining treatments. Since the eight treatment combinations are highly structured, it is unlikely that the experimenter will select the four combinations to be included in both treatment sets randomly. If he did so, and then found that the four common treatments were all those with the lower heating level (only a 1 in 70 chance, but with five other similar unfortunate patterns, not a negligible probability), then he would probably wish to abandon that random selection and start again. It would be better to recognise initially that the four common treatments should be representative of both heating levels, both lighting levels and both carbon dioxide levels, and to choose as the common treatments a set such as:

upper level of heating	upper level of lighting	CO_2 1
upper level of heating	lower level of lighting	CO_2 2
lower level of heating	upper level of lighting	CO_2 2
lower level of heating	lower level of lighting	CO_2 1

Such a systematic set is easily constructed from the three stated requirements. Methods of selecting such 'representative' fractional subsets of treatments are discussed in Chapters 13 and 14. Our present purpose is to suggest simply that random selection of treatment sets for incomplete blocks is not always a good idea. The allocation of two more treatments to each block set offers less scope for systematic selection, but the allocation of heating and lighting levels to the two blocks can be made orthogonal to blocks. After selecting the two sets of treatments systematically there should be two stages of randomisation. First the random allocation of sets to blocks, and secondly the random allocation of treatments to units within each block.

The details of the randomisation procedure for any design employing a single blocking system are extremely simple. In all stages of randomisation we have a set of items to be allocated to an equi-numeric set of objects. In a randomised complete block design there are t treatments to be allocated to t units in each block. We may either allocate single units randomly to treatment A, treatment B, ..., treatment T in turn, or we may allocate individual treatments to unit 1, unit 2, ..., unit t in turn; at each stage of either allocation all remaining units (treatments) are equally likely to be selected. In incomplete block designs the allocation of treatment sets to blocks, or of treatments to units within a block, is in exactly the same one-to-one form.

The random allocation of treatments to units in designs with two blocking criteria, such as a Latin square design, is more complex. Unlike the randomised block design, Latin squares cannot be randomised sequentially, for example by randomising treatments in each row subject to the requirement that no treatment appears twice in a column, because such a procedure generally leads to a position from which the column requirements cannot be satisfied. It is therefore necessary to adopt a different approach to randomisation.

If the set of possible Latin squares can be tabulated or indexed in some way, then the random selection of a Latin square can be achieved by simply using the structure of the tabulation. This has been the basis of classical methods for selecting a random Latin square. To understand methods of tabulation, it is important to realise that, for a Latin square, (i) permutations of the letters, or (ii) permutations of the rows, or (iii) permutations of the columns, leaves a design which is still a Latin square. Although we have not used this approach the randomisation procedure for a randomised complete block design can also be recognised as a permutation of treatments within the set of units (or equivalently as a permutation of the units within the set of treatments). For any randomisation test criterion which is symmetric in the treatments, permutation of the letters does not alter the value of the test statistic. We therefore consider only those Latin squares which are distinct in the sense that one cannot be transformed to another simply by relabelling the treatments. The number of distinct squares is small for small t (one for $t = 2$, two for $t = 3$, 24 for $t = 4$) but rapidly becomes too large for tabulation of all distinct squares.

Two methods of indexing the available Latin squares have been devised. The first uses the fact that, for any Latin square, it is possible by reordering first the rows and then columns to produce a Latin square in which the first row and the first column both contain the letters in the alphabetic order (A, B, C, ...). A square with the first row and first column 'in order' is called a *standard square*. Latin squares can be divided into sets such that all the squares in one set derive from a single standard square by the reordering of rows and columns. Since each Latin square has a unique associated standard square, a proper method of randomly selecting a Latin square is to select randomly a standard square and then randomly reorder the rows and the columns. However, although the number of standard squares is quite small for small values of t (one for $t = 2$ and for $t = 3$, four for $t = 4$, 56 for $t = 5$), for $t = 6$ the number is too large for easy tabulation.

The second effort at indexing involves the grouping of standard squares into sets called *transformation sets*, where each standard square in a transformation set can be transformed into any other standard square in the set by first permuting the letters, for example (A → B, B → D, D → C), and then reordering the rows and columns to produce a standard square. There are two difficulties associated with transformation sets. The first is that the number of standard squares in a transformation set varies between sets so that, in selecting a random Latin square by first selecting a transformation set, then a standard square and then a Latin square, the initial selection of a transformation set must take account of the different sizes (number of standard squares) of transformation sets. However, the second difficulty is that even the number of transformation sets becomes too large as t increases and this means that the indexing of transformation sets is not a complete solution to the search for a method of randomly selecting a Latin square.

The advice about the practical randomisation of a Latin square design will vary according to the size of square being considered. The basic principle on which the choice should be made is that the set of possible designs considered for selection must be large enough to allow a large number of values in the randomisation distribution.

For $t = 3$ or 4, there really are not enough different Latin squares available (two for $t = 3$ and 24 for $t = 4$) for a randomisation test to be meaningful. Nonetheless, the Latin square should be chosen from all the available squares by randomly selecting a standard square and then randomly permuting the rows and the columns.

For $t = 5$, a standard square should be selected at random from the 56 possible standard squares. This may be done directly or through selection of a transformation set. Subsequently, the rows and columns should be randomly reordered and, if the initial selection was by means of a transformation set, the letters should be randomly permuted.

For $t = 6$, a standard square for any of the larger transformation sets should be selected and the rows, columns and letters should be randomly permuted.

For $t = 7$ or more, it is sufficient to write down any Latin square and randomly permute rows, columns and letters.

For multiple Latin squares the randomisation procedure depends on the form of multiple Latin square design. For the design in which the component Latin squares are entirely separate, the randomisations of each square are independent. If a Latin rectangle design, using Latin squares with common row effects, is required then the reasonable randomisation procedure would be:

(*a*) select $n\,t \times t$ standard Latin squares,
(*b*) randomise the rows within each square,
(*c*) randomise the entire set of nt columns.

The procedure is illustrated for two 5×5 Latin squares in Figure 10.2.

Finally we consider the random allocation of treatments for more general row-and-column designs. If the rows and columns are orthogonal (i.e. each row-and-column combination has the same number of units, usually 1) then the randomisation procedure is essentially exactly

(a) A B C D E   A B C D E
     B C E A D   B E A C D
     C D A E B   C A D E B
     D E B C A   D C E B A
     E A D B C   E D B A C

(b) A B C D E   C A D E B
     E A D B C   E D B A C
     B C E A D   A B C D E
     D E B C A   B E A C D
     C D A E B   D C E B A

(c) B E A D B E C C A D
     C A D B A C D E E B
     E D B C C D E A B A
     D C E A E A B B D C
     A B C E D B A D C E

Figure 10.2 Randomisation of multiple Latin squares: (a) two standard squares,
(b) rows of each randomised, (c) columns randomised.

the same as for Latin squares of size $t \geq 7$. Any initial allocation of treatments to the row and column structure, such that the joint occurrence restrictions are satisfied, may be used as the starting point. The rows are randomly reordered; then the columns are randomly reordered; finally, if the treatments are regarded as equally important, the treatment letters are randomly reordered. The procedure is illustrated in Figure 10.3 for a design to compare ten treatments in a 6×5 array.

When rows and columns are not orthogonal the restrictions on possible randomisation are severe. Any set of rows (or columns) which have exactly the same pattern of occurrence in the columns (rows) may be permuted. And, of course, the treatment letters may be permuted. Other randomisation possibilities may be deduced from the occurrence of arbitrary choices within the design construction procedure.

Exercises

10.1 In an experiment on human–computer interactions, five subjects are selected randomly from a group of eight and are given a standard screen interface, the other three receiving a new style of screen. The scores on a particular test of accuracy in carrying out their tasks are given below. Use a randomisation test to test whether there is a genuine difference between the effects of the two screen types, defining carefully the criterion you use.

$$Scores \begin{cases} \text{standard} & 45, 85, 55, 80, 90 \\ \text{new} & 75, 95, 80. \end{cases}$$

10.2 A randomisation test is required of the null hypothesis that the five treatment effects are equal for the experimental data for five rice spacing treatments in four blocks given in Table 10.7. By sampling from the full randomisation distribution of $(5!)^3$ values to obtain a sample of 10 000 values, perform the required randomisation test at the 5% significance level. Think carefully about how you take your random sample of 10 000 points from the randomisation distribution.

Table 10.7. *Rice spacing data*

Treatment	Block			
	I	II	III	IV
1	5.9	5.3	6.5	6.3
2	7.1	6.4	6.6	5.8
3	7.0	6.5	6.3	8.9
4	8.8	7.6	7.0	7.9
5	7.8	6.7	8.2	7.2

(a)	A	B	C	D	E	F
	I	H	B	G	D	A
	C	E	G	A	H	J
	D	C	J	F	I	H
	J	I	F	E	G	B
(b)	D	C	J	F	I	H
	J	I	F	E	G	B
	A	B	C	D	E	F
	C	E	G	A	H	J
	I	H	B	G	D	A
(c)	I	F	J	C	H	D
	G	E	F	I	B	J
	E	D	C	B	F	A
	H	A	G	E	J	C
	D	G	B	H	A	I
(d)	E	F	A	J	H	D
	G	B	F	E	I	A
	B	D	J	I	F	C
	H	C	G	B	A	J
	D	G	I	H	C	E

Figure 10.3 Randomisation of design for ten treatments in 5 rows × 6 columns: (a) initial, (b) reorder rows, (c) reorder columns, (d) reorder letters.

10.3 Show that there are only 12 distinct 3 × 3 Latin squares. Show further that, if the ratio of treatment MS/error MS for a particular square is R, then the other possible values under randomisation are all either R or $1/R$.

10.4 Three treatments, A, B and C, are compared in an experiment for which the design consisted of two quite separate 3 × 3 Latin squares. Show that under randomisation, and assuming the null hypothesis that the three treatments have identical effects, the distribution of the ratio treatment MS/error MS contains 24 distinct values. For the data given in Table 10.8 test the null hypothesis that the treatments have identical effects.

10.5 The randomisation model for a completely randomised design is

$$y_{ik} = \mu + t_k + \varepsilon_i$$

Table 10.8. *Data from separate Latin squares*

A	C	B		B	A	C
4	2	2		4	6	3
B	A	C		A	C	B
3	4	3		4	5	2
C	B	A		C	B	A
3	4	5		4	4	4

Table 10.9. *Numbers of babies not crying when rocked or not*

Day	Number of control babies		Number of rocked babies	
	Total	Not crying	Total	Not crying
1	8	3	1	1
2	6	2	1	1
3	5	1	1	1
4	6	1	1	0
5	5	4	1	1
6	9	4	1	1
7	8	5	1	1
8	8	4	1	1
9	5	3	1	1
10	9	8	1	0
11	6	5	1	1
12	9	8	1	1
13	8	5	1	1
14	5	4	1	1
15	6	4	1	1
16	8	7	1	1
17	6	4	1	0
18	8	5	1	1

if treatment k is applied to unit i with

$$y_{.k} = \mu + t_k + \sum_i \delta_i^k \varepsilon_i / n_k,$$

where δ_i^k is the probability that treatment k is allocated to unit i. Identify the probability structure for δ_i^k if treatment k is to be allocated to n_k units and the total number of units is N. Obtain the variance of a difference between two treatment means. Show that the expected value of the residual MS in the analysis of variance is $\sum \varepsilon_i^2 / (n-1)$ and that the variance of the difference between two treatment means may be estimated by $s^2(1/n_1 + 1/n_2)$, where s^2 is the residual MS in the analysis of variance.

10.6 How many sample points are there in the randomisation distributions for:
   (*a*) a randomised complete block design for seven treatments in three blocks,
   (*b*) a BIB design for seven treatments in seven blocks of three units,
   (*c*) a Youden square for seven treatments in a 7×3 array?

10.7 Gordon and Foss (1966) reported the results of an experiment on the effect of rocking on the crying of very young babies. On each of 18 days, the babies not crying at a certain instant in a hospital nursery served as subjects. One baby selected at random was rocked for a set period, the remainder serving as controls. The numbers not crying at the end of a specified period were as in Table 10.9. Construct a randomisation test of the hypothesis that rocking has no effect on the propensity of babies to cry.

11

Restricted randomisation

11.0 Preliminary example

An experiment to compare four varieties of tomato is to be run in a greenhouse at a horticultural research station. The greenhouse has 16 compartments in a 4×4 array. The initial proposal is to use a Latin square design, such as that shown in Figure 11.1(a), which is randomised, as in Chapter 8 by randomly permuting rows and columns, to give the square shown in Figure 11.1(b). On seeing the randomised design, the experimenter recalls that previous experiments in this greenhouse showed unusually high yields along the top-right to bottom-left diagonal and is concerned that this might bias the results in favour of variety D. She also notes that the situation is even worse in the unrandomised design.

One possible solution is to abandon the natural seeming 4×4 row-and-column structure and define blocks according to the distance from the top-right to bottom-left diagonal. However, this is not satisfactory either, since the experimenter has also previously observed row and column trends. Another possible solution is to restrict the randomisation more than usual to ensure that the design has all of the required properties.

11.1 Time-trend resistant run orders and designs

In industrial experiments, the experimental units are often sequential runs of the same process. The possibility of time trends needs to be allowed for, but resources are often too scarce to benefit from using very small blocks. Rather than using very large blocks within which time trends are still likely, sometimes the time effect is modelled as a low-order, usually linear or quadratic, polynomial. This is particularly appropriate when something is known about the likely causes of the time trend, such as tool wear or ambient temperature, which would suggest that it is likely to be smoothly changing over the course of the experiment, rather than something like batch differences which could be expected to lead to step changes in response. Properly allowing for linear or quadratic trends also requires the time points at which the runs are made to be carefully controlled and they are usually used when the runs will be equally, or at least regularly, spaced in time.

Designing industrial experiments to be orthogonal, or nearly orthogonal, to time trends is an area of much research. With unstructured treatments, it is relatively easy to find a trend-resistent run order. For example, for a factor with t levels, the order $0\,1\,\cdots\,t\,t\,\cdots\,1\,0$ (repeated as many times as necessary) gives estimates of the main effect comparisons orthogonal to a linear trend, since the runs for any treatment are symmetrical around the midpoint in time (assuming equally spaced runs). For a quadratic time trend, things are in general more

(a)

	(a)					(b)		
A	B	C	D		D	A	B	C
B	C	D	A		B	C	D	A
C	D	A	B		C	D	A	B
D	A	B	C		A	B	C	D

Figure 11.1 Two 4×4 Latin squares: (a) standard square; (b) after randomisation.

complicated, but for two levels (with which these methods are most often used), the order $0\ 0 \cdots 0\ 1\ 1 \cdots 1$ is orthogonal to a pure quadratic time trend, since there is a comparison of treatments at each level on the quadratic surface. To be orthogonal to both linear and quadratic time trends, an order such as $0\ 1\ 1\ 0 \cdots 0\ 1\ 1\ 0\ 1\ 0\ 0\ 1 \cdots 1\ 0\ 0\ 1$ is required.

With two treatment factors, P and Q, in order to ensure that both main effects and the interaction are orthogonal to a linear time trend, a sequence such as $p_0 q_0\ p_0 q_1\ p_1 q_0\ p_1 q_1\ p_1 q_1\ p_1 q_0\ p_0 q_1\ p_0 q_0$ can be used. Finding designs for two factors in which main effects and the interaction are orthogonal to both linear and quadratic time trends is already quite difficult and requires either 32 runs or unequally spaced time points. In the former case, one might already question the adequacy of a quadratic assumption on the time trend over such a long period of time. With more factors, it becomes completely impracticable to find run orders which are orthogonal to linear and quadratic time trends. Instead, in small experiments we recommend using orders which are nearly orthogonal to these time trends. These can be found by trial and error, or by computer search, in exactly the same way as block designs.

For larger experiments, it is almost certainly preferable to use reasonably small blocks, since these are more robust to many different types of time effect and have the additional support of a randomisation-based justification for the analysis. Nevertheless, for small experiments, the idea of ensuring that the run order is nearly orthogonal to time trends is certainly worth pursuing and probably deserves further research. In particular, for the commonly used fractional factorial designs, to be discussed in Chapter 14, with between 8 and 32 runs, it should be possible to carry out exhaustive searches of all run orders to discover those which are best for different kinds of time trend.

11.2 Modelling spatial variation

Some agricultural crop experiments, involving large numbers of experimental plots, are performed on large areas of land. Typically, such an experiment may involve a rectangular array of between 60 and 600 plots, the plots being of identical size and shape, the shape being most frequently a long, thin rectangle. Blocking of such an experiment is often viewed as an attempt to insure against unforeseen patterns of variation in the field. There may be some probable directional trend of fertility, which would suggest that blocks should be long in the direction perpendicular to the fertility trend. If blocks are being designed to insure against unforeseen patterns, then two, perpendicular, blocking systems may be used in a row-and-column design involving incomplete sets of treatments in each row and in each column. The most appropriate designs for this purpose are discussed in Chapter 8.

Because the pattern of spatial yield variation is often largely unknown before the experiment, it might be expected that covariance analysis could be employed to utilise information

	(a)				(b)			
9	3	−3	−9		A	B	C	D
3	1	−1	−3		D	C	B	A
−3	−1	1	3		C	D	A	B
−9	−3	3	9		B	A	D	C

Figure 11.2 (a) Coefficients of the linear × linear row–column interaction; (b) Latin square design robust to row–column interaction.

on yield variation that becomes available during the experiment. For any specific variable that might be observed during the experiment and thought to be affecting yield, such as pest damage or waterlogging, this can be done in the normal way. However, the information about likely yield patterns may sometimes emerge from the yields themselves. The pattern in the strawberry experiment (Example D in Section 4.1) is an example, though in that case an obvious physical explanation of the pattern was readily available. In many cases, however, although clear yield patterns exist, no direct physical explanation is apparent. The methods described in this section have been developed for agricultural crop experiments. However, there are applications also in laboratory experiments where many samples are tested in moulded plastic plates containing two-dimensional arrays of wells for sample material.

There are two approaches to using covariance analysis to adjust treatment yields to allow for the underlying but unidentified field yield variation. The first assumes that there is a general two-dimensional response surface of yield which can be represented by a second-degree polynomial or two-dimensional Fourier series. The general covariance model discussed in Section 4.8 can allow a response surface model of this kind to be included with the experimental design model. Perhaps more usefully, this idea can be used in the row-and-column designs described in Chapter 8, to allow for the linear × linear component of the row–column interaction, in addition to the usual discrete row-and-column effects. This is particularly useful in small experiments in controlled environments such as greenhouses, when it would be unwise to assume random row and column effects, but the set of possible randomisations is not large enough for the finite population model to be reliable. When designing such experiments, the relevant treatment contrasts can be made orthogonal, or nearly orthogonal, to the linear interaction covariate. This would usually be a secondary consideration, so that we restrict the randomisation after choosing an optimal, or at least good, row-and-column design.

An example of such a design can be constructed for the introductory example. The coefficients of the linear × linear component of the row–column interaction are proportional to those shown in Figure 11.2(a). It is clear that it is not possible to construct a design for which treatment comparisons are orthogonal to this interaction contrast. By trial and error, or exhaustive search, it becomes clear that the best pattern which can be achieved is that of the design shown in Figure 11.2(b), in which all treatment comparisons are estimated nearly orthogonally to this interaction component. The only randomisation of this design which can be done is of treatments to letters and this is clearly insufficient for any sensible randomisation based inference. Of course, even the fully randomised 4×4 Latin square is barely sufficient in this regard, as discussed in Chapter 8. Equally, it would seem absurd to assume that the 16 plots in the greenhouse are in any sense a random sample from a population of such plots. Hence restricting the randomisation in the manner described has considerable attractions for experiments of this type.

In recent decades, there has been an explosion in the use of models for local spatial variation first suggested over 70 years ago, but largely neglected for 50 years, possibly because the models require substantial computational facilities. In this form of covariance model, it is assumed that the best information on the level of yield to be expected on a particular plot is contained in the yields from the immediately neighbouring plots. The covariance model (omitting treatment effects) would then take the form

$$y_{ij} = \mu + \beta \sum_{k(ij)} (y_k - \mu) + \varepsilon_{ij}, \qquad (11.1)$$

where $k(ij)$ denotes the set of neighbours of y_{ij}.

This form of model cannot be fitted in the usual manner for covariance models because the observed values of the covariate are also values of the principal variate, and all 'dependence' between variables is reciprocal. The model (11.1) is an example of a class of spatial effect models, for which maximum likelihood fitting methods have been developed by Besag (1974) and, for field crop experiments, by Bartlett (1978). If these fitting methods are used when the spatial effect model is the covariance component of an experimental design model, then an iterative procedure is necessary. The covariance effect estimation for the model in Section 4.8 is in terms of the residuals after fitting the design parameters. However, the fitted covariance effects imply patterns of adjacent plot yields and, when these are allowed for, the estimation of treatment effects will change. For example, it may become apparent that an unusually high plot yield for a particular treatment was associated with high yields for the four adjacent plots and, that allowing for those neighbouring yields, the initial yield was much less unusual. This leads to a revised covariance estimate. The general covariance model of Section 4.8 does not give rise to such problems.

This covariance adjustment method is highly effective in some situations (Kempton and Howes, 1981), but it can give ridiculous results if applied inappropriately. These difficulties arise when the experiment contains only two or three replicate plots for each treatment. The iterative method may then be strongly influenced by a small number of pairs of plots, and a form of positive feedback can occur, leading to estimates of the covariance coefficient, and treatment effects, which diverge increasingly from sensible values.

A series of important papers developing alternative methods using structured nearest neighbour models for the analysis of experimental data includes Wilkinson *et al.* (1983), Green, Jennison and Seheult (1985), Besag and Kempton (1986), Gleeson and Cullis (1987), Cullis and Gleeson (1989, 1991), Gilmour, Cullis and Verbyla (1997). Although there is continuing discussion about the merits of different methods and they will not be described in detail here, the last group of papers listed, all developed from the seminal discussion paper of Wilkinson *et al.*, have become particularly popular due to the possibility of fitting them using residual maximum likelihood, as discussed in Section 9.4. There is a recognisable unity in that all the methods involve the consideration either of first-order differences between adjacent plots, assuming that there is a dominant direction of plot correlation, or of second-order differences. The assumption in all methods is that the pattern of variation of fertility is sufficiently smooth that first- or second-order differences are largely unaffected by the fertility trend effect.

Column

		1	2	3	4	5
Row	1	A	B	C	D	E
	2	B	E	D	A	C
	3	C	D	B	E	A
	4	D	A	E	C	B
	5	E	C	A	B	D

Figure 11.3 Neighbour-balanced Latin square.

Column

		1	2	3	4	5	6	7	8	9	10
Row	1	C	A	B	C	D	A	D	C	B	D
	2	D	B	D	A	C	D	B	A	C	A
	3	B	C	A	D	B	C	A	B	D	C
	4	A	D	C	B	A	B	C	D	A	B

Figure 11.4 Neighbour-balanced Youden square.

Over a very large number of applications of these spatial methods, efficiency gains of greater than 30% have consistently been claimed compared with traditional methods, such as randomised complete block designs. Some care is needed in interpreting this, however, as there is sometimes a considerable amount of model selection done in order to find the highest efficiency gain. While the model selected might indeed be the best, its calculated efficiency will be seriously positively biased, due to the model selection. However, the gains are so large that they do suggest that the traditional methods can be substantially improved. One general implication of these developments is either that blocking is an inappropriate form of control of variation or at least that the block sizes commonly used in field crop experiments may be considerably too large. If conventional blocking designs are to be used, then blocks should be small and compact. Nearest neighbour adjustment implies that each pair of adjacent plots should be considered as a block because each neighbour supplies strong evidence of the expected yield level for the neighbouring plot. More specific implications of any of the forms of analysis for nearest neighbour models are that adjacent pairs of treatments should include all possible treatment pairs as nearly equally frequently as is possible. This should improve the precision of comparisons for essentially the same reason that having treatment pairs occur equally frequently together in a block improves precision. The choice of such designs will be discussed in the next section.

11.3 Neighbour balance

Some of the ideas discussed in Chapter 8 on designs to allow for residual effects are relevant to the problem of constructing designs so that all pairs of treatments occur adjacently equally frequently. Such designs have been developed for one- or two-dimensional adjacency by various authors, notably by Williams (1952) and Freeman (1979). Examples of neighbour-balanced designs using a basic Latin square and an extended Youden square are shown in Figures 11.3 and 11.4.

If only one-dimensional neighbour effects, or trend effects, are deemed important, then it is possible to arrange designs in randomised complete blocks with the blocks in sequence so

Block

	I	II	III	IV	V	VI	VII	VIII	IX	
C	BCDA	DBCA	DABC	BACD	CDBA	BDAC	ACBD	BDCA	BADC	B

Figure 11.5 Neighbour-balanced design in nine consecutive blocks of four plots plus two end plots.

(a)
A	B	C	D	E	F	G	H	I
J	K	L	A	H	C	I	E	B
D	G	E	J	I	K	A	F	L
C	J	G	K	B	L	H	D	F

(b)
F	C	J	D	G	H	I	A	L
K	A	F	H	L	E	J	G	B
L	D	B	I	E	A	H	K	C
I	G	E	K	J	F	B	C	D

Figure 11.6 Experimental plan with neighbour effect 'balance' for 12 treatments in 4 rows × 9 columns: (a) initial plan, (b) randomised.

that each treatment has every possible pair of neighbouring treatments. An example given by Dyke and Shelley (1976) which employs nine blocks of four treatments with two end plots is shown in Figure 11.5. This example was originally proposed to investigate the possible transfer effects of treatments from adjacent plots on the yield of a plot. However, the design philosophy is just as appropriate when attempting to allow for a smooth trend.

Doubtless much effort and ingenuity can be devoted to identifying the existence of, and devising, designs providing exact neighbour balance for those situations for which such balance is possible. However, in the same way that the construction of incomplete block designs which are as efficient as possible was shown in Chapter 7 to be possible without mathematical theory, so in designs for neighbour effects it is possible to construct designs giving near equality of adjacent occurrence of treatment pairs, especially when no exactly neighbour-balanced designs exist. Consider the allocation of 12 treatments within a 9×4 array with the intention of using nearest neighbour analysis in the direction in which there are rows of nine plots. Each treatment will occur three times and should have five or six different neighbouring treatments in the occurrences. This is a minimal level of requirement and the range of possible designs is enormous. If we add a further restriction that treatments should not be repeated in a row (to be able to allow efficiently for differences between rows), then the design shown in Figure 11.6(a) is quite easily constructed, there still being many points of choice. The randomisation of this design could involve permutation of the four rows, 'rotation' of the nine columns keeping adjacent columns together, and permutation of letters. Such a randomised version is shown in Figure 11.6(b). Further randomisation could be built into the initial construction of the design.

11.4 Advantages and disadvantages of restricting randomisation

In Chapter 10, the benefits of randomisation were emphasised, namely the removal of subjectivity from the allocation of treatments to experimental units and the justification it gives for the usual form of analysis based on very minimal assumptions about the experimental units.

In this chapter, in contrast, we have emphasised the benefits of restricting the randomisation to improve the efficiency of estimated treatment comparisons by removing the effects of known spatial or temporal covariates. Clearly there is some conflict here.

Considering first the removal of subjectivity in the allocation of treatments to units, this is useful but should not be overemphasised. In particular, it cannot be stressed too much that randomisation should be used to decide on the allocation, after taking account of all known or expected patterns of variation among the experimental units. In other words, it is usually appropriate to restrict the randomisation in some way and the arguments in favour of randomisation do not in any way contradict this. The overwhelming benefits of randomisation should never be misunderstood to be arguments in favour of complete, rather than restricted, randomisation.

In a general sense, therefore, restricted randomisation should always be considered and usually used. Of course, the most commonly used form of restricted randomisation is blocking. Appropriate blocking allows the design to be efficient with respect to many expected patterns of variation, while still retaining both of the main benefits of randomisation. Other forms of restricted randomisation, as discussed in the preceding sections, might gain some efficiency over the best blocking system for very specific forms of patterns of variation, but are usually less efficient with respect to other forms. They still allow some randomisation, so the loss of objectivity in the allocation is not a major problem. However, except in a few special cases which have been devised, restricted randomisations do not allow a valid analysis based only on the finite population and the randomisation. They thus depend much more heavily on the assumptions made about the experimental units.

In general, we would tend towards designs with appropriate block structures included to account for all expected sources of variation. Even if a continuous pattern, such as a linear time trend, is strongly expected, small blocks, for example of two runs each, will allow for efficient treatment comparisons with these unit effects removed. In large experiments, or those which require few blocks to allow for the expected variation, this argument is pretty conclusive. It remains only to emphasise again that small blocks might be needed to properly account for the expected variation among units.

In small experiments, or those in which the appropriate blocking structure would take up too many df, however, the forms of restricted randomisation described in this chapter provide sensible alternatives to block designs. This is particularly true when the appropriate form of blocking would leave little scope for any randomisation on which the analysis could rest, or when it would leave insufficient residual df to allow any sensible inferences to be drawn. The resource equation described in Chapter 1 is a useful guide in these circumstances.

For example in a 16-run industrial experiment, we might be reluctant to use more than 1 df for control and it might be expected that there will be a linear trend. Rather than use the df to define two blocks of size 8, which might be too large to give homogeneous responses, it will often be sensible to sacrifice the randomisation basis for the analysis (which might anyway be too approximate) and choose a design which is robust to the linear time trend, which can then be included in the analysis. In interpreting the experiments, however, it must be kept in mind that all conclusions are conditional on these assumptions. In such circumstances, there is a strong case for carrying out a Bayesian analysis, so that uncertainty about the assumptions made can be allowed for to avoid overoptimistic conclusions being drawn. For example in the 16-run industrial experiment, a quadratic trend could be allowed for with a fairly small prior probability. Designing for such analysis is an interesting area for further research.

11.5 Ignoring blocking in the data analysis

In some applications, blocks are used not because of expected patterns of variation in the response, but purely to simplify the management of the experiment. For example, an industrial experiment to be run on a process believed to be homogeneous might be split into two blocks as an insurance against early stopping of the experiment for unforeseen reasons. The argument is that the single block completed should include an informative set of treatments – for example, with complete blocks, we at least have a single replicate of each treatment. In clinical trials blocks are frequently used to ensure near balance between the treatments at all stages of recruitment. In many agricultural field trials, blocks are introduced for management purposes, for example to allow each data manager to look after a single block. In these cases, there is nothing about the runs of the process, the trial subjects, or the field plots which would lead us to expect systematically different responses from different blocks.

In circumstances like these, it is tempting to see the inclusion of the block effects in the analysis as simply wasting df. Under the believable assumption that the true block means are all equal, the expected block MS is σ^2 and it can be argued that the block SS should be pooled with the RSS in order to improve the estimates of SEs. In other words, the randomised block design should be analysed as if it had been completely randomised. Indeed, in many data sets, such an analysis can show an apparent improvement in precision.

We object to such a practice in the strongest possible terms. The first reason is that the analysis suggested is not justified by the randomisation and hence is absolutely dependent on the block effects really being zero. From a more practical point of view, the idea that sets of units which are managed together are not expected to be more similar than other units is extremely far-fetched. If industrial experimenters are concerned that the second half of the experiment might be lost due to extreme unexpected occurrences, surely less extreme unexpected occurrences are likely to lead to differences in response between the halves of the experiment. If balance throughout a clinical trial matters, it can only be because of the possibility of variation throughout the trial. If it is convenient for field trial data managers to work with a block each, this ease of working must surely lead us to expect the results from a single manager to be more homogeneous than those from different managers.

The concept of dropping block effects from the analysis seems to be based on a misunderstanding of what is being analysed. If the process, subjects, or plots were in normal use without any intervention in the form of an experiment, it might well be true that they would be expected to give similar responses. However, the very act of experimenting in the way we do changes the objects of interest. The experimental units are not just the runs of a process, the clinical subjects or parts of a field; they are these objects as they are used in the experiment, i.e. with some intervention to their normal condition. It is indefensible to intervene in a way that ensures that there will block effects, but to then analyse the data assuming there will be no block effects.

A variant on excluding block effects from the analysis which is even more objectionable is to exclude them only if they turn out to be small, for example not significant at 10%. This, of course, leads to an utterly meaningless analysis of the data. In Chapter 4, we noted that tests of the block effect are usually irrelevant and unbelievable and they do not arise from the finite population model of Chapter 10 under randomisation. Further, even if one believes this test, the impact of carrying it out for model selection cannot be ignored when assessing the treatment effects. For example, if the block SS are pooled with the RSS only when they are small, then if the block effects really are zero, the residual error estimator we use is clearly negatively biased. This in turn biases the estimated SEs of the treatment effects and therefore gives incorrect inferences. The impacts of these biases on the conclusions are, in general, very difficult to assess. Why any experimenter would want to use wrong inferences, with no idea of how wrong they can be, when correct inferences are available, is a question which we think cannot be answered.

The discussion in this section has been about a primary analysis of the data to draw inferences from the experiment. Sometimes, the data are also used to provide clues about how future experiments can be better designed. In this case, there really can be no objection to pooling block SS with RSS in order to estimate σ^2 for the purposes of designing a new experiment. However, since the main purpose is only to get an order of magnitude idea of what σ^2 might be, it also does not have any obvious advantages. The only exception is perhaps when there are very few residual df and a surprisingly small estimate of σ^2 is obtained. Then, if the block MS is not very large, one might consider pooling SS to get a very rough estimate of σ^2 for use in designing the next experiment. It would be important to recognise, however, that such an estimate cannot be considered very reliable.

11.6 Covariance or blocking

The ideas of covariance analysis appear to offer an alternative to blocking. Since we can adjust comparisons between treatments to allow for variation in a covariate, we can use as many covariates as may be thought appropriate and adjust treatment comparisons simultaneously for all the covariates considered. Why then should we bother with blocking at all? Is there any reason why we should not simply allocate units to treatments quite arbitrarily, and allow for any disparities between the units in respect of measurable covariates by a multiple covariance adjustment? Are blocking and covariance adjustment simple alternatives?

The arguments of Section 4.8 about the difficulty of treatment comparison when the values of the covariate differ substantially between treatments suggest that total abandonment of blocking in favour of covariance adjustment could lead to unfortunate patterns of covariate values which could very much reduce the precision of treatment comparisons. Hence, some degree of prior control of treatment allocation would seem desirable, but how much? What should the balance be between blocking and covariance?

In many forms of experiment the primary control of variation is attempted by blocking; adjustment by covariates is introduced to improve precision, usually as an afterthought prompted by some unusual circumstances. There will usually be only one, or occasionally two, covariates, and these will most frequently occur as variables measured after the commencement of the experiment, to record additional information not available or not considered before the experiment. In such experiments, the intention to use covariance adjustments has no implications for the design of the experiment. However, in clinical trials,

Table 11.1. *Numbers of patients in 16 combinations from four blocking factors*

		X_1		X_2	
		W_1	W_2	W_1	W_2
Y_1	Z_1	3	5	4	4
	Z_2	6	4	8	1
Y_2	Z_1	3	7	2	5
	Z_2	8	0	3	1

covariance has come to be regarded as the principal method of controlling variation and this has an important influence on experimental design.

In clinical trials, a great deal of information is available on each individual experimental unit, which is usually an individual patient. Each patient can be classified by age, sex, occupation, various physical characteristics and by factors appropriate to the particular experiment. There are many potential blocking factors, or covariates. Even if the clinical trial is planned with full information about all patients to be included in the trial available at the outset of the trial, then the use of blocking will not be simple. Patients will not occur in equal numbers for each class of each blocking factor. Typically, the pattern for four blocking factors each with two classes might be as shown in Table 11.1. In most clinical trials, moreover, patients enter the trial sequentially, so that it is impossible to designate a precise blocking structure in advance of the experiment. Methods of sequential allocation intended to avoid extreme forms of non-orthogonality by using restricted, or partially determined, randomisation will be discussed in Section 11.7, but these have not yet been widely accepted. This situation has led some medical statisticians to advocate covariance adjustments as the principal control technique for achieving precision in clinical trials with blocking, or stratification, relegated to a minor, or even non-existent, role.

To assess the relative importance of blocking and covariance, we must reconsider the advantages and disadvantages of each.

Blocking is highly effective when the pattern of variation between experimental units is recognisable in terms of a set of distinct groups. If potential variation between units corresponds to qualitative differences between units, then blocking is natural. The use of blocking is most advantageous when treatment effects are orthogonal to block effects. The benefits of orthogonality are partly in making the analysis of results easy, but principally in allowing interpretation of treatment differences to be independent of block effects.

If the blocks are not orthogonal to treatments, then, provided the non-orthogonality is not too large, the interpretation of treatment differences is not substantially affected by block differences. Orthogonal designs give independence of interpretation, but impose severe restrictions on block sizes. Mildly non-orthogonal designs allow less restrictive conditions on block sizes, with only a small amount of ambiguity in the interpretation of treatment effects. A further major advantage of blocking or stratification is that it allows the possibility of detecting stratum × treatment interactions when the block size is greater than the number of treatments as could often be the case in clinical trials.

A final advantage of blocking is for the organisation and administration of the experiment. If it is necessary for the experimental units to be assessed on different days or by different assessors the block structure provides an ideal structure for the control of any such management variation.

Covariance allows adjustment for as many covariate factors as are thought appropriate. If the form of the covariance relationship is properly identified, then a covariance adjustment for variation between units caused by quantitative differences between units will produce more precise results than the approximation which blocking offers to a continuously varying pattern of yield over the available units. However, if the occurrences of treatments in different classes of a covariate factor are uneven the interpretation of treatment effects will not be independent of the covariate effect. When the unevenness is severe the adjustment of treatment comparisons by the covariate may make the interpretation of treatment comparisons impossible because the precision of comparisons is extremely poor.

Relying solely on covariance adjustment, and allowing totally random allocation of patients to treatments across the whole experiment or even within a few large strata, assumes an avoidance, by chance, of unfortunate allocations. The argument for attempting to avoid unfortunate allocations by design through choosing appropriate blocking factors, with the probability of further adjustment for variation in the blocking factors or other variables by covariance, seems overwhelming.

11.7 Sequential allocation of treatments in clinical trials

For clinical trials in which the number of treatments is small and for which all the patients are available simultaneously at the beginning of the trial the requirements of randomisation and blindness can be satisfied sufficiently by the use of blocks, each of which contains several replicates of each treatment. Thus, patients can be grouped into appropriate blocks such that each block would contain several replicates of each treatment and treatments can be allocated randomly within blocks. The requirement of blindness may lead to the use of larger blocks than would normally be recommended for other forms of experiment. In practice, however, it is common for patients to be admitted to a clinical trial sequentially, and the allocation of treatments must be random and 'blind' while at the same time achieving, as nearly as possible, balance of the treatments over several different blocking factors. Methods for achieving this sequential balance, while retaining the element of randomisation, are discussed by Pocock (1979). An alternative scheme using traditional blocking methods can also be devised.

The principles of the restricted randomisation methods of Pocock and others are presented here for the two-treatment experiment which is an extremely common form of clinical trial. Each patient is classified by each of several blocking factor classifications, and the information on the numbers previously allocated to each treatment from each class of each blocking factor is available. Thus, for example, it may be that patients are classified by age (four classes), sex (two classes) and occupation (four classes). The numbers of patients for each class for each of two treatments, A and B, at a particular stage of the trial are shown in Table 11.2. If a new patient, male, under 30, in occupation class IV becomes available

Table 11.2. *Numbers of patients on each treatment, classified by age, sex and occupation*

		Treatment A	Treatment B
Age	<30	10	6
	30–50	12	12
	50–70	4	5
	>70	4	7
Sex	M	17	14
	F	13	16
Occupation	I	5	8
	II	9	13
	III	7	2
	IV	9	7

Table 11.3. *Expected frequencies in different class classifications*

Patient	A : B	Modification to P_A
1 : male	17 : 14	14/17
<30	10 : 6	6/10
IV	9 : 7	7/9
2 : female	13 : 16	16/13
50–70	4 : 5	5/4
III	7 : 2	2/7

for the trial, then allocation to treatment B would improve the balance of all three blocking factors. A female patient between 50 and 70 in occupation III could not be allocated so as to improve the balance for all three blocking factors. Even in the most clear cut case, however, some element of uncertainty must remain, so that clinicians involved with the trial cannot deduce the identity of the treatment, or even deduce that a patient will be allocated a treatment different from that received by a particular previous patient. Hence, treatment allocation must still include a random component.

The restricted randomisation method defines each allocation as a random event, the probability of allocating a patient to treatment A being P_A. This probability is modified to take account of the degree of imbalance of previous treatment allocations for the class of the new patient for each blocking factor in turn. The simplest form of modification assumes a basic probability of 0.5 and modification by the ratio of previous allocations for each blocking factor, n_B/n_A. For the two examples (male, <30, IV) and (female, 50–70, III) assuming the previous allocation patterns shown in Table 11.2 the calculations would be as shown in Table 11.3. The probability that patient 1 will be allocated to treatment A would be

$$P_A = 1/2 \times 14/17 \times 6/10 \times 7/9 = 49/255 = 0.19.$$

Table 11.4. *Expected frequencies*
of numbers of patients in different
classification combinations

		Occupation			
Sex	Age	I	II	III	IV
M	<30	4	4	4	0
	30–50	4	4	4	4
	50–70	2	2	4	4
	>70	0	2	2	4
F	<30	2	4	6	2
	30–50	2	4	6	4
	50–70	0	4	6	0
	>70	0	4	4	0

The probability of allocation to A for patient 2 is

$$P_A = 1/2 \times 16/13 \times 5/4 \times 2/7 = 20/91 = 0.22.$$

There are several other aspects of the restricted randomisation approach which deserve mention. First, more complex methods of modifying the P_A values may be used and the reader is referred to Pocock (1979) for a discussion of these. Second, the restricted randomisation method cannot be used until at least one patient has been allocated to each treatment from each class of each classification (this may not be necessary with modified allocation rules). Third, balance is considered separately for each factor in turn. To allow for all the various different combinations of sex, age and occupation class, balance must be considered for each of the 32 combinations separately. This would usually be too fine a subdivision of the population of patients, giving very small numbers in each combination. However, if the possibility of a very unbalanced allocation for, say, female patients in occupation class I worries the experimenter, then the blocking factors may be redefined to allow balance within this particular combination of classes. Finally, the need for the balancing mechanism to be a probabilistic one is a direct consequence of the need for treatment allocation to be unidentifiable by those directly involved in the trial. This form of secrecy is usually required in medical trials, but not in animal drug trials, and certainly not in agricultural crop experiments or industrial process experimentation. Although the ideas of sequential balancing and restricted randomisation are ingenious, they are, of course, inefficient when the secrecy element is not required, since blocking allocation could then be selected to minimise the treatment comparison variance.

An alternative approach to the problem of simultaneously controlling the variation of patients and achieving blind and random allocation is to define blocks of patient types such that each block may be expected to include, by the end of the sequential arrival of patients, at least three patients per treatment. A random allocation system is then constructed for each block using a block size bigger than that expected. The excess of, say, two replicates per treatment will allow for excess numbers of patients in any block class. To illustrate the procedure suppose that for the trial discussed earlier the expected frequencies for numbers of

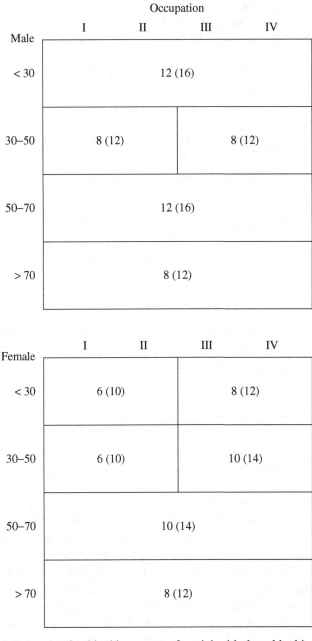

Figure 11.7 Prospective blocking system for trial with three blocking factors.

patients in the combinations of all three classifications are those shown in Table 11.4. Such figures will be approximate, and no effort to make them more exact is useful. If sex is believed to be the most important blocking factor, age the next most important, and occupation least important, then a blocking system as shown in Figure 11.7 might be appropriate, each block being expected to include at least six patients.

Table 11.5. *Probabilities of patients being in different categories*

Blocking factor	Category	Expected proportion
Sex	Male	0.5
	Female	0.5
Age	18–39	0.2
	40–59	0.4
	60+	0.4
Condition	Mild	0.5
	Moderate	0.3
	Severe	0.2
Weight	Underweight	0.1
	Normal	0.4
	Overweight	0.3
	Obese	0.2

Random allocation plans would be prepared for each block by the statistician without the knowledge of the clinician, assuming block sizes four more than those expected. The random allocation for the (male, <30) block would assume a possible block size of 16 ($= 12 + 4$) and might be

A A A B A B B B A B A B B B A A.

If now only nine patients in this category occur in the trial, then there will be five As and four Bs in the block. If 14 patients occur there will be six As and eight Bs. Since the clinician need not know the block size chosen the allocation is effectively blind.

Although the ideas of sequential randomised allocation are usually related to experiments to compare just two treatments, there is no difficulty in principle in adapting the methods for use in experiments with three or more treatments. Indeed, when more than two treatments are used, the dangers of the clinician penetrating the blindness of either random allocation procedure are much reduced simply because of the multiplicity of treatments.

Absolute blindness in the sense of the clinician being completely incapable of making a better-than-average prediction not only of whether the treatment to be allocated to the current patient is more likely to be A or B, but even of whether the treatment will be different to that applied to the last similar patient, is philosophically impossible. The objective of a trial is to compare two treatments, and any system of allocation must tend to equalise the overall replication of the two treatments: hence, the allocation of treatment A to one patient makes the allocation of treatment B to any other particular patient more likely and hence a limited level of prediction must be feasible. What can and should be attempted in any trial is a degree of randomness within a system of control of variation, retaining near blindness. Either the sequential restricted random allocation or the use of large multi-replicate blocks can achieve this combination of objectives.

Exercises

11.1 An experiment to compare five treatments is to be carried out in a greenhouse which contains 20 compartments in a 4×5 array. Write down the coefficients of the linear $\times$ linear component of the row–column interaction, and hence find a row-and-column design which is nearly orthogonal to this covariate, as well as to row and column effects.

11.2 In a clinical trial to compare a new drug with a standard (control) drug, 200 patients will be recruited sequentially. It is expected that the responses of patients will depend on their sex, their age (categorised as 18–39, 40–59 and 60 and above), the severity of their pre-existing condition (classified as mild, moderate, or severe) and their weight (categorised as underweight, normal, overweight or obese). The probabilities of patients being in any category for each blocking factor are assumed to be independent of the other blocking factors and are given in Table 11.5.

Suggest an appropriate blocking scheme:

(i) so that each block might be expected to include at least six patients;

(ii) so that each block might be expected to include at least eight patients.

Part III

Second subject

12

Experimental objectives, treatments and treatment structures

12.0 Preliminary examples

(*a*) An experiment is proposed to examine the effects of water stress on plant growth and development. It is already determined that the experiment will include plants of three varieties sown at two different sowing dates. The experimental unit is a single plant in a pot. About 50 plants for each variety/sowing date combination are available. The remaining design question concerns how to define the set of treatments for assessing the effects of water stress. Periods when water stress can be applied can be defined in terms of calendar date or of physiological stage of the plants. Several stress periods can be used – should the number be two, three, four or even more? For a given number of stress periods the number of experimental treatments remains to be chosen. Thus, for example, with three stress periods (A, B, C), there are eight possible experimental treatments as in Table 12.1.

From this complete set of possible experimental treatments, sensible choices of the set of treatments to be used in the experiment could be:

  (i) (1), (2), (3), (4) and (8), or
 (ii) (2), (3), (4) and (8), or
(iii) (1), (5), (6), (7) and (8), or
(iv) all 8, or
 (v) (8), (2), (3), (5) and (1)

or possibly other subsets.

(*b*) An experiment to compare different methods of control of an agricultural crop pest has 240 plants in individual pots available as experimental units. Two chemicals, O and E, are to be compared with a standard control treatment, and with an untreated control. O is an oil which requires a surfactant for application and two surfactants, S_1 and S_2, are to be compared in the experiment. Three forms of sprayer (Ed, U1 and Con) are available for comparison, but only Ed and U1 can be used with O and only U1 and Con with E. The experimenter would like to compare three concentration rates for each of the two new chemicals, O and E, but a single rate is sufficient for the standard control treatment. What is the treatment structure and what replication is appropriate for each treatment combination?

12.1 Different questions and forms of treatments

We have already mentioned, in Chapter 3, that there are many different kinds of experimental objectives, and that the choice of treatments to be used in an experiment requires careful

Table 12.1. *Treatments in water stress experiment*

	A	B	C
(1)	Stress	Stress	Stress
(2)	Stress	No stress	No stress
(3)	No stress	Stress	No stress
(4)	No stress	No stress	Stress
(5)	Stress	Stress	No stress
(6)	Stress	No stress	Stress
(7)	No stress	Stress	Stress
(8)	No stress	No stress	No stress

consideration. Although it might seem that the statistician, in attempting to emphasise the importance of relating the choice of experimental treatments to the objective of the experiment, is providing quite superfluous advice, practical experience suggests otherwise. Many experiments are conducted without any precise description of the purpose of the experiment; the treatments are chosen first and the questions that can be answered by comparison of the effects of those treatments are allowed to emerge. As one paper in a reputable journal put it (in 1979), 'The purpose of this experiment is ... to compare the experimental treatments.'

There are many different questions that can be asked before an experiment is performed to answer the questions; some examples of these questions are:

 (i) Which is the best method (within a defined set of options) of growing a particular crop, or of operating a chemical plant?
 (ii) What is the pattern of yield response to a stimulus whose level can be varied?
(iii) Which are the main causes of variation in the yield, or output, from a biological or chemical process?
(iv) What are the advantages, including perhaps the economic advantage, of a new form of medical treatment?
 (v) To what extent are the effects of different levels of one treatment factor dependent on the particular level of another treatment factor?
(vi) Which, if any, of a large array of chemicals have any effect on a disease?

Each form of question implies a rather different form of experimental treatment. In addition, there are three major distinctions which should be drawn between types of treatment or of treatment structures which will be discussed further in later sections. These are:

 (i) the distinction between treatment factors whose levels are qualitative (different diets, different breeds of animal), and those which are quantitative (drug concentration, diet amount);
(ii) the distinction between treatments which are a representative sample from a population (varieties from a breeding programme), and those which are selected individually because they specifically are of interest. The latter are sometimes referred to as fixed, in contrast to the former, which are referred to as random;

(iii) the distinction between factorial structure for treatments (introduced in Chapter 3) and treatments with a lesser degree of structure or even with no deliberate structure. Sometimes the treatments may be simply a collection of alternatives with no apparent structure. Intermediate structures could involve partial factorial or grouped structures.

Regardless of the precise nature of the treatments, it is important to emphasise that the practical operation of the experiment, including randomisation and the initial form of the analysis, the analysis of variance, is exactly the same once the set of experimental treatments has been chosen, independent of all these distinctions of treatment types. The initial analysis is determined only by the design structure of the experimental units and of the allocation of treatments to those units. The ways in which subsequent analysis will be modified for each of the three treatment types in turn will be discussed in Sections 12.4, 12.5 and 12.6, with consideration in Section 12.7 of the special problems of screening and selection. First, we consider the general problems of interpreting comparisons between treatments, and the presentation of results.

12.2 Comparisons between treatments

The theory of linear contrasts, including the concept of orthogonal contrasts, was developed in Chapter 4. We now consider the use of orthogonal contrasts and other comparisons in interpreting the effects of treatments from experiments.

The classical statistical approach to the interpretation of experimental results advocates the specification, before the experiment is started, of a set of p orthogonal comparisons, $L_1, \ldots, L_p$. Orthogonality of comparisons enables each comparison to be interpreted independently. Statistical theory suggests that there are two approaches to interpreting a set of comparisons, either by considering estimates of the comparisons, or through testing the significance of the comparisons. In general, in reporting the results of experiments there is much more emphasis on testing the significance of comparisons than on estimates of treatment comparisons. The emphasis on testing rather than estimation is surprising for two reasons: (i) the questions exemplified in the previous section are much more concerned with estimation than with testing; and (ii) experimenters do not in general conduct experiments to compare treatments which they expect to have identical effects. Hence testing a null hypothesis that treatments all have the same effect is not likely to be a sensible approach to interpreting the results of an experiment.

Nevertheless, because most experimenters seek to present their results through a series of significance statements it is necessary to spend some time explaining why some of the most frequently used methods of testing multiple null hypotheses are to be avoided. When a set of orthogonal comparisons has been defined, the fact that several significance tests are made simultaneously, and the probabilistic nature of the significance test, leads to consideration of the probability of a false significance, or type I error, over the whole set of comparisons. If we assume that s^2 is approximately equal to σ^2, then for p orthogonal comparisons,

prob ($\geq$ one significant result in the situation where all null hypotheses are true)

$$= 1 - (1 - \alpha)^p \simeq p\alpha,$$

where α, the type I error probability of each test, is assumed small (for example 0.05 or 0.01). The overall probability of a type I error is then

$$1 - (1 - \alpha)^p,$$

which will be much larger than α if p is substantial.

One approach to this problem has been to define an experimentwise error rate as the overall probability of a type I error if all the null hypotheses are true (which as has already been pointed out is improbable in most experiments). If this experimentwise error rate is to be a prescribed level, α, then the type I error probability for each orthogonal comparison test, α', must satisfy

$$1 - (1 - \alpha')^p = \alpha,$$

or

$$\alpha' = 1 - (1 - \alpha)^{1/p}.$$

The classical approach to interpreting the results of experiments has another, practical, difficulty, and this is that the appearance of the results may suggest additional questions beyond the specified p orthogonal comparisons. The classical response to this situation is to suggest a further experiment to test the hypotheses generated by these new questions. This is a theoretical approach to research which the shortage of resources will rarely permit and, in attempting to provide assistance to the justifiable desire of experimenters to interpret their results using information suggested by those results, statisticians have developed a wide range of *multiple comparison methods* or *automatic testing methods*.

12.2.1 *Problems of multiple comparison methods*

Although each of the methods for multiple comparisons was developed for a particular, usually very limited, situation, in practice these methods are used very widely with no apparent thought as to their appropriateness. For many experimenters, and even for editors of journals, they have become automatic in the less desirable sense of being used as a substitute for thought. The scale of the misuse and the narrowness of the situations for which each method is appropriate leads one to agree with Nelder (then Head of the Statistics Department at Rothamsted), who said in the discussion of a review paper by O'Neill and Wetheril (1971) on multiple comparison methods that 'In my view multiple comparison methods have no place at all in the interpretation of data'. We recommend strongly that multiple comparison methods be avoided unless, after careful thought to identify the situation for which the test you are considering was proposed, you decide that the method is exactly appropriate.

To illustrate why we make this recommendation, we shall consider one particular test as an example. The test is Duncan's test, and it was originally proposed for interpretation of a set of treatments which have no structure or pattern; the treatments can be identified simply as A, B, C, ..., where even the ordering of the letters has no implication of ordering for the corresponding treatments. Duncan's test considers the ordered set of treatment mean yields, and tests each group of two, three, four, or more, successive means to assess whether the spread between the highest and lowest of each group is significantly large in terms of the experimental error MS, allowing for the number of treatments in the group being tested.

Table 12.2. *Results using Duncan's multiple range test*

Substrate potassium concentration (mM/1)	Foliage dry weight	Roots dry weight
0	4.47d	0.67e
0.25	18.74c	4.51d
0.50	26.40bc	6.30c
1.00	27.34ab	7.48bc
1.50	31.87ab	9.07ab
3.00	32.31ab	8.91ab
6.00	31.68ab	8.46ab
9.00	34.85a	9.39a

The result of using the test is the identification of, possibly overlapping, groups of treatments, such that all the treatments in each group might reasonably be presumed not to differ in their effect on yields.

Examples where Duncan's test is inappropriately applied are many, and occur in almost every volume of every applied science journal in which any statistical analysis is used except where they are replaced by an alternative test of the same general intention. The unthinking use of tests such as Duncan's has been attacked in the very journals where examples of such use occur frequently, by various statisticians including Bryan-Jones and Finney (1983), Maindonald and Cox (1984), Morse and Thompson (1981). However, these tests are embedded in many computer packages and it is important to continue to point out why they are so misleading and inappropriate. A few examples are given here to illustrate some of the forms of misuse.

12.2.2 First example

The simple example shown in Table 12.2 is taken from an experiment to investigate the relationship between potassium level and growth of plants. Any attempt, such as was made by the authors of the paper from which Table 12.2 was taken, to interpret differences between treatments in terms of the Duncan's test information rapidly leads to confusion. Thus for foliage the potassium levels from 1.0 to 9.0 are not significantly different. Level 9.0 is signif-icantly different from 0.5 and lower levels but levels 1.0, 1.5, 3.0 and 6.0 are not significantly different from 0.5. The foliage weight for level 0.25 is also not significantly different from that for level 0.5 but is different from the weights of all higher levels. Fortunately everything is different from level zero. This last statement is true also for root weights but all the other statements have to be modified (slightly).

The propagation of such nonsense may be described as unfortunate or in stronger terms. However, the real wickedness of the use of Duncan's test in this situation, as in many others, is that the true interpretation of the data was submerged in the swamp of significance statements. The true interpretation is obtained from a graph of the results as shown in Figure 12.1. There is

Table 12.3. *Factorial means with Duncan's multiple range test*

| Methionine | Sodium sulphate | | | | |
	0%	0.08%	0.16%	0.24%	0.32%
0%		53a	54a	56a	56a
0.08%	114b	147d	141cd	145d	119bc
0.16%	170e	216fg	217fg	223fg	208f
0.24%	235gh	266ij	266ij	281jk	264i
0.32%	249hi	288jk	300kl	309l	304kl

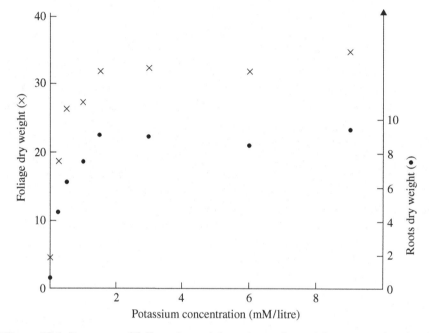

Figure 12.1 Response of foliage dry weight and root dry weight to potassium level of the substrate.

a clear asymptotic response pattern for increasing potassium concentration, both for foliage and root dry weights. The superimposed plots also demonstrate the rather more gradual nature of the approach to the asymptote for root dry weight.

12.2.3 Second example

An even more impressive piece of misrepresentation is shown in the next example in which a factorial structure for two components of the diet of chicks requires 12 letters to summarise the results of Duncan's test for 24 treatments, shown in Table 12.3.

Again graphical representation makes everything clear, which certainly cannot be said for the poor authors. In Figure 12.2 we can see that the effect of methionine is very consistent,

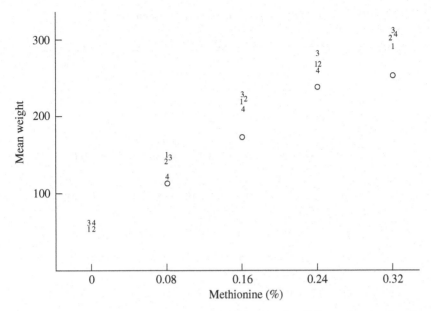

Figure 12.2 Response of chick growth to methionine for different levels of sodium
sulphate 0% (0), 0.08% (1), 0.16% (2), 0.24% (3), 0.32% (4).

for all levels of sodium sulphate. The response pattern is rather weaker for the zero sodium
sulphate level and is slightly reduced also for the 0.32% sodium sulphate level. There is a
slight suggestion that the effect of methionine might continue to increase further at this level
of sodium sulphate. Further insight might be gained by fitting response functions. Some
might argue that if the obtrusive letters are removed from Table 12.3 the results are clear
enough, though we prefer the graphical representation.

12.2.4 Third example

Our third example illustrates an alternative form of presentation of the results of a Duncan's
test in which the treatments are ordered so that the means increase consistently across the
page. The effects of performing this illusion for measurements at four different dates, for
the five treatment levels being compared, are shown in Table 12.4. In this case the attempt
to unravel the mystery will be left to the reader, though it should be pointed out that for day
25 the author of the results has confused himself into making two incompatible significance
statements.

The indictment of multiple comparison methods is on three grounds:

(i) they are regularly used in inappropriate situations,
(ii) they are used to test hypotheses known to be false, on the evidence of the overall
treatment F test, and
(iii) most importantly, they are used in a routine fashion, and thereby divert many experimenters from a proper analysis of their data.

Table 12.4. *Liver/body weight* × *100 for five levels on each of four measurement dates*

Day		Doses and yields					Overall significance of treatments (*F* test)
11	dose	0.8	0.4	0.0	1.6	3.2	
	yield	2.26	2.29	2.36	2.42	3.19	$p \leq 0.01$
18	dose	0.8	0.4	1.6	0.0	3.2	
	yield	2.35	2.87	3.10	3.23	4.03	$p < 0.01$
25	dose	0.8	3.2	0.4	1.6	0.0	
	yield	2.66	2.72	2.76	2.80	2.85	$p < 0.01$
32	dose	0.8	1.6	3.2	0.4	0.0	
	yield	1.85	2.18	2.38	2.50	2.50	$p < 0.05$

Treatment means not significantly different at 5% using Duncan's test are indicated by a common underline.

Even the simple idea of experimentwise error rates is difficult to justify in its simplest form. The experimentwise error rate depends on the idea that it is useful to consider the situation where all null hypotheses are true. However, not only is this situation improbable because experimenters rarely investigate, experimentally, situations where it is credible that all treatments have identical effects, but, in practice, an experimenter only looks at his or her treatment mean yields if the overall treatment *F* ratio gives some indication that there are some genuine differences between treatment mean yields. What might be practically useful is some guide to the real significance of apparent effects, given the acceptance that some effects are genuine and that some other effects are negligible. But it would be impossibly complex to devise a general procedure for such situations.

We recommend therefore: that multiple comparison methods be avoided; that the idea of experimentwise error rates be retained, but only as a general principle, helpful in the practical interpretation of data which involves the use of significance tests for several orthogonal comparisons.

12.3 Presentation of results

In the later sections of this chapter, we consider some particular approaches to the presentation of results, but it is useful first to consider some general principles of presentation. We believe the first principle should be that the reader of the presented results should be given information in a form which makes it easy to make his or her own interpretation of the results. This means, for example, that estimates for all *relevant* treatment comparisons should usually be provided, together with SEs for all comparisons and df for the SEs. Ideally the computer package should allow or encourage the experimenter to define their own comparisons to match their questions. The alternative approach of quoting least significance differences or equivalently the smallest difference which, if tested, would be found to be significant at a particular level, reduces the information available to the reader by forcing the use of the particular significance level chosen by the experimenter, restricts any attempt to extend the comparisons beyond a simple comparison of two means, and should be avoided in a responsible package.

Sometimes, the need for a statistical analysis to provide a concise summary of information will result in the presentation of only a subset of means or possibly of means for groups of treatments. It is still important to present SEs and their df, but it is also then desirable to provide a summary of the pattern of between-treatment variation, and this will usually be most effective in the form of an analysis of variance. Such a summary should at least include all MS and df. The practice of quoting only *p*-values is indefensible.

Presentation in the form of diagrams is often effective, but it is extremely easy to mislead either by design or inadvertently. Figures should not be used when the same information can be conveyed in tabular form, because the tabulated information is usually more precise. The principal benefit of diagrams is undoubtedly for displaying quantitative relationships. In addition to the greater precision of tabulation, it is also much easier to provide information on precision for means in tables than in figures. The use of vertical bars to indicate standard errors in a diagrammatic representation is fraught with possibilities of confusion. In many cases, we believe that the wish to include precision indicators in a diagrammatical representation should suggest that tabular presentation is a superior alternative.

One practice which should be avoided is the expression of results for all treatments as percentages of the result for a single treatment, this being usually some form of control treatment. Although from one viewpoint this is only a change of scale, the fact that the divisor is an estimate, and is therefore uncertain, makes statements of precision much more difficult. In particular, if SEs are scaled by the same factor that is used for the treatment means, the resulting SEs will be underestimated so that differences between means will be interpreted as more extremely significant than they should be. A further implication of presenting mean yields relative to a control mean yield is that the yields should be considered on a relative basis which is best achieved by transforming yields to a logarithmic scale.

The last general point about presentation must be a reminder that interpretation must match the objectives of the experiment. There is no such thing as a correct standard form of presentation for experimental data.

12.4 Qualitative or quantitative factors

The distinction between qualitative and quantitative factors is usually quite clear. In using levels of a quantitative factor, we are concerned with the relationship of output with the varying levels of the factor. The interest in the relationship may be direct – we may wish to establish the form of relationship – or it may be indirect. An example of an indirect use of a relationship is an attempt to identify the optimum level of the factor, for which it is necessary to make some assumption about the form of the relationship, at least in the neighbourhood of the optimum. However, the fact that there is a numerical component to the definition of a factor level does not necessarily make the factor quantitative. If we are considering various additions to a chemical process, the amount of the additive must be specified but, if the only two alternatives considered for this additive are (*a*) none, or (*b*) some specified amount, then the additive is essentially a qualitative factor.

Most results for qualitative factors will be presented simply as mean yields, with SEs. When there are only two levels presentation will often be more effective in the form of differences, particularly where there are other factor levels or treatments. For example in the experiment on water uptake of amphibia, discussed in Chapter 3, when the analysis of variance had demonstrated which main effects and interactions of the three qualitative factors were important, the interpretation was based primarily on differences between the two levels of the hormone factor, or on differences of these differences.

With more than two qualitative treatments there will usually be a structure for interpretation implied by the choice of treatments – otherwise we would have to deduce that the experimenter had no reason for selecting the treatments! Comparisons may be in the orthogonal form discussed in Section 4.7, but this may not always be appropriate, the most frequent example being when a control treatment is used. A more subtle example is shown in the following.

Example 12.1 An experiment to compare the effects of four diets on the growth of young chicks included 32 cages arranged in eight columns (brooders) of four tiers with each cage containing 15 chicks. The treatments were:

C: a control diet including 0.8% lysine,
L: a lysine enriched diet, being C + 0.2% of lysine,
W: a wheat diet in which some starch and sugar of diet C is replaced by wheat to produce an extra 0.2% lysine,
A: an amino acid diet in which some other amino acids present in W but not in C or L are added to L.

To understand the interpretation of the experimental results it is necessary to consider the background to the choice of experimental treatments. Previous research had shown that the addition of lysine to the diet of young chicks enhanced their growth and that there was an apparently linear relationship between growth response and the intake of lysine. A diet in which wheat was included to provide the desired additional lysine had been found not to promote growth as effectively as the equivalent lysine-enriched diet. The present experiment was intended to provide information on why the wheat diet was not as effective as expected.

Two possible explanations were to be investigated. First, the wheat diet might include other amino acids which are deleterious to growth. To investigate this explanation the fourth treatment (A) is included. This adds to the lysine diet some of the amino acids present in wheat, which it was thought might cause the relative failure of the wheat diet. The second idea was that the defect of the wheat diet might be that chicks do not eat so much of it as of the lysine diet and so in practice do not get the intended amount of lysine. It is not possible to monitor the amount of each diet eaten by individual birds or by the set of birds in each individual cage, but the total intake for each diet can be recorded. The lysine intake per chick can be estimated for each of the four diets and these intake figures can be used in the assessment of the treatment effects. The design and mean weight data are shown in Table 12.5(a). The results of the initial analysis of variance are shown in Table 12.5(b).

The treatment means are

C: 166,
L: 240,
W: 190,
A: 218,   SED = 5.9.

The control diet, C, is not the primary point of reference, being included merely as an origin. Both W and A might be expected to have effects equivalent to L and so the basic comparisons

Table 12.5. *Experiment of chick diets: (a) treatment design and mean weights; (b) analysis of variance*

(a)

Tier	Column							
	1	2	3	4	5	6	7	8
1	161C	246L	186W	229A	235L	164C	196A	182W
2	228L	199W	251A	170C	186W	233A	171C	235L
3	224A	169C	241L	183W	159C	191W	251L	175A
4	190W	210A	175C	247L	222A	236L	201W	158C

(b)

Source	SS	df	MS
Brooders	1756	7	251
Tiers	943	3	314
Diets	22473	3	7491
Error	2553	18	142
Total	27725	31	

$L - W$ and $L - A$ are important. A further comparison of interest is to examine whether the reduced benefit of the W and A treatments can be explained by variation in the food intake.

Figure 12.3 shows the treatment means plotted against lysine intake (milligram per chick per day) and provides the basis for testing whether the reduced performance of the W diet is attributable to the reduced intake of food (and hence of lysine). Assuming, as previous results show is reasonable, that the increased performance due to lysine is linearly related to actual intake of lysine, the expected mean yield for either of the W or A treatments would be given by the value on the straight line joining points C and L in Figure 12.3, at the corresponding food intake value.

The observed intakes are

C: 14.7,
L: 22.6,
W: 20.9,
A: 21.2.

The additional food intake for W compared with C is 0.78 of the food intake increase from C to L. The comparison of observed and expected performance for W is therefore

$$W - [C + 0.78(L - C)]$$

or

$$W - 0.22C - 0.78L,$$

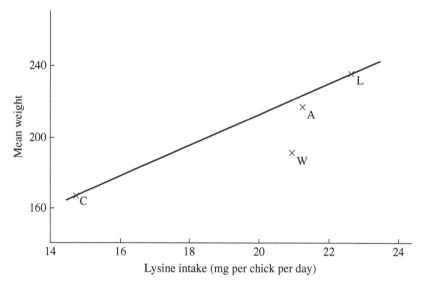

Figure 12.3 Mean weight of chicks plotted against lysine intake for the four experimental treatments.

for which the estimated effect is

$$190 - 0.22(166) - 0.78(240) = -33.7.$$

The SE of this estimate is

$$\{(s^2/8)(1 + 0.22^2 + 0.78^2)\}^{\frac{1}{2}} = 5.4.$$

Since the deficit of the W treatment compared with its 'expected' value based on linear interpolation between C and L is between five and six times its SE it is clear that the reduction of yield for W is not attributable solely to the reduced intake.

If we perform the same comparison for diet A the comparison is

$$A - [C + \{(21.2 - 14.7)/(22.6 - 14.7)\}(L - C)]$$

or

$$A - 0.18C - 0.82L.$$

The estimated effect is

$$218 - 0.18(166) - 0.82(240) = -8.7$$

and the SE of the estimate is

$$[(s^2/8)\{1 + (0.18)^2 + (0.82)^2\}]^{\frac{1}{2}} = 5.5.$$

The deficit for A below its 'expected' value is clearly not significant.

The conclusions from the experiment are therefore clearly that the failure of the wheat diet is not explained either by the reduced intake of chicks on that diet, or by the extra amino acids in the wheat diet which were included in the A diet.

Table 12.6. *(a) Analysis of variance and*
(b) treatment means for rice spacing experiment

(a)

	SS	df	MS
Blocks	5.88	3	1.9
Treatments	23.14	9	2.57
Error	12.42	27	0.46

(b)

Treatment	Mean yield	Density (plants/m^2)	Area per plant (cm^2)
30 × 30	6.02	11.1	900
30 × 24	6.48	13.9	720
30 × 20	7.19	16.7	600
30 × 15	6.92	22.2	450
24 × 24	7.85	17.4	576
24 × 20	7.78	20.8	480
24 × 15	7.50	27.8	360
20 × 20	7.71	25.0	400
20 × 15	8.70	33.3	300
15 × 15	8.20	44.4	225

When the treatments are quantitative, the questions being asked by the experimenter should also be quantitative. Such questions are rarely appropriately answered by simple comparisons of pairs of treatment means. The use of three, or more, levels of a quantitative factor implies an interest in the response pattern of yield to increasing amounts of the factor. The experimenter may be interested in estimating the level of the factor which produces maximum yield, or in some other specific characteristic of the response function. Alternatively, he or she may be interested in a more general description of the forms of response, examining the fit of several different functions to the data. In all cases, a proper first step is to plot the treatment mean yields, y, against the quantitative levels, x.

Example 12.2 In Chapter 2 we examined the analysis of an experiment on rice spacing from which the analysis of variance and the set of treatment means are reproduced in Table 12.6. The yields clearly increase as the density increases, or equivalently as the area per plant decreases. Both forms of relationship are displayed in Figure 12.4. The reader may like to assess which of the two plots appears to give the more linear relationship.

The treatment SS can be split into components for regression and residual using the contrast SS method of Chapter 4:

$$\text{SS (regression)} = \sum_j T_j (x_j - \bar{x}) \bigg/ \left\{ 4 \sum_j (x_j - \bar{x})^2 \right\}.$$

Table 12.7. *Two alternative analyses of variance for the rice spacing experiment*

Source	SS	df	MS
Regression on area	16.79	1	
Residual for area	6.35	8	0.79
Regression on density	13.40	1	
Residual for density	9.74	8	1.22
Error	12.42	27	0.46

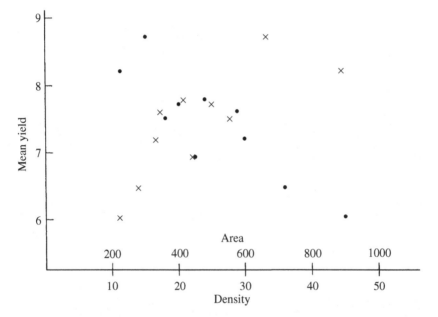

Figure 12.4 Mean yield of rice plotted against plant density ($\times$) and against area per plant ($\bullet$).

The two alternative explanatory variables give the SS summarised in the analysis of variance shown in Table 12.7.

The regression SS is very large in either case and there is no benefit from formally examining the ratio of regression MS to error MS. The residual MS is of more interest because it provides information on whether the fit is adequate. Clearly the area regression is considerably better than the density regression. However, the F ratio 0.79/0.46 suggests that there might be further small systematic treatment effects. To examine these we use the regression on area to calculate the deviations of observed yield from fitted yield shown in Table 12.8.

The regression equation is

$$y = 9.11 - 0.0034(\text{area}).$$

Examination of the deviations provides some evidence that the fitted regression overestimates yield when the two spacing dimensions are disparate. In other words better yields are

Table 12.8. *Deviations of yields from fitted regression*

Spacing	Observed mean	Fitted yield	Deviation
30 × 30	6.02	6.07	−0.05
30 × 24	6.48	6.66	−0.18
30 × 20	7.19	7.07	+0.12
30 × 15	6.92	7.58	−0.66
24 × 24	7.85	7.35	+0.50
24 × 20	7.78	7.64	+0.14
24 × 15	7.50	7.89	−0.39
20 × 20	7.71	7.75	−0.04
20 × 15	8.70	8.09	+0.61
15 × 15	8.20	8.35	−0.15

obtained when the spacing is more nearly square, a conclusion which would hardly surprise a biologist.

12.5 Treatment structures

Most sets of experimental treatments will include some structure. A major exception is experiments for screening or selecting from large numbers of drugs or new varieties of field crops, and this situation is discussed in Section 12.7. The other common situation where there is no structure is where only two or three treatments are being compared.

The most important concept in treatment structures is that of factorial structures, introduced in Chapter 3, which will be the basis for the ideas in the next three chapters. Here we reiterate and extend the rather simplified description of factorial treatments in Chapter 3. The basic components of factorial structure are:

factor	set of treatments of a single type,
level of a factor	particular treatment from the set,
experimental treatment	a combination of one level from each factor included in the experiment,
main effect	comparison between levels of a single factor, averaging over levels of all other factors,
interaction	comparison between levels of one factor of the comparison between levels of a second factor (of a comparison . . .) averaging over levels of all other factors.

A complete set of factorial combinations includes all possible combinations including one level from each factor. Within a complete set of factorial combinations, all main effect and interaction effect comparisons can be defined so as to be orthogonal. The set of main effects and interactions then provides a basis for the interpretation of the results.

For each factor, the main effect comparisons will be defined as is most appropriate along the lines described in the preceding section for either qualitative or quantitative factors. Interaction effects for two factors are derived directly from the main effect definitions for those factors. Thus, if an experiment includes four factors, A, B, C and D, with two, three, two and four levels, respectively, and treatment effects and mean yields are denoted by t_{ijkl} and $y_{.ijkl}$, then the main effect of A must be

$$A = t_{1...} - t_{0...},$$

estimated by

$$y_{.1...} - y_{.0...}.$$

A suitable pair of main effects for B would be

$$B = t_{.1..} - t_{.0..}$$

and

$$B' = t_{.2..} - \tfrac{1}{2}(t_{.1..} + t_{.0..}).$$

Then the interaction effects for factors A and B corresponding to these main effect definitions would be

$$AB = (t_{11..} - t_{10..}) - (t_{01..} - t_{00..})$$

and

$$AB' = \left\{ t_{12..} - \tfrac{1}{2}(t_{11..} + t_{10..}) \right\} - \left\{ t_{02..} - \tfrac{1}{2}(t_{01..} + t_{00..}) \right\}.$$

Although these interaction effects have been written in terms of a B effect varying over levels of A, they are symmetrical and can equivalently be written as

$$AB = (t_{11..} - t_{01..}) - (t_{10..} - t_{00..})$$

and

$$AB' = (t_{12..} - t_{02..}) - \tfrac{1}{2}\left\{ (t_{11..} - t_{01..}) + (t_{10..} - t_{00..}) \right\}.$$

The generalisation to the three-factor interaction effects is straightforward:

$$ABC = (t_{111.} - t_{101.} - t_{011.} + t_{001.}) - (t_{110.} - t_{100.} - t_{010.} + t_{000.}).$$

and

$$AB'C = [(t_{121.} - t_{120.}) - \tfrac{1}{2}\{(t_{111.} - t_{110.}) + (t_{101.} - t_{100.})\}]$$
$$- [(t_{021.} - t_{020.}) - \tfrac{1}{2}\{(t_{011.} - t_{010.}) + (t_{001.} - t_{000.})\}],$$

where ABC has been written as the contrast of the AB effect between levels of C, and AB'C as the variation of the C effect over AB combinations to illustrate the symmetry of the various possible forms of writing interactions.

Now, since each interaction effect may be interpreted as a modification of a main effect or of a lower level interaction effect, the factorial structure provides a sequential structure for interpretation. The main effects should be examined first, followed by two-factor interactions, and then three-factor interactions. It will only very rarely be worthwhile proceeding to four-factor interactions, and sometimes the interpretation of three-factor interactions may be ignored. Also, following from the form of definition of effects, there is a general expectation that main effects will be larger than two-factor interactions, and so on, size being most simply measured by MS in the analysis of variance. Where this expectation is fulfilled the interpretation should consist of the consideration first of the important main effects, followed by the discussion of the variation of those effects, as measured by the interactions involving those main effects. If the interactions are of similar size to their related main effects, then the main effects have little importance (being as variable as they are large!). If interactions are negligible, then the interpretation should ignore them.

Factorial treatment structure provides the most influential statistical contribution to the experimental component of scientific methodology. The classical scientific approach to a comparison of two treatments was to attempt to control all other sources of variation, so that the only difference was the comparison of interest. In contrast, factorial structure provides two complementary advantages. First, the main effect of each factor has been examined over a range of conditions, and is of greater validity if the interactions are negligible, but more importantly the possibility of interaction, i.e. variation of the effect, has been allowed for and examined. Second, and even more important, if interactions are negligible, factorial structure allows the effects of several factors to be assessed independently within the same set of resources. Thus, suppose that three changes to standard conditions are to be assessed, and to examine each change would require 12 observations for the standard, and 12 for the changed condition. Then Figure 12.5(a) shows the minimum number of observations (48) required for the classical scientific approach of changing one factor at a time, controlling all others. The factorial approach, shown in Figure 12.5(b) requires only 24 observations – only half the resources. The secret of the factorial advantage lies in the fact that each observation is included in each comparison. This provides 'hidden' replication in the variation of levels of other factors.

The hidden replication applies even if there are some large interactions, provided there are factors not interacting with the factors involved in these interactions. The advantages of hidden replication also apply if the factorial structure is incomplete (because some combinations of levels are impossible or prone to produce quantitatively different results), though the hidden replication will then be reduced for some main effect comparisons (illustrated in the next section) and the range of validity will also be reduced.

Although factorial structure is the most frequently useful form of structure, it is not the only useful form. Sometimes treatments form natural groups and the relevant treatment comparisons consist of an initial subdivision of the overall variation into between and within groups. Within groups, there may be a further structure; often one of the groups may consist of a control treatment or a small set of control treatments.

Example 12.3 Nine formulations of detergent were compared by washing plates one at a time until they were clean. The experimental procedure required that three basins be used at each time so that the observations were made in threes, and a BIB design with 12 blocks was used. The three operators wash at the same rate during the test and the 'yield' is the number of plates washed before the foam disappears. The results are listed in Table 12.9(a), and the within-block analysis of variance, Table 12.9(b), was obtained using the methods of Section 7.5.

(a)

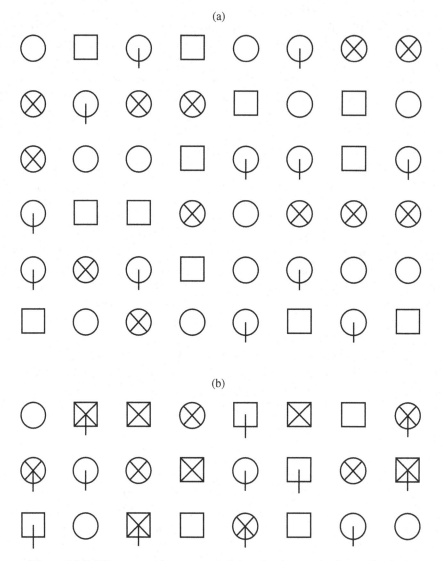

(b)

Figure 12.5 Diagrammatic representations of units required to make three comparisons with specified precision: (a) classical scientific approach, (b) factorial approach.

The treatments and the adjusted treatment means were:

A: base I + three parts additive = 19.8,
B: base I + two parts additive = 17.2,
C: base I + one part additive   = 13.2,
D: base I   = 6.5,
E: base II + three parts additive = 25.3,
F: base II + two parts additive = 23.0,
G: base II + one part additive   = 21.1,
H: base II   = 19.2,
J: control   = 29.5.

Table 12.9. *Data and analysis of variance for experiment on washing plates: (a) data; (b) analysis of variance*

(a)

Block	Treatments and numbers of plates washed					
1	A	19	B	17	C	11
2	D	6	E	26	F	23
3	G	21	H	19	J	28
4	A	20	D	7	G	20
5	B	17	E	26	H	19
6	C	15	F	23	J	31
7	A	20	E	26	J	31
8	B	16	F	23	G	21
9	C	13	D	7	H	20
10	A	20	F	24	H	19
11	B	17	D	6	J	29
12	C	14	E	24	G	21

(b)

Source	SS	df
Blocks (fitted first)	412.25	11
Detergents (fitted second)	1087.30	8
Error	13.20	16
Total	1512.75	35

The comparison of the control with the remaining eight combinations is a natural first step, followed by the subdivision of the SS for the eight combinations into main effect and interaction effect SS.

The subdivision of the treatment SS into components can be calculated directly, because the balance of the design means all component SS are modified by the same factor. Alternatively, it can be calculated by fitting three factors in a general program, the first factor having level 1 for J and level 2 for all other treatments, the second having level 1 for J, 2 for A, B, C and D, and 3 for E, F, G and H, and the third having level 1 for J, 2 for A and E, 3 for B and F, 4 for C and G, and 5 for D and H. The resulting extended analysis of variance is shown in Table 12.10. The very high precision of this experiment makes it inevitable that all treatment effects, including the interaction between base and additive, are highly significant. The pattern of interaction should be examined to assess whether the interaction is large enough to be practically important, as well as significant. Inspection shows that the beneficial effect of the additive in allowing more plates to be washed is much greater for base I than for base II, particularly for the lower levels of additive. Alternatively we can interpret the interaction in terms of the better performance of base II than base I, the difference being large when no additive is used and gradually diminishing as more additive is used.

Table 12.10. *Extended analysis of variance for detergent experiment*

Source	SS	df	MS
Blocks (fitted first)	412.25	11	
Control v. rest	344.29	1	344.29
I v. II	381.60	1	381.60
Additives	311.39	3	103.80
Base × additives	50.02	3	16.67
Error	13.20	16	0.82

12.6 Incomplete structures and varying replication

In developing the ideas of factorial structure later in this book we shall predominantly discuss the case of complete factorial structures. There are several reasons for doing so. The complete structure displays most clearly the benefits of factorial structure, i.e. the capacity to assess interaction and the increased precision due to hidden replication. The mathematical models are neatest when a complete factorial structure is used. And, at a purely practical level, the number of possible incomplete structures is enormous and a systematic discussion of incomplete factorial structures would be impossible.

It is, however, important to stress that, although complete factorial structure represents a powerful ideal, the benefits of factorial structure still hold for incomplete factorial structures. It is foolish to use a complete factorial simply to comply with the simple theory. If particular factorial combinations are practically ridiculous or unrealistic, then they should not be included. Consider the case of a set of treatments derived from three factors each at two levels, for which the combination of all three lower levels is considered impractical. The set of treatments is

$$a_0 b_0 c_1 \quad a_1 b_0 c_0 \quad a_1 b_0 c_1$$
$$a_0 b_1 c_0 \quad a_0 b_1 c_1 \quad a_1 b_1 c_0 \quad a_1 b_1 c_1.$$

Suppose an experiment consists of four blocks of these seven treatments. Then both main benefits of factorial structure are retained at a reduced level. Consider first the information on interaction. Considering only observations at level c_1 we clearly have some information on the AB interaction; using level b_1 there is information on the AC interaction; and similarly for BC at level a_1. In each case we have no hidden replication for the interaction as we would have had with the complete structure but there is still information on all two-factor interactions. Where interactions are negligible, then we still have most of the hidden replication for main effect comparisons. For the main effect of $A(a_1 - a_0)$ we have estimates based on the three combinations $(b_0 c_1)$, $(b_1 c_0)$, $(b_1 c_1)$, giving an effective replication of 12 for the A main effect. Similarly, in the absence of interactions, the effective replication for the two other main effects is 12.

The analysis of the main effects and interactions is computationally more complex because of the non-orthogonality resulting from the incompleteness of the structure. Further, there will be a small ambiguity in the interpretation of effects but this is of negligible practical importance provided the analysis is correctly produced. Essentially as the full factorial structure is allowed to crumble slowly so the analysis becomes increasingly non-orthogonal and the order of fitting of effects and the resultant interpretation requires more care, though it is unlikely that the non-orthogonality will become so large as to produce a practically serious level of ambiguity, as may often occur in multiple regression.

Another situation where the experimenter should consider an incomplete factorial structure is where the complete factorial structure would include identical treatments. A common example is where two factors represent:

 (i) alternative forms of a stimulus, and
(ii) different amounts of the stimulus.

If one of the amounts is zero, then the zero levels for the alternative forms will be identical. It would be absurd to include a whole set of zero treatments simply for completeness.

12.6.1 Different replication patterns

The idea of a zero level, which would be identical for alternative levels of another factor, prompts discussion of another completeness pattern which has been implicitly assumed, namely that all treatments, or treatment combinations, should normally be equally replicated. Suppose that we have three alternative formulations of a chemical stimulus and are to use four amounts including a zero (control) with three replications. Then we have $3 \times 3 = 9$ non-zero treatments plus a control making a total of 30 observations. If the alternative formulations do not have different effects, then we will wish to interpret comparisons between the four means for the four quantitative levels. The three non-zero amount means will be based on nine replications while the zero amount mean is based on three replications. We might therefore consider additional replication of the control treatment. However, if the alternative formulations are very different, then, if we have increased the replication of the control, the means for the non-zero levels for a particular alternative formulation will have less replication than the control.

The choice becomes more clear if we restate the problem in the form of specific alternative designs. Suppose we have 60 experimental units available and two alternative designs are considered:

design I

(three forms $\times$ three amounts + zero amount) $\times$ six replicates = 60 units

design II

three forms $\times$ (three amounts + zero) $\times$ five replicates = 60 units.

The replication of the means for the control treatment, for a particular combination of form and (non-zero) amount, and for the average over forms for a (non-zero) amount, will be as in Table 12.11. For each design we may calculate the variances of different comparisons, and these are also shown in the table. Thus design I gives greater precision for comparisons not involving the control, whereas design II gives greater precision for comparisons involving the control, the greatest difference between the precision achieved in the two designs being for the comparison of an overall amount mean with the control.

Table 12.11. *Different replication patterns and variances*

	Design I	Design II
Replications		
Control mean	6	15
Combination mean	6	5
Amount mean	18	15
Variances		
Control–Combination	$2\sigma^2/6 = 0.33\sigma^2$	$\sigma^2/15 + \sigma^2/5 = 027\sigma^2$
Control–Amount	$\sigma^2/6 + \sigma^2/18 = 0.22\sigma^2$	$2\sigma^2/15 = 0.13\sigma^2$
Combination 1–Combination 2	$2\sigma^2/6 = 0.33\sigma^2$	$2\sigma^2/5 = 0.40\sigma^2$
Amount 1–Amount 2	$2\sigma^2/18 = 0.11\sigma^2$	$2\sigma^2/15 = 0.13\sigma^2$

Table 12.12. *Treatments in pest control experiment*

Chemical	Surfactant added	Sprayer	Concentration 1	2	3
O	S_1	Ed	X	X	X
O	S_1	Ul	X	X	X
O	S_2	Ed	X	X	X
O	S_2	Ul	X	X	X
E		Ul	X	X	X
E		Con	X	X	X
Control		Standard			
Control		Untreated			

12.6.2 *Replication for a more complicated treatment structure*

To illustrate again the way in which decisions about replication for incomplete factorial structures may be considered, we return to the second problem posed in the introduction to this chapter. Two chemicals, O and E, are to be compared with a standard control (S) and an untreated control (C). Chemical O is an oil requiring a surfactant for application and two surfactants, S_1 and S_2, are included. Different sprayers are also to be compared, Ed and Ul for O, and Ul and Con for E. Three concentrations are to be used for the two new chemicals, but not for the controls. The set of treatments is set out in Table 12.12. The total number of distinguishable treatments is 20 and the number of experimental plants available is 240. It might seem that the obvious design would use 12 replications of each of the 20 treatments in a blocked design to control major patterns of variation within the set of available plants. However, such a design is only one of many reasonable alternatives and it ignores the hidden replication implicit in the partial factorial structure. For example, O will be replicated twice as often as E and therefore will be more precisely assessed (unless one of the surfactants

Table 12.13. *Replication for pest control experiment*

Chemical	Surfactant added	Sprayer	Concentration		
			1	2	3
O	S_1	Ed	×1	×1	×1
O	S_1	Ul	×1	×1	×1
O	S_2	Ed	×1	×1	×1
O	S_2	Ul	×1	×1	×1
E		Ul	×2	×2	×2
E		Con	×2	×2	×2
Control		Standard			×5
Control		Untreated			×1

renders chemical O ineffective). Also, the comparison with the controls will be poor because of the relatively low replication of the controls.

Probably the untreated control is included in the experiment merely to assess the general level of pest incidence for this particular experiment, and not for quantitative comparisons with other treatments. This is frequently true for untreated control treatments and it is usually wise not to include such treatments in an analysis of variance with the other treatments, since the error structure for the untreated control is likely to be quite different from that for the other treatments. The replication for the untreated control need not be increased and could even be decreased. But the standard control will be used for comparison with the new chemicals, and will need additional replication. We might therefore consider n_1 replicates of each of the twelve O treatment combinations, n_2 for the six E combinations, and n_3 of the standard control. Precise choice of n_1, n_2 and n_3 will depend on the relative precision required for different comparisons, and must be discussed with the experimenter. At one extreme, $n_2 = 2n_1$, $n_3 = 12n_1$ gives equal overall replication of O, B and S. Since the optimal concentration is not known for O or for E, it is likely that the comparisons of each new chemical with the standard control will involve only one of the three concentrations, and the replication of the control can be reduced accordingly.

The replication pattern actually employed for this particular problem was to use eight blocks with 30 plants per block with the within-block replication structure as shown in Table 12.13.

The anticipated treatment contrasts would include

(*a*) between the standard control and the weighted average of the 18 chemical treatments,
(*b*) between the three concentrations,
(*c*) between O and E,
(*d*) main effects and interaction of (S_1, S_2) and (Ed, Ul) for chemical O,
(*e*) interactions of the (*d*) contrasts with the three concentrations,
(*f*) (Ul–Con) for E and interaction with the three concentrations.

The important conclusion to be drawn from these examples, but applicable much more widely, is that choice of replication is the joint responsibility of the experimenter and statistician, and the choice will determine the relative precision with which answers are provided to different questions. Invariable use of equal replication is not a proper philosophy nor is 'equal replication' always simply definable as this previous example has shown.

12.7 Treatments as a sample

One of the topics in the design of experiments which has generated a great deal of often rather heated discussion is the distinction between 'fixed' and 'random' effects. In our opinion, the distinction, though it is clearly necessary, is a very minor one, and we shall discuss it only sufficiently to explain the distinction and the consequent change in the interpretation of the experimental results.

For some experiments specific treatments are selected at the outset of the planning of the experiment. A new drug to be compared with a control is of interest because it, specifically, has produced some previous good results. The levels of a chemical constituent in a complex chemical process are chosen because it is expected that the range of values will cover the optimum level symmetrically with sufficient range to identify the optimum clearly. In general, the levels of a quantitative factor will be chosen exactly and deliberately to give maximum information in answer to a specific question.

Sometimes, however, the treatments are selected to be a representative sample of some population, and the primary purpose of the comparison of the treatments is not to draw conclusions about these particular treatments but rather to assess the variability of the population. This use of representative treatments occurs widely in genetics and breeding experiments, where it is important to assess the levels of variance in different subpopulations, or attributable to different sources. Thus, in a pig breeding experiment, the offspring from several sows for each of several boars may provide data on characteristics measured on individual piglets. These data may then be used to assess the relative levels of variation between (i) different boars, (ii) different sows and (iii) different piglets within the same litter. The use of representative treatments is described as a random effects model because, in the model for experimental data, the treatment effects, t_j, may be regarded, at the stage of planning the experiment, as being unknown and randomly selected from a definable set of possible treatment effects.

Now the reason why there is no necessity to make a major distinction between fixed, or specifically selected, treatments, and random, or representative, treatments is that the experimental design and the initial analysis of data are identical for the two situations. Essentially, once the particular representative treatments have been chosen, the experimenter considers only those treatments, and the details of the experimental design and the initial analysis of results are conditional on the treatments chosen. The random allocation of treatments to experimental units is the same as for specifically selected treatments. The analysis of variance divides variation into components representing block differences, treatment differences, and residual variation in the usual way. The change which arises because the treatments are a representative sample comes in the interpretation which follows the initial analysis of variance.

When the experimental treatments are a representative sample from a population, comparisons between any two particular treatments are of no interest, because the occurrence of those two particular treatments in the experiment was due to chance. Instead, we wish to estimate the variance of the population from which the treatments are drawn. Obviously, the variance of the potential treatment population, σ_t^2, can be estimated from the treatment MS. Equally obviously, the treatment MS is partly attributable to the inherent variation between units. We therefore reconsider the expected value of the treatment MS for the case when the t_j are to be regarded as random variables with mean zero and variance σ_t^2. We assume a randomised complete block design or any design which has equal treatment replication and no non-orthogonal block effects:

$$\text{treatment SS} = \sum_j Y_{.j}^2/b - Y_{..}^2/(bt).$$

Now

$$Y_{.j}^2/b = b\left\{\mu + t_j + (1/b)\sum_i \varepsilon_{ij}\right\}^2,$$

$$\mathrm{E}(Y_{.j}^2)/b = b\mu^2 + b\sigma_t^2 + \sigma^2$$

and

$$Y_{..}^2/(bt) = bt\left\{\mu + \sum_j t_j/t + \sum_{ij}\varepsilon_{ij}/(bt)\right\}^2,$$

$$\mathrm{E}\{Y_{..}^2/(bt)\} = bt\mu^2 + b\sigma_t^2 + \sigma^2,$$

$$\mathrm{E}\left\{\sum_j Y_{.j}^2/b - Y_{..}^2/(bt)\right\} = \left(bt\mu^2 + bt\sigma_t^2 + t\sigma^2\right) - \left(bt\mu^2 + b\sigma_t^2 + \sigma^2\right)$$

$$= b(t-1)\sigma_t^2 + (t-1)\sigma^2.$$

Hence,

$$\mathrm{E}(\text{treatment MS}) = b\sigma_t^2 + \sigma^2,$$

and we can estimate σ_t^2 by

$$(\text{treatment MS} - \text{error MS})/b.$$

More generally, if the treatments are unequally replicated, with n_j observations for treatment j, then, assuming a completely randomised design,

$$\mathrm{E}(\text{treatment MS}) = \sigma^2 + \left\{\left(\sum_j n_j\right)^2 - \sum_j (n_j^2)\right\}\sigma_t^2 \Big/ \left\{\sum_j n_j(t-1)\right\}.$$

For more complex situations, where the experimental treatments are combinations or hierarchies of 'random' treatments, it is also possible to determine expected values for components of the treatment SS and to estimate the different population variances.

12.8 Screening and selection experiments

There is one substantial area of experimentation which does not fit into the classifications so far considered in this chapter. This is the screening of large numbers of compounds to discover the few which produce a desired effect, or the selection of the best of a large number of alternatives. These two types of investigation often occur as different stages of the same research project, and are met most frequently in breeding programmes or in searches for appropriate chemical compounds for controlling disease. We shall consider screening and selection sequences of experiments in more detail later, in Chapters 19 and 20, but an initial consideration is appropriate to this chapter.

In both types of investigation, the only structure usually adopted for the experimental treatments is the inclusion of a standard, or control, treatment. Apart from the control treatment, all other treatments are regarded identically, and there is no structure within the non-control set from which the treatments come. The treatments are, of course, a qualitative set.

The essential difference between screening and selection is that, in screening trials, the response variable is qualitative, with each compound giving either a positive or a negative result whereas, in selection, the response variable is quantitative. Both forms of investigation usually involve several stages in a sequential structure. Initially, the research scientist may have many thousands or hundreds of thousands of candidate drugs or breeding lines. The end point of the investigation is the identification of a small number of these candidates, either for immediate use, or, more commonly, for further investigation of properties other than those of primary interest.

12.8.1 *Selection*

Selection has been discussed from two different philosophical viewpoints. For each it is assumed that there is a true set of expected yields for the candidates available for selection, and that observations in any particular trial can be modelled as

$$y_{ij} = \mu_i + \varepsilon_{ij},$$

where μ_i is the true yield for candidate i, and ε_{ij} represents the response variability inherent in the experimental units. This variability may be subdivided into block variation and random variation, but both components are assumed to be independent of the particular candidate.

One selection philosophy assumes that the objective is to select candidates so as to maximise the probability of including in the selection those candidates which have the highest values of true yield, μ_i. The other philosophy attempts to maximise the average mean value of the μ_i for the set of selected candidates. Obviously the perfect achievement of either objective is the same. The difference comes in the relative value placed on different imperfect achievements. The first objective accepts a small probability of selecting a candidate with a very poor μ_i as necessary to achieve a high probability of including in the selection those candidates with the best μ_i. The second objective accepts a reduced probability of including all the best μ_i candidates to ensure that all selected candidates have μ_is which are nearly as good as the best.

The choice of philosophy should depend on the particular objective of the research programme. If the selected candidates will inevitably be used, as for example in a breeding programme for cereals, then the interest of the user, the farmer, is to maximise his future yield and this leads to the second philosophy. If, on the other hand, the chance of the selected candidates being ultimately successful is rather small, as in searches for cures for diseases, then maximising the probability of success is important, and the first philosophy would seem appropriate.

One aspect of the design of selection programmes for which intuitive results are available is the division of a selection research programme into stages. If there are to be r stages of selection, and an initial set of n candidates is to be reduced, through selection, to c candidates, then under some rather restrictive assumptions of equal replication within a stage for all candidates included in that stage, the optimal design, using the second philosophy, is to select at the same intensity

$$p = (c/n)^{1/r},$$

at each stage and to allocate equal resources to each stage. Thus if it is required to select four from 100 over three stages of selection the numbers selected after the first and second stages should be 35 and 12, respectively, and the relative replication of observations per candidate in the three stages should be in the ratios 1 : 3 : 8. Although this design principle was discovered by considering a very precisely specified situation, it is an appropriate general principle for a much wider range of situations.

12.8.2 Screening

Screening problems can be divided into two groups, according to the probability of detection of those individuals which possess the required characteristic. The extreme case is where the detection of the characteristics is certain. The classic example of this is the screening of blood samples for a particular constituent which led Dorfman (1943) to propose a method of group screening whereby many candidates can be assessed simultaneously using a single test. The principle of group screening assumes that the cost depends only on the number of tests regardless of how many candidates are assessed in each test. In a group test which includes several candidates the test result will be positive if any candidate included in the group would provide a positive result if tested separately. If all candidates are negative, then the test result will be negative.

The methodology of group screening has been developed notably by Sobel and Groll (1959), considering only strategies in which, after a positive result for a group test that group is divided into two subgroups which are tested independently (unless logically unnecessary). The optimal design strategy depends on the overall proportion of candidates with the required characteristic. However, the broad pattern of conclusions from Sobel and Groll's results is that for a wide range of, relatively small, values of p a strategy of testing groups of size 8 with successive subgroups of 4, 2 and 1 is very close to optimal.

If the probability of detection of a characteristic is uncertain, then the strategies for screening experiments are not so well understood. It is not clear whether it is useful to pursue experiments involving simultaneous testing of several candidates on the same experimental unit, with the implication of further testing for those individuals which appear, as a group, to have the required characteristic. The ideas of interblock information from incomplete block designs considered in Chapter 7 may have some relevance here but this aspect of design will not be considered further here.

Table 12.14. *Design and responses for bioassay experiment*

		Block				
Drug	Concentration[a]	I	II	III	IV	V
1	1	1.7	1.7	1.8	1.9	1.6
	2	2.0	1.7	2.2	2.1	1.7
	4	2.1	1.9	2.1	2.1	2.1
2	0.8	2.0	1.7	2.1	1.8	1.6
	1.6	2.0	2.0	2.1	1.9	2.0
3	0.8	1.8	1.7	1.7	1.9	1.9
	1.6	2.1	1.8	1.8	2.1	1.9
4	0.6	2.0	1.8	1.8	1.9	1.7
5	0.6	2.2	2.0	2.0	2.0	2.1

[a] No units available.

Exercises

12.1 In a randomised block experiment on potato scab, eight treatments were replicated four times. Two treatments were identical controls, the other six consisted of three amounts of sulphur (3, 6 and 12 units) applied either in spring or autumn. The totals for the eight treatments were:

$$C_1 \quad C_2 \quad S_3 \quad A_3 \quad S_6 \quad A_6 \quad S_{12} \quad A_{12}$$
$$75 \quad 106 \quad 38 \quad 67 \quad 62 \quad 73 \quad 23 \quad 57$$

Think about the questions that the experimenter should have been asking to have chosen these treatments. Define a set of practically sensible orthogonal comparisons and partition the treatment SS into appropriate components. If the error MS is 30, summarise the results of the experiment.

12.2 In a bioassay experiment, five drugs were compared. Drug 1, a control, was applied at three concentrations, 1, 2, 4; drugs 2 and 3 were each applied at two concentrations, 0.8 and 1.6, and drugs 4 and 5 each at only one concentration 0.6. The experiment was arranged in five randomised blocks and the yield responses for one variable are given in Table 12.14.

Carry out an analysis of variance, splitting the treatment SS into components each with a single df, where these are meaningful. Two questions you should attempt to answer are:

(i) Is the relationship between yield and log concentration linear?

(ii) Is the average slope of the relationship between yield and log concentration consistent over drugs?

Thinking about the concepts of main effects and interactions may help clarify your choice of contrasts. Briefly summarise the conclusions to be drawn, including comments about the questions that cannot be answered.

12.3 An incomplete block design was used to investigate the effects on oxygen uptake in microlitres per milligram of poultry spermatozoa dry tissue per hour of various media, the restriction to four observations per block being due to the laboratory equipment available. The ten treatments were two controls and the eight combinations of two forms of buffering (b_1 = phosphate, b_2 = glutamate), two dilutents (d_1 = sodium chloride, d_2 = mixture of salts) and two nutrients

Table 12.15. *Design for oxygen uptake experiment*

Block	Treatments				Block	Treatments			
I	1	2	3	4	IX	2	5	8	10
II	1	2	5	6	X	2	7	8	9
III	1	3	7	8	XI	3	5	9	10
IV	1	4	9	10	XII	3	6	7	10
V	1	5	7	9	XIII	3	4	5	8
VI	1	6	8	10	XIV	4	5	6	7
VII	2	3	6	9	XV	4	6	8	9
VIII	2	4	7	10					

Table 12.16. *Analysis of variance*

Source	SS
Blocks (ignoring treatment)	124.5
Treatment (eliminating blocks)	292.7
Error	175.2
Total	592.4

(n_0 = nil, n_1 = glucose):

1	2	3	4	5
C_1	C_2	$b_1 d_1 n_0$	$b_1 d_1 n_1$	$b_1 d_2 n_0$
6	7	8	9	10
$b_1 d_2 n_1$	$b_2 d_1 n_0$	$b_2 d_1 n_1$	$b_2 d_2 n_0$	$b_2 d_2 n_1$

The block design is shown in Table 12.15.

The initial within-block analysis of variance and the treatment totals adjusted for blocks $(T_j - \sum_{i(j)} B_i/4)$ are given in Tables 12.16 and 12.17. Complete the analysis by splitting the treatment SS into suitable component SS, or by calculating treatment contrasts and their SE. Summarise the conclusions to be drawn from the results.

12.4 In an experiment to examine leakage of propellate from aerosol cans the loss of weight (in grams) over eight weeks was measured for three different propellates X, Y and Z from cans with seals made by different methods. Four factors involved in the manufacture of seals were examined at the following levels:

depth of operation:	A, B, C;
tool diameter:	1, 2;
weight of sealing compound:	0.60, 0.55, 0.50 g;
position of sealing compound:	standard (S), non-standard (T).

The non-standard position of sealing compound could only be used when the weight of sealing compound was 0.50 g. Each observation in Table 12.18 is the mean weight loss for ten cans.

Table 12.17. *Adjusted treatment totals*

1	−12.3
2	−13.0
3	−7.2
4	+9.8
5	−9.6
6	+2.6
7	−14.9
8	+16.4
9	+12.6
10	+15.7
Overall mean = 13.5	

Table 12.18. *Weight loss data*

Propellate	Weight of sealing compound (g)	Position of sealing compound	A		B		C	
			1	2	1	2	1	2
X	0.60	S	0.25	0.41	0.40	0.37	0.41	0.49
	0.55	S	0.36	0.29	0.30	0.41	0.29	0.60
	0.50	S	0.38	0.60	0.33	0.41	0.32	0.54
	0.50	T	0.78	0.91	2.10	0.73	0.47	0.32
Y	0.60	S	1.20	1.09	1.13	0.91	0.74	1.00
	0.55	S	1.55	1.33	1.19	1.27	0.75	1.05
	0.50	S	1.80	1.68	1.45	1.27	1.04	1.55
	0.50	T	2.09	2.03	2.08	1.86	1.27	1.35
Z	0.60	S	1.07	1.44	0.66	0.63	0.62	0.28
	0.55	S	1.01	0.62	0.32	0.40	0.81	0.45
	0.50	S	0.64	1.58	0.38	0.36	0.33	0.31
	0.50	T	3.54	3.26	1.37	1.68	1.53	0.83

First consider what difficulties there might be in the assumptions required for analysis of the data. You may find it helpful initially to regard the different propellates as three 'blocks' within which the different methods for constructing the seals are to be compared. Complete an analysis of the data and present your conclusions.

13

Factorial structure and particular forms of effects

13.0 Preliminary example

Suppose eight treatments comprising all combinations of three two-level factors P, Q and R are compared and the resulting treatment means are shown in Table 13.1. What is the best estimate of the difference between the two treatment combinations $p_1q_1r_1$ and $p_1q_1r_0$? Or, in practical terms, if you have decided to recommend the upper levels of factors P and Q, what is the benefit of using r_1 rather than r_0? If you think there is only one possible answer, then you have not yet fully grasped the benefits of factorial structure. With a little effort, you should be able to find at least six different conceivable answers.

13.1 Factors with two levels only

In Chapters 3 and 12, we started to examine the general advantages of using experimental treatments with a factorial structure. In this chapter, we look in more detail at special forms of factorial structure to examine how the general advantages manifest themselves in particular cases, and we develop methods for using further advantages of these special forms.

The first special form of factorial to consider is that in which all factors have two levels. There are many reasons why this is an interesting form to examine in detail. First, it is a sensible practical solution for many scientific situations. We can think of one level of each factor as being the normal mode and the other level as a new, alternative mode. In a classic agricultural example, advisory officers found that two farmers growing mushrooms on a large scale in the same locality were getting very different yields, though much of their management procedure was identical. On careful examination, the advisors distinguished four ways in which the two farmers' procedures differed. The four differences were identified as four factors, and an experiment was planned in which each of the four factors appeared at two levels, the first level being the mode adopted by farmer 1 and the second that of farmer 2. All 16 combinations of levels for the four factors were included in the experiment to determine whether the differences between the two farmers' yields were due to a single factor or to a combination of factors.

Another area where experiments with two-level factors are important is in industrial research where many different facets of the industrial process might be changed. In attempting to discover which changes are beneficial and which changes produce interdependent effects, an experiment with a number of factors, each at two levels (normal and changed), is often an effective first stage of an investigation.

Table 13.1. *Treatment*
means for preliminary
example

	r_0	r_1
$p_0 q_0$	17	23
$p_0 q_1$	18	28
$p_1 q_0$	14	22
$p_1 q_1$	19	30

A third important area for two-level factor experimentation is the investigation of predictions from complex computer simulation models for ecological, chemical or agricultural systems. These models are being extensively developed but the investigation of the implications of such models is often hardly planned at all. Within any such model, there will be many parameter values which could be altered in the practical situation which the model is intended to represent. As an initial stage in the investigation of the properties of the model, consideration of the effects of changing each of six parameters in the model can be most effectively achieved by using a factorial experiment with six factors. For each factor one level is the value originally assumed for the parameter, and the second level is an alternative value.

One other reason for considering two-level factors in some detail is that they are particularly simple to manipulate mathematically. Factorial structures in which all factors have two levels are referred to as 2^n factorials. In a simplified form of the more general notation the levels are defined as 0 and 1, the upper level of a factor, P, is represented by the lower case letter, p, and the lower factor level by the absence of the letter. Thus the treatment combination with all factors at their upper level is represented by pqr $\cdots$; the combination with factors P, R and S at their upper levels and all other factors at their lower levels by prs; the combination with only factor R at its upper level by r. The combination with all factors at their lower levels is a special case and conventionally this is represented by (1).

A general factorial treatment combination is written $p_i q_j r_k \ldots$, where each of $i, j, k, \ldots$ may be 0 or 1. Effects are represented by capital letters, P, Q, PQ, PQR, ... It is convenient to use p, pq, pqr, . . . to represent both the treatment combinations and the mean yield for that treatment combination, and to use P, PQ, PQR, . . . to represent both effects and the least squares estimators of these effects. This might appear to lead to ambiguity and confusion but, in practice, does not, principally because of the simple unbiased form of least squares estimators.

We consider the detailed properties of the two-level factorial structure first for three factors because this enables us to illustrate all the important ideas without employing too large an array of symbols. If the three factors are P, Q and R, then the main effect of P is most simply defined as

$$P = (p + pq + pr + pqr)/4 - \{(1) + q + r + qr\}/4, \tag{13.1}$$

the simple difference between the mean yield for all observations at the upper level of P and the corresponding mean yield for the lower level of P. This effect can be rewritten as a pseudo-algebraic expression

$$P = \frac{1}{4}(p-1)(q+1)(r+1),$$

the advantage of which will be seen later.

The interaction between two factors, P and Q, is most simply expressed as

$$\{(pq+pqr)/2 - (p+pr)/2\} - [(q+qr)/2 - \{(1)+r\}/2], \tag{13.2}$$

showing clearly that it is the comparison of the (upper – lower) Q comparison between upper and lower levels of P. This can be rewritten either as

$$\{pq + pqr + (1) + r\}/2 - (p + pr + q + qr)/2$$

or as

$$\frac{1}{2}(p-1)(q-1)(r+1).$$

We can now see that, whereas the use of 2 as the divisor in (13.2) seems logical when thinking about the PQ effect as a difference of difference of mean yields, the resulting estimate is clearly not measured on the same scale as the effect P (13.1). It is usual therefore to define the PQ effect on a scale directly comparable to the P effect:

$$PQ = \frac{1}{4}(p-1)(q-1)(r+1).$$

Similarly, the natural expression of the PQR interaction 'a difference of a difference of a difference' is

$$\{(pqr - pq) - (pr - p)\} - [(qr - q) - \{r - (1)\}]$$

but, to make the PQR effect directly comparable with those for P and PQ, we define

$$PQR = \frac{1}{4}(p-1)(q-1)(r-1).$$

Two related patterns to note for future reference are:

(i) In the algebraic summary form for an effect each factor included in the effect contributes an $(x-1)$ term and each excluded factor an $(x+1)$ term, the latter representing the averaging of the effect over levels of factors not directly involved in the effect.
(ii) If the effects are rewritten as differences between two groups of treatment combinations, the two groups are, respectively, those with an even number of letters in common with the effect, and those with an odd number of letters in common with the effect. This is a

direct result of the definition of interactions as 'differences of differences of differences of ...'. The patterns may be confirmed in the further examples below:

$$Q = \frac{1}{4}(p+1)(q-1)(r+1)$$

$$= \frac{1}{4}[(pqr + qr + pq + q) - \{pr + r + p + (1)\}]$$

$$QR = \frac{1}{4}(p+1)(q-1)(r-1)$$

$$= \frac{1}{4}[\{pqr + qr + p + (1)\} - (pq + pr + q + r)]$$

$$PR = \frac{1}{4}(p-1)(q+1)(r-1)$$

$$= \frac{1}{4}[\{pqr + pr + q + (1)\} - (pq + p + qr + r)].$$

If we define an eighth effect, the overall mean

$$M = \frac{1}{8}\{pqr + pq + pr + qr + p + q + r + (1)\}$$

$$= \frac{1}{8}(p+1)(q+1)(r+1),$$

then we have eight effects and eight treatment combinations, and the relationship between the two can be expressed in matrix form by defining

$$\mathbf{x}' = ((1), p, q, pq, r, pr, qr, pqr),$$
$$\mathbf{y}' = (8M, 4P, 4Q, 4PQ, 4R, 4PR, 4QR, 4PQR)$$

and writing

$$\mathbf{y} = \mathbf{U}\mathbf{x},$$

where the elements of U are all ± 1 and the rows of U are orthogonal:

$$\mathbf{U} = \begin{bmatrix}
+1 & +1 & +1 & +1 & +1 & +1 & +1 & +1 \\
-1 & +1 & -1 & +1 & -1 & +1 & -1 & +1 \\
-1 & -1 & +1 & +1 & -1 & -1 & +1 & +1 \\
+1 & -1 & -1 & +1 & +1 & -1 & -1 & +1 \\
-1 & -1 & -1 & -1 & +1 & +1 & +1 & +1 \\
+1 & -1 & +1 & -1 & -1 & +1 & -1 & +1 \\
+1 & +1 & -1 & -1 & -1 & -1 & +1 & +1 \\
-1 & +1 & +1 & -1 & +1 & -1 & -1 & +1
\end{bmatrix}.$$

13.1.1 Relationships between effects and treatment combinations

Many elegant mathematical patterns may be observed in the expression of the relationship between effects **y** and treatment combinations **x**, some of which are statistically useful. For example the elements of the PQ row of **U** are obtained by multiplying the corresponding elements in the rows for P and Q; similarly, the row for PQR is the product of the rows for P and for QR, or of those for Q and for PR.

Also, notice that the relationship between effects, **y**, and treatment combinations, **x**, is the same as that between least squares estimators of effects, **y**, and the mean yields of treatment combinations, **x̄**. If the experiment has n observations of each treatment combination, then each element of **x̄** has variance σ^2/n, and each element of **y** has variance $8\sigma^2/n$. Hence the variance of each comparison effect is

$$\mathrm{Var}(\mathrm{P}) = \sigma^2/2n.$$

Also, from the general result for the SS for treatment contrasts in Chapter 4,

$$\mathrm{SS(contrast)} = \left(\sum_j l_{kj}Y_{.j}\right)^2 \bigg/ \left(n\sum_j l_{kj}^2\right)$$

and so

$$\mathrm{SS(P)} = \frac{\mathrm{P}^2}{n\{8(1/4n)^2\}} = 2n\mathrm{P}^2,$$

with corresponding results for other effects.

All these results can be generalised from three to m factors:

$$\mathrm{effect} = (1/2^{m-1})(p \pm 1)(q \pm 1)(r \pm 1)\dots(m \pm 1),$$
$$\mathrm{Var(effect)} = \sigma^2/n2^{m-2},$$
$$\mathrm{SS(effect)} = n2^{m-2}(\mathrm{effect})^2.$$

Although modern computational facilities have made the method practically obsolete, no discussion of 2^m factorials could be complete without reference to Yates' algorithm for obtaining estimates of effects. This is defined as follows:

(i) Write the mean yields for treatment combinations in a column in standard order ((1), p, q, pq, r, pr, qr, pqr, s, ps, ...).

(ii) The first half of a new column consists of the sums of consecutive values from the previous column, (1) + (2), (3) + (4), etc.

(iii) The second half of a new column consists of the differences of consecutive pairs of values from the previous column, (2) − (1), (4) − (3), etc.

(iv) Repeat steps (ii) and (iii) forming a series of new columns from previous columns until m new columns have been completed after the initial column of mean yields.

(v) Then the final column gives the effect vector **y** in standard order (2kM, kP, kQ, kPQ, kR, kPR, kQR, kPQR, kS, kPS, ...), where $k = 2^{m-1}$.

Table 13.2. *Calculations for Yates' algorithm*

Combination	Species	Moisture	Hormone					Effect
(1)	Toad	Wet	Control	0.36	2.24	33.80	97.10	
s	Frog	Wet	Control	1.88	31.56	63.30	−45.38	= 4S
m	Toad	Dry	Control	21.46	24.60	−9.84	43.42	= 4M
sm	Frog	Dry	Control	10.10	38.70	−35.54	−12.58	= 4SM
h	Toad	Wet	Hormone	21.26	1.52	29.32	29.50	= 4H
sh	Frog	Wet	Hormone	3.34	−10.36	14.10	−25.70	= 4SH
mh	Toad	Dry	Hormone	28.16	−17.92	−12.88	−15.22	= 4MH
smh	Frog	Dry	Hormone	10.54	−17.62	0.30	13.18	= 4SMH

The transformation from each column to the succeeding column is represented by the matrix:

$$
\mathbf{V} = \begin{bmatrix}
+1 & +1 & 0 & 0 & 0 & 0 & 0 & 0 & \dots \\
0 & 0 & +1 & +1 & 0 & 0 & 0 & 0 & \dots \\
0 & 0 & 0 & 0 & +1 & +1 & 0 & 0 & \dots \\
0 & 0 & 0 & 0 & 0 & 0 & +1 & +1 & \dots \\
\vdots & \vdots & \vdots & \vdots & \vdots & \vdots & \vdots & \vdots & \\
-1 & +1 & 0 & 0 & 0 & 0 & 0 & 0 & \dots \\
0 & 0 & -1 & +1 & 0 & 0 & 0 & 0 & \dots \\
0 & 0 & 0 & 0 & -1 & +1 & 0 & 0 & \dots \\
0 & 0 & 0 & 0 & 0 & 0 & -1 & +1 & \dots \\
\vdots & \vdots & \vdots & \vdots & \vdots & \vdots & \vdots & \vdots & \dots
\end{bmatrix}.
$$

It is simple to verify for any particular value of m that $\mathbf{V}^m = \mathbf{U}$ but a general proof is outside the scope of this book.

We illustrate the use of Yates' algorithm with the data from Chapter 3 on the water uptake of amphibia. The factors were S = species (toad or frog), M = moisture (wet or dry), H = hormone (control or hormone). The mean results for the eight treatment combinations with the Yates' algorithm calculations are given in Table 13.2.

From the final column of Table 13.2, we can calculate estimates for the effects and corresponding SS shown in Table 13.3. The SE of each effect estimate requires the estimate of variance $s^2 = 34.56$ from the analysis of variance in Chapter 3. The SE for each effect estimate is then $(34.56/4)^{1/2} = 2.94$.

13.2 Improved yield comparisons in terms of effects

Consider again the relationship $\mathbf{y} = \mathbf{U}\mathbf{x}$ between the vector of effects, $\mathbf{y}$, and the vector of treatment combinations, $\mathbf{x}$. Since the rows of $\mathbf{U}$ are orthogonal, $\mathbf{U}$ is a scalar multiple of an orthogonal matrix $\mathbf{L}$. For the three-factor structure $\mathbf{U} = \sqrt{8}(\mathbf{L})$. Hence,

$$
\mathbf{x} = \mathbf{U}^{-1}\mathbf{y} = \frac{1}{\sqrt{8}}\mathbf{L}'\mathbf{y} = \frac{1}{8}\mathbf{U}'\mathbf{y},
$$

Table 13.3. *Estimates and sums of squares for experiment on amphibia*

Effect	Estimate	SS
S	−11.34	514.38
M	10.86	471.74
SM	−3.14	39.44
H	7.38	217.86
SH	−6.42	164.87
MH	−3.80	57.76
SMH	3.30	43.56

where $\mathbf{L}'$, $\mathbf{U}'$ are the transposed forms of $\mathbf{L}$ and $\mathbf{U}$. Hence, we can write

$$
\mathbf{x} = \frac{1}{8}
\begin{bmatrix}
+1 & -1 & -1 & +1 & -1 & +1 & +1 & -1 \\
+1 & +1 & -1 & -1 & -1 & -1 & +1 & +1 \\
+1 & -1 & +1 & -1 & -1 & +1 & -1 & +1 \\
+1 & +1 & +1 & +1 & -1 & -1 & -1 & -1 \\
+1 & -1 & -1 & +1 & +1 & -1 & -1 & +1 \\
+1 & +1 & -1 & -1 & +1 & +1 & -1 & -1 \\
+1 & -1 & +1 & -1 & +1 & -1 & +1 & -1 \\
+1 & +1 & +1 & +1 & +1 & +1 & +1 & +1
\end{bmatrix}
\mathbf{y}.
$$

Hence,

$$
x_{ijk} = M + (1/2)\{(-1)^{i-1}P + (-1)^{j-1}Q + (-1)^{i+j-2}PQ + (-1)^{k-1}R
$$
$$
+ (-1)^{i+k-2}PR + (-1)^{j+k-2}QR + (-1)^{i+j+k-3}PQR\}. \tag{13.3}
$$

Thus each treatment combination may be expressed in terms of the set of seven treatment effects and, if numerical values are substituted for the effects, estimates of the treatment combination mean yields may be calculated. Of course, if all seven effect estimates are included the estimate of treatment combination means will be simply the original mean yields. For the frog–toad experiment, the predicted yield for toads in dry prior conditions with no hormone is

$$
x_{010} = 97.10/8 + (1/2)(11.34 + 10.86 + 3.14 - 7.38 - 6.42 + 3.80 + 3.30)
$$
$$
= 12.14 + (1/2)18.64 = 21.46.
$$

However, suppose we had decided that the three-factor interaction should be ignored, then we would obtain a different estimate, since the three-factor interaction effect would be set equal to zero instead of 3.30. The revised estimate of x_{010} would be 19.81. If we ignore other interaction effects, we would obtain other estimates.

Why should we expect any benefits from using estimates which are different from those actually observed? If we make an assumption, such as $PQR = 0$, which is true, or almost true, then the resulting estimates will be more precise because of the additional information contributed by the assumption. This trade-off of assumptions and precision is the basis of any statistical analysis of data. The present form is a particularly blatant example, but is not unusual. To examine the benefit in terms of improved precision, consider again the expression of x_{ijk} in terms of effects (13.3). The eight effects are orthogonal and each element of $\mathbf{y}$ has variance $8\sigma^2/n$ and each comparison effect has variance $\sigma^2/2n$. The estimate of x_{ijk}, setting PQR to zero, includes only seven of the eight terms in the $\mathbf{y}$ vector and therefore the variance of the estimate assuming PQR $= 0$ is

$$(1/8)^2 7(8\sigma^2/n) = (7/8)(\sigma^2/n),$$

a reduction of $1/8$ from the original variance of the observed mean, σ^2/n.

Usually, we are most interested in comparisons between means for two treatment combination means, for example the comparison of hormone and control treatments for frogs in dry preexperiment conditions. We require an estimate of $x_{011} - x_{010}$, and from the Expression (13.3) we obtain

$$x_{011} - x_{010} = R - PR + QR - PQR.$$

To illustrate the effects of making assumptions about interactions, we shall consider five different estimates of $x_{011} - x_{010}$, with the SEs of those estimates and confidence intervals. The assumptions we shall consider are

  (i) no additional assumptions (use observed means),
 (ii) assume PQR $= 0$,
(iii) assume PQR $= 0$, QR $= 0$,
(iv) assume PQR $= 0$, PR $= 0$,
 (v) assume PQR $= 0$, PR $= 0$, QR $= 0$.

Since each effect has variance $\sigma^2/2n$, the variances of the alternative estimates of $(x_{111} - x_{110})$ will be

  (i) σ^2,
 (ii) $3\sigma^2/4$,
(iii) $\sigma^2/2$,
(iv) $\sigma^2/2$,
 (v) $\sigma^2/4$.

The results for the frog data are tabulated in Table 13.4 in terms of factors S($=$ P), M($=$ Q) and H($=$ R). The estimates of $x_{011} - x_{010}$ shown in Table 13.4 vary considerably, and so do the 95% confidence intervals.

Each point and interval estimate is valid, given the assumptions made, and each could be criticised, not least the estimate (i), which is the value we would use if we had not considered the factorial structure of the treatments. From the analysis of variance for this experiment in Chapter 3, there is little evidence of a three-factor interaction or of an interaction (MH) between the hormone/control factor and the moisture factor. However, the SS for the interaction (SH) between the hormone factor and the species factor is not negligible and this should make us wary about using estimates based on assumptions (iv) or (v). The evidence for an MH interaction is slightly greater than that for the SMH interaction, so the choice of an estimate for $x_{011} - x_{010}$ lies between (ii) and (iii). The best estimate is probably that

Table 13.4. *Estimates and confidence intervals for amphibia data*

Assumptions	Estimate of $x_{011} - x_{010}$	Standard error of estimate	95% confidence interval
(i) All effects non-zero	6.70	5.88	(−6.88 to 20.28)
(ii) SMH = 0	10.00	5.09	(−1.76 to 21.76)
(iii) SMH = MH = 0	13.80	4.16	(4.19 to 23.41)
(iv) SMH = SH = 0	3.58	4.16	(−6.03 to 13.19)
(v) SMH = MH = SH = 0	7.38	2.94	(0.59 to 14.17)

Table 13.5. *Treatment means for amphibia data*

Species	Hormone	Wet	Dry	Mean
Toad	Control	0.36	21.46	10.91
Frog	Control	1.88	10.10	5.99
Toad	Hormone	21.26	28.16	24.71
Frog	Hormone	3.34	10.54	6.94
Mean		6.71	17.56	12.14

from (iii), which is the same as was implied in summarising the results of this experiment in Chapter 3.

13.2.1 What estimates of treatment means should we use?

It may seem that we have spent a lot of time in discussing a relatively trivial example. However, the central point in this example is the benefit of factorial structure, and in particular the advantages of hidden replication. Using a factorial structure for experimental treatments does provide better information about treatment effects than is implied by the explicit replication. To use this benefit, it is essential that the estimates of treatment mean yields derived and quoted from the experiment should use the advantages of hidden replication and, to achieve this, we must be prepared to estimate treatment mean yields, ignoring interaction effects for which there is little evidence.

To illustrate this argument, consider first the presentation of treatment means as they would be calculated ignoring the factorial structure, but with a presentation emphasising the species × hormone interaction which the analysis of variance suggests is important. The eight means are shown in Table 13.5.

Now consider the presentation of the results if we accept the full implications of the analysis of variance that the species × moisture, hormone × moisture and the three-factor interaction may properly be ignored. The main effect means remain unaltered, but the eight treatment combination mean yields are re-estimated using (13.3), ignoring SM, MH and SMH, or equivalently using a model which includes only the effects S, M, H and SH, and are shown in Table 13.6. The differences are not large but the estimates and differences between

Table 13.6. *Treatment means from model with a single interaction*

Species	Hormone	Wet	Dry	Mean
Toad	Control	5.48	16.33	10.91
Frog	Control	0.56	11.41	5.99
Toad	Hormone	19.28	30.13	24.71
Frog	Hormone	1.51	12.36	6.94
Mean		6.71	17.56	12.14

Table 13.7. *Analysis of variance for amphibia data*

	SS	df	MS
Species	515	1	515
Moisture	471	1	471
Hormone	218	1	218
SM	40	1	40
SH	165	1	165
MH	58	1	58
SMH	44	1	44
Error	276	8	34.5
Total	1786	15	

estimates in Table 13.6 are arguably more appropriate and definitely more precise than those in Table 13.5. The crux of the argument is that we are using a simpler model which we believe is an adequate representation of situation. The more complex model implied by presentation of the observed means is unnecessarily complicated and gives less accurate estimates.

13.2.2 *What estimate of variance should be used?*

One question which arises when the possibility of ignoring certain interactions is being discussed is which estimate of σ^2 to use. Consider again the analysis of variance for the frog–toad experiment shown in Table 13.7.

If we decide to ignore the SM, MH and SMH interactions, then we might consider omitting the SS of these effects from the analysis of variance as separate components, grouping the SS with the error SS as in Table 13.8.

This procedure is one which has been much discussed by statisticians and research scientists, and it requires careful thought, and will be considered in more depth later in this chapter. Our belief is that combining insignificant effect SS with the error SS should be avoided unless there are clear benefits to be obtained. There may be advantages from pooling MS when there are few df for the error MS. However, the major philosophical difficulty is that, whereas the original error MS provides an unbiased estimator of the variance of

Table 13.8. *Reduced analysis*
of variance for amphibia data

	SS	df	MS
Species	51	1	515
Moisture	471	1	471
Hormone	218	1	218
SH	165	1	165
Error	418	11	38
Total	1786	15	

an individual observation, the MS from the combined SS will not be unbiased. Consider the situation when the effect, whose SS may be pooled, is actually zero. Then that effect SS will be pooled with the error SS, whenever the effect SS is not significantly larger than the error MS. This will, on average, reduce the error MS, since when the error MS might be most increased we will not pool the SS. In practice, with very large factorial structures, we would be able to examine large numbers of SS for individual interaction effects and, by pooling all the smallest ones, could substantially reduce the error MS.

Later, in Section 13.6, we consider the use of interaction SS for the estimation of σ^2. Our arguments for this will be that the interactions concerned are of little interest, being those which involve many factors and for which the practical interpretation is difficult and uninformative. However, we must decide *before* doing the experiment which interaction SS to use to estimate σ^2. Choosing which interaction SS to use in the light of the observed values of the various interaction SS is a dangerous procedure and should be avoided.

13.3 Analysis by considering sums and differences

One other facet of factors with only two levels which may give further insight into the results of 2^m experiments is the consideration of sums and differences of pairs of yields. In considering the definitions of main effects and interactions earlier in this chapter, it is clear that all the effects involving a particular factor, P, include the component $(p - 1)$. Sometimes in a 2^m experiment one factor is particularly important. This importance may be recognised before the experiment is performed, or it may become apparent only after the results are obtained. In either case, there are often advantages in considering the analysis of the experimental data in two parts, analysing separately the sums and differences of pairs of yields, each pair differing only in the level of this dominant factor.

The two principal advantages of this approach are first that the interpretation of the results should usually be in terms of the pattern of variation of the effect of this important factor over levels of the other factors. This, of course, is essentially a matter of presentation and does not require alteration of the analysis. However, the second advantage is that, if one factor has a very substantial effect, then the variability of sums and of differences of pairs of observations for the two levels of that factor might be rather different, contrary to the usual assumptions of the analysis of variance.

Table 13.9. *Yields from maize trial*

Variety	Infestation	Treatment	Block I	Block II
A	I	O	10.4	10.0
		P	12.5	10.1
		PM	15.8	13.6
		PNK	11.3	11.4
B	I	O	9.5	9.6
		P	11.8	9.5
		PM	13.5	13.4
		PNK	11.6	9.2
A	U	O	14.8	14.0
		P	19.7	18.0
		PM	24.9	22.0
		PNK	19.9	19.2
B	U	O	12.8	13.0
		P	16.9	16.0
		PM	22.3	20.0
		PNK	17.1	16.6

The analysis of sums and differences is only practicable if each pair of yields is uniquely defined, and this will be possible only when a randomised block design is used. The frog–toad experiment was not blocked, and therefore it is not possible, for example, to establish which of the two yields for frog/wet/hormone should be paired with which of the frog/wet/control yields.

Example 13.1 We therefore use some published data (Rayner, 1969, p. 456) from an experiment on maize in which there were three factors:

 (i) two varieties (A and B),
 (ii) infestation of plots with witch weed (I) or no infestation (U),
(iii) four fertiliser treatments:
    O = control
    P = 400 lb superphosphate per morgen,
    PM = P + 10 tons farmyard manure per morgen,
    PNK = P + 150 lb sulphate of ammonia per morgen
            + 100 lb muriate of potash per morgen.

The yields (lb per 1/200 morgen) from two randomised blocks are shown in Table 13.9.

It is clear immediately that the (U − I) difference is substantial and so we examine the separate analyses of sums (U + I) shown in Table 13.10, and differences (U − I) shown in Table 13.11. Note that the 'correction factor SS' for the mean in the analysis of differences is the SS for the main effect (U − I) in the original analysis. The two analyses confirm the

Table 13.10. *Analysis of sums of infected and uninfected yields: (a) sums* (U + I); *(b) analysis of variance*

(a)

Variety	Treatment	Block		Total
		I	II	
A	O	25.2	24.0	49.2
	P	32.2	28.1	60.3
	PM	40.7	35.6	76.3
	PNK	31.2	30.6	61.8
B	O	22.3	22.6	44.9
	P	28.7	25.5	54.2
	PM	35.8	33.4	69.2
	PNK	28.7	25.8	54.5
	Total	244.8	225.6	470.4

(b)

	SS	df	MS
Blocks	23.04	1	23.04
Varieties	38.44	1	38.44
Fertilisers	335.49	3	111.83
VF	1.41	3	0.47
Error	11.52	7	1.65
Total	409.90	15	

suspicion that the very large effect of infestation might distort the assumption of homogeneous variances. The error MS for differences is markedly lower than that for sums, though the ratio is not significant at 5%. This would suggest that variances of interaction effects involving infestation may be smaller than variances of main effects averaging over infestation levels.

In view of the substantial difference in error variances we may choose to present the results in terms of the separate analyses. The major effect, as we have already observed, is that of infestation. The infestation effect is modified by fertiliser and, to a lesser extent, by variety (difference analysis). There are also main effects of fertiliser and variety (sum analysis). The significant interaction effects are two or three times smaller than the corresponding main effects, and the interpretation should therefore include both main effect means and two-way tables of means. The appropriate means and SEs are therefore shown in Table 13.12.

The SEs given in Table 13.12 are for individual quoted values not for differences between these values. Comparison of the values in the difference column with those in the column of means indicates that the infestation effect $(U - I)$ tends to increase with increasing mean value, but, both for fertilisers and for varieties, the differences change relatively more than the

Table 13.11. *Analysis of differences of uninfected and infected yields: (a) differences* (U − 1); *(b) analysis of variance*

(a)

| | | Block | | |
| | | I | II | Total |
Variety	Treatment			
A	O	4.4	4.0	8.4
	P	7.2	7.9	15.1
	PM	9.1	8.4	17.5
	PNK	8.6	7.8	16.4
B	O	3.3	3.4	6.7
	P	5.1	6.5	11.6
	PM	8.8	6.6	15.4
	PNK	5.5	7.4	12.9
	Total			104.0

(b)

	SS	df	MS
Varieties	7.29	1	7.29
Fertilisers	44.45	3	14.82
VF	0.66	3	0.22
Error	6.10	8	0.76
Total	58.50	15	
Correction factor (infestation)	676.00	1	676.00

Table 13.12. *Means and standard errors for important effects*

| | Treatment mean yields | | Difference | Mean |
	Infested	Uninfested	U − I	(U + I)/2
Mean	11.45	17.95	6.50	
			SE = 0.22	
O	9.88	13.65	3.78	11.76
P	10.98	17.65	6.67	14.31
PM	14.08	22.30	8.22	18.19
PNK	10.88	18.20	7.32	14.54
			SE = 0.44	SE = 0.32
A	11.89	19.06	7.17	15.47
B	11.01	16.84	5.83	13.92
			SE = 0.31	SE = 0.23

means. In practical terms the absolute differences between fertiliser treatments or between varieties are reduced under witch-weed infestation compared with uninfested plots, and the differences are also proportionally smaller for the infested plots.

13.4 Factors with three or more levels

The simplicity of the effects for factors with two levels does not extend to factors with more levels. With three levels, there are two comparisons to be made, and while there are, theoretically at least, many different forms of comparison, there is no pair of comparisons which retains symmetry between all three levels. Consequently, the three levels are treated differently and, particularly as more factors or more levels are included, the effects defined for main effects and interactions become more complex. The range of possible comparisons corresponds to a range of possible questions which may be appropriate to the objectives of the experiment, and it is important that the choice of main effect and interaction comparisons be made in the context of the particular experiment rather than always using a standard set of effects.

In this section we consider some definitions of effects that may be useful. With three levels of a single factor two orthogonal effects may be defined in a multitude of ways, some of which are shown below:

$$\begin{array}{lll} (1, -1, 0) & \text{and} & (1, 1, -2) \\ (1, 2, -3) & \text{and} & (-5, 4, 1) \\ (2, 3, -5) & \text{and} & (-8, 7, 1) \\ (1, 3, -4) & \text{and} & (-7, 5, 2). \end{array}$$

Of these, the first is much the simplest and most frequently useful. The two comparisons split the three levels into two groups, one with two levels and the other with a single level, and the two effects are, respectively, a comparison between the two levels of the first group, and a comparison of the third level with the average of the first two. Thus the crucial choice is which level should be the odd one.

One particular practical situation where the two comparisons $(1, -1, 0)$ and $(1, 1, -2)$ have a simple interpretation is when the three levels of the factor are equally spaced quantitative values. The comparisons

$$t_2 - t_0,$$

and

$$t_2 - 2t_1 + t_0$$

represent the linear and quadratic regression effects of the quantitative factor. This may be recognised directly from the analogy with a regression variable, taking values $-1, 0$ and $+1$. Less directly, by considering the two differences $(t_1 - t_0)$ and $(t_2 - t_1)$ which measure the slopes of the response for the two half intervals, $(t_2 - t_0)$ may be identified as the sum of these two slopes (representing average slope) and $(t_2 - 2t_1 + t_0)$ as the difference (change of slope).

The extension of effects for three level factors to allow for two or more factors tends to increase the complexity of the set of effects, but it is useful to consider in some detail at least the 3^2 factorial structure. The standard set of effects used are:

$$P' = (1/3)(p_2 - p_0)(q_2 + q_1 + q_0),$$
$$P'' = (1/6)(p_2 - 2p_1 + p_0)(q_2 + q_1 + q_0),$$
$$Q' = (1/3)(p_2 + p_1 + p_0)(q_2 - q_0),$$
$$Q'' = (1/6)(p_2 + p_1 + p_0)(q_2 - 2q_1 + q_0),$$
$$P'Q' = (1/2)(p_2 - p_0)(q_2 - q_0),$$
$$P'Q'' = (1/4)(p_2 - p_0)(q_2 - 2q_1 + q_0),$$
$$P''Q' = (1/4)(p_2 - 2p_1 + p_0)(q_2 - q_0),$$
$$P''Q'' = (1/8)(p_2 - 2p_1 + p_0)(q_2 - 2q_1 + q_0).$$

Note that the interaction effects are all obtained as direct combinations of main effects and their interpretation follows the same pattern. Thus, $P'Q'$ represents the linear $\times$ linear interaction, the change of the linear effects of P between the extreme levels of Q (by symmetry it is also the change of the linear effect of Q between the extreme levels of P). The $P'Q''$ effect is the linear $\times$ quadratic interaction and has two rather different interpretations. It can be thought of either as the comparison of the linear effects of P between the central and extreme levels of Q; or as the change in the quadratic effect of Q between the extreme levels of P. These interpretations are, of course, all in terms of quantitative factors, but similar interpretations will exist when qualitative factor levels are used, determined by the particular levels used. A final point to note about the effects defined above is the set of multiplicative constants used. In each case, the multiplier is chosen so that the effect is measured as a difference between mean yields per experimental unit.

If we define a ninth effect, the mean,

$$M = (1/9)(p_2 + p_1 + p_0)(q_2 + q_1 + q_0)$$

then, as for two-level factors, there is a relationship between the effects, $\mathbf{y}$, and the treatment combinations, $\mathbf{x}$, which in matrix form is

$$\mathbf{y} = \mathbf{Ux}$$

and is written out in full, as

$$
\begin{bmatrix} 9M \\ 3P' \\ 6P'' \\ 3Q' \\ 6Q'' \\ 2P'Q' \\ 4P'Q'' \\ 4P''Q' \\ 8P''Q'' \end{bmatrix}
=
\begin{bmatrix}
1 & 1 & 1 & 1 & 1 & 1 & 1 & 1 & 1 \\
-1 & -1 & -1 & 0 & 0 & 0 & 1 & 1 & 1 \\
1 & 1 & 1 & -2 & -2 & -2 & 1 & 1 & 1 \\
-1 & 0 & 1 & -1 & 0 & 1 & -1 & 0 & 1 \\
1 & -2 & 1 & 1 & -2 & 1 & 1 & -2 & 1 \\
1 & 0 & -1 & 0 & 0 & 0 & -1 & 0 & 1 \\
-1 & 2 & -1 & 0 & 0 & 0 & 1 & -2 & 1 \\
-1 & 0 & 1 & 2 & 0 & -2 & -1 & 0 & 1 \\
1 & -2 & 1 & -2 & 4 & -2 & 1 & -2 & 1
\end{bmatrix}
\begin{bmatrix} p_0q_0 \\ p_0q_1 \\ p_0q_2 \\ p_1q_0 \\ p_1q_1 \\ p_1q_2 \\ p_2q_0 \\ p_2q_1 \\ p_2q_2 \end{bmatrix}.
$$

The matrix $\mathbf{U}$ can be expressed as the product of two matrices

$$\mathbf{U} = \mathbf{VW},$$

where $\mathbf{W}$ is orthogonal and $\mathbf{V}$ is diagonal with

$$v_{ii}^2 = \sum_j u_{ij}^2,$$

giving squared diagonal elements (9, 6, 18, 6, 18, 4, 12, 12, 36) for $\mathbf{V}$. Again, just as for two-level factors, the relationship between effects and combinations can be reversed:

$$
\begin{aligned}
x &= \mathbf{U}^{-1}\mathbf{y} \\
&= (\mathbf{VW})^{-1}\mathbf{y} \\
&= \mathbf{W}^{-1}\mathbf{V}^{-1}\mathbf{y} \\
&= \mathbf{W}'\mathbf{V}^{-1}\mathbf{y} \\
&= \mathbf{W}'\mathbf{V}\mathbf{V}^{-1}\mathbf{V}^{-1}\mathbf{y} \\
&= \mathbf{U}'\mathbf{V}^{-2}\mathbf{y}.
\end{aligned}
$$

Hence, using the squared diagonal elements of $\mathbf{V}$ and the transpose of $\mathbf{U}$ we can derive comparisons between the treatment combinations in terms of effects:

$$
\begin{aligned}
p_0 q_0 &= 9M/9 - 3P'/6 + 6P''/18 - 3Q'/6 + 6Q''/18 + 2P'Q'/4 \\
&\quad - 4P'Q''/12 - 4P''Q'/12 + 8P''Q''/36 \\
&= M - P'/2 + P''/3 - Q'/2 + Q''/3 + P'Q'/2 - P'Q''/3 \\
&\quad - P''Q'/3 + 2P''Q''/9.
\end{aligned}
$$

Similarly

$$p_0 q_1 = M - P'/2 + P''/3 - 2Q''/3 + 2P'Q''/3 - 2P''Q''/9,$$

and hence

$$p_0 q_0 - p_0 q_1 = -Q'/2 - Q'' + P'Q'/2 - P'Q'' - P''Q'/3 + 2P''Q''/3.$$

If all interactions are assumed zero, this comparison reduces to

$$-Q'/2 - Q''.$$

If the linear $\times$ linear interaction is assumed to be non-zero but other interaction effects are assumed zero, then our estimate of $p_0 q_0 - p_0 q_1$ is

$$-Q'/2 - Q'' + P'Q'/2.$$

As with the two-level factors, estimates of comparisons between treatment combinations assuming certain interactions to be zero are more precise provided the assumptions are correct. Again the details are more complicated than for two-level factors because the different effects have different variances. Each element of the effect vector, $\mathbf{y}$, has variance $\sigma^2 v_{ii}^2/n$, where n is the replication of each treatment combination. The variances of effects can be calculated as

$$
\begin{aligned}
\mathrm{Var}(P') \quad &= \mathrm{Var}(Q') \quad = (1/3)^2 6\sigma^2/n = 2\sigma^2/3n, \\
\mathrm{Var}(P'') \quad &= \mathrm{Var}(Q'') \quad = (1/6)^2 18\sigma^2/n = \sigma^2/2n, \\
\mathrm{Var}(P'Q') \quad & \qquad\qquad\quad = (1/2)^2 4\sigma^2/n = \sigma^2/n, \\
\mathrm{Var}(P'Q'') &= \mathrm{Var}(P''Q') = (1/4)^2 12\sigma^2/n = 3\sigma^2/4n, \\
\mathrm{Var}(P''Q'') \quad & \qquad\qquad\quad = (1/8)^2 36\sigma^2/n = 9\sigma^2/16n.
\end{aligned}
$$

Hence the variance of the three alternative estimates of $(p_0q_0 - p_0q_1)$ considered previously are:

(a) making no assumptions,

$$(1/4)(2\sigma^2/3n) + (\sigma^2/2n) + (1/4)(\sigma^2/n) + (3\sigma^2/4n)$$
$$+ (1/9)(3\sigma^2/4n) + (4/9)(9\sigma^2/16n) = 2\sigma^2/n,$$

which is, of course, the variance for comparing two treatment means each based on n observations;

(b) assuming all interactions are zero:

$$(1/4)(2\sigma^2/3n) + (\sigma^2/2n) = 2\sigma^2/3n;$$

(c) assuming all interactions except the linear $\times$ linear are zero:

$$(1/4)(2\sigma^2/3n) + (\sigma^2/2n) + (1/4)(\sigma^2/n) = 11\sigma^2/12n.$$

The variance for (b) is simply interpreted as the variance for comparing two treatment means each based on $3n$ observations and, since we are assuming zero interactions, the comparison $(p_0q_1 - p_0q_1)$ is estimated by the main effects comparison $(q_0 - q_1)$. The variance for (c), which might often be the most reasonable assumption, is much closer to that for (b) than to that for (a), illustrating again the benefit of factorial structure when some interaction effects are negligible.

The extension of the set of factorial effects from the 2^m to the 3^m case can be continued to allow for more than three levels of a factor and for mixed levels. For three-level factors, although there is theoretically an infinite choice of possible pairs of effects, we have argued that in practice much the most important pair of effects are $(1, -1, 0)$ and $(1, 1, -2)$. With four levels, there are three orthogonal effects to be defined and there is not a single dominant set of three effects. If the levels are equally spaced levels of a quantitative factor, then the three effects may be defined to represent linear, quadratic and cubic effects, as follows:

$$\begin{array}{ll} \text{linear} & (-3, -1, 1, 3), \\ \text{quadratic} & (-1, 1, 1, -1), \\ \text{cubic} & (1, -3, 3, -1). \end{array}$$

Alternatively, the effects can represent successive comparisons of one level compared with the remainder:

$$\begin{array}{l} (3, -1, -1, -1), \\ (0, 2, -1, -1), \\ (0, 0, 1, -1). \end{array}$$

A third alternative involves a hierarchical structure of the levels with two main pairs, each split:

$$\begin{array}{l} (1, 1, -1, -1), \\ (1, -1, 0, 0), \\ (0, 0, 1, -1). \end{array}$$

A small modification of this last set of comparisons leads us back to the 2^2 structure for four levels:

$$(\ 1, \ \ 1, -1, -1),$$
$$(\ 1, -1, \ \ 1, -1),$$
$$(-1, \ \ 1, \ \ 1, -1).$$

However, the set of effects are defined the effects vector, $\mathbf{y}$, may be written in terms of the set of treatment combination means, $\mathbf{x}$, through an equation in the form

$$\mathbf{y} = \mathbf{Ux},$$

where $\mathbf{U} = \mathbf{VW}$ and $\mathbf{W}$ is orthogonal and $\mathbf{V}$ diagonal, exactly as for the 3^m or 2^m structure.

The same structural relationship of effects and treatment combinations holds for mixtures of levels. Thus, for three factors, $\mathbf{P}, \mathbf{Q}$ and $\mathbf{R}$ with three, two and two levels, respectively,

$$\begin{aligned}
\mathbf{P}' &= (1/4)(p_2 - p_0)(q_1 + q_0)(r_1 + r_0), \\
\mathbf{P}'' &= (1/8)(p_2 - 2p_1 + p_0)(q_1 + q_0)(r_1 + r_0), \\
\mathbf{Q} &= (1/6)(p_2 + p_1 + p_0)(q_1 - q_0)(r_1 + r_0), \\
\mathbf{R} &= (1/6)(p_2 + p_1 + p_0)(q_1 + q_0)(r_1 - r_0), \\
\mathbf{P}'\mathbf{Q} &= (1/4)(p_2 - p_0)(q_1 - q_0)(r_1 + r_0), \\
\mathbf{P}''\mathbf{Q} &= (1/8)(p_2 - 2p_1 + p_0)(q_1 - q_0)(r_1 + r_0), \\
\mathbf{P}'\mathbf{QR} &= (1/4)(p_2 - p_0)(q_1 - q_0)(r_1 - r_0).
\end{aligned}$$

The remaining effects may be completed by the reader or read from the matrix $\mathbf{U}$ given below:

$\mathbf{y}$	=				$\mathbf{U}$									$\mathbf{x}$
12M		1	1	1	1	1	1	1	1	1	1	1	1	$p_0q_0r_0$
4P'		−1	−1	−1	−1	0	0	0	0	1	1	1	1	$p_0q_0r_1$
8P''		1	1	1	1	−2	−2	−2	−2	1	1	1	1	$p_0q_1r_0$
6Q		−1	−1	1	1	−1	−1	1	1	−1	−1	1	1	$p_0q_1r_1$
6R		−1	1	−1	1	−1	1	−1	1	−1	1	−1	1	$p_1q_0r_0$
4P'Q	=	1	1	−1	−1	0	0	0	0	−1	−1	1	1	$p_1q_0r_1$
8P''Q		−1	−1	1	1	2	2	−2	−2	−1	−1	1	1	$p_1q_1r_0$
4P'R		1	−1	1	−1	0	0	0	0	−1	1	−1	1	$p_1q_1r_1$
8P''R		−1	1	−1	1	2	−2	2	−2	−1	1	−1	1	$p_2q_0r_0$
6QR		1	−1	−1	1	1	−1	−1	1	1	−1	−1	1	$p_2q_0r_1$
4P'QR		−1	1	1	−1	0	0	0	0	1	−1	−1	1	$p_2q_1r_0$
8P''QR		1	−1	−1	1	−2	2	2	−2	1	−1	−1	1	$p_2q_1r_1$

As for the three-level factors $\mathbf{U} = \mathbf{VW}$, where $\mathbf{W}$ is orthogonal and $\mathbf{V}$ is diagonal with

$$v_{ii}^2 = \sum_j u_{ij}^2,$$

giving squared diagonal elements of $\mathbf{V}$:

$$(12, 8, 24, 12, 12, 8, 24, 8, 24, 12, 8, 24).$$

Table 13.13. *Analysis of variance for three factors*

Source	df
Blocks	1
P	3
Q	2
R	1
PQ	6
PR	3
QR	2
PQR	6
Error	23
Total	47

And once again we can express the vector of treatment combination yields, $\mathbf{x}$, in terms of the effect vector,

$$\mathbf{x} = \mathbf{U}'\mathbf{V}^{-2}\mathbf{y}.$$

13.5 The use of only a single replicate

The logic of the economy afforded by factorial structure leads naturally to including more and more factors, gradually replacing explicit replication by hidden replication. If we completely replace direct replication by hidden replication, then we would use just a single complete replicate of a factorial structure. There are two obvious problems about this possible design. The first is how to utilise relevant blocking structures and this is a major subject which will be developed at length in Chapter 15. The second problem which might seem to make the single-replication design inadmissible is the requirement to estimate σ^2 so that SEs of treatment effects may be calculated; this will be discussed in some detail in the next section. First we look at the possibility of using a single replicate.

Consider, for example, a four-factor experiment with factors P (four levels), Q (three levels), R (two levels) and S (two levels), a total of 48 experimental units in a single replicate. If the factor S were not included, we could have two blocks of 24 units each, with the analysis of variance structure as in Table 13.13, and σ^2 can be estimated from the error MS.

But 23 is a luxuriously large number of df for estimating σ^2 and, by including S, we could obtain additional information not only about S but also about the two-factor interactions of S with P, Q and R.

For the four-factor experiment, without blocks, the full analysis of variance structure will be as shown in Table 13.14, leaving no df for estimating σ^2. Now, in most factorial experiments, the numerical size of effects tends to decline as the number of factors involved in the effect increases. That is two-factor interaction effects are usually smaller than the corresponding main effects for the two factors; three-factor interaction effects are usually smaller than the corresponding two-factor interaction effects, and so on. There is, of course,

Table 13.14. *Analysis of variance for four factors*

Source	df	Source	df
P	3	QS	2
Q	2	RS	1
R	1	PQR	6
S	1	PQS	6
PQ	6	PRS	3
PR	3	QRS	2
PS	3	PQRS	6
QR	2	Total	47

no absolute rule that this trend shall occur but statistically such a pattern is often observed. It might be anticipated theoretically by arguing that interactions are essentially modifications of main effects. It is also sometimes argued that the response to changing various quantitative factors could be modelled by a Taylor series expansion, and the terms in that expansion, with more variables included, which represent higher-order interactions, must diminish for convergence of the series.

The general drift of the arguments in the last paragraph is that we may expect to find that three-factor and higher-order interactions are negligibly small. Although only a wide experience can be convincing, most examples in this book do support this expectation as do most examples in a wide range of published books and journals. The conclusion to be drawn is that the three- and four-factor interactions will rarely be of importance, and that the expectations of the MS for these effects will often be only slightly larger than σ^2. Hence, we can obtain a sensible estimate of σ^2 by pooling the MS for some, or all, three- and four-factor interactions.

Example 13.2 To illustrate the use of a single replicate consider an experiment on the survival of *Salmonella typhimerium*, in which three factors were examined. Three levels of sorbic acid, six levels of water activity and three pH levels were combined to give 54 combinations. The data, presented in Table 13.15 are log (density/ml) measured seven days after treatments started.

The analysis of variance, including SS for all main effects and two- and three-factor interactions is shown in Table 13.16.

The design of the experiment clearly implies the intention of using the three-factor interaction MS as an estimate of error, and the variance ratios have been calculated using $s^2 = 0.03$ as the divisor. The single replicate clearly provides effective replication for all main effects and the two-factor AS interaction. The major effect is that of water activity, and comparison of the six water activity means is based on an effective nine-fold replication. The sorbic acid means and the pH means have an effective replication of 18, the means for water activity $\times$ sorbic acid combinations have effective replication of 3 and the two-way means for pH $\times$ sorbic acid have effective replication of 6. The presentation of results should be as in Table 13.17.

Table 13.15. *Data for salmonella experiment*

Sorbic acid (ppm)	pH	Water activity					
		0.78	0.82	0.86	0.90	0.94	0.98
0	5.0	4.20	4.52	5.01	6.14	6.25	8.33
	5.5	4.34	4.31	5.35	5.98	6.70	8.37
	6.0	4.31	4.85	5.06	5.87	6.65	8.19
100	5.0	4.18	4.18	4.29	5.78	6.51	7.59
	5.5	4.39	4.43	4.95	5.28	6.19	7.79
	6.0	4.13	4.29	4.85	5.01	6.52	7.64
200	5.0	4.15	4.37	4.79	5.43	6.43	7.19
	5.5	4.12	4.27	4.40	5.10	6.18	6.92
	6.0	3.93	4.26	4.41	5.20	6.33	7.14

Table 13.16. *Analysis of variance for salmonella experiment*

	SS	df	MS	F
Water activity (A)	81.57	5	16.31	473
Sorbic acid (S)	2.76	2	1.38	40
pH	0.01	2	0.01	0.2
AS	1.32	10	0.13	3.8
ApH	0.45	10	0.04	1.3
SpH	0.23	4	0.06	1.7
ASpH (error)	0.69	20	0.03	
Total	87.03	53		

Table 13.17. *Means for important differences in salmonella experiment*

Sorbic acid (ppm)	Water activity						Mean
	0.78	0.82	0.86	0.90	0.94	0.98	
0	4.28	4.56	5.14	6.00	6.53	8.30	5.80
100	4.23	4.30	4.77	5.36	6.41	7.67	5.44
200	4.07	4.30	4.53	5.24	6.31	7.08	5.26
Mean	4.19	4.39	4.79	5.53	6.42	7.68	

SE difference (water activity means) = 0.105
SE difference (sorbic acid means) = 0.075
SE difference (combination means) = 0.183

13.6 Analysis of unreplicated factorials

As we have seen in Example 13.2, the analysis of results from a single replicate is straight-forward except in one respect. All SS in the analysis for main effects and for interactions are calculated in the usual way, and the only question is 'How do we estimate σ^2?' Obviously an estimate of σ^2 is needed both for the calculation of SEs for estimates of treatment effects and comparisons, and for the calculation of F-ratios in the analysis of variance.

Clearly, because there is no replication there is no possibility of defining an unbiased estimator of σ^2. Equally clearly the hidden replication must have some potential to provide an estimate of σ^2 which will often be only a little biased, and which will allow inferences about the effect estimates to be made.

In Example 13.2 we used the three-factor interaction MS, for which there are 20 df, as an estimate for σ^2; in the earlier four-factor example in Section 13.5, if all three- and four-factor interactions SS are used to estimate σ^2 there are 23 df for the estimate. In both cases there are ample df for estimating σ^2.

In general it will usually be sensible to use MS from high-order interactions to estimate σ^2. However, the decisions about which interaction SS to use for the estimation of σ^2 require careful thought. (i) Ideally we would prefer to decide which interaction SS to use before the experiment, (ii) although in general we would usually expect to be able to use three-factor interaction SS there might be good reason to expect one or more three-factor interaction SS to be substantial, (iii) after the experiment analysis is completed some high-order interaction SS might appear large, and we might then wish to exclude them from the SS to estimate σ^2. (iv) Similarly some lower-order interaction SS in the completed analysis might appear small, and there could be the temptation to include them in the SS for estimating σ^2. (v) In general, to what extent can we choose which SS to include in the SS from which σ^2 will be estimated in the light of the results, when the choice would be better informed?

We have already noted in Section 13.2 that including interaction SS because they are small will result in biasing the estimator of σ^2. In principle, if we only included interaction SS when they were very small, this would produce a very small estimate of σ^2, and using such an estimate could lead us to claim that the main effects and two-factor interactions were much more precise than they really are. On the other hand, to always use only those interaction SS determined before the experiment could lead to sometimes having an inflated estimate of σ^2, or to using a less precise estimate of σ^2 than could be available.

If we consider a complete set of single df SS, then when all treatment effects are zero we would have a distribution of SS (chi-squared on one df) in which most SS are less than the expected value, but with a few larger values; combining all SS would provide the best estimate of σ^2.

If we further consider a situation when a small subset of the single-df treatment effects are definitely non-zero, then the distribution of SS would appear similar to that if all effects are zero, except that there would be a few much larger SS. The question is how is it possible to detect at what point the larger SS have departed from the pattern of distribution defined by the majority of (smaller) SS.

Although such a detection might appear very difficult, there are graphical methods which are practically useful, namely normal and half-normal plots.

Table 13.18. *Factors for washing powders experiment*

	Levels	
Factor	0	1
A: Bleach level	Low	High
B: Surfactant type	Standard	New
C: Surfactant level	Low	High
D: Catalyst	Absent	Present
E: Wash temperature	Low	High
F: Water hardness	Soft	Hard

13.6.1 The (half-)normal plot method

The principle of the normal plot is that we consider the distribution of the size of effects. However, instead of the distribution of SS, we use the equivalent distribution of actual effects. If all effects are really zero, then the distribution of the effect sizes would be normal. If, however, we consider the absolute sizes of effects, ignoring the + or − signs, then the effects will be ordered in increasing absolute size, and the resulting distribution should look like the half-normal distribution.

It is relatively simple to calculate the distribution we should expect for a set of values, evenly spaced, from a normal or half-normal distribution, and if we plot the sequence of observed effect values (from smallest to largest) against the expected values from the normal or half-normal distribution, then if all the effect values are really zero the graph should show an approximately straight line. If there are some non-zero effects, then the points corresponding to those effects should show a clear deviation at the upper end of the straight line provided by the other effects. The half-normal plot, rather than the normal plot, is recommended when the number of effects estimated is fewer than about 16, since in this case it is difficult to get a good picture at the tails of a normal plot. With fewer than about eight runs, the half-normal plot is also difficult to interpret, although huge effects can still be detected easily enough.

The simplest form of experiment for the use of the normal or half-normal plot method is when all factors are at two levels, since then all effects inevitably have a single df. The use of the method is illustrated in the following example.

Example 13.3 An experiment on the factors affecting the performance of washing powders was run using a single replicate of the combinations of two levels of each of six factors. The factors and their levels are described in Table 13.18. The performance was measured by recording the light fluorescence from stained test cloths after a wash in a standard household washing machine.

A normal plot of the estimated effects is shown in Figure 13.1. It shows clearly that the main effects of surfactant type, bleach level and probably wash temperature are larger than

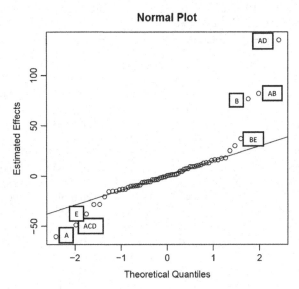

Figure 13.1 Normal plot of effects.

could be expected from random variation, while several interactions are also large, including that between bleach level and catalyst, as well as two involving the large main effects. There is also a fairly large three-factor interaction, between bleach level, surfactant level and catalyst, which makes interpretation difficult. In addition to the effects labelled, several others fall off the straight line of estimates enough to possibly merit further investigation, although effects clearly less than 10 in magnitude were considered too small to be of practical importance.

Interpretation using the normal plot seems clearer than that based on an analysis of variance constructed using some of the high-order interactions to estimate σ^2. The difficulty with the latter approach in this example is that the interpretation depends quite strongly on which interactions are included. Assuming all three-factor and higher-order interactions are negligible gives the analysis of variance shown in Table 13.19, while if we include only four-factor and higher-order interactions, we get that in Table 13.20. Note that the mean square error has halved. The conclusions change somewhat, for example for the AE and CD interactions, although the qualitative interpretation is not greatly affected if done carefully.

This example illustrates another advantage of the normal plot over analysis of variance tables for experiments of this type, namely that the graphical display encourages the user to focus on the relative sizes of effects, rather than their formal statistical significance. When studying many factors in one experiment, especially in industrial experiments, the objective is often to find a few factors whose manipulation will allow substantial improvements in quality, rather than to find the absolute optimum combination of factor levels. The analysis of variance in Table 13.20 identifies so many significant effects that the main message, i.e. that we should use the high level of bleach, the new surfactant and the catalyst to improve wash performance, is almost hidden by the sea of p-values. This also might suggest that the use of even a single replicate was excessive for this experiment.

Table 13.19. *Analysis of variance assuming three-factor interactions are negligible*

	df	SS	MS	F-value	p-value
A	1	3643.4	3643.4	25.0294	0.000
B	1	5886.0	5886.0	40.4362	0.000
C	1	11.4	11.4	0.0786	0.781
D	1	104.6	104.6	0.7185	0.401
E	1	1435.3	1435.3	9.8604	0.003
F	1	36.9	36.9	0.2537	0.617
AB	1	6744.9	6744.9	46.3363	0.000
AC	1	173.3	173.3	1.1907	0.281
AD	1	18343.8	18343.8	126.0193	0.000
AE	1	810.1	810.1	5.5652	0.023
AF	1	14.0	14.0	0.0959	0.758
BC	1	0.5	0.5	0.0032	0.955
BD	1	240.1	240.1	1.6497	0.206
BE	1	1373.7	1373.7	9.4373	0.004
BF	1	87.9	87.9	0.6042	0.441
CD	1	633.3	633.3	4.3509	0.043
CE	1	42.4	42.4	0.2913	0.592
CF	1	804.1	804.1	5.5239	0.024
DE	1	98.8	98.8	0.6789	0.415
DF	1	3.6	3.6	0.0245	0.876
EF	1	902.4	902.4	6.1992	0.017
Residuals	42	6113.7	145.6		

Exercises

13.1 A 2^4 experiment on yield from a chemical reaction gave the treatment totals (from two replicates) shown in Table 13.21. Estimate the effects and calculate the analysis of variance. If the error MS is 100 (based on 15 df), draw your conclusions from the experiment.

   If it could be assumed that only main effects and two-factor interactions were non-zero, find estimates of the comparisons:

$$(a_1 b_0 c_0 d_0 - a_0 b_0 c_0 d_0),$$
$$(a_0 b_1 c_0 d_0 - a_0 b_0 c_0 d_0),$$
$$(a_0 b_0 c_1 d_0 - a_0 b_0 c_0 d_0)$$

and

$$(a_0 b_0 c_0 d_1 - a_0 b_0 c_0 d_0);$$

also find the variance of any one of these estimates.

13.2 Write out the matrix relationship between the effects vector, **y**, and the observations vector, **x**, for an experiment with six treatments comprising three levels of factor A with two levels of factor B. The usual form of effects for A should be assumed $((a_2 - a_0)$ and $(a_2 - 2a_1 + a_0))$ with corresponding interaction effects.

Table 13.20. *Analysis of variance assuming four-factor interactions are negligible*

	df	SS	MS	F-value	p-value
A	1	3643.4	3643.4	48.9291	0.000
B	1	5886.0	5886.0	79.0473	0.000
C	1	11.4	11.4	0.1537	0.699
D	1	104.6	104.6	1.4046	0.249
E	1	1435.3	1435.3	19.2758	0.000
F	1	36.9	36.9	0.4959	0.489
AB	1	6744.9	6744.9	90.5813	0.000
AC	1	173.3	173.3	2.3276	0.141
AD	1	18343.8	18343.8	246.3508	0.000
AE	1	810.1	810.1	10.8792	0.003
AF	1	14.0	14.0	0.1875	0.669
BC	1	0.5	0.5	0.0062	0.938
BD	1	240.1	240.1	3.2249	0.086
BE	1	1373.7	1373.7	18.4486	0.000
BF	1	87.9	87.9	1.1811	0.289
CD	1	633.3	633.3	8.5054	0.008
CE	1	42.4	42.4	0.5694	0.459
CF	1	804.1	804.1	10.7986	0.003
DE	1	98.8	98.8	1.3271	0.262
DF	1	3.6	3.6	0.0478	0.829
EF	1	902.4	902.4	12.1186	0.002
ABC	1	31.4	31.4	0.4222	0.523
ABD	1	2.3	2.3	0.0305	0.863
ABE	1	11.0	11.0	0.1480	0.704
ABF	1	1.3	1.3	0.0171	0.897
ACD	1	2377.2	2377.2	31.9252	0.000
ACE	1	22.9	22.9	0.3079	0.585
ACF	1	306.5	306.5	4.1161	0.055
ADE	1	53.2	53.2	0.7143	0.407
ADF	1	430.0	430.0	5.7742	0.025
AEF	1	257.2	257.2	3.4542	0.077
BCD	1	31.2	31.2	0.4195	0.524
BCE	1	85.4	85.4	1.1470	0.296
BCF	1	85.9	85.9	1.1541	0.294
BDE	1	89.2	89.2	1.1973	0.286
BDF	1	79.7	79.7	1.0702	0.312
BEF	1	53.5	53.5	0.7178	0.406
CDE	1	158.7	158.7	2.1314	0.158
CDF	1	126.9	126.9	1.7039	0.205
CEF	1	228.3	228.3	3.0659	0.094
DEF	1	43.8	43.8	0.5876	0.452
Residuals	22	1638.2	74.5		

Table 13.21. *Treatment totals for chemical reaction experiment*

	d_0		1_1	
	c_0	c_1	c_0	c_1
a_0b_0	8	31	79	77
b_1	53	12	73	49
a_1b_0	4	9	68	38
b_1	43	36	8	23

Table 13.22. *Potato fertiliser experiment: (a) treatment means (tons/acre); (b) analysis of variance*

(a)

	m_0p_0	m_0p_1	m_1p_0	m_1p_1
n_0	4.0	4.2	5.4	6.4
n_1	4.8	5.5	6.2	6.4
n_2	4.4	4.9	5.2	5.9

(b)

	SS	df	MS
Blocks	18.17	2	
Treatments	23.29	11	2.12
Error	7.04	22	0.32
Total	48.5		

Derive the reverse relationship expressing $\mathbf{x}$ in terms of $\mathbf{y}$. Express the variances of the comparisons $(a_1b_1 - a_0b_0)$, $(a_0b_1 - a_0b_0)$, $(a_1b_1 - a_1b_0)$ and $(a_2b_0 - a_0b_0)$ in terms of the variances of the five effects. Examine how these variances are reduced when

  (i) interactions are assumed zero,

 (ii) the quadratic main effect and interaction are assumed zero,

(iii) only the quadratic interaction is assumed zero.

13.3 The analysis of variance and treatment means in Table 13.22(a) and (b) are taken from a fertiliser experiment on potatoes conducted in three randomised blocks of 12 plots. The 12 treatments were all combinations of three nitrogen levels (n_0 = none, n_1 = sulphate of ammonia, n_2 = ammonia bicarbonate), two manure levels (m_0 = none, m_1 = some) and two phosphate levels (p_0 = none, p_1 = some).

Define a complete set of single df contrasts and calculate the corresponding SS.

Obtain estimates of the differences of $n_1m_0p_0$, $n_2m_0p_0$, $n_0m_1p_0$ and $n_0m_0p_1$, from $n_0m_0p_0$ assuming

Table 13.23. *Data from experiment on aircraft components*

Treatment	Score	Treatment	Score
(1)	1.4	e	1.7
a	1.2	ae	2.0
b	3.6	be	3.1
ab	1.2	abe	1.2
c	1.5	ce	1.9
ac	1.4	ace	1.2
bc	1.5	bce	1.0
abc	1.6	abce	1.8
d	5.0	de	9.5
ad	9.0	ade	5.9
bd	12.0	bde	12.6
abd	5.4	abde	6.3
cd	4.2	cde	8.0
acd	4.4	acde	4.2
bcd	9.3	bcde	7.7
abcd	2.8	abcde	6.0

 (i) all interactions are zero,
(ii) the three factor interaction is zero.

What estimates would you give a farmer who wanted to know what benefits could be expected from adding each one of the fertilisers to the no fertiliser control treatment?

13.4 An experiment was concerned with the problem of trying to obtain a new type of coating material for metals used in making aircraft components. The purpose of coating the metal is to increase its strength. The coating powder is shot, using a spray gun, at extremely high velocity through a gas flame onto the metal or substrate. When the particles of coating hit the surface they are squashed and stick to the material. A measure of abrasion loss was then obtained by rubbing the substrate with a rough diamond pin over a fixed period of time.

The effects of five factors were studied, with each factor at two levels. The factors were:

factor A: method of cleaning or preparing the substrate;

factor B: composition of the substrate;

factor C: preheat temperature of the substrate before applying the coating powder;

factor D: type of coating powder;

factor E: size of the powder particles.

The 2^5 experiment was completely randomised using just one replicate. The scores were as shown in Table 13.23.

Analyse these data paying particular attention to the possibilities that some responses may be outliers, that the residual variance about an appropriate linear model may not be homogeneous over the whole experiment, and that a more parsimonious factorial model may be achieved by first transforming the observed responses.

Summarise the conclusions from the experiment.

14

Fractional replication

14.0 Preliminary examples

(*a*) An industrial experiment is to be planned to investigate the effects of varying seven factors in a chemical process. It is decided to use two levels of each factor, and there is sufficient material and time to make observations on 64 treatment combinations. How should the 64 treatment combinations to be included in the experiment be chosen?

(*b*) An experiment on competition between grass and legume species is to be planned to investigate the effects and pairwise interactions of eight factors. The eight factors are:

(1) two different cutting regimes,
(2) two different grass species,
(3) two different legume species,
(4) two levels of phosphate,
(5) two levels of potassium,
(6) two levels of nitrogen,
(7) root competition between grass and legume allowed or restricted,
(8) shoot competition between grass and legume allowed or restricted.

The total number of experimental units that can be managed is 48. How should the 48 combinations be chosen?

(*c*) An experiment on making hot cross buns is to be planned to investigate the effects, both main effects and two-factor interactions, of six factors. The factors to be varied are: (i) three different forms of flour (F), (ii) three different temperatures for baking (T), (iii) three levels for the amount and pattern of dried fruit in the mixture (D) and (iv) two levels of each of three additives related to improving storage performance (A, B and C). It is decided that it is possible to observe up to 60 combinations. How can a fractional replicate be constructed to estimate all main effects, the two-factor interactions for the three-level factors (F $\times$ T, F $\times$ D and T $\times$ D), and the interactions between the two-level factors and the baking temperature (T $\times$ A, T $\times$ B and T $\times$ C)? What other two-factor interactions could be estimated (possibly only partial information)?

(*d*) An experiment system is required to screen a large number of chemical compounds for a particular biological response. The requirement of the system is that many compounds can be screened with only a small number of runs. It is assumed that only a tiny proportion of the compounds will have any effect, that any interactions can be ignored and that the signal to noise ratio is high so that there is no requirement to estimate the error variation. What kind of designs are available for this situation?

14.1 The use of a fraction of a complete factorial experiment

In the previous chapter we have seen that it is possible to use a single replicate of a factorial set of treatments to obtain estimates of main effects and two-factor interactions, using high-order interaction MS to estimate σ^2 and obtaining sufficient effective replication from the hidden replication of the factorial structure. The full set of requirements which need to be considered when proposing the use of a single-replicate design are:

(i) main effects are always important, and must be estimated;
(ii) two-factor interactions should usually be estimated;
(iii) three-factor or higher-order interactions may occasionally need to be estimated;
(iv) it should be possible to obtain an estimate of σ^2 using higher-order interaction MS for interactions likely to have only small effects; and
(v) sufficient replication of the important treatment effects must be achieved without explicit replication.

In this chapter we will develop designs using only a fraction of the complete set of factorial combinations, most of which will enable most of these five crucial requirements to be satisfied. A design using only a fraction of the possible factorial treatment combinations is called a fractional replicate. For example suppose that we wish to investigate six factors, P, Q, R, S, T and U, each at two levels but that an experiment of 64 observations is too large for the available resources. To assess all the main effects and two-factor interactions would require 6 and 15 df, respectively. Using half of the 64 combinations would give a total of 31 df so that after estimating the main effects and two-factor interactions there could be 10 df available to estimate σ^2. Therefore the question is whether we can identify a suitable set of 32 combinations from the total 64 combinations, which allows us to estimate the effects and σ^2.

Fractional replicates can be useful in a variety of forms ranging from using half of the possible combinations and achieving all five requirements to using a tiny proportion of the possible combinations in a saturated design, and being able to satisfy only items (i) and (v). In these designs we shall ignore (iii) the estimation of three-factor interactions; in practice it is not often relevant to look for interactions involving three factors, and it is also often difficult to construct a sensible interpretation when such interactions appear to be large. In each fractional replicate design we shall examine how the particular requirements for the particular experiment can be achieved using a fraction of the possible factorial combinations, and show how the relevant fraction can be chosen.

The potential for theoretical developments for fractional replicate designs is very large. Here we shall consider only those aspects which are necessary for practically useful designs. This will mean that we do not discuss in great detail some important theoretical concepts and ideas. In most cases practically useful designs will be neither too small nor too large, and we shall concentrate on designs which would use between 25 and 80 observations.

In the first two sections we shall consider the simplest form of fractional replicate designs, looking at half-replicates of 2^n structures, and one-third-replicates of 3^n structures. The designs are very easy to construct and we shall also consider the analysis of a quarter-replicate of a 4^4 structure.

In the following two sections we shall look at smaller fractions for 2^n structures. These start by looking at how to construct one-quarter- and one-eighth-replicates, which leads us into consideration of how to choose which contrasts should be used to define the fractional replicate we wish to construct. We then go on to consider more irregular fractions such as three-quarters, three-eighths and five-eighths, again using the concepts of defining contrasts and the way they determine which effects can be estimated.

Then we return to 3^n structures and consider smaller fractions than the one-third, and also look at fractional replicate designs when a mixture of two-level and three-level factors are used.

Finally we consider designs where only main effects are of interest, but where the number of potentially interesting main effects can be very large, while the number of observations is quite small. The designs are generally referred to as saturated and supersaturated and are essentially screening designs to sort out a small number of important factors from a population of factors, most of which are expected to have no effect at all. These designs are very different from those discussed elsewhere in this chapter, and the concept may seem rather unlikely at first. However, the potential gain in screening programmes can be very substantial.

As in the previous chapter, we shall not be considering questions of blocking control of variation in this chapter. Obviously in many situations the number of combinations required will be large and then it would usually be necessary to think about dividing the combinations into blocks of a manageable size. The problems of blocking design for replicated, unreplicated and fractional factorial designs will be considered in Chapter 15.

14.2 Half-replicates of 2^n factorials

The simplest form of fractional replicate is a half-replicate of a 2^n factorial and the properties of half-replicates illustrate most of the important aspects of fractional replicates. To illustrate the construction and analysis of a fractional replicate we first consider a trivial example of a half-replicate of a 2^4 factorial structure with factors P, Q, R and S. The combinations to be included in the experiment should satisfy the two criteria that:

 (i) to estimate a main effect efficiently the two levels of the factor must be equally replicated;
(ii) to estimate main effects independently each pair of main effects should be orthogonal, and to achieve this all four combinations of the levels of the two factors should be equally replicated.

These two criteria can be satisfied by several sets of combinations, of which four are shown in Figure 14.1. In fact the number of possibilities is very limited. Either each set of three factors (PQR, PQS, PRS, QRS) has all eight of the possible combinations present (set I) or there is exactly one set of three factors which does not have all eight combinations present (sets II(PRS), III(PQS), IV(PQR)). All designs satisfying (i) and (ii) are in the form of set I or set II.

If we adopt the arbitrary, but intuitively satisfying, additional requirement that for each set of three factors all combinations shall be equally replicated, then we would select set I as our design; note that for each set there is a complementary design, consisting of the other eight combinations from the total set of 16, which possesses exactly the same properties. The estimation of effects can be represented as in Table 14.1.

Table 14.1. *Coefficients for estimating effects in half-replicate of 2^4 design*

Combination	P	Q	R	S	PQ	PR	PS	QR	QS	RS
$p_0q_0r_0s_0$	-1	-1	-1	-1	$+1$	$+1$	$+1$	$+1$	$+1$	$+1$
$p_0q_0r_1s_1$	-1	-1	$+1$	$+1$	$+1$	-1	-1	-1	-1	$+1$
$p_0q_1r_0s_1$	-1	$+1$	-1	$+1$	-1	$+1$	-1	-1	$+1$	-1
$p_0q_1r_1s_0$	-1	$+1$	$+1$	-1	-1	-1	$+1$	$+1$	-1	-1
$p_1q_0r_0s_1$	$+1$	-1	-1	$+1$	-1	-1	$+1$	$+1$	-1	-1
$p_1q_0r_1s_0$	$+1$	-1	$+1$	-1	-1	$+1$	-1	-1	$+1$	-1
$p_1q_1r_0s_0$	$+1$	$+1$	-1	-1	$+1$	-1	-1	-1	-1	$+1$
$p_1q_1r_1s_1$	$+1$	$+1$	$+1$	$+1$	$+1$	$+1$	$+1$	$+1$	$+1$	$+1$

Set I	Set II	Set III	Set IV
$p_0q_0r_0s_0$	$p_0q_0r_0s_0$	$p_0q_0r_0s_0$	$p_0q_0r_0s_0$
$p_0q_0r_1s_1$	$p_0q_0r_1s_1$	$p_0q_0r_1s_0$	$p_0q_0r_0s_1$
$p_0q_1r_0s_1$	$p_0q_1r_0s_0$	$p_0q_1r_0s_1$	$p_0q_1r_1s_0$
$p_0q_1r_1s_0$	$p_0q_1r_1s_1$	$p_0q_1r_1s_1$	$p_0q_1r_1s_1$
$p_1q_0r_0s_1$	$p_1q_0r_0s_1$	$p_1q_0r_0s_1$	$p_1q_0r_1s_0$
$p_1q_0r_1s_0$	$p_1q_0r_1s_0$	$p_1q_0r_1s_1$	$p_1q_0r_1s_1$
$p_1q_1r_0s_0$	$p_1q_1r_0s_1$	$p_1q_1r_0s_0$	$p_1q_1r_0s_0$
$p_1q_1r_1s_1$	$p_1q_1r_1s_0$	$p_1q_1r_1s_0$	$p_1q_1r_0s_1$

Figure 14.1 Subsets of the 2^4 factorial combinations for factors P, Q, R, S satisfying equal replication conditions given in the text.

Clearly all main effects are defined to be orthogonal to each other. Equally clearly it is not possible for all ten effects represented to be orthogonal since with eight combinations there can only be seven orthogonal combinations. By inspection there are pairs of two-factor interactions for which the estimates are identical. The estimate of PS is exactly the same as that for QR; $PQ \equiv RS$ and $PR \equiv QS$. In each case the linear contrast of combinations estimates not one parameter of the effects model but two. If the quantity

$$(p_0q_0r_0s_0 - p_0q_0r_1s_1 - p_0q_1r_0s_1 + p_0q_1r_1s_0$$
$$+ p_1q_0r_0s_1 - p_1q_0r_1s_0 - p_1q_1r_0s_0 + p_1q_1r_1s_1)$$

is large, then we can only interpret this as evidence that the factors P and S interact if we assume that the effect QR is negligible. This is expressed by saying that PS and QR are aliases.

For an alternative view of the aliasing concept consider the two sets of combinations defined in Table 14.2. All 16 treatment combinations are included in one set or the other, and set A is our previous set I. The effect PS is

$$PS = (1/8)(p_1 - p_0)(q_1 + q_0)(r_1 + r_0)(s_1 - s_0),$$

Hence,

$$8PS = (y_1 - y_2 - y_3 + y_4 + y_5 - y_6 - y_7 + y_8)$$
$$- (y_9 - y_{10} - y_{11} + y_{12} + y_{13} - y_{14} - y_{15} + y_{16}),$$

Table 14.2. *Sets of combinations forming separate half-replicates*

Set A	Set B
$y_1 = p_0 q_0 r_0 s_0$	$y_9 = p_0 q_0 r_0 s_1$
$y_2 = p_0 q_0 r_1 s_1$	$y_{10} = p_0 q_0 r_1 s_0$
$y_3 = p_0 q_1 r_0 s_1$	$y_{11} = p_0 q_1 r_0 s_0$
$y_4 = p_0 q_1 r_1 s_0$	$y_{12} = p_0 q_1 r_1 s_1$
$y_5 = p_1 q_0 r_0 s_1$	$y_{13} = p_1 q_0 r_0 s_0$
$y_6 = p_1 q_0 r_1 s_0$	$y_{14} = p_1 q_0 r_1 s_1$
$y_7 = p_1 q_1 r_0 s_0$	$y_{15} = p_1 q_1 r_0 s_1$
$y_8 = p_1 q_1 r_1 s_1$	$y_{16} = p_1 q_1 r_1 s_0$

while

$$8QR = (y_1 - y_2 - y_3 + y_4 + y_5 - y_6 - y_7 + y_8)$$
$$+ (y_9 - y_{10} - y_{11} + y_{12} + y_{13} - y_{14} - y_{15} + y_{16}).$$

Thus PS is estimated by the difference of two contrasts, one from each of the two sets, while QR is estimated by the sum of the same two contrasts. When the complete set of 16 observations is available, PS and QR are orthogonal and may each be estimated separately. When only one set of observations is available, then the estimates of PS and QR will be identical (except possibly in sign).

We may consider whether the main effects also have aliases since, on the argument of the last paragraph, the estimate of P is only 'half' of the normal estimate and might be expected to be 'half' of the normal estimate of some other effect. If we consider the three-factor interaction effects, then the estimate of PQR from the half-replicate (set A) would be

$$-y_1 + y_2 + y_3 - y_4 + y_5 - y_6 - y_7 + y_8,$$

which is identical with the estimate for S. Further investigation reveals the complete set of aliases to be

$$P \equiv QRS, \qquad PQ \equiv RS,$$
$$Q \equiv PRS, \qquad PR \equiv QS,$$
$$R \equiv PQS, \qquad PS \equiv QR.$$
$$S \equiv PQR,$$

The reader will perceive a pattern in these aliases, and mathematical theory provides a basis for deducing such patterns, but we shall defer discussion of this theory until Section 14.4. The aliasing of P and QRS is not likely to cause problems of interpretation. If the estimate, which could be attributable to either effect, is large, then there will rarely be any hesitation in assuming that QRS is negligible and interpreting the estimate as the effect P.

We have spent some time on this trivial example to display the way in which the estimates of different effects are composed, and how estimates of different effects are related. We will now look at the more practically realistic example of a half-replicate for a 2^6 factorial structure, considered briefly in the previous section.

P	Q	R	S	T	U		P	Q	R	S	T	U
0	0	0	0	0	0		0	0	0	0	1	1
0	0	0	1	0	1		0	0	0	1	1	0
0	0	1	0	0	1		0	0	1	0	1	0
0	0	1	1	0	0		0	0	1	1	1	1
0	1	0	0	0	1		0	1	0	0	1	0
0	1	0	1	0	0		0	1	0	1	1	1
0	1	1	0	0	0		0	1	1	0	1	1
0	1	1	1	0	1		0	1	1	1	1	0
1	0	0	0	0	1		1	0	0	0	1	0
1	0	0	1	0	0		1	0	0	1	1	1
1	0	1	0	0	0		1	0	1	0	1	1
1	0	1	1	0	1		1	0	1	1	1	0
1	1	0	0	0	0		1	1	0	0	1	1
1	1	0	1	0	1		1	1	0	1	1	0
1	1	1	0	0	1		1	1	1	0	1	0
1	1	1	1	0	0		1	1	1	1	1	1

Figure 14.2 A half-replicate of a 2^6 factorial structure, including all combinations of levels for every set of five factors.

Example 14.1 As noted earlier, a half-replicate of a 2^6 requires 32 observations, towards the lower end of our practically sensible range of design sizes. For the purposes of estimating all main effects and interactions we require

(i) for each factor the two levels are equally replicated,
(ii) for each pair of factors the four combinations are equally replicated.

To ensure that the estimates are all orthogonal we require

(iii) for any pair of interactions all combinations of levels for all factors involved are equally replicated. This requires that for each set of four factors all 16 combinations of levels are equally replicated.

These requirements are not sufficient to produce a unique design and, as with the simpler example, we adopt the general requirement that as far as is possible, for each set of factors, all combinations shall be equally replicated. In this case the requirement leads to having all combinations for each set of five factors occurring in the design. The resulting design is shown in Figure 14.2. The analysis of variance structure, as discussed earlier, is shown in Table 14.3. The alias structure can be determined to be

$$
\begin{array}{llll}
P \equiv QRSTU, & PQ \equiv RSTU, & QR \equiv PSTU, & RT \equiv PQSU, \\
Q \equiv PRSTU, & PR \equiv QSTU, & QS \equiv PRTU, & RU \equiv PQST, \\
R \equiv PQSTU, & PS \equiv QRTU, & QT \equiv PRSU, & ST \equiv PQRU, \\
S \equiv PQRTU, & PT \equiv QRSU, & QU \equiv PRST, & SU \equiv PQRT, \\
T \equiv PQRSU, & PU \equiv QRST, & RS \equiv PQTU, & TU \equiv PQRS. \\
U \equiv PQRST, & & & \\
\end{array}
$$

Table 14.3. *Analysis of variance structure for half-replicate*

	df
Six main effects	6
Fifteen two-factor interactions	15
Error (three-factor interactions)	10
Total	31

These two examples illustrate the general principle that the obvious half-replicate of a 2^n factorial structure to use is the set of combinations defined by the n-factor interaction, being either those combinations which are positive in the definition of the interaction, or those which are negative. It will be obvious that the set of combinations to be used as a half-replicate will have either all those combinations which have an even number of factors at the lower level, or all the combinations which have an odd number of factors at the lower level.

We can recognise that these two sets of combinations are the positive and negative groups of combinations in the definition of the n-factor interaction. Thus the natural half-replicate could also be defined as either the positive or negative set of combinations in the highest-order interaction, and this interaction is referred to as the Defining Contrast for the half-replicate. In our two examples the Defining Contrast is the interaction effect PQRS in the four-factor example, and PQRSTU in the six-factor example. These ideas are developed further in Section 14.4 where we consider more complex fractional structures.

14.3 Simple fractions for factors with more than two levels

When the factorial structure which we wish to investigate has factors with three or four levels it is also possible to construct fractional factorials, though the theory is not as simple as in two-level factorials.

For three-level factors, the simplest fraction is one-third of the full set of factorial combinations. For our range of practical experiment sizes there are two possible designs which are:

 (i) a one-third replicate of a 3^4 factorial with 27 observations, and
(ii) a one-third replicate of a 3^5 factorial with 81 observations.

We shall consider the first, smaller, design to demonstrate the principles and practice of the design in detail, and then briefly discuss how the latter design could be constructed.

For three-level factors, we have considered in the previous chapter possible definitions of the two single-df contrasts for each factor main effect. However, there is no simple way of using practically useful contrasts in the construction of an aliasing system for single-df contrasts. This means that design construction looks rather different from the simple methods for two-level factors. Instead we construct designs in terms of estimating the two-df effects rather than working with single-df contrasts.

P	Q	R	S	P	Q	R	S	P	Q	R	S
0	0	0	0	0	0	1	1	0	0	2	2
0	1	0	1	0	1	1	2	0	1	2	0
0	2	0	2	0	2	1	0	0	2	2	1
1	0	0	1	1	0	1	2	1	0	2	0
1	1	0	2	1	1	1	0	1	1	2	1
1	2	0	0	1	2	1	1	1	2	2	2
2	0	0	2	2	0	1	0	2	0	2	1
2	1	0	0	2	1	1	1	2	1	2	2
2	2	0	1	2	2	1	2	2	2	2	0

Figure 14.3 Levels of factors P, Q, R and S for the combinations of a one-third replicate of a 3^4 factorial structure satisfying the equal replication conditions specified in the text.

The fundamental principle (as with the two-level factors) is that each main effect will have a simple estimate provided the three factor levels are equally replicated. Two-factor interactions will have simple estimates provided the nine combinations of levels are equally replicated, and this requirement also ensures that the two related main effects will be estimated orthogonally. For a two-factor interaction effect and an unrelated main effect to be estimated orthogonally we require that the 27 combinations of levels for the three factors should be equally replicated.

To construct a one-third replicate of a 3^4, which will have 27 observations, we would wish that for each set of three factors all 27 combinations should appear once. This is quite easy to achieve by writing down the 27 combinations for the first three factors and then adding the level of the fourth factor by simple logical argument, with some choice for some of the first few combinations. One form of the design is shown in Figure 14.3, in which the levels of factor S are shown for each combination of levels of P, Q and R.

The logic of construction is the same as that used in the construction of a 3×3 Latin square. Thus the choice of S for $(P,Q,R) = (0,0,0)$ is free and we have chosen 0; the choice for $(0,0,1)$ is also free, but not 0, and we have chosen 1; the choice for $(0,0,2)$ is then inevitable. The choice for $(0,1,0)$ is again free, but not 0 and we have again chosen 1, after which the levels of S for the remaining combinations with P = 0 are inevitable. There is then one more free choice, for the level of S (1 or 2) to go with $(1,0,0)$, after which all further choices are inevitable. Thus there are exactly four free choices.

The design shown in Figure 14.3 is the result of choosing S = 0 for $(P,Q.R) = (0,0,0)$, and thereafter S = 1 for $(0,0,1)$, and for $(0,1,0)$ and for $(1,0,0)$. We could have chosen either S = 1 or S = 2 for each of the last three choices and eight different designs would have resulted from the eight different combinations of choices. If we had started with S = 1 or S = 2 for $(0,0,0)$ we would have produced sixteen further different designs.

The aliasing structure is difficult to express and is not particularly helpful in assessing the merits of the design. Clearly each two-factor interaction must be partially aliased with the interaction of the two remaining factors; in fact half (two df) of each two-factor interaction is aliased with half of the complementary two-factor interaction: however, the only way of expressing the aliasing requires definitions of contrasts which have no practical interpretation, and it does not seem helpful to discuss this further. There is some discussion of a related subject in Section 15.6.

The power of modern computer programs for analysing experimental data ensures that all available information about the practically interesting components of the two-factor interactions can be extracted from the data, with supporting information about the variance–covariance structure of the estimates of effects. Thus it is easy to obtain estimates of linear × linear interactions when the factors are quantitative, though the correlation of effects can be quite high, up to about 0.7.

The construction of a one-third replicate for a 3^5 factorial is very similar to the process just discussed. It is useful to decide that for each set of four factors all 81 combinations should occur once. Then the level of the fifth factor, T, can be determined for each of the 81 combinations of P, Q, R and S with five free choices for (0,0,0,0), (0,0,0,1), (0,0,1,0), (0,1,0,0) and (1,0,0,0), and the other levels of T inevitable.

14.3.1 Design and analysis for an experiment with four four-level factors

Designs for four-level factors are not commonly used, mainly because the number of combinations becomes large very quickly as the number of factors increases. There is one design which has a size within the range we are considering in this chapter and that is the quarter-replicate of a 4^4 factorial structure. We will consider the design and the analysis for a practical example of this design.

Example 14.2 Four four-level factors were investigated in a chemical process experiment using 64 observations, a quarter-replicate. The four experimental factors were

    MC = moisture content,
    TE = temperature,
    TI = duration (time),
    PH = pH of solution.

The design, which included all possible combinations of levels from three factors, for each possible set of three factors, involved some aliasing of components of several two-factor interactions. The full design together with the results for one variable, TSS, is displayed in Table 14.4. For each combination of MC, TE, TI there is an observation for just one pH level, and in Table 14.4 the coded pH values and observed TSS values are placed adjacently. The first impression from the data is of a major effect of pH with possibly some modification by MC. There appears to be a possibility that the variability is not homogeneous but the absence of explicit replication makes this difficult to assess formally. However, for such situations a log transformation will often provide a valid analysis and logged data are used for the analysis in this example.

The first stage of the analysis is to calculate the analysis of variance for main effects only and this is shown in Table 14.5. Clearly PH provides the dominant contribution to the variation, with MC and TE showing smaller but not negligible effects.

The aliasing structure involves the pairing of some contrasts from different two-factor interactions. Thus part of the MCTE interaction is aliased with part of the TIPH interaction.

Table 14.4. *Design and data for chemical process experiment*

MC	TE	TI = 0		TI = 1		TI = 2		TI = 3	
		PH	TSS	PH	TSS	PH	TSS	PH	TSS
0	0	0	23.70	2	41.44	3	63.34	1	24.13
0	1	2	41.25	0	22.66	1	24.19	3	65.74
0	2	3	65.78	1	22.35	0	22.99	2	42.84
0	3	1	22.89	3	73.28	2	40.52	0	22.03
1	0	1	32.40	0	20.39	2	48.50	3	75.37
1	1	3	75.24	2	47.87	0	19.10	1	26.87
1	2	2	46.83	3	71.51	1	19.01	0	19.07
1	3	0	19.69	1	19.54	3	69.66	2	29.13
2	0	3	70.62	1	21.26	0	21.58	2	41.68
2	1	1	19.84	3	71.89	2	43.86	0	20.82
2	2	0	21.14	2	44.52	3	61.91	1	20.03
2	3	2	35.90	0	20.29	1	18.96	3	27.48
3	0	2	44.14	3	72.84	1	20.00	0	19.99
3	1	0	21.66	1	19.76	3	69.82	2	58.99
3	2	1	19.14	0	20.57	2	19.28	3	21.93
3	3	3	31.99	2	18.59	0	21.25	1	21.60

Table 14.5. *Analysis of variance for main effects*

	SS	df	MS
MC	0.629	3	0.210
TE	0.754	3	0.251
TI	0.071	3	0.024
PH	11.567	3	3.856
Residual	2.805	51	0.055
Total	15.826	63	

Similarly there is some aliasing between MCTI and TEPH and between MCPH and TETI. In view of the patterns of main effects, it would be reasonable to expect MCPH and TEPH to be more important than their aliases but the distinction between MCTE and TIPH is not clearly predictable. We therefore calculate two analyses of variance, shown in Table 14.6(a) and (b). In neither analysis do the interactions appear significant. The TIPH interaction accounts for rather more variation than does MCTE (remember the aliasing is only partial).

If the analysis is taken further and the main effects and interactions are split into linear, quadratic and cubic effects, then the analysis suggests that the linear × linear component of the MCTE interaction is important. In contrast, simple components of the TIPH interaction show no substantial effect. Consequently it was decided to interpret the result in terms of the MCTE interaction rather than the TIPH interaction. The detailed analysis of variance is given in Table 14.7.

Table 14.6. *Two analyses of variance for chemical process experiment, including (a) interactions MCTE, MCPH and TEPH and (b) MCPH, TEPH and TIPH*

(a)

	SS	df	MS
MC	0.629	3	0.210
TE	0.754	3	0.251
TI	0.071	3	0.024
PH	11.567	3	3.856
MCTE	0.705	9	0.078
MCPH	0.446	9	0.050
TEPH	0.436	9	0.048
Residual	1.218	24	0.051

(b)

	SS	df	MS
MC	0.629	3	0.210
TE	0.754	3	0.251
TI	0.071	3	0.023
PH	11.567	3	3.856
MCPH	0.446	9	0.050
TEPH	0.436	9	0.048
TIPH	0.809	9	0.090
Residual	1.114	24	0.046

Table 14.7. *Analysis of variance for the model with fitted linear and quadratic effects*

Terms	SS	df	MS
MC linear	0.576	1	0.576
other	0.045	2	0.022
TE linear	0.672	1	0.672
other	0.279	2	0.140
TI	0.071	3	0.023
PH linear	10.603	1	10.603
quadratic	0.537	1	0.537
cubic	0.427	1	0.427
MCTE linear × linear	0.285	1	0.285
MCPH linear × linear	0.203	1	0.203
TEPH linear × linear	0.294	1	0.294
Other interaction contrasts	0.805	24	0.033
Residual	1.218	24	0.051

Table 14.8. *Tables of means for the chemical process experiment*

		PH				TE			
		0	1	2	3	1	2	3	4
MC	1	3.13	3.15	3.75	4.20	3.56	3.55	3.57	3.56
	2	2.97	3.17	3.74	4.29	3.67	3.61	3.50	3.39
	3	3.04	3.00	3.72	3.99	3.53	3.52	3.49	3.21
	4	3.04	3.00	3.44	3.77	3.52	3.60	3.01	3.13
TE	1	3.06	3.18	3.78	4.25				
	2	3.04	3.11	3.86	4.26				
	3	3.04	3.00	3.62	3.92				
	4	3.04	3.03	3.39	3.83				
Mean		3.04	3.08	3.66	4.06				
TI	1	3.07	3.14	3.73	4.06				
	2	3.04	3.03	3.58	4.28				
	3	3.05	3.02	3.58	4.19				
	4	3.02	3.14	3.76	3.73				

Standard error of differences between two means within a two-way table = 0.159.
SED between two PH means = 0.080.

The important conclusion from the design point of view is that the use of a quarter-replicate enabled the experimenter to examine all four main effects and to obtain some indications about the pattern of two-factor interactions. The two-way tables of means are given in Table 14.8. For completeness, the PHTI two-way table, in which no simple pattern has been found, is also shown in Table 14.8.

14.4 Smaller fractions for 2^n structures

With 2^n factorial structures it is possible and often practically useful to construct fractions smaller than one-half of the complete replicate. Compared with the simple half-replicate, a smaller fraction has a more complex system of aliasing effects, and requires more than one Defining Contrast. If we continue to restrict our attention to designs with between 25 and 80 observations, then the fractional replicates which could be used include: (i) a quarter-replicate of a 2^7 structure, using 32 observations, (ii) a quarter-replicate of a 2^8 structure, using 64 observations, (iii) a one-eighth-replicate of a 2^8 structure, using 32 observations, (iv) a one-eighth-replicate of a 2^9 structure, using 64 observations, and even smaller fractions of structures with more factors (the pattern should be obvious).

We shall examine the quarter-replicate of the eight-factor structure in some detail, in order to demonstrate the properties of smaller fractions and then consider briefly the potential for other designs to provide information about the effects in which we are likely to be interested.

For an eight-factor structure we have factors P, Q, R, S, T, U, V and W. We would hope to be able to obtain independent (orthogonal) estimates of all eight main effects and of the 28 two-factor interactions. The restrictions which need to be satisfied to achieve these objectives are that each level of each factor is replicated 32 times, each of the four combinations of levels of two factors should be replicated 16 times, and, to achieve orthogonal estimation of all the 36 effects of interest, each of the 16 combinations of levels of four factors should be replicated exactly four times, for each of the sets of four factors. This is a very large set of requirements, and even though all the requirements are interrelated it is not easy to achieve, and indeed it is not obvious how to start to construct the design. For example it is not possible to require that each combination of levels of six factors occurs just once, as might appear desirable. We need to develop an alternative method of construction, and this is possible through the concept of Defining Contrasts introduced at the end of Section 14.2.

Defining Contrasts are very important in developing fractional replicates and it is useful to examine the definitions and consequences in more detail. In the half-replicates in Section 14.2 the highest-order interaction is used as the Defining Contrast and the half-replicate used will be either the set of positive terms for the Defining Contrast, or the set of negative terms. An alternative characterisation of the two sets is as the combinations which are even or odd for the Defining Contrast. The even combinations are all those for which the number of factors in the Defining Contrast with upper levels in each combination is an even number (0 or 2 or 4 or . . .). The odd combinations are those for which the number of factors in the Defining Contrast with upper levels in each combination is an odd number (1 or 3 or . . .).

In Figure 14.2, the half-replicate of the 2^6 structure, with the Defining Contrast PQRSTU, includes all the combinations which are even for PQRSTU; this includes the single combination with no upper levels, the 15 combinations with two upper levels, the 15 combinations with four upper levels, and the single combination with all six upper levels.

For the present example we are constructing a quarter-replicate and we need to split the complete factorial set into half by one Defining Contrast and then split that half in half by a second Defining Contrast. The resulting quarter will be the set of combinations which are positive in the definitions of both Defining Contrasts. Suppose that we decide to use PQRSTV and PQRUW as the two Defining Contrasts. Then the set of observations in the quarter-replicate will be those that are even for PQRSTV and also even for PQRUW. Now if a combination is even for PQRSTV and also even for PQRUW, it will be found that it is also even for the contrast STUVW, which includes all the factors in one, but not both, of the two Defining Contrasts. Hence STUVW is also a Defining Contrast.

In general the total number of Defining Contrasts for a fraction $1/n$ will be $(n - 1)$. This should not be a surprise since the Defining Contrast system defines not one unique $1/n$ fraction but n different fractions. In our quarter-replicate example there could be four possible fractions which are characterised as: (i) even for PQRSTV, even for PQRUW, and hence even for STUVW, (ii) even for PQRSTV, odd for PQRUW and odd for STUVW, (iii) odd for PQRSTV, even for PQRUW and odd for STUVW, (iv) odd for PQRSTV, odd for PQRUW and even for STUVW. We can observe that the three Defining Contrasts occur in a symmetric overall pattern for the four possible quarters, with observations being either even for all three or even for one and odd for the other two. The logic of all these patterns is relatively simple, because of the nature of two-level factors.

We can now consider further how the choice of a Defining Contrast system should be made. For any quarter-replicate we will have three contrast effects in the Defining Contrast system. By the nature of the relationship between the three contrasts, each factor will occur in either two of the contrasts or in none of them. We would wish to choose the set of contrasts so that all three are high-order interactions. Hence we will not wish that any factor is omitted from all three contrasts. However, if we choose very-high-order interaction effects for two of our three contrasts the third contrast will be a low-order interaction or even a main effect. Thus if we select PQRSTUVW and PQRSTUV as our first two Defining Contrasts, then the third contrast will be W, a main effect. Logically we would wish to select the set of three Defining Contrasts so that all three include as nearly the same number of factors as is possible. For our example with eight factors this implies that we will use one six-factor and two five-factor interactions. Thus the choice we considered earlier of PQRSTV, PQRUW and STUVW is in fact the best we could have chosen.

We can now return to the construction of a quarter-replicate design for a 2^8 structure. Any set of six factors which is not a Defining Contrast will have all 32 possible combinations of levels each occurring once. Hence we can write down the 32 combinations of levels for factors P, Q, R, S, T and U, and then add the levels of V and W to satisfy the requirements that each combination shall be even for the two contrasts PQRSTV and PQRUW. The resulting design is shown in Figure 14.4. An additional check that we have selected the correct set of combinations is that each combination must be even for STUVW, the last five factors, which is a very easy check to make visually.

The set of Defining Contrasts ensures that no two-factor interaction or main effect has an alias which is either a main effect or a two-factor interaction. Hence the estimates of all eight main effects and 28 two-factor interactions can be obtained independently, leaving 27 df for estimating the error variance. This is more df than would normally be thought ideal but there is no obvious alternative use for these df. The design is efficient at providing the information required.

The construction and potential benefits of the other designs mentioned at the beginning of this section can be developed as follows:

(i) The quarter-replicate of a 2^7 structure using 32 observations will have a set of Defining Contrasts in the pattern, PQRST, PQUV, RSTUV. This means that the aliases for the seven main effects are all interactions involving at least three factors. However, some of the two-factor interactions will be aliased with other two-factor interactions, because of the four-factor Defining Contrast PQUV. In fact, six of the 21 two-factor interactions will be aliased in pairs, PQ and UV, PU and QV, and PV and QU. The other 15 two-factor interactions can be estimated independent of the main effects and all other two-factor interactions.

The 31 df in the analysis of variance are therefore allocated 7 for the main effects, 15 for unaliased two-factor interactions and 9 for estimating the error variance. Three of the error df correspond to the aliased pairs of two-factor interactions, but there is no advantage from separating these out, because 9 is already on the low side for estimating the error variance. Using 22 of the 31 df is a very high level of information and makes this design very efficient.

(ii) The one-eighth replicate of a 2^8 structure, using 32 observations, is really rather too small for all the effects we would hope to estimate. The Defining Contrast system will be in the pattern PQRS, PQTU, PRTVW, RSTU, QSTVW, QRUVW and PSUVW. This means that the aliases for the eight main effects are all interactions involving at least three factors. However, there are many aliased pairs of two-factor interactions. For the particular Defining Contrasts listed the aliasing patterns are PQ with both RS and TU,

00000000	00000101	00001010	00001111
00010010	00010111	00011000	00011101
00100011	00100110	00101001	00101100
00110001	00110100	00111011	00111110
01000011	01000110	01001001	01001100
01010001	01010100	01011011	01011110
01100000	01100101	01101010	01101111
01110010	01110111	01111000	01111101
10000011	10000110	10001001	10001100
10010001	10010100	10011011	10011110
10100000	10100101	10101010	10101111
10110010	10110111	10111000	10111101
11000000	11000101	11001010	11001111
11010010	11010111	11011000	11011101
11100011	11100110	11101001	11101100
11110001	11110100	11111011	11111110

Figure 14.4 Combinations of levels for eight factors, P, Q, R, S, T, U, V and W for the 64 observations of a quarter-replicate of a 2^8 structure.

PR with QS, PS with QR, PT with QU, PU with QT, RT with SU and RU with ST. Thus of the total 28 two-factor interactions, only 13 are not aliased with other two-factor interactions.

The 31 df in the analysis of variance are therefore allocated 8 for the main effects, 13 for unaliased two-factor interactions and 10 for estimating the error variance. Note, however, that the SS of seven of the ten error df are in fact possibly inflated by the effects of the two-factor interactions to which they correspond, and that there is a strong probability that the error variance will be considerably overestimated. Hence this design has problems, which can be partially overcome by a judicious choice of which two-factor interactions are to be ignored; if the two-factor interactions most likely to be large can be guessed in advance, then the particular set of Defining Contrasts can be chosen so as to permit estimation of the more important two-factor interactions, with the consequent further benefit that the remaining two-factor interactions, which are included in the error variance estimation, should be small. Of course, if the prior identification of which two-factor interactions are large is wrong then everything about the design will be worse!

(iii) The one-eighth-replicate of a 2^9 structure, using 64 observations, has more potential for a successful design largely because it uses twice as many observations and therefore allows the estimation of more possibly important effects without aliasing. The Defining Contrast system can be selected in the pattern PQRST, PQRUV, RSUWX, STUV, PQTUWX, PQSVWX and RTVWX. It is not possible to construct a system which includes only five- and six-factor interactions, and one four-factor interaction must be included, which means that there will be some aliased two-factor interactions. For the particular set of Defining Contrasts listed the aliased two-factor interactions are ST and UV, SU and TV, and SV and TU. All the other 30 two-factor interactions are unaliased.

The 63 df in the analysis of variance are therefore allocated 9 for the main effects, 30 for unaliased two-factor interactions and 24 for the estimation of the error variance. Using 24 df to estimate error is rather more than is necessary, but since most of these df correspond to three-factor interactions which are aliased with other three-factor interactions, there is little to be gained from looking at other single-df contrasts, with the possible exception of the three df for the aliased pairs of two-factor interactions; if one of these were to be relatively large it would tell us that at least one of the two-factor interactions is important, and excluding them from the estimation of error would reduce the tendency to overestimate the error variance.

14.5 Irregular fractions for 2^n structures

Thus far in our discussion of fractional replicate designs for two-level factorial structures we have considered only fractions in which the total number of observations is a power of 2. This restricts not only the designs in our self-imposed range of practical designs but also the effective replication of main effects and two-factor interactions. Also, as we have seen in some of the later examples in the previous section the quarter- or one-eighth-replicates can give a considerable reduction in the number of two-factor interactions which can be estimated without aliases.

Other fractions can also be constructed where the total number of observations is 24, 48 or 72. This expands the range of useful designs considerably, giving more possible levels of replication of effects. These designs allow more estimates of two-factor interactions, with some partial aliasing, the consequence of which is that we can obtain estimates of all the two-factor interactions but with some degree of non-orthogonality of estimation. The effects whose estimates are not orthogonal have correlations which in most cases are small; this is a useful development in the use of fractional designs, and one which can further extend the range of designs, as we shall see in a more extreme form in the final section of this chapter.

When we use more complex fractions the aliasing properties change so that while there are more contrasts in the Defining Contrast system, many of the aliased effects are only partially aliased. This means that pairs, or triples, of effects have to be estimated in a joint estimation process, and the estimates of effects will have a (slightly) smaller effective replication and will be correlated, usually with quite a low level of correlation, but the total number of

000000	000011	001100	001111
010101	010110	011001	011010
101010	101001	100110	100101
110000	110011	111100	111111

Figure 14.5 Sixteen combinations of six factors P,Q,R,S,T,U forming the quarter-replicate of the 2^6 structure using the Defining Contrasts PQRS and PQTU.

effects for which some estimate is possible will be much increased. Among many possible interesting designs are:

 (i) a three-quarters-replicate for a 2^6 structure, using 48 observations,
 (ii) a three-eighths-replicate for a 2^7 structure, using 48 observations,
(iii) a three-quarters-replicate for a 2^5 structure, using 24 observations,
(iv) a three-quarters-replicate for a 2^7 structure, using 72 observations,
 (v) a three-eighths-replicate for a 2^8 structure, using 72 observations,
(vi) a five-eighths-replicate for a 2^6 structure, using 40 observations.

We will consider the first two in some detail and leave the others for the reader to investigate.

(i) The three-quarters-replicate for a 2^6 structure, using 48 observations, is a very useful design. The Defining Contrast system is essentially the same as would be used for a quarter-replicate, the actual design consisting of all the observations not included in the quarter-replicate. However, whereas in the quarter-replicate all two-factor interactions are fully aliased in pairs or triples, in the three-quarters-replicate, the aliasing is partial, and we can obtain estimates for all two-factor interactions. The Defining Contrast system has the pattern PQRS, PQTU and RSTU. All two-factor interactions are partially aliased, twelve in pairs: PR and QS, PS and QR, PT and QU, PU and QT, RT and SU, RU and ST, and three in a triple, PQ and RS and TU.

The design can be most easily constructed by omitting a quarter-replicate from the complete factorial structure. The 16 combinations of the quarter-replicate are listed in Figure 14.5.

The three-quarters-replicate design with 48 observations is then all the combinations not listed in Figure 14.5. For completeness the 48-observation design is listed in Figure 14.6, in a pattern that hopefully shows the logic of construction. The first 16 observations shown are even for PQRS and odd for PQTU (and therefore odd for RSTU); the next 16 are odd for PQRS, even for PQTU, and odd for RSTU; the last 16 are odd for both PQRS and PQTU, and even for RSTU.

The estimates of main effects are independent of estimates of all other main effects and two-factor interactions, and are based on the full replication of 24. For the two-factor interactions which are partially aliased in pairs, the least squares equations for effects E_1 and E_2 occur in the form:

$$24\,(E_1) - 8\,(E_2) = \text{total (even for } E_1) - \text{total (odd for } E_1),$$
$$-8\,(E_1) + 24\,(E_2) = \text{total (even for } E_2) - \text{total (odd for } E_2)$$

000001	000010	001101	001110
	PQRS even		
010100	010111	011000	011011
	PQTU odd		
101000	101011	100100	100111
	RSTU odd		
110001	110010	111101	111110

000100	000111	001000	001011
	PQRS odd		
010001	010010	011101	011110
	PQTU even		
101001	101010	100101	100110
	RSTU odd		
110100	110111	111000	111011

000101	000110	001001	001010
	PQRS odd		
010000	010011	011100	011111
	PQTU odd		
100000	100011	101100	101111
	RSTU even		
110101	110110	111001	111010

Figure 14.6 The three-quarters-replicate design for the 2^6 factorial structure with the three separate quarters shown as separate groups of observations.

and the estimates of each interaction effect are based on an effective replication of 64/3 (11% less than the full replication), and the correlation between the two estimates is 1/9. For the three two-factor interactions which are partially aliased in a triple, the least squares equations occur in the form:

$$24 (E_1) - 8 (E_2) - 8 (E_3) = \text{total (even for } E_1) - \text{total (odd for } E_1),$$

$$-8 (E_1) + 24 (E_2) - 8 (E_3) = \text{total (even for } E_2) - \text{total (odd for } E_2),$$

$$-8 (E_1) - 8 (E_2) + 24 (E_3) = \text{total (even for } E_3) - \text{total (odd for } E_3)$$

and the estimate of each interaction effect is based on an effective replication of 16 (33% less than the full replication), the correlation between any two estimates being 3/16. Thus the information on the estimates of main effects and two-factor interactions is at least as good as, and in most cases much better than, that from the half replicate, as would be hoped.

(ii) The three-eighths-replicate for a 2^7 structure, using 48 observations, is also a potentially useful design. The Defining Contrast system will be the same as would be used for a one-eighth-replicate. The design is constructed in two stages: first we construct a half-replicate; then we identify one-quarter of that half, which is the one-eighth-replicate, and then use the remaining three-quarters of the half-replicate to form the desired three-eighths replicate. The Defining Contrast system requires a set of seven interrelated contrasts, one of which (the one involving most factors) defines the half-replicate, and the remainder defining the one-eighth-replicate to be omitted from the half.

One obvious possible set of contrasts, avoiding including any interaction of fewer than four factors, would be.

PQRS, PQTU, PRTV, RSTU, QSTV, QRUV and PSUV;

this, however, will have the disadvantage that since all the contrasts are four-factor interactions, the contrast defining the half-replicate will be a four-factor contrast and therefore three pairs of two-factor interactions will be fully aliased and there will be no information at all on those six two-factor interactions. Also because the other two-factor interactions will be aliased in triples, the effective replication will be the same as the lowest level in the previous example, 16, with correlations of 3/16. Hence, comparing the advantages of the 48-observation three-eighths-replicate with the 32-observation quarter-replicate, we get 50% more information on all the main effects, but for two-factor interactions we have no information on six interactions in each design, and for the other interactions we have the same effective replication but the estimates from the three-eighths-replicate are correlated, so that the overall information on two-factor interactions is actually less using the three-eighths-replicate than with the quarter-replicate.

If we consider other sets of Defining Contrasts, with some interactions involving more than four factors, and inevitably some interactions involving only three factors, then we can obtain better information, but also less even, and rather more complicated, patterns of information. In general we will get some information on all two-factor interactions, but at the cost of some main effects having two-factor interactions as aliases. A possible set of Defining Contrasts of this, less even, form would be: PQR, PSTU, QRSTU, PRTV, QTV, RSUV, and PQSUV (the last being the Defining Contrast for the half-replicate). Because the set of Defining Contrasts includes two three-factor interactions, the alias patterns will involve some main effects, the full set of aliased effects being: P with QR, Q with PR and TV, R with PQ, T with QV, V with QT, PS with TU, PT with SU and RV, PU with ST, PV with RT, RS with UV, and RU with SV; the remaining two-factor interactions, QS, QU and RT are aliased only with three-factor or higher-order interactions.

The construction of the design is more difficult than in the previous example, and is most easily achieved through the use of the Defining Contrast system. The Contrast defining the half-replicate is PQSUV, and all combinations in the three-eighths-replicate will have factor levels so that the combinations are even for PQSUV. This means that for any combination of levels of P, Q, R, S, T, U, the level of factor V is uniquely determined. In addition the combinations must not be even for both of PQR and PSTU. We can therefore construct the design by choosing three subsets which are: (i) even for PQR and odd for PSTU, (ii) odd for PQR and even for PSTU, and (iii) odd for PQR and odd for PSTU. The resulting design is shown in Figure 14.7.

The reduction in information for effects aliased in pairs or in triples will be essentially the same as in the previous example. Thus of the main effects S and U will have full information based on an effective replication of 24, with no correlation, P, R and T will be partially aliased with QR, PQ and QT, respectively, and will be estimated with effective replication 64/3 and a 1/9 correlation with the corresponding two-factor interaction, and Q, being partially aliased with PR and TV will have an effective replication of 16, with correlations of 3/16 with the two two-factor interactions. Most two-factor interactions will have an effective replication of 64/3, but QS, QU and RT will have the full replication of 24 and a few will have the effective replication of 16.

0000011	0000100	0001001	0001110
	PQR even		
0110010	0110101	0111000	0111111
	PSTU odd		
1010001	1010110	1011011	1011100
	PQSUV even		
1100000	1100111	1101010	1101101

0010000	0010111	0011010	0011101
	PQR odd		
0100001	0100110	0101011	0101100
	PSTU even		
1000010	1000101	1001000	1001111
	PQSUV even		
1110011	1110100	1111001	1111110

0010011	0010100	0011001	0011110
	PQR odd		
0100010	0100101	0101000	0101111
	PSTU odd		
1000001	1000110	1001011	1001100
	PQSUV even		
1110000	1110111	1111010	1111101

Figure 14.7 Levels of the seven factors P, Q, R, S, T, U, V for a three-eighths-replicate of a 2^7 structure.

Compared with the quarter-replicate using 32 observations, there is 50% more information in the estimates of S and U, 33% more information about P, R, S and V, rather less information on Q, because of the correlations, and a lot more information about almost all two-factor interactions.

Particularly from the second example we can see that using irregular fractions can provide improved information on main effects and two-factor interactions by using more replication and consequently more observations. We can choose the most suitable size of experiment, but the detailed choice of design can require more careful thought than the relatively straightforward choice of Defining Contrast systems for the simpler fractions. Partial aliasing is quite simple to handle, and, of course, with modern computing packages for the analysis of results, there is no difficulty with the analysis.

14.6 Other fractions for three-level factors and for mixed levels

Fractional replicate designs for three-level factors are generally less effective than those for two-level factors because:

(i) the number of observations increases much quicker as the number of factors is increased,
(ii) the aliasing and Defining Contrast systems are not directly related to the contrasts which are of practical interest, and
(iii) the numbers of df required for two-factor interactions when many factors are used becomes large quickly.

Although the Defining Contrast systems are more complicated to define precisely, compared with two-level factorial structures, they have some benefits in causing only partial aliasing of effects of interest, rather than the full aliasing which occurs in quarter or one-eighth fractions of 2^n structures. This is because with three level factors each Defining Contrast corresponds to two df from an interaction effect which will have 8 df if it is a three-factor interaction, or 16 if it is a four-factor interaction. This contrasts with the two-level structures where every effect has a single df which is either aliased fully or not at all.

The construction of designs is relatively straightforward through direct construction methods, using the objective of having all possible combinations included once each, for the highest-order interactions for which this is feasible. The pattern of which sets of all possible combinations can be made to occur once each is determined by the particular pattern of Defining Contrasts.

Although the implications for aliasing from Defining Contrasts are different for three-level factors compared with those for two-level factors, the patterns for the numbers of factors involved in the interactions which occur in the Defining Contrast system are essentially the same. That is if we have a five-factor structure, and more than one Defining Contrast, then the Defining Contrasts must be components of at least one three-factor interaction and at least one four-factor interaction. Consequently in the design construction we will encounter restrictions on the possible sets of factors for which all possible combinations can be included once.

Possible fractional replicate designs within our range of practical experiment sizes which could provide solutions for some practical design requirements include:

 (i) the one-ninth-replicate of a 3^5 factorial structure, using 27 observations,
 (ii) the one-ninth replicate of a 3^6 structure, using 81 observations,
(iii) the two-ninths-replicate of a 3^5 structure, using 54 observations.

We will discuss each of these designs briefly.

(i) The one-ninth-replicate of a 3^5 factorial structure, using 27 observations is a useful design in providing information about the main effects of the five factors but is not capable of providing much information about two-factor interactions.

To construct the design we attempt to ensure that, as far as is possible, all 27 combinations for each set of three factors occur once each. We know that this is not possible for all ten sets of three factors, but trying to achieve the objective for as many sets as possible will give us the best available design. The result is shown in Figure 14.8.

We can see that all 27 combinations do occur once each for the three-factor sets, PQR, PQS, PQT, PRS, PRT, PST, QRS, QRT and QST, but not for RST.

The design will provide estimates of all five main effects, each with two df. This uses up 10 of the 26 df, and most of the rest are needed for estimating the error variance. The two-factor interactions are estimated independently of the main effects, except for the partial aliasing caused by the fact that only 9 of the 27 combinations for RST are included in the design; R is partially aliased with ST and similarly S with RT, and T with RS. If we ignore two-factor interactions in the fitted model, then the analysis is simple and straightforward.

However, it is possible to look for some information about the two-factor interactions. Quite apart from the aliasing of some main effects and two-factor interactions it would clearly be impossible, with only 27 observations, to estimate all two-factor interactions

P	Q	R	S	T	P	Q	R	S	T	P	Q	R	S	T
0	0	0	0	0	0	0	1	1	2	0	0	2	2	1
0	1	0	1	1	0	1	1	2	0	0	1	2	0	2
0	2	0	2	2	0	2	1	0	1	0	2	2	1	0
1	0	0	1	1	1	0	1	2	0	1	0	2	0	2
1	1	0	2	2	1	1	1	0	1	1	1	2	1	0
1	2	0	0	0	1	2	1	1	2	1	2	2	2	1
2	0	0	2	2	2	0	1	0	1	2	0	2	1	0
2	1	0	0	0	2	1	1	1	2	2	1	2	2	1
2	2	0	1	1	2	2	1	2	0	2	2	2	0	2

Figure 14.8 The set of factor levels for a one-ninth-replicate of a 3^5 structure.

(which would require 10 sets of 4 df). Nevertheless it is possible to calculate the SS for fitting each interaction as the only term to be fitted in addition to the five main effects. This could give information about which interaction effects would be the largest. If the factors are quantitative it would also be possible to try fitting a few linear × linear interaction effects on one df each. None of the information about interaction effects can be regarded as conclusive, but it may be possible to get some ideas about which interaction effects would be worth further investigation.

(ii) The one-ninth-replicate of a 3^6 factorial structure, using 81 observations obviously has the potential to provide information about a much larger range of effects. There are six main effects each on two df, and fifteen two-factor interactions, each on four df, so that potentially the total number of df (80) could allow all these effects to be estimated, using 72 df, and leaving 8 df for error variance estimation.

The design construction to enable the best estimation of main effects and as many two-factor interactions as possible can be approached again by direct selection of combinations. To start we can write down all the 81 combinations of levels for the four factors P, Q, R and S. Then we can add levels for factors S and T as for the previous example, using the same patterns as in the previous example, and remembering the patterns used in 3 × 3 Latin squares and superimposed Latin squares. The general philosophy about the number of factors in the Defining Contrast system resulting from the direct approach will again be broadly the same as for two-level factors, so that we should expect to find that various components of several four-factor interaction effects will be included in the Defining Contrasts.

As in the previous example many two-factor interaction effects will be partially aliased, with, in general, fairly low levels of correlation between the effects estimates.

(iii) The two-ninths-replicate for the 3^5 structure can be most easily constructed by using, for each PQR combination, the six ST combinations which are not included in Figure 14.8. The resulting design has very similar properties to the one-ninth-replicate but, with twice as many observations, the estimation of effects is more precise and the aliasing patterns involve lower correlations of effect estimates. It is possible to obtain estimates of all five main effects, each on two df, and of all ten two-factor interactions, each on four df. However this occupies a total of 50 df, leaving only three for estimating the error variance.

Hence the model to be fitted to the data should probably be developed sequentially:

 (i) first fit the model with all main effects;
 (ii) the main effects will all remain in the model, regardless of their size;
(iii) then add to the model some two-factor interactions, choosing first those for which at least one factor has an important main effect, and also choosing those interactions which are not partially aliased with main effects (i.e. not RS, RT or ST);
 (iv) at this stage it is possible to omit some two-factor interaction terms from the model, if they appear to be uninteresting, possibly trying some other two-factor interactions;
 (v) another possibility if the factors are quantitative is to include some particular single df interaction terms, most obviously the ten linear × linear interaction terms.

The gradual development of a fitted model in the third example, which may also be useful for the previous two examples, must be undertaken carefully. There are dangers in trying very many sets of terms, because considering many alternatives increases the chance that the terms finally selected are chance extreme values. There are also dangers in discarding too many terms, because of the tendency to arrive at an error variance estimate which is biased downwards because of the long decision process and the inclusion in error of many non-significant terms. Essentially the process of looking for the best fitted model is similar to the selection of the best model in a multiple regression investigation with the same opportunities and dangers.

14.6.1 *Fractions for mixed-level factorial structures*

The other area of factorial structures where fractional replicates can be useful is for mixed level structures. Fractional replicate designs for mixtures of two- and three-level factors can be constructed quite easily by direct methods.

Suppose we are considering a $2^3 \times 3^2$ treatment structure, and wish to use fewer than 72 observations. The most likely sizes of experiment smaller than 72 are 36 or 48 observations. The latter can be constructed most simply by considering the design for 24 observations and using the remaining 48 observations, since any contrasts which are aliased in the 24 observations will also be at least partially aliased in the 48 observations.

Example 14.3 The design with 36 observations can clearly include all combinations of levels from four of the factors Q(2), R(2), S(3) and T(3), and the crucial stage of the design is the choice of the level of P(2) to combine with each of the 36 combinations. We might reasonably expect that it should be possible to choose the levels of P to include all combinations of levels of

 (i) PQST
(ii) PRST

and to include each of the PQR level combinations four or five times. This is equivalent to using the PQR three-factor interaction as a Defining Contrast, but using each of the two halves of the contrast combinations with four or five of the ST combinations. This results in another, different, form of partial aliasing. We could, of course, have simply defined PQR

P	Q	R	S	T	P	Q	R	S	T	P	Q	R	S	T
0	0	0	0	0	1	0	0	0	1	0	0	0	0	2
1	0	0	1	0	0	0	0	1	1	1	0	0	1	2
0	0	0	2	0	1	0	0	2	1	0	0	0	2	2
1	0	1	0	0	0	0	1	0	1	1	0	1	0	2
0	0	1	1	0	1	0	1	1	1	0	0	1	1	2
1	0	1	2	0	0	0	1	2	1	1	0	1	2	2
1	1	0	0	0	0	1	0	0	1	1	1	0	0	2
0	1	0	1	0	1	1	0	1	1	0	1	0	1	2
1	1	0	2	0	0	1	0	2	1	1	1	0	2	2
0	1	1	0	0	1	1	1	0	1	0	1	1	0	2
1	1	1	0	2	0	1	1	1	1	1	1	1	1	2
0	1	1	2	0	1	1	1	2	1	0	1	1	2	2

Figure 14.9 A half-replicate of a $2^3 \times 3^2$ structure.

as a Defining Contrast, and have used the same set of four PQR combinations with all the nine ST combinations, but this would completely alias the main effects for P, Q and R with the two-factor interaction of the other two factors, and would lose most of the information on two factor effects. The resulting set of treatment combinations for the design, with the partial aliasing pattern is shown in Figure 14.9.

The only unusual aspect of the analysis for this design is the unequal replication of the combinations of the three-factor effect PQR. This results in some partial aliasing of

$$P \equiv QR, Q \equiv PR \text{ and } R \equiv PQ,$$

and both effects from each pair can be estimated from a non-orthogonal analysis. The effects P and QR, for example, are estimated with a correlation of 4/36 because the degree of departure from equal replication of all PQR combinations amounts to four observations in 36. This may be seen directly from the least squares equations for the effects P and QR:

$$36\hat{P} - 4\widehat{QR} = (\text{total for } p_1) - (\text{total for } p_0)$$
$$-4\hat{P} + 36\widehat{QR} = (\text{total for } q_0 r_0 \text{ and } q_1 r_1) - (\text{total for } q_0 r_1 \text{ and } q_1 r_0).$$

Example 14.4 The design with 24 observations is too small to allow estimation of all the effects in which we would be likely to be interested, but, as mentioned earlier, it is useful as a means to constructing the design with 48 observations. The design problem is simply how to choose the levels of factors S and T for each of the eight combinations of PQR. Again there is a simple solution, using a component of the two-factor interaction, ST, as a Defining Contrast, but this results in quite a high level of correlation between the main effects of S and of T, even when the 48-observation complement of the 24-observation design is used. More usefully, we could allocate three sets of ST combinations each to two or three of the PQR combinations. For the 48-observation design this would result in omitting three sets of ST combinations for each of the eight PQR combinations, and the resulting design is shown in Figure 14.10.

Combinations of levels of S and T

P	Q	R						
0	0	0	(0,1)	(0,2)	(1,0)	(1,2)	(2,0)	(2,1)
0	0	1	(0,0)	(0,1)	(1,1)	(1.2)	(2,0)	(2,2)
0	1	0	(0,0)	(0,2)	(1,0)	(1,1)	(2,1)	(2,2)
0	1	1	(0,1)	(0,2)	(1,0)	(1,2)	(2,0)	(2,1)
1	0	0	(0,0)	(0,1)	(1,1)	(1,2)	(2,0)	(2,2)
1	0	1	(0,0)	(0,2)	(1,0)	(1,1)	(2,1)	(2,2)
1	1	0	(0,1)	(0,2)	(1,0)	(1,2)	(2,0)	(2,1)
1	1	1	(0,0)	(0,1)	(1,1)	(1,2)	(2,0)	(2,2)

Figure 14.10 The combinations of factor levels P, Q, R, S and T for a two-thirds replicate of a $2^3 \times 3^2$ structure.

The construction of designs for larger structures involving a mixture of two- and three-level factors uses essentially the same procedures. That is we look for a way of combining subsets of the two-level factors with each combination of three-level factors, or subsets of the three-level factors with each combination of two-level factors, or combining subsets of the two-level factors with subsets of the three-level factors.

Suppose that we have a $2^3 \times 3^3$ structure and wish to use fewer than 216 observations. The possible choices for the number of observations within our practical range of experiment sizes are 36, 54 or 72. For the 36-observation design, a one-sixth-replicate, there are various ways of combining a half-replicate of the 2^3 structure with a one-third replicate of the 3^3 structure.

The simplest is to choose one half-replicate (P,Q,R) $=(1,0,0)$, $(0,1,0)$, $(0,0,1)$ and $(1,1,1)$ and one one-third-replicate (S,T,U) $=(0,0,0)$, $(0,1,1)$, $(0,2,2)$, $(1,0,1)$, $(1,1,2)$, $(1,2,0)$, $(2,0,2)$, $(2,1,0)$ and $(2,2,1)$ and to use each combination from the half-replicate with each combination from the one-third-replicate. However, the use of the same half-replicate throughout the design means that main effects for the two-level factors are all fully aliased with two-level interactions, and a similar problem occurs for the three-level factors.

We can use all the two-level factor combinations and all the three-level factor combinations in the overall design if we split the three-level factor combinations into nine one-ninth-replicates of the three-level factor combinations and then combine each one-ninth-replicate with a different single combination of the two-level factors.

A possible set of one-ninth-replicates is

$$(000, 112, 221), \ (001, 110, 222), \ (002, 111, 220)$$
$$(010, 122, 201), \ (011, 120, 202), \ (012, 121, 200)$$
$$(020, 102, 211), \ (021, 100, 212), \ (022, 101, 210).$$

There are many possible different ways of selecting sets of fractional replicates with two-level factor combinations, and the only way of determining which is best would be to try all the different ways, and for each one to calculate an appropriate criterion of optimality of the total estimation process. The definition of the criterion is not simple, and the search procedure to find the optimum design for that criterion is also not simple. However, it is possible to construct reasonably efficient designs through a direct approach, trying to avoid

P Q R	S T U		P Q R	S T U		P Q R	S T U
0 0 0	0 0 0		0 0 0	1 1 2		0 0 0	2 2 1
0 0 1	1 1 0		0 0 1	2 2 2		0 0 1	0 0 1
0 1 0	2 2 0		0 1 0	0 0 2		0 1 0	1 1 1
1 0 0	1 2 2		1 0 0	2 0 1		1 0 0	0 1 0
1 1 1	2 0 2		1 1 1	0 1 1		1 1 1	1 2 0
1 1 0	0 1 2		1 1 0	1 2 1		1 1 0	2 0 0
1 0 1	2 1 1		1 0 1	0 2 0		1 0 1	1 0 2
0 1 1	0 2 1		0 1 1	1 0 0		0 1 1	2 1 2
0 0 0	1 0 1		0 0 0	2 1 0		0 0 0	0 2 2
0 0 1	2 2 1		0 0 1	0 0 0		0 0 1	1 1 2
0 1 0	0 0 1		0 1 0	1 1 0		0 1 0	2 2 2
1 0 0	1 1 1		1 0 0	2 2 0		1 0 0	0 0 2

Figure 14.11 A one-sixth-replicate of a $2^3 \times 3^3$ factorial structure based on combining quarter-replicates and one-ninth-replicates.

any particular combinations occurring unevenly. One possible design obtained by combining elements of the two sets is shown in Figure 14.11.

We cannot claim that this design is optimal, but the method of construction does ensure that all PQR combinations occur at least three times, that all STU combinations occur at least once, and there is very little repetition of four-factor combinations involving a mixture of two- and three-level factors. Similar approaches are likely to produce designs for other sets of factors or different sizes of experiments.

14.7 Very small fractions for main effect estimation

Finally we consider a very different form of design, which is also a fraction of a full factorial structure, but in a very different sense from the fractions considered in the previous sections. Not only are the fractions very different but the purposes and sets of important effects are also very different.

The fractions used in this section are very small (often less than 10%), and the full factorial structures are usually very large, including very many two-level factors. The only effects of interest are main effects, all interactions being tentatively assumed to be zero, and the objective is essentially to screen a large number of factors to discover the small number which have any effect at all. It is assumed that any effect which is non-zero will be clearly identifiable, being much larger than any random variation. Equivalently it is assumed that the error variance is small, and that there is no need to estimate it.

The consequence of these assumptions is that all df will be used to estimate main effects, and hence we can expect to estimate many main effects. However, if we are estimating many main effects, while using a relatively small fraction of the full set of possible observations, then we should not expect that the estimates of different main effects will be orthogonal or independent. We have seen that having correlated estimates need not be a problem for some of the smaller fractions with many effects to be estimated. In the screening designs considered in this section, there is the further benefit that we are expecting that only a small

```
Block     1   2   3   4   5   6   7

Treatments  1   2   3   4   5   6   7
            2   3   4   5   6   7   1
            4   5   6   7   1   2   3
```

Figure 14.12 A cyclic design to compare seven treatments in seven blocks of three units per block, where the columns are the blocks and the rows indicate the treatments included in each block.

```
Factor    1    2    3    4    5    6    7
Observations
1        +1   -1   -1   -1   +1   -1   +1
2        +1   +1   -1   -1   -1   +1   -1
3        -1   +1   +1   -1   -1   -1   +1
4        +1   -1   +1   +1   -1   -1   -1
5        -1   +1   -1   +1   +1   -1   -1
6        -1   -1   +1   -1   +1   +1   -1
7        -1   -1   -1   +1   -1   +1   +1
8        +1   +1   +1   +1   +1   +1   +1
```

Figure 14.13 Design matrix for eight observations on seven factors derived from the cyclic design in Figure 14.12, with rows for the observations and columns for the factors.

fraction of the factors screened will have any effect. Hence designs which provide correlated estimates are acceptable, provided the correlations are not too high, such that it is possible to separate out which factor or factors actually have an effect.

The designs in this section are described as saturated designs, or supersaturated designs, and we shall give only a brief idea of the potential of such designs. Most of the designs use considerably fewer than our previous lower limit of 25 observations for a practical design. This is mainly because it is not necessary to have any df to estimate an error variance, and because relatively low replication is needed to identify non-zero effects when there is little or no error variation.

We will look at two examples to illustrate the benefits of these saturated designs. The first design uses the ideas of simple cyclic designs, previously discussed in Chapter 7, to produce a design to estimate the main effects for seven factors using just eight observations. The design shown in Figure 14.12 is in the standard pattern of a cyclic incomplete block design, to compare seven treatments in seven blocks of three units per block. Obviously this design is a BIB design.

We now rewrite the design of Figure 14.12 so that each block (column) corresponds to a factor, and construct rows which will define the observations for our saturated design, where the factor level in row i and column j will be $+1$ if treatment i occurred in block j and -1 if treatment i did not occur in block j. We add an eighth row with all factor levels at $+1$ to produce a set of eight observations with each factor having four observations for each of the two levels. The design is shown in Figure 14.13.

This is obviously a saturated design since there are seven main effects, all orthogonal, using completely the seven df for comparisons between the eight observations. It is a very small fraction using only eight of the possible 128 factor level combinations, and provided

Block

1	2	3	4	5	6	7	8	9	10	11	12	13	14	15	16	17	18	19	20	21	22	23	24	25	26	27
1	2	3	4	5	6	7	8	9	1	2	3	4	5	6	7	8	9	1	2	3	4	5	6	7	8	9
2	3	4	5	6	7	8	9	1	2	3	4	5	6	7	8	9	1	2	3	4	5	6	7	8	9	1
3	4	5	6	7	8	9	1	2	4	5	6	7	8	9	1	2	3	3	4	5	6	7	8	9	1	2
6	7	8	9	1	2	3	4	5	8	9	1	2	3	4	5	6	7	5	6	7	8	9	1	2	3	4

Figure 14.14 Three cyclic designs each with nine blocks of four units, to compare nine treatments.

Factor

1	2	3	4	5	6	7	8	9	10	11	12	13	14
+1	-1	-1	-1	+1	-1	-1	+1	+1	+1	-1	+1	-1	-1
+1	+1	-1	-1	-1	+1	-1	-1	+1	+1	+1	-1	+1	-1
+1	+1	+1	-1	-1	-1	+1	-1	-1	-1	+1	+1	-1	+1
-1	+1	+1	+1	-1	-1	-1	+1	-1	+1	-1	+1	+1	-1
-1	-1	+1	+1	+1	-1	-1	-1	+1	-1	+1	-1	+1	+1
+1	-1	-1	+1	+1	+1	-1	-1	-1	-1	-1	+1	-1	+1
-1	+1	-1	-1	+1	+1	+1	-1	-1	-1	-1	-1	+1	-1
-1	-1	+1	-1	-1	+1	+1	+1	-1	+1	-1	-1	-1	+1
-1	-1	-1	+1	-1	-1	+1	+1	+1	-1	+1	-1	-1	-1
+1	+1	+1	+1	+1	+1	+1	+1	+1	+1	+1	+1	+1	+1

Factor

15	16	17	18	19	20	21	22	23	24	25	26	27
-1	+1	-1	+1	+1	-1	-1	-1	-1	+1	-1	+1	+1
-1	-1	+1	-1	+1	+1	-1	-1	-1	-1	+1	-1	+1
-1	-1	-1	+1	+1	+1	+1	-1	-1	-1	-1	+1	-1
+1	-1	-1	-1	-1	+1	+1	+1	-1	-1	-1	-1	+1
-1	+1	-1	-1	+1	-1	+1	+1	+1	-1	-1	-1	-1
+1	-1	+1	-1	-1	+1	-1	+1	+1	+1	-1	-1	-1
+1	+1	-1	+1	-1	-1	+1	-1	+1	+1	+1	-1	-1
-1	+1	+1	-1	-1	-1	-1	+1	-1	+1	+1	+1	-1
+1	-1	+1	+1	-1	-1	-1	-1	+1	-1	+1	+1	+1
+1	+1	+1	+1	+1	+1	+1	+1	+1	+1	+1	+1	+1

Figure 14.15 Supersaturated design based on the cyclic designs in Figure 14.14, with ten observations and 27 factors, each at two levels.

that the assumption that the error variance is small is correct (which cannot be tested) it is a very efficient design for estimating main effects.

The second example is a rather larger, supersaturated, design assessing the main effects of 27 factors using 10 observations. Inevitably not all main effects can be orthogonally estimated and the design is constructed so that the correlations between the main effect estimates are minimised. There are several possible criteria for minimising the set of correlations and we shall not discuss this aspect of design estimation further here, except to note that the subject of supersaturated designs has a large literature and further information can be obtained from papers by Butler *et al.* (2001) and Eskridge *et al.* (2004) – see also Gilmour (2006).

The supersaturated design is again based on cyclic designs, this time three designs, each to compare nine treatments in nine blocks of four units, and is taken from Eskridge *et al.* (2004). The set of three cyclic designs is shown in Figure 14.14, and the derived supersaturated design is shown in Figure 14.15.

The analysis of results from a supersaturated design requires careful thought. Because there are many more main effects than the number of observations any model fitting can include at most nine main effect terms in the model. Because the effects are not orthogonal some effects can appear to be bigger than they really are because of the size of other effects with which they are correlated. An initial set of analyses would fit each main effect separately to get a quick idea of which effects are important, then fit groups of four or five effects to try to sort out the bogus apparent effects.

With only 27 effects the chance of a completely ineffective factor being believed to have a substantial effect after a careful analysis should not be large. However, the class of supersaturated designs does include designs which are very much larger, with up to 12 000 factors compared in no more than 24 observations. For designs with really large numbers of factors there is a high risk of not detecting important factors, due to non-orthogonality and correlation between columns of the design and it is important to use caution in the application of these designs and to use follow-up experimentation to validate results wherever possible.

Exercises

14.1 In a 2^6 experiment with factors A, B, C, D, E and F all three-factor and higher interactions may be assumed negligible as may the two-factor interactions involving either A or C. Construct a quarter-replicate which allows main effects and other two-factor interactions to be estimated.

14.2 It is proposed to use a one-third-replicate to investigate the main effects and two-factor interactions for a 3^5 structure. Construct the design and check that all the effects of interest can be estimated independently of each other.

14.3 A research scientist wants to investigate a 2^7 structure. It is important that all main effects and two-factor interactions should be estimated. Several fractional replicate designs are being considered. These include a half-replicate, a quarter-replicate, a three-eighths-replicate and a five-sixteenths-replicate. The experimenter would like to use as few observations as possible for reasons of cost. Provide advice on how the best design can be identified, making sure that you can explain how to construct each of the designs.

14.4 Using the design for a one-ninth-replicate for a 3^5 structure in Section 14.6, construct a two-ninths-replicate for the same structure and identify the advantages of the larger design compared with the smaller design. You may find it useful to use a computer analysis program for the two designs to examine the precision and correlation of estimates of main effects and interactions. If the factors are quantitative and only the linear $\times$ linear two-factor interaction effects need to be considered how does this change the relative benefits of the two designs?

14.5 Consider again the one-sixth-replicate of a $2^3 \times 3^3$ factorial structure, discussed in Section 14.6. Can you devise alternative ways of combining the two sets of fractions for the two-level and three-level factors which might give different, or better, information about two-factor interaction effects. It may be useful to use computer analysis programs, and to consider linear effects for the three-level factors.

14.6 Find some cyclic designs with between eight and fifteen blocks from which sensible saturated designs to investigate the main effects of between eight and fifteen factors could be constructed.

15

Incomplete block size for factorial experiments

15.0 Preliminary examples

(*a*) In an animal feeding experiment, six dietary treatments are to be compared. The diets are all possible combinations of three different levels of molasses at two energy levels. Twenty-four sheep are available and each sheep can be fed a different diet in each of two time periods. It is expected that there will be large differences in nutritional performances between sheep and some systematic differences between the results for the first and second periods. The structure of the experimental units therefore has two blocking classifications, giving a 24 × 2 row-and-column structure. The treatments have a 3 × 2 structure and main effect comparisons, particularly between the three levels of molasses, are the principal area of interest.

(*b*) An industrial experiment is to be planned to investigate the effects of varying seven factors in a chemical process. Eight treatment combinations can be tested using the same batch of basic material. Eight different batches of material will be available, and it is expected that there may be substantial differences in output for sample units from the different batches. If it is decided to use two levels of each factor how shall the 64 treatment combinations to be included in the experiment be chosen, and how shall they be allocated to the eight blocks, or batches?

(*c*) An experiment on absorption of sugar by rabbits is to be designed to compare eight experimental treatments. The treatments have a 2^3 factorial structure, the three factors being

(i) concentrations of sugar in the diet,
(ii) chemical forms of sugar,
(iii) with or without an additive.

The experimental resources available for the experiment are five rabbits, each of which can be observed during five successive time periods. The design problem is therefore to allocate a 2^3 treatment structure within a 5 × 5 row-and-column structure, recognising the implied requirement for maximum precision of main effects and minimum precision for the three-factor interaction.

15.1 Small blocks and many factorial combinations

There is an apparent incompatibility between two fundamental principles discussed in earlier chapters. The idea of blocking is to group together experimental units which are likely to give

Block

I	II	III	IV
$n_1 s_1 m_1$	$n_1 s_1 m_1$	$n_1 s_1 m_1$	$n_1 s_1 m_1$
$n_1 s_1 m_2$	$n_1 s_1 m_2$	$n_1 s_1 m_2$	$n_1 s_1 m_2$
$n_1 s_2 m_1$	$n_1 s_2 m_1$	$n_1 s_2 m_1$	$n_1 s_2 m_1$
$n_1 s_2 m_2$	$n_1 s_2 m_2$	$n_1 s_2 m_2$	$n_1 s_2 m_2$
$n_1 s_3 m_1$	$n_1 s_3 m_1$	$n_1 s_3 m_1$	$n_1 s_3 m_1$
$n_1 s_3 m_2$	$n_1 s_3 m_2$	$n_1 s_3 m_2$	$n_1 s_3 m_2$
$n_2 s_1 m_1$	$n_2 s_1 m_1$	$n_2 s_1 m_1$	$n_2 s_1 m_1$
$n_2 s_1 m_2$	$n_2 s_1 m_2$	$n_2 s_1 m_2$	$n_2 s_1 m_2$
$n_2 s_2 m_1$	$n_2 s_2 m_1$	$n_2 s_2 m_1$	$n_2 s_2 m_1$
$n_2 s_2 m_2$	$n_2 s_2 m_2$	$n_2 s_2 m_2$	$n_2 s_2 m_2$
$n_2 s_3 m_1$	$n_2 s_3 m_1$	$n_2 s_3 m_1$	$n_2 s_3 m_1$
$n_2 s_3 m_2$	$n_2 s_3 m_2$	$n_2 s_3 m_2$	$n_2 s_3 m_2$

Figure 15.1 Experimental plan for a $2 \times 3 \times 2$ structure in four blocks of 12 units.

similar yields if treated with the same treatment, and to base comparisons between treatments on the differences between observations on units within a group. The general implication of the principle of blocking is that blocks will contain small numbers of experimental units, rarely more than 15 and frequently 8 or fewer. The principle of factorial structure, on the other hand, leads to experiments with many different experimental treatment combinations. When discussing blocking in Chapter 7, we developed a philosophy for designs using blocks with fewer units than the total number of treatments, but only the case where all treatment comparisons were of equal importance was considered. With factorial structures, there is an implied ordering of importance of effects with, in general, main effects being more important than interactions, two-factor interactions being more important than higher-level interactions and so on. In the context of a priority ordering of effects, and with block sizes less than the total number of treatment combinations, how do we allocate treatment combinations to blocks?

An alternative situation producing the same problem is when additional treatments are to be added to an existing experiment, either when the experiment is a continuing one and the additional treatments are for the second stage, or when the additional treatments are conceived after the initial block–treatment pattern is determined. This second situation is simpler than the more general case, and we consider an initial example of this type to illustrate the principles of confounding treatment comparisons.

Example 15.1 An experiment to compare two nitrogen levels (n_1 and n_2), three spatial arrangements (s_1, s_2 and s_3) and two management systems (m_1 and m_2) for a maize crop is planned in four randomised blocks of 12 plots each. The experimenter wishes to include also two different genotypes of maize, but does not wish to have more than 12 plots per block. The initial (unrandomised) design of the experiment is shown in Figure 15.1, and the analysis of variance structure is given in Table 15.1. It is worth noting that 33 df are amply sufficient for the estimation of σ^2 and that the resources which are represented in these 33 df could be more effectively used to estimate further treatment effects.

To introduce two different genotypes, while (*a*) retaining all the information on the first three treatment factors, (*b*) keeping a block size of 12 plots, and (*c*) obtaining information on the genotype main effects and the interactions of genotypes with each other factor, certain

Table 15.1. *Analysis of variance for nitrogen experiment*

Effect	df
Blocks	3
Nitrogen (N)	1
Spacings (S)	2
Management (M)	1
NS interaction	2
NM interaction	1
SM interaction	2
NSM interaction	2
Error	33
Total	47

simple restrictions on the design can be specified. First, all 24 treatment combinations should be replicated equally so that all the main effects and interactions of the four factors are mutually orthogonal. Second, each block of 12 units already includes all 12 combinations of the three original factors so that all estimates of effects for these three factors are unaffected by block differences.

Finally, the two genotypes (g_1 and g_2) should be allocated in such a way that each effect involving the factor G is not affected by block differences. A sufficient requirement for a treatment effect to be estimated independently of block effects is that each level or combination of levels included in the definition of the effect shall occur equally frequently in each block. We can now consider the allocation restrictions for g_1 and g_2 in terms of the main effects of G and the interactions involving G.

(i) To be able to estimate the main effect of G — each genotype must occur six times in each block and twice with each of the original 12 treatment combinations.

(ii) To be able to estimate the GN interaction — each of the four (g, n) combinations must occur thrice in each block and twice with each of the six (s, m) combinations.

(iii) To be able to estimate the GS interaction — each of the six (g, s) combinations must occur twice in each block and twice with each of the four (n, m) combinations.

(iv) To be able to estimate the GNS interaction — each of the 12 (g, n, s) combinations must occur once in each block, and twice with each level of m.

(v) To be able to estimate the GMS interaction — each of the 12 (g, m, s) combinations must occur once in each block, and twice with each level of n.

(a)   Block

                 I   II

$n_1 s_1 m_1 g_1$ $n_1 s_1 m_1 g_2$
$n_1 s_1 m_2 g_2$ $n_1 s_1 m_2 g_1$
$n_1 s_2 m_1 g_1$ $n_1 s_2 m_1 g_2$
$n_1 s_2 m_2 g_2$ $n_1 s_2 m_2 g_1$
$n_1 s_3 m_1 g_1$ $n_1 s_3 m_1 g_2$
$n_1 s_3 m_2 g_2$ $n_1 s_3 m_2 g_1$
$n_2 s_1 m_1 g_2$ $n_2 s_1 m_1 g_1$
$n_2 s_1 m_2 g_1$ $n_2 s_1 m_2 g_2$
$n_2 s_2 m_1 g_2$ $n_2 s_2 m_1 g_1$
$n_2 s_2 m_2 g_1$ $n_2 s_2 m_2 g_2$
$n_2 s_3 m_1 g_2$ $n_2 s_3 m_1 g_1$
$n_2 s_3 m_2 g_1$ $n_2 s_3 m_2 g_2$

(b)   Block

                 I   II

$n_1 s_1 m_1 g_1$ $n_1 s_1 m_1 g_2$
$n_1 s_1 m_2 g_2$ $n_1 s_1 m_2 g_1$
$n_1 s_2 m_1 g_2$ $n_1 s_2 m_1 g_1$
$n_1 s_2 m_2 g_1$ $n_1 s_2 m_2 g_2$
$n_1 s_3 m_1 g_1$ $n_1 s_3 m_1 g_2$
$n_1 s_3 m_2 g_2$ $n_1 s_3 m_2 g_1$
$n_2 s_1 m_1 g_2$ $n_2 s_1 m_1 g_1$
$n_2 s_1 m_2 g_1$ $n_2 s_1 m_2 g_2$
$n_2 s_2 m_1 g_1$ $n_2 s_2 m_1 g_2$
$n_2 s_2 m_2 g_2$ $n_2 s_2 m_2 g_1$
$n_2 s_3 m_1 g_2$ $n_2 s_3 m_1 g_1$
$n_2 s_3 m_2 g_1$ $n_2 s_3 m_2 g_2$

Figure 15.2 Two alternative designs for a $2 \times 3 \times 2 \times 2$ structure in two blocks of 12 units: (a) with four (g, n, m) combinations thrice in a block; (b) with all eight (g, n, m) combinations once or twice in a block.

(vi) To be able to estimate the GNM interaction — each of the eight (g, n, m) combinations must occur equally frequently in each block and thrice with each level of s.

(vii) To be able to estimate the GNSM interaction — each of the 24 (g, n, s, m) combinations must occur equally frequently in each block.

Clearly the requirements for (vi) and (vii) are impossible in blocks of 12 units. The other requirements are all individually possible, and the design problem is simply a question of whether they are simultaneously possible. The best way to convince oneself that, in fact, the requirements (i)–(v) can be achieved simultaneously is to try adding the genotypes in two blocks of the original design given in Figure 15.1 (with the treatments written in a systematic order instead of a randomised order to clarify the logic of the allocation). It will be found that choice is very limited, and that the design 'constructs itself'.

Two essentially different designs are possible, and these are shown in Figure 15.2(a) and (b). The difference between the two designs is in the pattern of occurrence of the eight (g, n, m) combinations. In Figure 15.2(a), four of these combinations occur thrice in the

Table 15.2. *Outline analysis of variance for nitrogen experiment*

Effect	df
Blocks	3
Nitrogen (N)	1
Spacing (S)	2
Management (M)	1
Genotype (G)	1
NS interaction	2
NM interaction	1
SM interaction	2
GN interaction	1
GS interaction	2
GM interaction	1
NSM interaction	2
NSG interaction	2
SMG interaction	2
Error	24
Total	47

first block, and the other four occur thrice in the second block. In Figure 15.2(b), all eight occur in each block, four occurring twice in block I, and the other four twice in block II. In either design the comparison of the eight combinations must involve the differences between blocks. The choice between the designs is considered further in Section 15.3.

Since the complete set of treatment effects is defined to be orthogonal, all effects which are orthogonal to blocks can be estimated independently of block effects and of each other. The calculation of effect estimates and SS in an analysis of variance is exactly the same as for any other factorial or blocked experiment, except for the GNM and GNSM interaction effects. The structure of the analysis of variance for the four-block design which consists of two replicates of either of the plans shown in Figure 15.2 can be specified by the df for the fitted terms; see Table 15.2. The GNM interaction, discussed earlier, is said to be confounded with blocks and is necessarily omitted from the analysis. The four-factor interaction GNSM is also omitted but would not usually be included in the analysis.

The experiment we have designed is called a confounded experiment. It is important, however, not to concentrate on the information which has been confounded in a confounded experiment, but rather to consider what has been gained. All the main effects and two-factor interactions for all four factors can be estimated, and all the three-factor interactions involving factor S. Compared with the original design involving only three factors, we have additional information on nine treatment comparisons, and we still have a very adequate estimate of σ^2, based on 24 df. Indeed, we could still be accused of not using the available resources efficiently because the 24 df for estimating σ^2 is more than sufficient.

We have discussed the design construction in the context of adding an extra factor to an existing experiment, but if the full design problem of a $2 \times 2 \times 2 \times 3$ factorial structure in four blocks of 12 units is considered directly, we should again arrive at the designs of Figure 15.2.

We have deliberately started the discussion of confounded experiments by considering an experimental problem with many factors, and with different numbers of levels per factor because the principle of avoiding confounding is a simple one which applies equally to large or small, complex or simple, experiments. We continue the discussion by considering two further, practical, large experiments, and then discuss some smaller factorial structures in which all factors have the same number of levels, before going on to consider extensions of the basic principle.

Example 15.2 Four factors, two (P, Q) with three levels and two (R, S) with two levels, are to be compared, again using blocks of 12 units, but this time having a total of only 36 observations so that each treatment combination occurs just once.

The effects for which it might be hoped that confounding could be avoided are those for which the number of combinations of levels is a divisor of 12, so that the combinations of levels can be equally replicated in a block of 12. These effects, and the necessary restrictions, are:

   (i) main effects of P and Q, each having three levels, to appear four times in each block;
  (ii) main effects of R and S, each with two levels, each level to occur six times per block;
 (iii) two-factor interaction, RS, four combinations of levels, each three times per block;
  (iv) two-factor interactions, PR, PS, QR, QS, each with six combinations of levels, each combination twice per block;
   (v) three-factor interactions, PRS, QRS, 12 combinations, all 12 to occur in each block.

Effects which will not allow equal occurrence of all combinations in each block are PQ, PQR, PQS, PQRS. Two possible designs are given in Figure 15.3, the design in (a) having all combinations of (p, q) appearing in each block, three duplicated in each block, while in (b) different groups of three (p, q) combinations appear in each block. The construction of the design may be followed more easily by first recognising that all QRS combinations occur in each block and that the levels of P are added to these combinations so as to satisfy the requirements.

For either design, and for the other designs possible under the restrictions (i)–(v), which are intermediate between (a) and (b), the analysis of variance structure is straightforward (see Table 15.3), including all effects for which the design has been constructed to provide estimates, and of course the block effects.

Note that the error df for the estimation of σ^2 are still adequate at 14, but are nearing the reasonable minimum, so that the experiment provides an efficient use of resources.

Example 15.3 A very large experiment was encountered in experimentation on intercropping in which there were five factors, three (P, Q, R) with three levels and two (S, T) with two levels. It was proposed to use 18 plots per block, which was probably too many. The factors were:

   P: three densities of maize,
   Q: three densities of beans,
   R: three genotypes of beans,
   S: two genotypes of maize,
   T: two row arrangements,

Table 15.3. *Outline analysis of variance for Example 15.2*

Effect	df
Blocks	2
P	2
Q	2
R	1
S	1
RS interaction	1
PR interaction	2
PS interaction	2
QR interaction	2
QS interaction	2
PRS interaction	2
QRS interaction	2
Error	14
Total	35

(a)

	Block	
I	II	III
$p_1q_1r_1s_1$	$p_2q_1r_1s_1$	$p_3q_1r_1s_1$
$p_2q_1r_1s_2$	$p_3q_1r_1s_2$	$p_1q_1r_1s_2$
$p_3q_1r_2s_1$	$p_1q_1r_2s_1$	$p_2q_1r_2s_1$
$p_1q_1r_2s_2$	$p_2q_1r_2s_2$	$p_3q_1r_2s_2$
$p_2q_2r_1s_1$	$p_3q_2r_1s_1$	$p_1q_2r_1s_1$
$p_3q_2r_1s_2$	$p_1q_2r_1s_2$	$p_2q_2r_1s_2$
$p_1q_2r_2s_1$	$p_2q_2r_2s_1$	$p_3q_2r_2s_1$
$p_2q_2r_2s_2$	$p_3q_2r_2s_2$	$p_1q_2r_2s_2$
$p_3q_3r_1s_1$	$p_1q_3r_1s_1$	$p_2q_3r_1s_1$
$p_1q_3r_1s_2$	$p_2q_3r_1s_2$	$p_3q_3r_1s_2$
$p_2q_3r_2s_1$	$p_3q_3r_2s_1$	$p_1q_3r_2s_1$
$p_3q_3r_2s_2$	$p_1q_3r_2s_2$	$p_2q_3r_2s_2$

(b)

	Block	
I	II	III
$p_1q_1r_1s_1$	$p_2q_1r_1s_1$	$p_3q_1r_1s_1$
$p_1q_1r_1s_2$	$p_2q_1r_1s_2$	$p_3q_1r_1s_2$
$p_1q_1r_2s_1$	$p_2q_1r_2s_1$	$p_3q_1r_2s_1$
$p_1q_1r_2s_2$	$p_2q_1r_2s_2$	$p_3q_1r_2s_2$
$p_2q_2r_1s_1$	$p_3q_2r_1s_1$	$p_1q_2r_1s_1$
$p_2q_2r_1s_2$	$p_3q_2r_1s_2$	$p_1q_2r_1s_2$
$p_2q_2r_2s_1$	$p_3q_2r_2s_1$	$p_1q_2r_2s_1$
$p_2q_2r_2s_2$	$p_3q_2r_2s_2$	$p_1q_2r_2s_2$
$p_3q_3r_1s_1$	$p_1q_3r_1s_1$	$p_2q_3r_1s_1$
$p_3q_3r_1s_2$	$p_1q_3r_1s_2$	$p_2q_3r_1s_2$
$p_3q_3r_2s_1$	$p_1q_3r_2s_1$	$p_2q_3r_2s_1$
$p_3q_3r_2s_2$	$p_1q_3r_2s_2$	$p_2q_3r_2s_2$

Figure 15.3 Two alternative designs for a $3 \times 3 \times 2 \times 2$ structure in three blocks of 12 units, (a) with all (p, q) combinations in each block; (b) with different groups of (p, q) combinations in each block.

PQS combinations in all blocks	I	II	III	IV	V	VI
			Block			
$p_1q_1s_1$	r_1t_1	r_1t_2	r_2t_1	r_2t_2	r_3t_1	r_3t_2
$p_1q_1s_2$	r_2t_2	r_2t_1	r_3t_2	r_3t_1	r_1t_2	r_1t_1
$p_1q_2s_1$	r_2t_1	r_2t_2	r_3t_1	r_3t_2	r_1t_1	r_1t_2
$p_1q_2s_2$	r_3t_2	r_3t_1	r_1t_2	r_1t_1	r_2t_2	r_2t_1
$p_1q_3s_1$	r_3t_1	r_3t_2	r_1t_1	r_1t_2	r_2t_1	r_2t_2
$p_1q_3s_2$	r_1t_2	r_1t_1	r_2t_2	r_2t_1	r_3t_2	r_2t_1
$p_2q_1s_1$	r_2t_2	r_2t_1	r_3t_2	r_3t_1	r_1t_2	r_1t_1
$p_2q_1s_2$	r_3t_1	r_3t_2	r_1t_1	r_1t_2	r_2t_1	r_2t_2
$p_2q_2s_1$	r_3t_2	r_3t_1	r_1t_2	r_1t_1	r_2t_2	r_2t_1
$p_2q_2s_2$	r_1t_1	r_1t_2	r_2t_1	r_2t_2	r_3t_1	r_3t_2
$p_2q_3s_1$	r_1t_2	r_1t_1	r_2t_2	r_2t_1	r_3t_2	r_3t_1
$p_2q_3s_2$	r_2t_1	r_2t_2	r_3t_1	r_3t_2	r_1t_1	r_1t_2
$p_3q_1s_1$	r_3t_1	r_3t_2	r_1t_1	r_1t_2	r_2t_1	r_2t_2
$p_3q_1s_2$	r_1t_2	r_1t_1	r_2t_2	r_2t_1	r_3t_2	r_3t_1
$p_3q_2s_1$	r_1t_1	r_1t_2	r_2t_1	r_2t_2	r_3t_1	r_3t_2
$p_3q_2s_2$	r_2t_2	r_2t_1	r_3t_2	r_3t_1	r_1t_2	r_1t_1
$p_3q_3s_1$	r_2t_1	r_2t_2	r_3t_1	r_3t_2	r_1t_1	r_1t_2
$p_3q_3s_2$	r_3t_2	r_3t_1	r_1t_2	r_1t_1	r_2t_2	r_2t_1

Figure 15.4 Design for a $3 \times 3 \times 2 \times 3 \times 2$ structure in six blocks of 18 units.

and the design in Figure 15.4 allows all main effects and two-factor interactions, except ST, to remain non-confounded, and also allows the estimation of three-factor interactions PQS, PQT, PRS, PRT, QRS. When the levels of T are added to complete the treatment combinations for each block a choice emerges. Either each block contains only two (S, T) combinations and all QRT combinations occur in each block, or all four (S, T) combinations occur in each block, although unevenly, but not all QRT combinations can be included in each block. The design in Figure 15.4 is the latter form based on the practical desire to obtain some information about the ST interaction.

15.2 Factors with a common number of levels

In the experiments considered in the previous section, the number of levels per factor varies within each experiment. This is likely to be true of many, possibly of most, practical experimental situations. However, when all factors have the same number of levels, the principles of design are simpler because of the symmetry of all factors.

15.2.1 Designs for factors with two levels

Example 15.4 Three factors in blocks of four. Main effects and two-factor interactions can be estimated independently of blocks by allocating combinations to blocks as shown in Figure 15.5(a). All four combinations of levels for factors P and Q are first allocated to each block. The levels of the third factor, R, are completely determined by the choice in the first combination in the block and the requirements not to confound main effects or two-factor interactions.

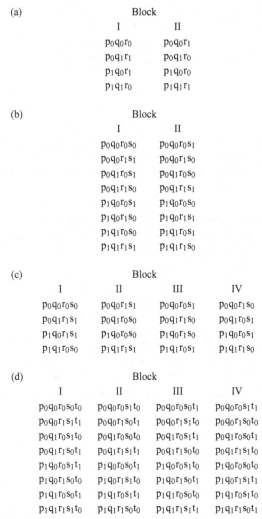

Figure 15.5 (a) $2 \times 2 \times 2$ in two blocks of four, (b) $2 \times 2 \times 2 \times 2$ in two blocks of eight, (c) $2 \times 2 \times 2 \times 2$ in four blocks of four, (d) $2 \times 2 \times 2 \times 2 \times 2$ in four blocks of eight.

Example 15.5 Four factors in blocks of eight. The combinations of levels can be allocated to blocks so that all effects up to three-factor interactions are independent of blocks. To construct the design first allocate the eight combinations of levels from the first three factors to each block. As in Example 15.4, after the arbitrary choice for the first combination in the block the levels of the last factor are completely determined by the requirements of avoiding confounding of interactions up to three factors. The resulting design is shown in Figure 15.5(b).

Example 15.6 Four factors in blocks of four. Main effects and most interactions between two factors can be kept clear of confounding as shown in Figure 15.5(c). All four combinations of levels from the first two factors occur in each block. The choice of levels for the third

Block

I	II	III
p_0q_0	p_0q_1	p_0q_2
p_1q_1	p_1q_2	p_1q_0
p_2q_2	p_2q_0	p_2q_1

Block

I	II	III
p_0q_0	p_0q_1	p_0q_2
p_1q_2	p_1q_0	p_1q_1
p_2q_1	p_2q_2	p_2q_0

Figure 15.6 The two possible designs for a 3×3 structure in three blocks of three units.

factor is determined as in Example 15.4. The choice of levels for the fourth factor produces the first failure of our optimistic assumption that, if the number of combinations required for an interaction effect is divisible into the block size, then it will be possible to arrange for that interaction to be not confounded. The allocation of levels of the fourth factor must coincide with the pattern of levels of one of the first three factors. In Figure 15.5(c), the two-factor interaction, RS, is confounded with blocks.

Example 15.7 Five factors in blocks of eight. All main effects and two-factor interactions can be estimated from the design shown in Figure 15.5(d). Obviously, with eight units per block, many three-factor interactions can be arranged to be orthogonal to blocks. Examination of the design shows that the only three-factor interactions which have been confounded are PRT and QST.

While the simple logical methods of design construction employed in the previous section are still effective for 2^n structures, the increasing number of desirable restrictions as n increases makes it necessary to develop a method based less on trial and error. In fact, for 2^n, a very simple method is available, and this is described in Section 15.6, which is concerned with mathematical theory for the construction of confounded designs. This theory demonstrates that we could not have improved on the four designs constructed in Figure 15.5.

15.2.2 Designs for factors with three levels

The two simplest possible cases for two and three factors demonstrate clearly the principles for confounding with three-level factors.

Example 15.8 Two three-level factors. To arrange main effects orthogonal to blocks it is necessary to have all three levels of each factor occurring equally frequently in each block, which requires blocks of three units or of six units. Since a design with three units per block implies a complementary design with six units per block, in which the six combinations in a block are simply those left out of the corresponding block in the three-unit block design, only the three-unit block design is examined. The problem then is to arrange the nine treatment combinations in three blocks of three so that each level of each factor occurs once in each block. There are two possible solutions shown in Figure 15.6. The problem is exactly

(a)

	p_0	p_1	p_2
q_0	r_0	r_1	r_2
q_1	r_1	r_2	r_0
q_2	r_2	r_0	r_1

(b)

	Block I	Block II	Block III	Block I	Block II	Block III	Block I	Block II	Block III	Block I	Block II	Block III
	I	II	III	I	II	III	I	II	III	I	II	III
p_0q_0	r_0	r_1	r_2	r_0	r_1	r_2	r_0	r_1	r_2	r_0	r_1	r_2
p_0q_1	r_1	r_2	r_0	r_1	r_2	r_0	r_2	r_0	r_1	r_2	r_0	r_1
p_0q_2	r_2	r_0	r_1	r_2	r_0	r_1	r_1	r_2	r_0	r_1	r_2	r_0
p_1q_0	r_1	r_2	r_0	r_2	r_0	r_1	r_1	r_2	r_0	r_2	r_0	r_1
p_1q_1	r_2	r_0	r_1	r_0	r_1	r_2	r_0	r_1	r_2	r_1	r_2	r_0
p_1q_2	r_0	r_1	r_2	r_1	r_2	r_0	r_2	r_0	r_1	r_0	r_1	r_2
p_2q_0	r_2	r_0	r_1	r_1	r_2	r_0	r_2	r_0	r_1	r_1	r_2	r_0
p_2q_1	r_0	r_1	r_2	r_2	r_0	r_1	r_1	r_2	r_0	r_0	r_1	r_2
p_2q_2	r_1	r_2	r_0	r_0	r_1	r_2	r_0	r_1	r_2	r_2	r_0	r_1

Figure 15.7 (a) Allocation pattern for levels of R to PQ combinations; (b) four alternative designs for a $3 \times 3 \times 3$ structure in three blocks of nine units.

equivalent to the 3×3 Latin square design, the rows, columns and treatments of the Latin square corresponding to levels of factor P, levels of factor Q, and blocks for the confounded design. The two solutions of Figure 15.6 correspond to the two distinct forms of the 3×3 Latin square.

Example 15.9 Three three-level factors in three blocks of nine units each. Again, the ideas of 3×3 Latin squares help to clarify the possible choices. To avoid confounding the two-factor interactions, each block must contain all nine combinations of levels for each pair of factors. The first requirement is that each block contains all nine (p, q) combinations. The levels of R must be allocated so that each (p, r) combination and each (q, r) combination occur once in a block. If the nine (p, q) combinations are written in a 3×3 square with levels of P corresponding to rows and levels of Q corresponding to columns, then each R level must occur once in each row and once in each column. There are four essentially different patterns. The first is shown in Figure 15.7(a), the other acceptable patterns are obtained by swapping either the p_1 and p_2 columns or the q_1 and q_2 rows, or both. Each of the four patterns produces a distinct block I and the other two blocks are totally determined by the first block. The four designs are shown in Figure 15.7(b), in which only the levels of r are shown, the nine combinations of (p, q) being in the same standard order for each design.

15.2.3 Analysis of results from confounded experiments

For the analysis of results from the experiments discussed in this section, it would be sufficient from a practical viewpoint to remark that all non-confounded treatment effects of interest can be estimated, and the corresponding SS calculated, in the usual way. The error SS, providing an estimate of σ^2, is calculated by subtracting those treatment SS included in the analysis from the total SS. A standard computer program for the analysis of experiments will need

Table 15.4. *Contrasts for 2^3 treatments in blocks of four*

	Contrasts						
	1(P)	2(Q)	3(R)	4(PQ)	5(PR)	6(QR)	7(PQR $\equiv$ block)
$p_0q_0r_0$	-1	-1	-1	$+1$	$+1$	$+1$	-1
$p_0q_0r_1$	-1	-1	$+1$	$+1$	-1	-1	$+1$
$p_0q_1r_0$	-1	$+1$	-1	-1	$+1$	-1	$+1$
$p_0q_1r_1$	-1	$+1$	$+1$	-1	-1	$+1$	-1
$p_1q_0r_0$	$+1$	-1	-1	-1	-1	$+1$	$+1$
$p_1q_0r_1$	$+1$	-1	$+1$	-1	$+1$	-1	-1
$p_1q_1r_0$	$+1$	$+1$	-1	$+1$	-1	-1	-1
$p_1q_1r_1$	$+1$	$+1$	$+1$	$+1$	$+1$	$+1$	$+1$

to be told only which treatment effects are to be fitted, and the analysis should be produced. However, further insight into the properties of confounded designs may be obtained through considering unit contrasts and the corresponding sums of squares.

For the 2^3 in blocks of four, shown in Figure 15.5(a), there are seven orthogonal comparisons between units within a single replicate. We know that these should include the seven treatment comparisons, and the block comparison, with three main effect comparisons and three two-factor interaction comparisons independent of the block comparisons. It follows that the three-factor interaction comparison and the block comparison must be identical, since each is orthogonal to all six other orthogonal treatment contrasts. This is confirmed by the set of contrasts displayed in Table 15.4, where it can be seen that column 7 is the only contrast which can be constructed to be orthogonal to the first six columns, and that it represents both the three-factor interaction $(p-1)(q-1)(r-1)$ and the difference between block I and block II.

Similar identification of contrasts can be written down for the other 2^n designs considered earlier and shown in Figure 15.5(b), (c) and (d). The reader is invited to confirm that the block contrasts in these designs are identical with treatment contrasts as follows:

(b) 2^4 in two blocks of eight units; the block difference is identical to the four-factor interaction PQRS.

(c) 2^4 in four blocks of four units; there are three block contrasts:
   (i) $(b_1 + b_2 - b_3 - b_4)$, which is identical with the three-factor interaction PQR;
   (ii) $(b_1 - b_2 + b_3 - b_4)$, which is identical with the three-factor interaction PQS; and
   (iii) $(b_1 - b_2 - b_3 + b_4)$, which is identical with the two-factor interaction RS.
       Note that the four-factor interaction is not confounded. The reason for this unsought and essentially useless 'benefit' will become clear in Section 15.6.

(d) 2^5 in four blocks of eight units; the pattern is similar to that for (*c*), the three block contrasts being:
   (i) $(b_1 + b_2 - b_3 - b_4)$, which is identical to the four-factor interaction PQRS;
   (ii) $(b_1 - b_2 + b_3 - b_4)$, which is identical to the three-factor interaction PQS; and
   (iii) $(b_1 - b_2 - b_3 - b_4)$, which is identical to the three-factor interaction RST.
       And again the highest-order interaction PQRST is not confounded.

Table 15.5. *Contrasts for* 3^2 *treatments in blocks of three*

	1	2	3	4	5	6	7	8	9	10	11	12
p_0q_0	-1	$+1$	-1	$+1$	-1	$+1$	-1	$+1$	$+1$	-1	-1	$+1$
p_0q_1	-1	$+1$	0	-2	$+1$	$+1$	$+1$	$+1$	0	0	$+2$	-2
p_0q_2	-1	$+1$	$+1$	$+1$	0	-2	0	-2	-1	$+1$	-1	-1
p_1q_0	0	-2	-1	$+1$	0	-2	$+1$	$+1$	0	$+2$	0	-2
p_1q_1	0	-2	0	-2	-1	$+1$	0	-2	0	0	0	$+4$
p_1q_2	0	-2	$+1$	$+1$	$+1$	$+1$	-1	$+1$	0	-2	0	-2
p_2q_0	$+1$	$+1$	-1	$+1$	$+1$	$+1$	0	-2	-1	-1	$+1$	$+1$
p_2q_1	$+1$	$+1$	0	-2	0	-2	-1	$+1$	0	0	-2	-2
p_2q_2	$+1$	$+1$	$+1$	$+1$	-1	$+1$	$+1$	$+1$	$+1$	$+2$	$+1$	$+1$

If we consider individual contrasts for the 3^2 design in three blocks of three units each, shown in Figure 15.6(a), then the eight orthogonal contrasts must include two main effect contrasts for each factor, and two block contrasts. Also, there are four interaction contrasts which are orthogonal to the four main effect contrasts. The set of eight orthogonal contrasts, shown in the first eight columns of Table 15.5, includes the four main effect contrasts (contrasts 1–4), and the two block contrasts (5 and 6) together with two contrasts (7 and 8) orthogonal to the previous six. Although there is an element of choice of contrasts for three-level factors as mentioned in Section 13.4, in most practical circumstances the contrasts $(1, -1, 0)$ and $(1, 1, -2)$ are the most relevant contrasts, and with this form of contrasts the choice for the contrasts in Table 15.5, including columns 7 and 8, is minimal. The last four columns of Table 15.5 comprise the normal form of the four interaction contrasts for PQ, and it can be seen that they bear no resemblance either to the block contrasts in columns 5 and 6 or to the two 'surplus' contrasts in columns 7 and 8.

Unlike the 2^n experiments, we are not able to identify in terms of practically interpretable effects either those effects which have been confounded or those effects which in addition to the main effects, remain unconfounded. The contrasts in columns 5–8 must represent interaction comparisons, since they are orthogonal to the main effect contrasts, but they are not apparently related to the form of interaction effect in which we have previously been interested. We return to this question of interpretation of the confounded effects in 3^n experiments in Section 15.4.

15.3 Incompletely confounded effects

In discussing the analysis of confounded experiments, we have considered only those effects which are either completely unaffected by block differences, or identical with block differences. Classical design theory has concentrated on designs for which all effects are either completely confounded or entirely unconfounded. However, with the computational difficulties previously associated with incompletely confounded designs now eliminated by modern computational practice, it is important also to consider a wider range of designs to identify which designs are practically appropriate. This leads to the consideration of

Block

I	II
$n_1 s_1 m_1 g_1$ (−)	$n_1 s_1 m_1 g_2$ (+)
$n_1 s_1 m_1 g_2$ (−)	$n_1 s_1 m_2 g_1$ (+)
$n_1 s_2 m_1 g_2$ (+)	$n_1 s_2 m_1 g_1$ (−)
$n_1 s_2 m_2 g_1$ (+)	$n_1 s_2 m_2 g_2$ (−)
$n_1 s_3 m_1 g_1$ (−)	$n_1 s_3 m_1 g_2$ (+)
$n_1 s_3 m_2 g_2$ (−)	$n_1 s_3 m_2 g_1$ (+)
$n_2 s_1 m_1 g_2$ (−)	$n_2 s_1 m_1 g_1$ (+)
$n_2 s_1 m_2 g_1$ (−)	$n_2 s_1 m_2 g_2$ (+)
$n_2 s_2 m_1 g_1$ (+)	$n_2 s_2 m_2 g_1$ (−)
$n_2 s_2 m_2 g_2$ (+)	$n_2 s_2 m_2 g_1$ (−)
$n_2 s_3 m_1 g_2$ (−)	$n_2 s_3 m_1 g_1$ (+)
$n_2 s_3 m_2 g_1$ (−)	$n_2 s_3 m_2 g_2$ (+)

Figure 15.8 Design for a $2 \times 3 \times 2 \times 2$ structure in two blocks of 12 units (Figure 15.2(b)) showing + and − terms for the GNM ($2 \times 2 \times 2$) interaction.

incompletely confounded effects, which are neither orthogonal to block differences nor identical with block differences.

Several designs with incompletely confounded effects have already been introduced. In the $3 \times 2 \times 2 \times 2$ example in the first section of this chapter, the $2 \times 2 \times 2$ interaction could not be arranged to be independent of blocks when blocks of 12 units were used. Two alternative designs were constructed in Figure 15.2. In the first (Figure 15.2(a)), the $2 \times 2 \times 2$ interaction was exactly identified with the differences between the two blocks and was completely confounded. However, in Figure 15.2(b), reproduced in Figure 15.8, all eight combinations of the three factors occurred, albeit unequally, in each block.

If we consider the GNM interaction effect

$$\text{GNM} \propto (g_2 - g_1)(n_2 - n_1)(m_2 - m_1)(s_1 + s_2 + s_3)$$
$$= \{(g_2 n_2 m_2 + g_2 n_1 m_1 + g_1 n_2 m_1 + g_1 n_1 m_2)(s_1 + s_2 + s_3)\}$$
$$- \{(g_2 n_2 m_1 + g_2 n_1 m_2 + g_1 n_2 m_2 + g_1 n_1 m_1)(s_1 + s_2 + s_3)\}$$

the pattern of + and − terms in the two blocks shown in Figure 15.8 is easily identified.

Clearly, the estimation of the GNM interaction effect is equivalent to the estimation of the difference between two treatments such that, in two blocks of 12, one treatment occurs eight times in one block and four times in the other, while the other treatment occurs four times in the first block and eight times in the other. Least squares equations for this situation are

$$12\hat{\mu} + 12\hat{b}_1 + 4\hat{t}_1 = B_1,$$
$$12\hat{\mu} + 12\hat{b}_2 + 4\hat{t}_2 = B_2,$$
$$12\hat{\mu} + 4\hat{b}_1 + 12\hat{t}_1 = T_1,$$
$$12\hat{\mu} + 4\hat{b}_2 + 12\hat{t}_2 = T_2,$$

and the variance of the estimate of $\hat{t}_2 - \hat{t}_1$ is $3\sigma^2/16$, compared with the usual $2\sigma^2/12$, giving an efficiency of $\frac{8}{9}$ for the effect GNM. In the design of Figure 15.2(b) the eight GNM combinations are allocated to blocks as evenly as possible within the constraints imposed by the requirement to completely avoid confounding all main effects and two-factor interactions.

Block

I	II	III
p_1q_1	p_1q_1	p_1q_1
p_2q_1	p_2q_1	p_2q_1
p_3q_1	p_3q_1	p_3q_1
p_1q_2	p_1q_2	p_1q_2
p_2q_2	p_2q_2	p_2q_2
p_3q_2	p_3q_2	p_3q_2
p_1q_3	p_1q_3	p_1q_3
p_2q_3	p_2q_3	p_2q_3
p_3q_3	p_3q_3	p_3q_3
p_1q_1	p_2q_1	p_3q_1
p_2q_2	p_3q_2	p_1q_2
p_3q_3	p_1q_3	p_2q_3

Figure 15.9 Levels of PQ for the design of Figure 15.3(a).

And the cost of using blocks of 12 units instead of 24 is to lose $\frac{1}{9}$ of the information about the GNM three-factor interaction. In the analysis of variance we can add a term for the GNM interaction with a consequent reduction in error df:

$$\begin{array}{ll} \text{GNM (allowing for blocks)} & \text{1 df} \\ \text{error} & \text{23 df.} \end{array}$$

Another example from Section 15.1 illustrates a slightly more complex case of incomplete confounding. The $3 \times 3 \times 2 \times 2$ design confounded in three blocks of 12 units in Figure 15.3(a) inevitably fails to avoid some confounding of the PQ (3×3) interaction. The levels of PQ for the design of Figure 15.3(a) are shown in Figure 15.9 without the levels of the other two factors. Note first that, if the problem has been presented as that of comparing nine treatments in three blocks of 12 assuming all treatment comparisons to be of equal importance, the design of Figure 15.9 would have been the obvious design, using the general philosophy of Chapter 7.

To assess the loss of efficiency of information on the PQ interaction, consider the nine combinations of PQ levels as nine treatments and the eight orthogonal contrasts between these nine treatments. The variance of the estimated difference between any two treatments which are duplicated in the same block is $2\sigma^2/4$ (efficiency of 1.0), and between two treatments which are duplicated in different blocks, the variance of the estimated difference is $23\sigma^2/45$ (efficiency of 0.98). The overall average variance of treatment comparisons for the design is therefore calculated to be

$$\{9(2\sigma^2/4) + 27(23\sigma^2/45)\}/36 = 61\sigma^2/120.$$

The main effects of P and Q are orthogonal to blocks, and therefore are estimated fully efficiently.

Since the eight orthogonal treatment comparisons between the nine PQ combinations consist of four main effects and four interactions, the average efficiency of the interaction comparisons is

$$(2\sigma^2/4)/(61\sigma^2/120) = 60/61 = 0.98.$$

Table 15.6. *Duplication of fractions for some common situations*

Treatment structure	Block size	Non-orthogonal interaction	
$2 \times 2 \times 3$	6	2×2	Combinations duplicated in pairs such that each pair includes each level of each factor
$2 \times 3 \times 3$	6	3×3	Combinations omitted in threes, including each level of each factor
$2 \times 3 \times 4$	12	2×4	Combinations duplicated in fours balanced over levels of each factor
$2 \times 2 \times 3 \times 3$	6	2×2	Duplicated in pairs
		3×3	Omitted in threes
$2 \times 3 \times 3 \times 3$	6	3×3	Omitted in threes
$2 \times 3 \times 3 \times 4$	12	2×4	Duplicated in fours
		3×3	Duplicated in threes
$2 \times 3 \times 4 \times 4$	12	2×4	Duplicated in fours
		4×4	Omitted in fours

The detailed arithmetic of efficiency is not important. The important information concerns the high efficiency of all treatment comparisons for the nine combinations which ensures that the efficiency of all effects estimates is high.

At this point it is useful to recognise that the general principles of block–treatment designs developed in Chapter 7 can be used to minimise the effect of confounding on the estimation of particular treatment interaction effects when it is impossible to arrange for those effects to be orthogonal to blocks. If the combinations for these interaction effects are treated as a set of equally important treatments, the principles of Chapter 7 will suggest a simple form of allocation of combinations to blocks. The allocation must, of course, take into account restrictions required for those effects which can be made orthogonal to blocks. In general the efficiency of estimation of effects which are necessarily non-orthogonal to blocks, but which are allocated as evenly as possible, will be high.

For factorial structures where all factors have two, three or four levels and for small block sizes which allow main effects to be orthogonal to blocks the allocation problem for all non-orthogonal two-factor interactions will always be simple. Either a simple fraction of the combinations has to be duplicated in each block, or a simple fraction of the combinations has to be omitted from each block. The possible situations are listed in Table 15.6 for some of the more common treatment structures.

15.4 Partial confounding

In some of the examples of Section 15.2, when the set of 2^n treatment combinations was divided into more than two blocks, it was not possible to avoid confounding a relatively low-level effect. In the 2^4 structure in four blocks of four units a two-factor interaction was inevitably confounded. In the 2^5 structure in four blocks of eight units two three-factor interactions were necessarily confounded. In practice, we would be most unlikely to use only a single replicate of a 2^4 structure and usually two or three replicates would be used. There

Replicate 1
Block

I	II	III	IV
$p_0q_0r_0s_0$	$p_0q_0r_1s_1$	$p_0q_0r_0s_1$	$p_0q_0r_1s_0$
$p_0q_1r_1s_1$	$p_0q_1r_0s_0$	$p_0q_1r_1s_0$	$p_0q_1r_0s_1$
$p_1q_0r_1s_1$	$p_1q_0r_0s_0$	$p_1q_0r_1s_0$	$p_1q_0r_0s_1$
$p_1q_1r_0s_0$	$p_1q_1r_1s_1$	$p_1q_1r_0s_1$	$p_1q_1r_1s_0$

Replicate 2
Block

V	VI	VII	VIII
$p_0q_0r_0s_0$	$p_0q_0r_1s_0$	$p_0q_0r_0s_1$	$p_0q_0r_1s_1$
$p_0q_1r_1s_1$	$p_0q_1r_0s_1$	$p_0q_1r_1s_0$	$p_0q_1r_0s_0$
$p_1q_0r_0s_1$	$p_1q_0r_1s_1$	$p_1q_0r_0s_0$	$p_1q_0r_1s_0$
$p_1q_1r_1s_0$	$p_1q_1r_0s_0$	$p_1q_1r_1s_1$	$p_1q_1r_0s_1$

Figure 15.10 Two replicates of a $2 \times 2 \times 2 \times 2$ structure each in four blocks of four units, with different two-factor interactions confounded in the two replicates.

is then a choice of either allowing the same two-factor interaction to be confounded between blocks in each replicate, or arranging that different two-factor interactions are confounded in different replicates. The former choice provides no information on one particular two-factor interaction. The latter allows each two-factor interaction effect to be estimated from information for those blocks in which the effect was not confounded. We would then have some information on each effect of interest but, for each *partially confounded* effect, we would have only partial information.

Consider the 2^4 design in four blocks of four units each. In the design in Figure 15.5(c) a two-factor interaction, RS, is confounded with blocks. A two-replicate design with a total of 32 experimental units would usually be practically relevant. Assume that the design from Figure 15.5(c) has been used for the first replicate, shown again in the first four blocks of Figure 15.10. The three effects confounded between blocks are PQR, PQS and RS.

In the second replicate, we must ensure that the two-factor interaction which is confounded should not be RS. We may choose which of the five two-factor interactions, other than RS, should remain unconfounded in both replicates. Suppose all interactions involving P are thought to be important, and that QS is also thought likely to be important. We can then choose that QR should be confounded in the second replicate. This is achieved by allocating the levels of R last, after arranging that each block includes all eight (P, Q, S) combinations. Alternatively we may recognise that confounding QR means that in two blocks the levels of R are identical to those of Q, and in the other two blocks the levels of R are the opposite of those of Q, The second replicate in Figure 15.10 is constructed so that all main effects and two-factor interactions other than QR are orthogonal to blocks. By inspection the two other effects confounded with blocks in the second replicate are seen to be PQS (again) and PRS.

15.4.1 Analysis for partially confounded effects

What form does the analysis of the two-replicate design of Figure 15.10 take? First consider the analysis from the intuitive approach. The four main effects and the four two-factor

Table 15.7. *Esimates of effects*

Effect	Estimate	Var(estimate)
P, Q, R, S	From data of both replicates	$2\sigma^2/16$
PQ, PR, PS, QS	From data of both replicates	$2\sigma^2/16$
QR, PRS	From replicate 1 data only	$2\sigma^2/8$
RS, PQR	From replicate 2 data only	$2\sigma^2/8$
PQS	None	
QRS, PQRS	From data of both replicates	$2\sigma^2/16$

interactions PQ, PR, PS, QS are all orthogonal to blocks in both replicates, and the estimates of these effects and the SS for the analysis of variance would be calculated in the usual way. So also, if it is of any interest, can the estimate and SS for the three-factor interaction QRS. For RS and QR, in each replicate one block contrast is identical with the estimate of one of these effects. The estimate of each effect is therefore calculated using only the data from the replicate in which that effect is not confounded. Similarly for the three-factor interactions PQR and PRS, each confounded in one replicate, estimates and SS could be calculated from the other replicate only.

Standard computer programs for analysis of designed experiments should produce the appropriate SS and estimates of effects and their SEs for the analysis of confounded experiments which include partially confounded effects, simply by specifying the effects to be fitted. The estimates of effects would then be as defined in Table 15.7, and the analysis of variance structure would be as shown in Table 15.8. The inclusion of separate SS for the three- and four-factor interactions is debatable. It will usually be sensible to pool them with the error to give 14 df for estimating σ^2.

15.4.2 *Partial confounding for other two-level factor designs*

Partial confounding can be used in any multireplicate 2^n experiment. It is most common when:

(i) the number of factors is three, four, or five since for more factors it is unlikely that an experiment will have more than one replicate;
(ii) the number of confounded effects in a single replicate is three, or possibly seven, since, if only one effect is confounded in a single replicate, it will almost always be the highest possible order interaction which will be of little interest, and which can therefore be confounded in all replicates, obviating any need for partial confounding.

If the number of replicates is more than two, then partial confounding will involve a set of effects in each replicate, and there will be a choice between confounding different effects in each replicate or having some effects confounded more than once. The important consideration is the relative information (precision) for each of the effects to be estimated. Thus, suppose it was decided to use four replicates of a 2^4 structure in blocks of four, then different strategies of design could be employed:

Table 15.8. *Basis of calculation for effect SS*

Source	SS	df
Blocks	Using block totals as usual	7
P	Using both replicates	1
Q	Using both replicates	1
R	Using both replicates	1
S	Using both replicates	1
PQ	Using both replicates	1
PR	Using both replicates	1
PS	Using both replicates	1
QS	Using both replicates	1
QR	Using replicate 1 only	1
RS	Using replicate 2 only	1
QRS	Using both replicates	1
PQR	Using replicate 2 only	1
PRS	Using replicate 1 only	1
PQRS	Using both replicates	1
Error	By subtraction as usual	10

(*a*) different confounding in each replicate:
    (i) PQR, PQS, RS,
    (ii) PRS, QRS, PQ,
    (iii) PQS, QRS, PR,
    (iv) PQR, PRS, QS;
(*b*) same confounding in all replicates:
    PQR, PQS, RS;
(*c*) two confounding patterns:
    (i), (iii) PQR, PQS, RS,
    (ii), (iv) PQS, PRS, QR;
(*d*) three confounding patterns avoiding confounding any two factor interaction involving P:
    (i), (iv) PQR, PQS, RS,
    (ii) PQS, PRS, QR,
    (iii) PQR, PRS, QS.

The information available on each interaction effect for each design strategy, relative to the precision for any effect unaffected by confounding, is summarised in Table 15.9. It can be seen that making the choice of the required relative information for each two-factor interaction has the apparent effect of producing levels of information for three-factor interactions in precisely the opposite pattern to that which would seem desirable from the selected pattern of information for two-factor interactions. Thus the reason for using design (d) should be that the principal interest is in factor P so that two-factor interactions involving P are estimated as precisely as possible. However, the consequence for three-factor interactions is that those involving P have the worst precision. This undesirable pay-off between the patterns of relative

Table 15.9. *Information available on interactions under different partial confounding schemes*

	Strategy			
Effect	a	b	c	d
PQ	75%	100%	100%	100%
PR	75%	100%	100%	100%
PS	100%	100%	100%	100%
QR	100%	100%	50%	75%
QS	75%	100%	100%	75%
RS	75%	0	50%	50%
PQR	50%	0	50%	25%
PQS	50%	0	0	25%
PRS	50%	100%	50%	50%
QRS	50%	100%	100%	100%

precision for two-factor interactions and for three-factor interactions is inevitable for reasons which will become apparent in Section 15.6. The precision for the two-factor interactions should usually be the dominant consideration in choosing patterns of partial confounding.

15.4.3 *Partial confounding for three-level factors*

Probably the best known classical partially confounded designs are those for two designs involving three-level factors. These two design problems were considered in Section 15.2 but the concept of partial confounding is necessary to obtain the maximum advantage from these designs.

First, we consider the 3^2 design using blocks of three. The two designs which allow main effects of each factor to be estimated with full precision are shown again in Figure 15.11. The consideration of confounded and non-confounded contrasts in Table 15.5 demonstrated that, in the design in Figure 15.11(a), the block differences are equivalent to comparisons between three sets of treatment combination

$$I_1 = p_0q_0 + p_1q_1 + p_2q_2,$$
$$I_2 = p_0q_1 + p_1q_2 + p_2q_0$$

and

$$I_3 = p_0q_2 + p_1q_0 + p_2q_1,$$

while, in the design in Figure 15.11(b), block differences are equivalent to comparisons between

$$J_1 = p_0q_0 + p_1q_2 + p_2q_1,$$
$$J_2 = p_0q_1 + p_1q_0 + p_2q_2$$

and

$$J_3 = p_0q_2 + p_1q_1 + p_2q_0.$$

(a)

	Block	
I	II	III
$p_0 q_0$	$p_0 q_1$	$p_0 q_2$
$p_1 q_1$	$p_1 q_2$	$p_1 q_0$
$p_2 q_2$	$p_2 q_0$	$p_2 q_1$

(b)

	Block	
I	II	III
$p_0 q_0$	$p_0 q_1$	$p_0 q_2$
$p_1 q_2$	$p_1 q_0$	$p_1 q_1$
$p_2 q_1$	$p_2 q_2$	$p_2 q_0$

Figure 15.11 The two alternative designs for a 3×3 structure in three blocks of three units: block differences equivalent to (a) differences between I_1, I_2 and I_3 and (b) differences between J_1, J_2 and J_3.

Any comparison between the three I groups of combinations is orthogonal to any comparison between the three J groups of combinations, since each I group contains one combination from each J group. Similarly, any comparison between the I groups, or between the J groups, is orthogonal to main effect comparisons of P or of Q, and hence I group comparisons and J group comparisons are orthogonal components of the PQ interaction. The orthogonality of the I group comparisons and the J group comparisons means that two orthogonal I group comparisons and two orthogonal J group comparisons constitute a set of four orthogonal comparisons, all of which are PQ interaction contrasts and, since there can be only four orthogonal PQ interaction comparisons, it follows that the set of I and J group comparisons contains complete information about the PQ interaction effects. However, as noted in Section 15.2, the I and J comparisons are not apparently related in any way to the PQ interaction effects suggested in Chapter 13 as being practically relevant.

To see how the I and J comparisons can be related to practical forms of PQ interaction effects, consider a particular PQ effect, from Chapter 13,

$$P'Q' = \frac{1}{2}(p_2 - p_0)(q_2 - q_0)$$

and two general contrasts from the I group and the J group

$$C_1 = \beta_1 I_1 + \beta_2 I_2 + \beta_3 I_3, \quad \text{where } \beta_1 + \beta_2 + \beta_3 = 0,$$

and

$$C_2 = \gamma_1 J_1 + \gamma_2 J_2 + \gamma_3 J_3, \quad \text{where } \gamma_1 + \gamma_2 + \gamma_3 = 0.$$

C_1, C_2 and $C_3 = C_1 + C_2$ can be represented as combinations of $p_i q_j$ as shown in Figure 15.12. For C_3 to be identical with $P'Q'$, we require, in addition to the general restrictions for contrasts,

$$\beta_1 + \beta_2 + \beta_3 = 0, \qquad \gamma_1 + \gamma_2 + \gamma_3 = 0,$$

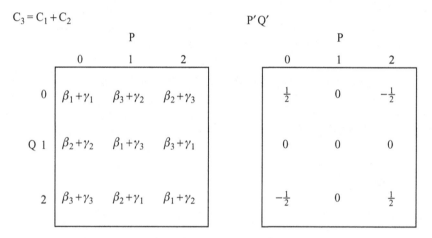

Figure 15.12 Representation of two contrasts of the I and J subsets and the conditions for a combination of contrasts to estimate $P'Q'$.

the additional restrictions

$$\beta_1 + \gamma_1 = \tfrac{1}{2}, \qquad \beta_2 + \gamma_1 = 0, \qquad \beta_3 + \gamma_1 = 0,$$
$$\beta_1 + \gamma_2 = \tfrac{1}{2}, \qquad \beta_2 + \gamma_2 = 0, \qquad \beta_3 + \gamma_2 = 0,$$
$$\beta_1 + \gamma_3 = 0, \qquad \beta_2 + \gamma_3 = -\tfrac{1}{2}, \qquad \beta_3 + \gamma_3 = -\tfrac{1}{2}.$$

The solution is

$$(\beta_1, \beta_2, \beta_3) = \left(\frac{2}{6}, -\frac{1}{6}, -\frac{1}{6} \right),$$

$$(\gamma_1, \gamma_2, \gamma_3) = \left(\frac{1}{6}, \frac{1}{6}, -\frac{2}{6} \right).$$

Block

I	II	III	IV	V	VI
$p_0 q_0$	$p_0 q_1$	$p_0 q_2$	$p_0 q_0$	$p_0 q_1$	$p_0 q_2$
$p_1 q_1$	$p_1 q_2$	$p_1 q_0$	$p_1 q_2$	$p_1 q_0$	$p_1 q_1$
$p_2 q_2$	$p_2 q_0$	$p_2 q_1$	$p_2 q_1$	$p_2 q_2$	$p_2 q_0$

Figure 15.13 Two replicates of a 3×3 structure in six blocks of three units with partial confounding of the PQ interaction.

Similar identification of the other usual PQ interaction effects,

$$P''Q' = \frac{1}{4}(p_2 - 2p_1 + p_0)(q_2 - q_0),$$

$$P'Q'' = \frac{1}{4}(p_2 - p_0)(q_2 - 2q_1 + q_0),$$

$$P''Q'' = \frac{1}{8}(p_2 - 2p_1 + p_0)(q_2 - 2q_1 + q_0),$$

with sums of contrasts of the I and J groups is possible, and the construction of the appropriate contrasts is left to the reader.

The practically sensible form of partial confounding for the 3^2 treatment structure in two replicates, each containing three blocks of three units, is to confound the J comparisons in one replicate and the I comparisons in the other replicate, i.e. use the design in Figure 15.11(a) for the first replicate and the design in Figure 15.11(b) for the second. The complete, two-replicate, design is shown in Figure 15.13. In practice, four or even six replicates might be used repeating the initial two replicates.

Combinations of comparisons between (I_1, I_2, I_3) and between (J_1, J_2, J_3) can provide estimates of the more usual form of interaction effects. Similarly, SS calculated from the totals for (I_1, I_2, I_3) and for (J_1, J_2, J_3) provide the interaction SS for the analysis of variance of the design shown in Figure 15.13. We have already remarked that the set of four comparisons, two between Is and two between Js, provides a complete set of interaction effects for the 4 df of the PQ interaction, and correspondingly the interaction SS can be calculated from the set of four comparisons, or equivalently from the two sets of totals for the Is and the Js. The argument can be supported by an algebraic proof based on the results of Section 4.7, and is illustrated in the following example.

Example 15.10 Suppose that a complete 3^2 factorial structure in two replicates, with no confounding, gives the treatment totals shown in Table 15.10. The simple calculation of main effect and interaction SS is

$$
\begin{aligned}
\text{SS(P)} &= (72^2 + 83^2 + 90^2)/6 - 245^2/18 &&= 27.44, \\
\text{SS(Q)} &= (75^2 + 82^2 + 88^2)/6 - 245^2/18 &&= 14.11, \\
\text{total treatment SS} &= (20^2 + 25^2 + \ldots + 30^2)/2 - 245^2/18 &&= 49.78, \\
\text{interaction SS} &= 49.78 - 27.44 - 14.11 &&= 8.23.
\end{aligned}
$$

Table 15.10. *Treatment totals for* 3^2
experiment

	p_0	p_1	p_2	Total
q_0	20	25	30	75
q_1	24	28	30	82
q_2	28	30	30	88
Total	72	83	90	245

The totals for the I and J groups of combinations are

$$I_1 = 78, I_2 = 84, I_3 = 83,$$
$$J_1 = 80, J_2 = 79, J_3 = 86,$$
$$SS(I) = (78^2 + 84^2 + 83^3)/6 - 245^2/18 = 3.45,$$
$$SS(J) = (80^2 + 79^2 + 86^2)/6 - 245^2/18 = 4.78,$$
$$SS(I) + SS(J) = 8.23 = \text{interaction SS.}$$

The importance, for partially confounded experiments, of this alternative method of calculating the interaction SS is that, in the standard partially confounded design of Figure 15.13, while the normal form of interaction effects are confounded, the I and J comparisons are each unconfounded in one replicate. In the first replicate (blocks I, II, III) the I comparisons are confounded with blocks, but the J comparisons are not, so that the SS(J) can be calculated from the first replicate data. Similarly, SS(I) can be calculated from the second replicate data (blocks IV, V, VI), in which J comparisons are confounded. The PQ interaction SS can therefore be calculated in two halves, each on 2 df, from the two separate replicates, and the overall interaction SS is assessed by combining SS(I) + SS(J) on 4 df. The *F*-test for significance of the PQ interaction will be less powerful than in an experiment where the interaction is not confounded, being based effectively on a single replicate compared with the two replicates for the main effects. The analysis of variance structure is outlined in Table 15.11. A standard analysis program should produce the analysis of variance and estimates of effects that would be calculated by putting together the SS from the two separate replicates. Note that, while the two-replicate partially confounded design has been used for illustration, the fact that there are only 4 df for the estimation of σ^2 makes the design unsuitable for practical use, and the four-replicate design which provides 16 df for error would usually be required.

The other design for factors with three levels which utilises the ideas of partial confounding is for three factors in blocks of nine units. The principles are exactly the same as for the two-factor design. As remarked in Section 15.2 the 27 combinations can be arranged in three blocks of nine, so that all main effects and two-factor interactions are not confounded. The four different designs which were summarised in Figure 15.7 each confound a part of the three-factor interaction. Since the three-factor interaction has 8 df, and a replicate in three

Table 15.11. *Analysis of variance structure for* 3^2
experiments

Source	SS calculation	df
Blocks	Usual	5
A main effect	Totals from both replicates	2
B main effect	Totals from both replicates	2
I comparison	totals from second replicate	2
J comparison	totals from first replicate	2
Interaction	$SS_2(I) + SS_1(J)$	4
Error		4
Total	Usual	17

blocks necessarily confounds two single df comparisons, it should not be surprising to find that

(i) the four different designs of Figure 15.7 each confound a different pair of comparisons;
(ii) the four pairs of comparisons are all orthogonal and together constitute a complete subdivision of the 8 df for the three-factor interaction; and
(iii) a four-replicate design can be used which provides information on each pair of comparisons in three out of four replicates.

Details of the four-replicate design and the calculations for the analysis of variance are given in other books, notably Cochran and Cox (1957). They will not be repeated here for three reasons. Such details add nothing further to the principles which are clearly displayed in the 3^2 design. If it is desired to use the four-replicate 3^3 design in blocks of nine, the analysis of variance should be done by computer and a modern computer program will give the correct analysis without special manipulation. However, the third, and most compelling, reason for not giving details of the calculations is that the design is not practically relevant. Four replicates will almost always be excessive. Two or three replicates will usually be found to give the most appropriate level of precision.

In addition, it will rarely be necessary to examine the three-factor interaction. Interactions of three factors rarely provide useful information. If a two-replicate design is used, then some information on the three-factor interaction is available, though not in a symmetrical fashion and this can be extracted by the use of a suitable computer package analysis.

15.4.4 Partial confounding in different fractions of a replicate

There is one further aspect of partial confounding that merits discussion, though the theory for this is not yet properly developed. We have discussed the use of different confounding systems, each in a complete replicate. However, it is possible to use different confounding systems in different fractions of a single replicate. Consider again the confounding of a 2^4 experiment in blocks of four units. If treatment levels are allocated sequentially, then all main effects are kept unconfounded, and one two-factor interaction is inevitably confounded

Block

I	II	III	IV
$p_0q_0r_0s_0$	$p_0q_0r_0s_1$	$p_0q_0r_1s_0$	$p_0q_0r_1s_1$
$p_0q_1r_1s_1$	$p_0q_1r_1s_0$	$p_0q_1r_0s_1$	$p_0q_1r_0s_0$
$p_1q_0r_1s_1$	$p_1q_0r_1s_0$	$p_1q_0r_0s_0$	$p_1q_0r_0s_1$
$p_1q_1r_0s_0$	$p_1q_1r_0s_1$	$p_1q_1r_1s_1$	$p_1q_1r_1s_0$

Figure 15.14 Partial confounding of different two-factor interactions in two half-replicates of a $2 \times 2 \times 2 \times 2$ structure.

Block

I	II	III	IV
$p_0q_0r_0s_0$	$p_0q_0r_0s_1$	$p_0q_0r_1s_0$	$p_0q_0r_1s_1$
$p_0q_1r_1s_1$	$p_0q_1r_1s_0$	$p_0q_1r_0s_1$	$p_0q_1r_0s_0$
$p_1q_0r_1s_1$	$p_1q_0r_1s_0$	$p_1q_0r_0s_0$	$p_1q_0r_0s_1$
$p_1q_1r_0s_0$	$p_1q_1r_0s_1$	$p_1q_1r_1s_1$	$p_1q_1r_1s_0$
V	VI	VII	VIII
$p_0q_0r_0s_0$	$p_0q_0r_1s_0$	$p_0q_0r_0s_1$	$p_0q_0r_1s_1$
$p_0q_1r_0s_1$	$p_0q_1r_1s_1$	$p_0q_1r_1s_0$	$p_0q_1r_0s_0$
$p_1q_0r_1s_1$	$p_1q_0r_0s_1$	$p_1q_0r_0s_0$	$p_1q_0r_1s_0$
$p_1q_1r_1s_0$	$p_1q_1r_0s_0$	$p_1q_1r_1s_1$	$p_1q_1r_0s_1$

Figure 15.15 Partial confounding in four half-replicates of a $2 \times 2 \times 2 \times 2$ structure.

between the first two blocks. Each block will include all four combinations of P and Q. The levels of R are added so that PQR is identical with the comparison between the first two blocks and the last two blocks. When the levels of S are allocated to the first two blocks one two-factor interaction involving S, say RS, must be confounded with the difference between the first two blocks. However, when the allocation of treatments in blocks III and IV is considered, there is no necessity to confound the RS interaction again. If, instead, the QS interaction is confounded the resulting design is shown in Figure 15.14. The design shown in Figure 15.14 provides full information on the four main effects and four of the two-factor interactions, and half-information on the other two interactions, RS and QS.

To illustrate further the use of partial confounding within different fractions of a replicate, consider three alternatives for the design of a 2^4 experiment using two replicates with eight blocks of four units each. In the first design one two-factor interaction, RS, is completely confounded in both replicates, so that there is no information on that interaction, but full information on main effects and other two-factor interactions. Alternatively, a different two-factor interaction may be confounded in different replicates, RS in replicate 1, and QS in replicate 2. In the third design we could partially confound RS between the first two blocks, QS between blocks III and IV, PR between blocks V and VI, and QR between blocks VII and VIII, the resulting design being shown in Figure 15.15. The resulting information on different treatment effects from the different designs is summarised in Table 15.12. The choice between these alternative designs, like the choice of which interactions to confound, depends on the prior assessment of the relative importance of the different effects. If the experimenter is quite certain that the RS interaction is of no importance, then the first, completely confounded, design would be appropriate. However, in general, it seems likely that spreading the information evenly over different interactions of a similar order will often prove attractive, in which case the third alternative should be chosen.

Table 15.12. *Treatment information from different replicates*

	Complete confounding (%)	Partial confounding in each replicate (%)	Partial confounding in half-replicates (%)
P	100	100	100
Q	100	100	100
R	100	100	100
S	100	100	100
PQ	100	100	100
PR	100	100	75
PS	100	100	100
QR	100	100	75
QS	100	50	75
RS	0	50	75

				Block				
I	II	III	IV	V	VI	VII	VIII
$p_0 q_0 r_0$	$p_0 q_0 r_0$	$p_0 q_0 r_0$	$p_0 q_0 r_1$	$p_0 q_0 r_1$	$p_0 q_0 r_1$	$p_0 q_1 r_0$	$p_0 q_1 r_0$
$p_0 q_1 r_1$	$p_1 q_0 r_1$	$p_1 q_0 r_0$	$p_0 q_1 r_1$	$p_0 q_1 r_0$	$p_1 q_0 r_0$	$p_0 q_1 r_1$	$p_1 q_0 r_1$
$p_1 q_1 r_0$	$p_1 q_1 r_1$	$p_1 q_1 r_1$	$p_1 q_1 r_0$	$p_1 q_0 r_1$	$p_1 q_1 r_1$	$p_1 q_0 r_0$	$p_1 q_1 r_0$

Figure 15.16 Three replicates of a $2 \times 2 \times 2$ structure in eight blocks of three units.

15.5 Confounding for general block size and factor levels

The discussion of confounding has been concerned solely with those designs in which factor levels or combinations of factor levels can appear equally frequently in each block, with a few exceptions where some effects are incompletely confounded with blocks. Consequently only block sizes which are products of the numbers of levels of the various factors have been considered for confounded experiments. This restriction may seem inappropriate, given the emphasis in Chapter 7 on always seeking the natural block size. Inevitably the theory and the recognition of suitable patterns for confounding are less advanced for the more general confounding situation and the philosophy of general confounding is presented only briefly, mainly through examples.

If block size is not related to the numbers of levels for the several factors, the estimation of main effects and interactions of different factors will not be orthogonal to blocks. Consequently, the estimation of different factorial effects will no longer be orthogonal between different effects, even if all treatments are replicated equally over the whole experiment. Hence, estimates of effects will depend on the set of effects which are fitted, and the SS attributed to different effects will depend on the order of fitting. The scale of variation caused by different orders of fitting is demonstrated in the following example.

Example 15.11 The design shown in Figure 15.16 includes three replicates of a 2^3 treatment set arranged in eight blocks of three units each. The selection of treatment combinations in each block is such that for each factor both levels occur in each block, and main effects should therefore be well estimated. The data are shown in Table 15.13.

Table 15.13. *Data from a 2^3 experiment*

	Block							
	I	II	III	IV	V	VI	VII	VIII
$p_0q_0r_0$	33.2	31.7	34.6					
$p_0q_0r_1$				44.1	43.2	40.1		
$p_0q_1r_0$					42.7		46.6	43.1
$p_0q_1r_1$	40.7			46.4			48.2	
$p_1q_0r_0$			43.0			47.0	51.3	
$p_1q_0r_1$		44.1			49.0			50.3
$p_1q_1r_0$	48.6			55.7				53.0
$p_1q_1r_1$		45.4	49.1			49.4		

Table 15.14. *SS for effects fitted in six different orders*

Blocks	265.5										
P	446.5	Q	88.0	R	19.6	Q	88.0	R	19.6	P	446.5
Q	65.3	R	14.9	P	446.6	P	423.8	Q	63.3	R	19.7
R	15.4	P	424.3	PR	2.2	PQ	0.1	QR	20.3	Q	61.0
PQ	0.4	QR	7.7	Q	66.3	R	15.8	P	411.8	PR	7.5
PR	7.4	PQ	0.0	QR	8.1	QR	7.4	PR	7.7	PQ	0.3
QR	7.8	PR	7.8	PQ	0.1	PR	7.8	PQ	0.1	QR	7.8
PQR	0.1										
Residual	6.5										

The logic of main effects and interactions requires that main effects must be fitted before interaction effects involving corresponding factors. The possible orders of fitting are numerous and sensible orders include:

$$\{P, Q, R\}, \quad \{PQ, PR, QR\}, \quad PQR,$$
$$\{P, Q\}, \quad PQ, \quad R, \quad \{PR, QR\}, \quad PQR,$$
$$\{P, R\}, \quad PR, \quad Q, \quad \{PQ, QR\}, \quad PQR,$$

and

$$\{Q, R\}, \quad QR, \quad P, \quad \{PQ, PR\}, \quad PQR,$$

where the brackets indicate that the terms within a bracket may be fitted in any order,

Analyses of variance for some of these orders are shown in Table 15.14 and this gives some idea of the variation in apparent information according to the order.

Although the degree of non-orthogonality for this design might be thought to be considerable the changes in the SS for effects, as the order of fitting is varied, are not substantial in relative terms so that the interpretation is not likely to be altered by the particular order, or orders, chosen. It is particularly notable that the information about the strength of the

Table 15.15. *Variances and efficiencies for the seven effects for the design of Figure 15.16*

Effect	Variance	Efficiency
P	0.1418	0.84
Q	0.1452	0.82
R	0.1370	0.87
PQ	0.1615	0.74
PR	0.1883	0.64
QR	0.2290	0.52
PQR	0.1885	0.63

interaction effects hardly varies at all, except where an interaction is fitted before a main effect.

When all terms are included in the model, the variances of the seven estimates of effects and their covariances and correlations can be estimated in terms of the error variance, s^2, which is 0.7176 for these data, and these are shown in Table 15.15. The variances can be compared with the variances from the three-replicate randomised block design to examine the efficiency of the design. The efficiencies tabulated in Table 15.15 are calculated as the ratio of the variance of the effect for the randomised complete block design to the variance of the effect for the incomplete block design (Figure 15.16), assuming that the variance, σ^2, is the same for both designs. This assumption is likely to be considerably wrong for the same reasons that led to the use of an incomplete block design in the first place.

From Table 15.15 it can be seen that the efficiencies for main effect estimation are between 0.82 and 0.87, but that the efficiencies for interactions are lower, and unevenly so, suggesting that a design with a more even spread of interaction information could be found. If the use of the small blocks reduces σ^2 by 20% then Figure 15.16 is a sensible design for main effect estimation. If the reduction of σ^2 is 50% then Figure 15.16 is a good design for all effects. The incomplete block design has the further disadvantage of some correlations between estimates of effects, though in this case those correlations are small. However, although the three-unit blocks do have some disadvantages, there will always be situations where the use of the reduced block size will clearly produce a more precise experiment.

15.5.1 General principles demonstrated through a range of examples

What general principles are necessary to guide the selection of appropriate designs for particular block sizes and treatment structures? Essentially, they are those developed at the beginning of this chapter and earlier, in Chapter 7, during the discussion of designs for comparing treatments in general block structures. For the 2^3 experiment just considered, the design principles are simply expressed, since each of the seven effects is a simple difference between two sets of treatment combinations. Therefore, in each block of three units, we should try to include three combinations such that for each effect the combinations include both positive and negative combinations. The correlations of estimates of different effects are determined by the interrelationship of effect estimates over the eight different blocks but, provided the joint pattern of $+$ and $-$ terms for pairs of different effects is varied as much as possible between blocks, the correlation should be minimised. Precise methods for optimising the design relative to a criterion defined in terms of the relative importance of the seven different effects are not yet fully developed, but we believe that it is possible to produce good designs essentially by 'trial and error' provided a general computer program is available to assess the efficiency of each possible design.

To illustrate our belief, we consider a range of examples of designs for experiments given

 (i) a particular factorial structure,
(ii) a pattern of 'natural' block sizes.

The general philosophy of block–treatment designs from Chapter 7 is applied to each effect in turn, usually starting with main effects, and then two-factor interactions. The levels, or combinations of levels, should be 'spread evenly' over blocks so as to make the number of concurrences of pairs of levels, or of pairs of combinations, as nearly equal as is possible, for all pairs of levels or all pairs of combinations. Where the interaction of two factors is judged to be of particular interest, all combinations of levels of the two factors should be treated as equally important in the sense of Chapter 7. It would not be realistic to treat interactions as *more* important than the related main effects because the interpretation of interactions requires the estimation of main effects. If block sizes vary, then, of course, we should consider the sum of concurrences inversely weighted by block size.

Example 15.12   In the second example at the beginning of this chapter the treatment structure is three levels of molasses (M) $\times$ two levels of energy (E). The unit structure has 24 blocks (sheep) with two units per block. The possibility of a systematic difference between the first and second observation across all blocks is a complicating factor, but with 24 sheep and six treatment combinations it is obviously easy to ensure that each treatment occurs four times in the first position and four times in the second position.

Clearly each sheep receives two treatments, which must include both energy levels, and two of the three molasses levels. There are six pairs of combinations which can be constructed:

$$(e_1m_0, e_2m_1), \quad (e_1m_0, e_2m_2), \quad (e_1m_1, e_2m_2),$$
$$(e_1m_1, e_2m_0), \quad (e_1m_2, e_2m_0), \quad (e_1m_2, e_2m_1).$$

Each pair can be applied to the sheep in either order, and two replicates of each of the two orders of each of the six pairs gives the design in Figure 15.17 which clearly arranges treatment effect comparisons to be orthogonal to time periods. Inevitably in the real life experiment everything did not go according to plan, and one sheep died leaving only 23 blocks and a non-orthogonal design. There was, however, no difficulty in calculating the analysis of variance and estimates of main effects and interaction effects from within-sheep variation, allowing for a systematic difference between the two time periods.

Example 15.13   In Chapter 1 we introduced the experimental problem of a microbiologist who wanted to investigate a set of treatments making up a $5 \times 4 \times 2 \times 2$ factorial structure

Sheep

Period	I	II	III	IV	V	VI	VII	VIII
1	e_1m_1	e_2m_1	e_1m_0	e_2m_0	e_1m_0	e_2m_1	e_2m_0	e_1m_1
2	e_2m_0	e_1m_2	e_2m_1	e_1m_2	e_2m_2	e_1m_0	e_1m_1	e_2m_2

	IX	X	XI	XII	XIII	XIV	XV	XVI
1	e_1m_2	e_2m_2	e_1m_2	e_2m_2	e_1m_0	e_1m_2	e_2m_2	e_2m_2
2	e_2m_1	e_1m_0	e_2m_0	e_1m_1	e_2m_2	e_2m_1	e_1m_1	e_1m_0

	XVII	XVIII	XIX	XX	XXI	XXII	XXIII	XXIV
1	e_2m_0	e_1m_1	e_1m_1	e_2m_1	e_2m_1	e_1m_0	e_1m_2	e_2m_0
2	e_1m_1	e_2m_0	e_2m_2	e_1m_2	e_1m_0	e_2m_1	e_2m_0	e_1m_2

Figure 15.17 Eight replicates of a 2×3 structure in an array of 24 sheep $\times$ 2 periods.

Day 1

$p_1q_1r_1s_1$	$p_2q_1r_1s_2$	$p_3q_1r_1s_1$	$p_4q_1r_1s_2$	$p_5q_1r_1s_1$
$p_1q_1r_2s_2$	$p_2q_1r_2s_1$	$p_3q_1r_2s_2$	$p_4q_1r_2s_1$	$p_5q_1r_2s_2$
$p_1q_2r_1s_2$	$p_2q_2r_1s_1$	$p_3q_2r_1s_2$	$p_4q_2r_1s_1$	$p_5q_2r_1s_2$
$p_1q_2r_2s_1$	$p_2q_2r_2s_2$	$p_3q_2r_2s_1$	$p_4q_2r_2s_2$	$p_5q_2r_2s_1$
$p_1q_3r_1s_2$	$p_2q_3r_1s_1$	$p_3q_3r_1s_2$	$p_4q_3r_1s_1$	$p_5q_3r_1s_2$
$p_1q_3r_2s_1$	$p_2q_3r_2s_2$	$p_3q_3r_2s_1$	$p_4q_3r_2s_2$	$p_5q_3r_2s_1$
$p_1q_4r_1s_1$	$p_2q_4r_1s_2$	$p_3q_4r_1s_1$	$p_4q_4r_1s_2$	$p_5q_4r_1s_1$
$p_1q_4r_2s_2$	$p_2q_4r_2s_1$	$p_3q_4r_2s_2$	$p_4q_4r_2s_1$	$p_5q_4r_2s_2$

Day 2

$p_1q_1r_1s_2$	$p_2q_1r_1s_1$	$p_3q_1r_1s_2$	$p_3q_1r_1s_1$	$p_5q_1r_1s_2$
$p_1q_1r_2s_1$	$p_2q_1r_2s_2$	$p_3q_1r_2s_1$	$p_4q_1r_2s_2$	$p_5q_1r_2s_1$
$p_1q_2r_1s_1$	$p_2q_2r_1s_2$	$p_3q_2r_1s_1$	$p_4q_2r_1s_2$	$p_5q_2r_1s_1$
$p_1q_2r_2s_2$	$p_2q_2r_2s_1$	$p_3q_2r_2s_2$	$p_4q_2r_2s_1$	$p_5q_2r_2s_2$
$p_1q_3r_1s_1$	$p_2q_3r_1s_2$	$p_3q_3r_1s_1$	$p_4q_3r_1s_2$	$p_5q_3r_1s_1$
$p_1q_3r_2s_2$	$p_2q_3r_2s_1$	$p_3q_3r_2s_2$	$p_4q_3r_2s_1$	$p_5q_3r_2s_2$
$p_1q_4r_1s_2$	$p_2q_4r_1s_1$	$p_3q_4r_1s_2$	$p_4q_4r_1s_1$	$p_5q_4r_1s_2$
$p_1q_4r_2s_1$	$p_2q_4r_2s_2$	$p_3q_4r_2s_1$	$p_4q_4r_2s_2$	$p_5q_4r_2s_1$

Figure 15.18 Design for a $5 \times 4 \times 2 \times 2$ structure in two days of 40 observations.

but who was restricted to sets of at most 40 observations per day. The simplest approach to this design problem, recognising that 40 is simply 5×8 and that this factorisation fits the treatment structure rather neatly, is to divide the 16 possible combinations of levels for the last three factors into two sets of eight combinations retaining estimation for main effects and two-factor interactions for these three factors. The five levels of the first factor are then allocated three to one set of eight combinations and two to the other set to give a set of 40 observations for a day, a possible example for two days being shown in Figure 15.18 (not randomised).

Example 15.14 An experiment on exudates from plants involved sections of individual leaves being treated with chemicals and placed (and observed for several days) in a quarter of a petri dish. The basic experimental unit is a section of a leaf in a quarter of a petri dish. The petri dishes are kept in piles of four. Four plastic bags, each containing four dishes, each dish divided into four quarters, provide a total of 64 experimental units. The plastic bags may be assumed to be different, providing one blocking system. A second system of

Layer	Bag			
	I	II	III	IV
	c_1v_1	c_1v_2	c_1v_3	c_1v_2
Top	c_2v_2	c_2v_3	c_2v_1	c_3v_2
	c_3v_3	c_3v_1	c_4v_2	c_4v_1
	c_4v_3	c_5v_2	c_5v_1	c_5v_3
	c_2v_3	c_1v_1	c_1v_2	c_1v_3
Upper	c_3v_2	c_2v_2	c_2v_1	c_2v_1
Middle	c_4v_1	c_4v_2	c_3v_1	c_3v_3
	c_5v_1	c_5v_3	c_4v_3	c_5v_2
	c_1v_2	c_1v_3	c_2v_2	c_1v_1
Lower	c_3v_1	c_2v_1	c_3v_3	c_2v_3
Middle	c_4v_3	c_3v_2	c_4v_1	c_4v_2
	c_5v_2	c_4v_3	c_5v_3	c_5v_1
	c_1v_3	c_1v_2	c_1v_1	c_2v_2
	c_2v_1	c_3v_3	c_2v_3	c_3v_1
Bottom	c_4v_2	c_4v_1	c_3v_2	c_4v_3
	c_5v_3	c_5v_1	c_5v_2	c_5v_1

Figure 15.19 Design for a 5×3 structure in four sets of four blocks of four units.

blocking is suggested as the position of each dish within the pile (top, middle, bottom). A lower level of blocking is the individual petri dish within which the four quarters should be similar.

The treatments proposed are three varieties $\times$ five chemical treatments of the plants. The design requirements would be

(i) in each dish all three varieties should occur (one twice);
(ii) in each dish four of the five chemicals should occur;
(iii) in each bag all 15 combinations should occur at least once;
(iv) at each layer all 15 combinations should occur at least once.

A possible design is shown in Figure 15.19; the reader is invited to try to improve this design.

Example 15.15 In Chapter 7 we considered the problem of comparing nine treatments using a block structure of 18 blocks with two units per block. This problem actually arose in the context of an experiment in which the nine treatments had a 3×3 factorial structure and the design question required allocation of pairs of treatments to blocks to maximise information about important treatment contrasts. A comprehensive investigation of the design possibilities and the choice of an optimal design is a substantial piece of research and the problem is not discussed in depth here. Instead we compare two alternative designs which illustrate well the principles of this chapter.

With nine treatments there are 36 possible pairs of treatments, and the block structure requires that at most only 18 different pairs can be used in an experiment. Two possible experimental designs with clearly different philosophies might seem appropriate. Either

Table 15.16. *Efficiencies of main effect and interaction estimates from two designs*

	Main effect	Interaction
Figure 15.20(a)	37.5%	75.0%
Figure 15.20(b)	75.0%	37.5%

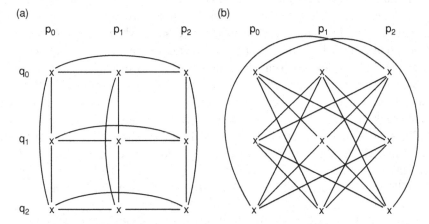

Figure 15.20 Two alternative designs for a 3×3 structure in 18 blocks of two units: (a) each block contains a common level for one of the factors, (b) each block pair of treatments has no common factor level.

(a) each block contains a pair of treatments with a common level for one of the factors, or
(b) each block contains a pair of treatments with no common level of either factor.

There are exactly 18 pairs of treatments of each type so that each experiment can contain 18 different blocks. Using a link to represent a pair of observations in a block, the two designs are represented in Figure 15.20(a) and (b). A simple question is which is the better design for estimating main effects. The reader is urged to answer before reading further.

The answer is that the efficiencies of main effect and interaction estimates, relative to the design with four randomised blocks of nine units each, are given in Table 15.16. Thus the second design gives twice as much information about main effects as about interactions, while the reverse is true for the first design. The reason why the second design is better for main effects is that in every block the levels of a factor are different and therefore each block contributes to the information on the main effect of the factor. In contrast, only half of the blocks contain information about the main effect of factor P, and the other half contain information about the main effect of Q. Because the main effects are more efficiently estimated in Figure 15.22(b), the interactions must be correspondingly less efficiently estimated.

15.6 The negative approach to confounding for two-level factors

In the earlier sections of this chapter the intuitive ideas of confounding, including partial confounding, were explored and the thesis was developed that designing appropriate experiments for many combinations of block size and factorial structure is straightforward. We now explore how mathematical theory can be developed to provide support for that thesis, not only by demonstrating why the methods of design construction used earlier should be effective, but also providing the means to construct designs when the size of the treatment set becomes so large that the intuitive approach becomes more difficult to organise.

Whereas the intuitive approach involves the construction of designs to allow all those effects which are regarded as important to be estimated without confounding, the classical, mathematical, approach to confounding is in terms of the effects which are confounded with blocks, and which therefore cannot be estimated. If, in an experiment with a number of blocks, certain effects are selected to be confounded with blocks, then the crucial mathematical theory identifies which other effects are also confounded, and hence which other effects are not confounded.

We shall illustrate this seemingly negative approach to confounding by considering a very simple example, which is almost invariably used in any presentation of the mathematical theory of confounding. Consider a 2^4 factorial structure with factors P, Q, R and S. Suppose it is initially decided that the experiment should be designed in blocks of eight units. Of all the possible treatment effects (four main effects, six two-factor interactions, four three-factor interactions and one four-factor interaction) it would seem reasonable to decide that the four-factor interaction is most easily ignored. Since the 15 treatment effects are orthogonal, if the treatment combinations are allocated so that the four-factor interaction is identical with the difference between blocks, then the other 14 effects will be orthogonal to blocks and therefore not confounded. The four-factor interaction is

$$
\begin{aligned}
PQRS \propto (p-1)(q-1)(r-1)(s-1) \\
= \{pqrs + pq + pr + ps + qr + qs + rs + (1)\} \\
- (pqr + pqs + prs + qrs + p + q + r + s),
\end{aligned}
$$

the familiar pattern of positive 'even' combinations and negative 'odd' combinations identified in Chapter 13. If in each replicate the even combinations are allocated to one block and the odd combinations to the other, then the PQRS effect is confounded with blocks in each replicate, and all other effects will be unconfounded. The design is shown in Figure 15.21(a), and it can quickly be verified that all other treatment effects are unconfounded, and that this is the same design as that in Figure 15.5(b), produced by the direct allocation approach.

Suppose, after further thought, it is decided that blocks of eight are too large, and that blocks of four must be used. The PQRS effect is already confounded. The most obvious candidate for a second confounded effect is one of the three-factor interactions. Suppose it is decided to confound PQR in addition to PQRS. Now PQR is also defined as the difference

(a)   Block

	I		II
	(1)		p
	pq		q
	pr		r
	ps		s
	qr		pqr
	qs		pqs
	rs		prs
	pqrs		qrs

(b)   Block

I_1	I_2	II_1	II_2
ps	pq	p	s
qs	pr	q	pqs
rs	qr	r	prs
pqrs	(1)	pqr	qrs

Figure 15.21 Dividing a $2 \times 2 \times 2 \times 2$ structure (a) into two blocks confounding PQRS, (b) into four blocks also confounding PQR.

between even and odd combinations:

$$PQR \propto (p-1)(q-1)(r-1)(s+1)$$
$$= (pqrs + pqr + ps + qs + rs + p + q + r)$$
$$- \{pqs + prs + qrs + pq + pr + qr + s + (1)\}.$$

If, from each of the blocks of Figure 15.21(a), two blocks are constructed, one containing even combinations wrt PQR and the other containing odd combinations wrt PQR, the four blocks shown in Figure 15.21(b) are obtained. However, since there are four blocks, there must be three contrasts between blocks and hence there should be three effects confounded with block comparisons (or possibly more than three effects not unconfounded). Inspection of the four blocks of Figure 15.21(b) reveals that the other confounded effect is the main effect of S, blocks I and IV having all the combinations with the upper level of S, and blocks II and III all the combinations with the lower level of S. Clearly a design in which the main effect of one factor is confounded is most unlikely to be suitable. It is necessary to devise a method of construction of designs, by selecting the effects to be confounded, which avoids confounding important effects.

The fundamental principle on which the theory of choosing confounded designs is based was established by Fisher. If, in an experiment for n factors, each with two levels, the 2^n treatment combinations are divided into 2^m blocks each having 2^{n-m} treatment combinations, then the subdivision into blocks can be conceived sequentially. The 2^n combinations are first split into two halves with one effect, X, confounded between the two halves; then the halves are split into quarters with a further effect, Y, confounded; then the quarters into eighths with a further effect, Z, confounded, and so on, until after m splits there are 2^m blocks each with 2^{n-m} combinations. In this procedure, m effects are selected to be confounded in the m splits. Since there are 2^m blocks, there are $(2^m - 1)$ block contrasts, and the same number of confounded effects. For a given set of chosen confounded effects, the complete set of confounded effects is determined by the following rule.

15.6.1 *Fisher's multiple confounding rule*

If, in a replicate of a 2^n factorial structure, two effects, X and Y, are both confounded, then the generalised interaction, $Z(= XY)$, is also confounded, where the set of factors involved in the interaction, Z, includes all factors involved in either X or Y, but not in both.

Thus, if X and Y are the interactions PQRS and PQR, then $Z(= XY)$ will be S. If X and Y are PQS and PRTU, then $Z(= XY)$ will be QRSTU. The validity of the rule may be demonstrated as follows: if the effect X is confounded with blocks, then the blocks are split into two groups, $\{X_1\}$ and $\{X_2\}$, such that the blocks in $\{X_1\}$ contain only treatment combinations which are even for X, i.e. with an even number of factors from X having their upper level. Correspondingly, the blocks in $\{X_2\}$ contain only treatment combinations odd for X, i.e. with an odd number of factors from X having their upper level. If Y is confounded, there is a similar division of blocks into two groups $\{Y_1\}$ which has even combinations, and $\{Y_2\}$, which has odd combinations.

Consider the four groups caused as a result of the two divisions and, in particular, the blocks which occur in both $\{X_1\}$ and $\{Y_1\}$. All treatment combinations in these blocks are even for X and even for Y, i.e. for each treatment combination, the number of factors from X at their upper level is even, and the number of factors from Y at their upper level is even. It follows that, for these combinations, the number of upper level factors in X or Y, but not both, must also be even, since it is equal to the sum of two even numbers minus twice the number of common upper level factors. Hence, all treatment combinations in both $\{X_1\}$ and $\{Y_1\}$ are even for Z.

Similarly, treatment combinations in $\{X_2\}$ and $\{Y_2\}$ are even for Z, and those in $\{X_1\}$ and $\{Y_2\}$ or in $\{X_2\}$ and $\{Y_1\}$ are odd for Z. To express the result more formally we use two mathematical concepts. First the idea of modulo arithmetic, where the expression

$$a = b(c)$$

means that a is the remainder when b is divided by c. The numbers a, b and c are integers and clearly a is an integer between 0 and $(c - 1)$. Thus, for example,

$$7(2) = 1, 10(3) = 1, 10(5) = 0.$$

The numbers, b, such that $b(2) = 0$ are the even numbers and those for which $b(2) = 1$ are the odd numbers.

Secondly we use the set theory concepts of union and intersection. For two sets, A and B the union $A \cup B$ comprises all elements which are in A or B or both: the intersection $A \cap B$ comprises all elements which are in both A and B. Hence the set of elements which are in exactly one of A and B is

$$A \cup B - A \cap B.$$

Formally, if two effects X and Y are to be confounded, and if for any treatment combination, $U(X)$ is the number of factors involved in the effect X which have their upper level in the treatment combination, then

for combinations in $\{X_1\}$ $\quad$ $U(X) = 0(2)$ $\quad$ (= even),
for combinations in $\{X_2\}$ $\quad$ $U(X) = 1(2)$ $\quad$ (= odd)
similarly, in $\{Y_1\}$ $\quad\quad\quad$ $U(Y) = 0(2)$
in $\{Y_2\}$ $\quad\quad\quad$ $U(Y) = 1(2)$.

But
$$U(Z) = U(X \cup Y) - U(X \cap Y)$$
and
$$U(X \cup Y) = U(X) + U(Y) - U(X \cap Y),$$
hence
$$U(Z) = U(X) + U(Y) - 2U(X \cap Y).$$

Since $2U(X \cap Y) = 0(2)$ for any value of $U(X \cap Y)$ it follows that

in $\{X_1 Y_1\}$ $\quad\quad\quad\quad\quad\quad$ $U(Z) = 0(2)$.

Similar arrangements lead to the corresponding results:

in $\{X_1 Y_2\}$ $\quad\quad\quad\quad\quad\quad$ $U(Z) = 1(2)$,
in $\{X_2 Y_1\}$ $\quad\quad\quad\quad\quad\quad$ $U(Z) = 1(2)$,
in $\{X_2 Y_2\}$ $\quad\quad\quad\quad\quad\quad$ $U(Z) = 0(2)$.

Hence, the interaction effect Z is such that the blocks in $Z_1 = \{X_1, Y_1\} + \{X_2, Y_2\}$ are 'even' for Z, and the blocks in $Z_2 = \{X_1, Y_2\} + \{X_2, Y_1\}$, are 'odd' for Z.

15.6.2 *Identifying the confounded effects*

Fisher's rule enables us to identify immediately all confounded effects when more than one effect is confounded. If three effects, W, X, Y, are chosen initially to be confounded, then the set of confounded effects will be W, X, Y, WX, WY, XY and WXY. The pattern of the limitations on confounding is quickly found by considering examples:

(i) 2^4 in four blocks of four:

$\quad\quad\quad\quad$ confound PQRS and PQR $\quad\quad$ $\Rightarrow$ S confounded;
$\quad\quad\quad\quad$ confound PQR and PQS $\quad\quad\;$ $\Rightarrow$ RS confounded.

This latter design was found in Figure 15.5(c).

(ii) 2^5 in four blocks of eight:

$\quad\quad\quad\quad$ confound PQRST and PQRS $\quad$ $\Rightarrow$ T confounded,
$\quad\quad\quad\quad$ confound PQRS and PQRT $\quad\;\;$ $\Rightarrow$ ST confounded,
$\quad\quad\quad\quad$ confound PQRS and PRT $\quad\quad\;$ $\Rightarrow$ QST confounded.

(See Figure 15.5(d).)

(iii) 2^5 in eight blocks of four:

$\quad\quad\quad$ confound PQRST, PQRS, QRST $\;\;$ $\Rightarrow$ T, P, PT, QRS also confounded,
$\quad\quad\quad$ confound PQRS, PRT, PQR $\quad\quad\;\;$ $\Rightarrow$ QST, S, QT, PRST also confounded,
$\quad\quad\quad$ confound PQRS, PRT, RST $\quad\quad\;\;$ $\Rightarrow$ QST, PQT, PS, QR also confounded.

(iv) 2^6 in eight blocks of eight:

confound PQRSTU, PQR, PST $\Rightarrow$ STU, QRU, QRST, PU also confounded,
confound PQRS, PQTU, PRT $\Rightarrow$ RSTU, QST, QRU, PSU also confounded.

In each of these examples, we have deliberately included in the possible alternatives a confounding system including the highest possible factor interaction, and also the optimal confounding system in the sense of avoiding confounding main effects as a first priority, and two-factor interactions as a second priority. It can be seen that confounding the highest possible interaction is never optimal.

15.6.3 Choosing the set of confounded effects

The guidelines for optimal systems of confounding three effects can be simply deduced by observing that no factor can be included in all three effects X, Y and Z $= $XY. Therefore, if three effects are to be confounded, each factor will be included in at most two of the confounded effects. The complete omission of any factor from all three confounded effects must involve unnecessarily-low-order interactions. Hence, in the set of three confounded effects, all factors must appear twice, and the most efficient confounding system will require confounded effects involving n_1, n_2 and n_3 factors, respectively, where n_1, n_2 and n_3 are the most nearly equal partition of $2n$ into three components.

More generally, for a 2^n set of treatment combinations, in 2^m blocks, there will be $2^m - 1$ confounded effects, and each factor should occur 2^{m-1} times. The optimal confounding system should be sought by considering the most nearly equal partition of $n(2^{m-1})$ into $2^m - 1$ components. Thus, for $n = 6$ and $m = 3$, we have to divide 24 into seven components, and the best conceivable solution would be (4, 4, 4, 3, 3, 3, 3), as was actually achieved in example (iv).

It may be helpful to think of an 'average' number of factors in the set of confounded interactions which will be

$$\frac{n2^{m-1}}{2^m - 1},$$

which, if m is 3 or more, is approximately $n/2$. Hence, if a replicate is divided into a large number of blocks, the confounding system should be such that most of the confounded effects are interactions of about half the factors in the experiment. If higher-order interactions are confounded, this necessitates confounding also interactions of fewer factors. For $m = 1$, of course, the higher-order interaction can be confounded. For $m = 2$, the 'average' number of factors in the confounded effects will be $2n/3$, and it is always possible to achieve the most nearly equal partition of $2n$ into three components.

15.6.4 Identifying the treatment combinations in each block

The selection of the set of confounded effects defines the design uniquely, but there remains the task of identifying the treatment combinations in each block. Fisher's rule for determining the set of confounded effects implies the pattern of 'even' and 'odd' treatment combinations in the different blocks. There must be one block, known as the principal block, in which all

			Block				
I	II	III	IV	V	VI	VII	VIII
Principal	×p	×q	×pq	×r	×pr	×qr	×pqr
(1)	p	q	pq	r	pr	qr	pqr
pst	st	pqst	qst	prst	rst	pqrst	qrst
qsu	pqsu	su	psu	qrsu	pqrsu	rsu	prstu
rstu	prstu	qrstu	pqrstu	stu	pstu	qstu	pqstu
pqtu	qtu	ptu	tu	pqrtu	qrtu	prtu	rtu
pru	ru	pqru	qru	pu	u	pqu	qu
qrt	pqrt	rt	prt	qt	pqt	t	pt
pqrs	qrs	prs	rs	pqs	qs	ps	s

Figure 15.22 A 2^6 design in eight blocks of eight units confounding PQRS, PQTU, PRT.

treatment combinations are even for all the confounded effects. That is for every treatment combination in the principal block and for each confounded effect the number of factors involved in the confounded effect having an upper level in the treatment combination will be even:

$$U(X) = 0(2),$$
$$U(Y) = 0(2),$$
$$\vdots$$

Each treatment combination is defined by the factors for which the upper level is used in the combination. If two treatment combinations occur in the principal block, then so also will the product combination defined by the factors having an upper level for either of the combinations, but not for both.

Any combination not in the principal block can be used as the seed of another block. The new block will contain the new treatment combination and all product combinations formed from the principal block combinations and the new combination. Since the principal block combinations are even for all confounded effects, all the other combinations in a block with this new combination will have the same 'even' or 'odd' class as the new combination for each confounded effect.

Example 15.16 2^6 in eight blocks of eight. The seven confounded effects will be interactions such that the total of the number of factors per interaction will be 24, and a possible set of confounded effects is

PQRS, PQTU, PRT, RSTU, QST, QRU, PSU.

The combinations which are even with respect to each of these seven effects are those which are even for the first three (the others are generalised interactions of the first three) and may be identified as pst, qsu, rstu and their products pqtu, pru, qrt and pqrs, together with the combination of lowest levels of all factors, (1).

The other blocks are constructed using any combination not already included and forming products with the principal block combinations. The resulting design is shown in Figure 15.22. Note that for any non-confounded three-factor interaction effect, such as PQR, all eight combinations ((1), p, q, r, pq, pr, qr and pqr) will occur in each of the eight blocks.

Table 15.17. *Analysis of variance structure for 2^6 treatments in blocks of 8*

Source	df
Blocks	7
Main effects	6
Two-factor interactions	15
Three-factor interactions (except for PRT, QST, PSU, QRU)	16
Error	19
Total	63

As for all other confounded experiments where effects are either completely confounded or not confounded, the analysis of variance includes SS for all important effects, except those which are confounded. For this example, the structure of the analysis of variance is as shown in Table 15.17.

15.7 Confounding theory for other factorial structures

All theory for confounding in factorial structures other than the 2^n involves either adapting the ideas of 2^n confounding or generalising those ideas in a mathematically neat, but practically unattractive, fashion. We shall discuss each approach briefly.

15.7.1 *Using dummy two-level factors to represent four-level factors*

The adaptation of the 2^n confounding ideas to factors with other numbers of levels can be achieved by defining dummy factors each with two levels, such that the combinations of levels of these dummy factors correspond to the actual levels of the real factors, although the dummy factors and their levels have no sensible practical interpretation. A four-level factor, P, may be represented by two two-level dummy factors, P' and P'', as follows:

$$
\begin{aligned}
p_0' + p_0'' &\Rightarrow p_0, \\
p_0' + p_1'' &\Rightarrow p_1, \\
p_1' + p_0'' &\Rightarrow p_2, \\
p_1' + p_1'' &\Rightarrow p_3.
\end{aligned}
\tag{15.1}
$$

The design of a 4^n experiment with confounding in blocks is achieved by first determining the set of confounded effects, and the allocation of treatment combination to blocks, in terms of the dummy factors, and then translating the combinations of levels of dummy factors to levels of the real factors through the definition (15.1). The choice of which interactions of the dummy factors to confound must, of course, be primarily based on the real factors and must involve consideration of the translation from the dummy factors to the genuine factors. The main effects and interactions of P′ and P″ are all main effects of the genuine factor P since

$$P' \propto (p_3 + p_2 - p_1 - p_0),$$
$$P'' \propto (p_3 - p_2 + p_1 - p_0),$$
$$P'P'' \propto (p_3 - p_2 - p_1 + p_0).$$

Hence the interaction of dummy factors P′P″ constitutes a main effect comparison of P, which for a quantitative factor we can recognise as the quadratic effect. However, the interaction of dummy factors P′Q″ is an interaction comparison for the two factors P and Q.

Example 15.17 To demonstrate the necessary steps consider a complete 4^3 experiment to be designed in eight blocks of eight units each. The corresponding dummy factor design has six two-level dummy factors in eight blocks of eight, for which an appropriate design has been constructed in the previous example. Let the four-level factors be A, B, C and the corresponding dummy factors, A′, A″, B′, B″, C′, C″. In the previous example, the set of confounded interactions was

(PQRS, PQTU, PRT, RSTU, QST, QRU, PSU).

We would like to allocate (A′, A″, B′, B″, C′, C″) to (P, Q, R, S, T, U) so that each of the seven confounded effects involves a dummy factor from each of the three genuine factors. This is not possible and the best that can be achieved is to have four of the confounded effects involving a dummy factor from each genuine factor, and the other three involving dummy factors from only two genuine factors. A typical solution is P = A′, Q = B″, R = B′, S = A″, T = C′, U = C″, giving confounded effects (A′A″B′B″, A′B″C′C″, A′B′C′, A″B′C′C″, A″B″C′, B′B″C″, A′A″C″). The confounded effects may be expressed in terms of the levels of the genuine factors, and two examples are

$$A'A''B'B'' = (a_3 - a_2 - a_1 + a_0)(b_3 - b_2 - b_1 + b_0)(c_3 + c_2 + c_1 + c_0),$$
$$A'B''C'C'' = (a_3 + a_2 - a_1 - a_0)(b_3 - b_2 + b_1 - b_0)(c_3 - c_2 - c_1 + c_0).$$

The treatment combinations are translated into levels of A, B, C by

$$ps \Rightarrow a_3, \quad p \Rightarrow a_2, \quad s \Rightarrow a_1, \quad (1) \Rightarrow a_0;$$
$$qr \Rightarrow b_3, \quad r \Rightarrow b_2, \quad q \Rightarrow b_1, \quad (1) \Rightarrow b_0;$$
$$tu \Rightarrow c_3, \quad t \Rightarrow c_2, \quad u \Rightarrow c_1, \quad (1) \Rightarrow c_0.$$

The resulting design is shown in Figure 15.23, with the combinations in the same order in each block as those in Figure 15.22.

In the analysis, the SS for each two-factor interaction are derived from only eight of the usual nine df, because of the one confounded component of each two-factor interaction. The

			Block				
I	II	III	IV	V	VI	VII	VIII
$a_0b_0c_0$	$a_2b_0c_0$	$a_0b_1c_0$	$a_2b_1c_0$	$a_0b_2c_0$	$a_2b_2c_0$	$a_0b_3c_0$	$a_2b_3c_0$
$a_3b_0c_2$	$a_1b_0c_2$	$a_3b_1c_2$	$a_1b_1c_2$	$a_3b_2c_2$	$a_1b_2c_2$	$a_3b_3c_2$	$a_1b_3c_2$
$a_1b_1c_1$	$a_3b_1c_1$	$a_1b_0c_1$	$a_3b_0c_1$	$a_2b_3c_1$	$a_3b_3c_1$	$a_1b_2c_1$	$a_3b_2c_1$
$a_1b_2c_3$	$a_3b_2c_3$	$a_1b_3c_3$	$a_3b_3c_3$	$a_2b_0c_3$	$a_3b_0c_3$	$a_1b_1c_3$	$a_3b_1c_3$
$a_2b_1c_3$	$a_0b_1c_3$	$a_2b_0c_3$	$a_0b_0c_3$	$a_1b_3c_3$	$a_0b_3c_3$	$a_2b_2c_3$	$a_0b_2c_3$
$a_2b_2c_2$	$a_0b_2c_2$	$a_2b_3c_2$	$a_0b_3c_2$	$a_1b_0c_2$	$a_0b_0c_2$	$a_2b_1c_2$	$a_0b_1c_2$
$a_0b_3c_1$	$a_2b_3c_1$	$a_0b_2c_1$	$a_2b_2c_1$	$a_0b_1c_1$	$a_2b_1c_1$	$a_0b_0c_1$	$a_2b_0c_1$
$a_3b_3c_0$	$a_1b_3c_0$	$a_3b_2c_0$	$a_1b_2c_0$	$a_3b_1c_0$	$a_1b_1c_0$	$a_3b_0c_0$	$a_1b_0c_0$

Figure 15.23 A $4 \times 4 \times 4$ design in eight blocks of eight units.

Table 15.18. *Analysis of variance structure for 4^3 treatments in blocks of 8*

Source	df
Blocks	7
A	3
B	3
C	3
AB	8
AC	8
BC	8
Error	23
Total	63

actual calculation of SS can be done by using a general analysis program which recognises the loss, through confounding, of the ninth component. Recognising that the confounded effect of the AB interaction is the quadratic $\times$ quadratic would suggest that calculating the SS for other particular components of that effect such as linear $\times$ linear and linear $\times$ quadratic would be useful. More generally, the fact that only one out of the nine df for each two-factor interaction is confounded enables the more important contrasts of the interaction to be arranged to be unconfounded. The structure of the analysis of variance is shown in Table 15.18.

15.7.2 *Using dummy factors to represent three-level factors*

To use the 2^n confounding systems for factors with three levels, it is necessary to accept unequal replication of the three levels. The presumption that this is a major disadvantage reflects the tendency to regard equal replication as inevitable without questioning whether all effects are equally important. Sometimes one level of a three-level factor should be regarded as more important than the other two and should be replicated more frequently. Alternatively, the level to have increased replication has to be arbitrarily chosen.

As with the four-level case, two dummy factors, P' and P'', with two levels are used to represent the levels of the genuine factor, **P**, as follows:

$$p_0' + p_0'' \Rightarrow p_0,$$
$$p_0' + p_1'' \Rightarrow p_1,$$
$$p_1' + p_0'' \Rightarrow p_1,$$
$$p_1' + p_1'' \Rightarrow p_2.$$

The choice of the level (p_1) to be duplicated is for the experimenter.

The translation from dummy factor levels to real factor levels for three-level factors has two other important differences from that for four-level factors. There is a symmetry of the two dummy factors in their relationship to the genuine factor, and the interpretation of the main effects and interaction of the dummy factors is directly relevant to the main effects of the genuine factor. The main effect of P' is

$$P' \propto (p_1' - p_0')(p_1'' + p_0'') \equiv p_2 - p_1 + p_1 - p_0 = (p_2 - p_0).$$

Similarly, P'' $\propto (p_2 - p_0)$, while the interaction is

$$P'P'' \propto (p_1' - p_0')(p_1'' - p_0'') = (p_2 - p_1 - p_1 + p_0) = (p_2 - 2p_1 + p_0).$$

Both P' and P'' provide estimates of the $(p_2 - p_0)$ main effect of P which, for quantitative levels, is the linear effect of P. The interaction P'P'' provides an estimate of the quadratic effect of P. More generally, as noted in Chapter 13, the pair of contrasts (i) between two levels, and (ii) between that pair and the third is, practically, almost the only useful form of contrasts for three levels.

The obvious advantage of having two estimates of the linear main effect of P is that it is possible to confound one of the dummy factor main effects and still have an estimate of the linear main effect of P from the other dummy factor main effect. This advantage is tempered by the reduced precision of each dummy factor effect as an estimator of the linear main effect of P. In fact, the duplication of the middle level changes the entire pattern of precision.

To examine this change of precision pattern, consider two sets of 12 observations. The first consists of four replicates of a single three-level factor; the second consists of three replicates of the 2×2 dummy factors, equivalent to three replicates of p_0 and p_2, and six replicates of p_1. Consider the estimates of the linear and quadratic effects of P:

(a) P_L estimated by $(\bar{p}_2 - \bar{p}_0)$ variance $2\sigma^2/4 = \sigma^2/2$, P_Q estimated by $1/2(\bar{p}_2 - 2\bar{p}_1 + \bar{p}_0)$ variance $1/4(6\sigma^2/4) = 3\sigma^2/8$.

(b) P_L estimated by $2(\bar{p}_1' - \bar{p}_0')$ variance $4(2\sigma^2/6) = 4\sigma^2/3$, also estimated by $2(\bar{p}_1'' - \bar{p}_0'')$ variance $4(2\sigma/6) = 4\sigma^2/3$, P_Q estimated by $1/2(\bar{p}_1' - \bar{p}_0')(\bar{p}_1'' - \bar{p}_0'')$ variance $1/4(4\sigma^2/3) = \sigma^2/3$.

If both dummy factor main effects are used to estimate P_L, the variance of the combined estimate is $2\sigma^2/3$, compared with $\sigma^2/2$ for the equal replication design, an efficiency of $3/4$. Correspondingly, the estimation of P_Q is more efficient with unequal replication by a factor of $9/8$. If only one of the dummy factor main effects is used to estimate P_Q, then the efficiency relative to the equal replication design is $3/8$. Although the estimation of the linear main effect has been discussed in terms of using one, or both, dummy main effect estimates, in practice of course we would not calculate the two separate estimates but would calculate, directly, a single estimate. Thus, the estimation of P_L is relatively better in the equal replicate design, but the estimation of P_Q is relatively better in the unequal replicate design. This obviously has some relevance to the subsequent consideration of alternative forms of confounding system.

The benefit of retaining some information about the linear effect, even if one dummy factor main effect is confounded, is not likely to be a usable advantage for main effects where the linear effect is already less precise than the quadratic because of the unequal replication. But, when interactions of the real factors are used as confounded effects the loss of one estimate of a linear × linear interaction effect can be tolerated because there are three other estimates. The basic design of the two previous examples for a 2^6 experiment confounded in eight blocks of eight units can be used again to illustrate the use of dummy factors for confounding 3^n experiments.

Example 15.18 Consider the 3^3 design using dummy factors A′, A″, B′, B″, C, C″. The six dummy factors must be equated to the six factors of Example 15.1 so that the confounded effects

$$PQRS, PQTU, PRT, RSTU, QST, QRU, PSU$$

represent effects where loss is not unacceptable. From the similar exercise in Example 15.17, some of the confounded effects must be components of two-factor interactions. A sensible solution is

$$P = A', Q = B', R = C', S = A'', T = C'', U = B'',$$

giving confounded effects

$$A'B'C'A'', A'B'C''B'', A'C'C'', C'A''C''B'', B'A''C'', B'C'B'', A'A''B''.$$

The third, sixth and seventh confounded effects would give estimates of the linear A × quadratic C, linear C × quadratic B, and linear B × quadratic A effects, but in each case there is a second estimator of the same effect (A″C′C″, B′B″C″ and A′A″ B′) which is not confounded.

The allocation of treatment combinations to blocks is determined by translation of the dummy factor levels to the real three-level factor levels:

$$
\begin{array}{llll}
ps \Rightarrow a_2, & p \Rightarrow a_1, & s \Rightarrow a_1, & (1) \Rightarrow a_0; \\
qu \Rightarrow b_2, & q \Rightarrow b_1, & u \Rightarrow b_1, & (1) \Rightarrow b_0; \\
rt \Rightarrow c_2, & r \Rightarrow c_1, & t \Rightarrow c_1, & (1) \Rightarrow c_0.
\end{array}
$$

The resulting design is shown in Figure 15.24. All components of all main effects and interactions of the real factors can be estimated, but the precision of different estimates will vary both because of the unequal replication and because of the loss of some dummy factor effects through confounding. The structure of the analysis of variance is inevitably very similar to that for an experiment without confounding (see Table 15.19).

Table 15.19. *Analysis of variance structure for 3^3 treatments in blocks of 8*

Source	df
Blocks	7
Three main effects	6
Three two-factor interactions	12
Three-factor interactions	8
Error	40
Total	63

Table 15.20. *Standard errors for confounded design and for full block design*

Effect	SE for confounded design	SE for full block design $(\times \sqrt{(27/64)})$
A linear	$\sigma\sqrt{(1/8)} = 0.35\sigma$	$\sigma\sqrt{(3/32)} = 0.31\sigma$
A quadratic	$\sigma/4 = 0.25\sigma$	$\sigma\sqrt{(9/128)} = 0.27\sigma$
AB lin × lin	$\sigma = 1.00\sigma$	$\sigma(3/4) = 0.75\sigma$
AB lin × quad	$\sigma\sqrt{(1/2)} = 0.71\sigma$	$\sigma\sqrt{(27/64)} = 0.65\sigma$
AB quad × lin	$\sigma = 1.00\sigma$	$\sigma\sqrt{(27/64)} = 0.65\sigma$

				Block			
I	II	III	IV	V	VI	VII	VIII
$a_0b_0c_0$	$a_1b_0c_0$	$a_0b_1c_0$	$a_1b_1c_0$	$a_0b_0c_1$	$a_1b_0c_1$	$a_0b_1c_1$	$a_1b_1c_1$
$a_2b_0c_1$	$a_1b_0c_1$	$a_2b_1c_1$	$a_1b_1c_1$	$a_2b_0c_2$	$a_1b_0c_2$	$a_2b_1c_2$	$a_1b_1c_2$
$a_1b_2c_0$	$a_2b_2c_0$	$a_1b_1c_0$	$a_2b_1c_0$	$a_1b_2c_1$	$a_2b_2c_1$	$a_1b_1c_1$	$a_2b_1c_1$
$a_1b_1c_2$	$a_2b_1c_2$	$a_1b_2c_2$	$a_2b_2c_2$	$a_1b_1c_1$	$a_2b_1c_1$	$a_1b_2c_1$	$a_2b_2c_1$
$a_1b_2c_1$	$a_0b_2c_1$	$a_1b_1c_1$	$a_0b_1c_1$	$a_1b_2c_2$	$a_0b_2c_2$	$a_1b_1c_2$	$a_0b_1c_2$
$a_1b_1c_1$	$a_0b_1c_1$	$a_1b_2c_1$	$a_0b_2c_1$	$a_1b_1c_0$	$a_0b_1c_0$	$a_1b_2c_0$	$a_0b_2c_0$
$a_0b_1c_2$	$a_1b_1c_2$	$a_0b_0c_2$	$a_1b_0c_2$	$a_0b_1c_1$	$a_1b_1c_1$	$a_0b_0c_1$	$a_1b_0c_1$
$a_2b_1c_1$	$a_1b_1c_1$	$a_2b_0c_1$	$a_1b_0c_1$	$a_2b_1c_0$	$a_1b_1c_0$	$a_2b_0c_0$	$a_1b_0c_0$

Figure 15.24 A $3 \times 3 \times 3$ design in eight blocks of eight units.

The SEs for the various linear and quadratic main effects and two-factor interactions can be obtained simply from a general analysis program, and are given in Table 15.20. For comparison with the unconfounded, equal replicate design, the SE for the same effects in the latter design have been calculated and are also given in Table 15.20, but with a multiplying factor of $\sqrt{(27/64)}$ to adjust for the different numbers of units used in the two designs. The comparison of the two designs is difficult because not only are the relative precisions of the two experiments different for different comparisons, but the value of σ^2 will be different for the two designs, since the smaller block size of 8 should give a much smaller σ^2 than the larger block size of 27. A choice between the two designs must therefore involve not only the consideration of the relative importance of different effects, but also the benefit of using smaller blocks.

The use of dummy factors to represent three-level factors can also be extended to factorial structures with a mixture of two- and three-level factors.

For 3^n designs, there are also other theoretical approaches to constructing designs which do not require the use of duplicated levels and these are discussed in the next section.

15.7.3 Classical confounding for 3^n

The classical approach to confounding for 3^n factorial structures extends the ideas of 2^n confounding in a different manner from the dummy factor approach. The basis of the 2^n confounding systems was the division of treatment combinations into two sets, characterised, with reference to a particular effect, as 'even' and 'odd'. More formally, the treatment combinations in the two sets, X_0 and X_1, are such that $U(X)$, the number of factors in effect X having their upper level in the treatment combination, must be

$$U(X) = 0(2) \text{ for all treatment combinations, X, in } X_0$$

and

$$U(X) = 1(2) \text{ for all treatment combinations in } X_1.$$

With three levels, any equal subdivision of treatment combinations will be into three sets, X_0, X_1 and X_2. The characterisation of the three sets must now involve all three levels and, instead of a simple count of factors with upper levels in treatment combinations, $U(X)$ is defined as a weighted sum of levels for the factors included in the effect, X. The combinations in the three treatment sets, X_0, X_1 and X_2, will be those for which $U(X) = 0(3)$, $1(3)$ and $2(3)$, respectively.

The actual subdivision of combinations into three sets is determined by the form of the weighted sum of levels $U(X)$. The general form is

$$\alpha i + \beta j + \gamma k + \cdots,$$

where $i, j, k, \ldots$ represent the levels of the different factors for a treatment combination $(p_i q_j r_k \ldots)$ and $\alpha, \beta, \gamma, \ldots$ represent the weights.

Consider the various possible subdivisions for two factors. There are nine treatment combinations,

$$p_0 q_0, p_0 q_1, p_0 q_2, p_1 q_0, p_1 q_1, p_1 q_2, p_2 q_0, p_2 q_1, p_2 q_2.$$

(1) If $\alpha = 1, \beta = 0$, $U(X) = i$.

Treatment combinations are allocated to sets according as

$$i = 0(3), i = 1(3) \text{ or } i = 2(3).$$

In other words,

set 1 includes all combinations with p_0,

set 2 includes all combinations with p_1,

set 3 includes all combinations with p_2.

(2) If $\alpha = 2, \beta = 0, \quad U(X) = 2i$.

This levels to exactly the same allocation except that set 2 has the p_2 combinations and set 3 the p_1 combinations. As a basis for a confounded design the difference is immaterial.

(3) If $\alpha = 0, \beta = 1, \quad U(X) = j$.

Now the allocations are based on q_0 (set 1), q_1 (set 2), q_2 (set 3).

(4) If $\alpha = 0, \beta = 2, \quad U(X) = 2j$.

Again this provides the same allocation as (3).

(5) If $\alpha = 1, \beta = 1, \quad U(X) = i + j$.

Treatment combinations are allocated to sets based on the sum of i and j. This gives

set 1	$p_0 q_0$,	$p_1 q_2$,	$p_2 q_1$;
set 2	$p_0 q_1$,	$p_1 q_0$,	$p_2 q_2$;
set 3	$p_0 q_2$,	$p_1 q_1$,	$p_2 q_0$.

(6) If $\alpha = 1, \beta = 2, U(X) = i + 2j$.

Now it is $i + 2j$ which determines the allocation

set 1	$p_0 q_0$,	$p_1 q_1$,	$p_2 q_2$,
set 2	$p_0 q_2$,	$p_1 q_0$,	$p_2 q_1$,
set 3	$p_0 q_1$,	$p_1 q_2$,	$p_2 q_0$.

(7) If $\alpha = 2, \beta = 1, U(X) = 2i + j$.

This provides the same allocation as (6).

(8) If $\alpha = 2, \beta = 2, U(X) = 2i + 2j = 2(i + j)$.

This, perhaps more obviously, provides the same allocation as (5).

The four alternative patterns of allocation can be interpreted in terms of the treatment effects as follows:

(1) or (2) confounds the main effect of P between sets,
(3) or (4) confounds the main effect of Q between sets.

In (5), (6), (7), (8) main effects are clearly not confounded and hence the confounded components are part of the PQ interaction. It should also be noted that the allocations (5) and (6) are precisely the two designs constructed in Section 15.2 (Figure 15.6).

To distinguish the four subdivisions which lead to part of the PQ interaction effect being confounded, the subdivisions are identified in the form $P^\alpha Q^\beta$. Thus subdivision (5) is $P^1 Q^1$ or PQ, and (6) is $P^1 Q^2$ or PQ^2. As already noted, (7), which is $P^2 Q^1$ or $P^2 Q$, gives the same allocation as PQ^2 (6), and (8), which is $P^2 Q^2$, is identical with PQ (5). Comparisons between totals for the three sets for any subdivision are contrasts for the corresponding effect. For either PQ or PQ^2 the comparisons between totals are orthogonal to main effect comparisons for P and for Q and must therefore be interaction effect contrasts. The comparisons between $(PQ)_0$, $(PQ)_1$, and $(PQ)_2$ are thus interaction effect contrasts (which we have met previously as contrasts between J_1, J_2, J_3 in Section 15.4). Similarly, comparisons between $(PQ^2)_0$, $(PQ^2)_1$ and $(PQ^2)_2$ are interaction effect contrasts (equivalent to those between I_1, I_2, I_3 in Section 15.4).

Although the contrasts between the three sets (I_1, I_2, I_3) or (J_1, J_2, J_3) are not directly practically meaningful, they can, as noted in Section 15.4, be used to estimate effects of practical interest. More importantly, they provide a basis for constructing confounding systems with several effects confounded, in the same way that the 'even' and 'odd' division did for 2^n treatment structures. The general statement providing for the implications of multiple confounding is exactly the same.

15.7.4 Fisher's multiple confounding rule

If, in a replicate of a 3^n factorial, two effects X and Y are both confounded, then the generalised interaction, $Z = XY$, is also confounded. The generalised interaction of two effects

$$X = P^\alpha Q^\beta R^\gamma \ldots \text{ and } Y = P^{\alpha'} Q^{\beta'} R^{\gamma'} \ldots$$

is

$$Z = P^{\alpha''} Q^{\beta''} R^{\gamma''} \ldots,$$

where

$$\alpha'' = (\alpha + \alpha')(3),$$
$$\beta'' = (\beta + \beta')(3).$$

The argument for this result is essentially the same as that for the earlier result for two-level factors, and we shall not give any details of the proof.

We should also note that since the effects X and XX or X^2 define the same division into three sets of treatment combinations, the simultaneous confounding of X (and X^2) and Y (and Y^2) implies the additional confounding not only of $Z = XY$, but also of $W = X^2Y$, $V = XY^2$ and $U = X^2Y^2$. Since confounding two effects, X and Y, implies nine blocks, and each confounded effect corresponds to two df, it is inevitable that the deliberate confounding of two effects will imply a total of eight confounded effects.

15.7.5 Practical construction of designs

The development of suitable systems of confounded effects is rather easier for 3^n experiments than for 2^n, essentially because the confounded component interactions comprise more than a single effect. However, just as in 2^n confounding, when we confound more than one effect, we should expect to be unable to restrict the confounded effects to the highest possible order interactions. An example shows this more clearly:

Example 15.19 Four factors, 81 combinations in nine blocks of nine units each. If a pair of four-factor interaction components are confounded this implies the additional confounding of a main effect or two-factor interaction. If PQRS and PQ^2RS^2 are confounded, then so are

$$(PQRS)(PQ^2RS^2) = P^2R^2 = PR$$

and

$$(PQRS)(PQ^2RS^2)^2 = Q^2S^2 = QS.$$

If two three-factor interaction component effects are confounded, then the other effects confounded can also be three-factor interaction component effects. If PQR and PR^2S are chosen to be confounded, then the other confounded effects are

$$(PQR)(PR^2S) = P^2QS \equiv PQ^2S^2$$

and

$$(PQR)(PR^2S)^2 = QR^2S^2.$$

This latter example illustrates the benefit of three-level factors instead of two-level factors for confounding. With four two-level factors, confounding two three-factor interactions implies that a two-factor interaction must be confounded. With three-level factors the existence of both PQR and P^2QR allows sufficient extra alternatives that we can avoid confounding any two-factor interaction.

The definition of the confounding system provides a unique design. The actual composition of treatment combinations in blocks is determined in exactly the same manner as for 2^n confounded designs. First, construct the principal block containing only treatment combinations in the first set of the three sets from each confounded effect. The required combinations $(p_i q_j r_k \ldots)$ satisfy

$$\alpha i + \beta j + \cdots = 0(3)$$
$$\alpha' i + \beta' j + \cdots = 0(3)$$
$$\vdots$$

for as many independent confounded effects as are defined. For example (2), the combinations would satisfy

$$i + j + k \equiv 0(3)$$

and

$$i + 2k + l \equiv 0(3).$$

Since PR is not confounded, all (i, k) combinations must occur in each block. For each (i, k) combination, there will be unique values of j and l. The principal block will therefore be

$$
\begin{array}{ccc}
p_0 q_0 r_0 s_0 & p_0 q_2 r_1 s_1 & p_0 q_1 r_2 s_2 \\
p_1 q_2 r_0 s_2 & p_1 q_1 r_1 s_0 & p_1 q_0 r_2 s_1 \\
p_2 q_1 r_0 s_1 & p_2 q_0 r_1 s_2 & p_2 q_2 r_2 s_0.
\end{array}
$$

The other blocks will satisfy other combinations of

$$i + j + k = 0, 1 \text{ or } 2(3),$$

and

$$i + 2k + l = 0, 1 \text{ or } 2(3),$$

and these are generated most easily by cyclic rotation of the levels of j ($0 \to 1 \to 2 \to 0$) or of l, or both. The resulting design is shown in Figure 15.25.

I	II	III	IV	V
$p_0q_0r_0s_0$	$p_0q_1r_0s_0$	$p_0q_2r_0s_0$	$p_0q_0r_0s_1$	$p_0q_0r_0s_2$
$p_0q_2r_1s_1$	$p_0q_0r_1s_1$	$p_0q_1r_1s_1$	$p_0q_2r_1s_2$	$p_0q_2r_1s_0$
$p_0q_1r_2s_2$	$p_0q_2r_2s_2$	$p_0q_0r_2s_2$	$p_0q_1r_2s_0$	$p_0q_1r_2s_1$
$p_1q_2r_0s_2$	$p_1q_0r_0s_2$	$p_1q_1r_0s_2$	$p_1q_2r_0s_0$	$p_1q_2r_0s_1$
$p_1q_1r_1s_0$	$p_1q_2r_1s_0$	$p_1q_0r_1s_0$	$p_1q_1r_1s_1$	$p_1q_1r_1s_2$
$p_1q_0r_2s_1$	$p_1q_1r_2s_1$	$p_1q_2r_2s_1$	$p_1q_0r_2s_2$	$p_1q_0r_2s_0$
$p_2q_1r_0s_1$	$p_2q_2r_0s_1$	$p_2q_0r_0s_1$	$p_2q_1r_0s_2$	$p_2q_1r_0s_0$
$p_2q_0r_1s_2$	$p_2q_1r_1s_2$	$p_2q_2r_1s_2$	$p_2q_0r_1s_0$	$p_2q_0r_1s_1$
$p_2q_2r_2s_0$	$p_2q_0r_2s_0$	$p_2q_1r_2s_0$	$p_2q_2r_2s_1$	$p_2q_2r_2s_2$

VI	VII	VIII	IX
$p_0q_1r_0s_1$	$p_0q_1r_0s_2$	$p_0q_2r_0s_1$	$p_0q_2r_0s_2$
$p_0q_0r_1s_2$	$p_0q_0r_1s_0$	$p_0q_1r_1s_2$	$p_0q_1r_1s_0$
$p_0q_2r_2s_0$	$p_0q_2r_2s_1$	$p_0q_0r_2s_0$	$p_0q_0r_2s_1$
$p_1q_0r_0s_0$	$p_1q_0r_0s_1$	$p_1q_1r_0s_0$	$p_1q_1r_0s_1$
$p_1q_2r_1s_1$	$p_1q_2r_1s_2$	$p_1q_0r_1s_1$	$p_1q_0r_1s_2$
$p_1q_1r_2s_2$	$p_1q_1r_2s_0$	$p_1q_2r_2s_2$	$p_1q_2r_2s_0$
$p_2q_2r_0s_2$	$p_2q_2r_0s_0$	$p_2q_0r_0s_2$	$p_2q_0r_0s_0$
$p_2q_1r_1s_0$	$p_2q_1r_1s_1$	$p_2q_2r_1s_0$	$p_2q_2r_1s_1$
$p_2q_0r_2s_1$	$p_2q_0r_2s_2$	$p_2q_1r_2s_1$	$p_2q_1r_2s_2$

Figure 15.25 A $3 \times 3 \times 3 \times 3$ design in nine blocks of nine units.

15.8 Confounding in fractional replicates

The development of the concept of fractional replication in Chapter 14 ignored blocking, and we shall now reexamine the ideas of fractional replication in the context of the various approaches to using small blocks developed in the earlier parts of this chapter. We have argued previously that it is rare for there to be no appropriate blocking structure among large sets of observations, and the examples of 27 and 36 sets of observations certainly come in the category of large sets of observations.

The use of blocking within a fractional replicate provides no new conceptual problems. The intuitive approach of Section 15.1 is entirely effective for treatment structures in which the total number of treatments is not large. For the important main effects and interactions each block of the fractional replicate must include equal replication of levels for each factor, and of combinations of levels for each important interaction. For small sets of experimental treatments there is therefore nothing further to be added.

For larger sets of treatment the more formal approach of Sections 15.6 and 15.7 provides the appropriate methodology. Using the classical theory of confounding for fractional replicates there are three distinct stages:

(i) Choose the fractional replicate without consideration of the later confounding choices. The fraction enabling the maximum number of relevant effects to be estimated with unimportant aliases should always be chosen.

(ii) Identify the group of effects which must be estimated together with the aliases of all such effects.

(iii) Choose a set of confounded effects such that none is an alias of an important effect and so that no generalised interaction of confounded effects is an alias of an important effect.

The choice of confounded effects is complicated by the number of implications of each possible choice. However, the design problem in practical situations is not as complex as might appear from a general descussion, since the choices are always limited, and a careful listing of the possibilities can usually be simply achieved. Once again the general philosophy of the two approaches is illustrated through a series of examples as follows.

Example 15.20 For a 2^8 treatment structure 64 observations (a quarter-replicate) are available and the block size is chosen to be 16 units. An initial consideration of the df structure for the analysis of variance reveals that the main effects and two-factor interactions account for 8 and 28 df respectively, leaving a further 27 df. After allowing for block and error df there are unlikely to be many df available for three-factor interactions, and it is probably sensible at this stage to abandon interest in three-factor interactions. With so many factors it is probably easier to design the experiment through the more formal approach.

First choose as the three defining contrasts the highest-order interaction effects possible:

PQRST, STUVW and the generalised interaction PQRUVW.

Examination of aliases for a subset of effects gives

$$
\begin{array}{llll}
P & \equiv QRST & \equiv PSTUVW & \equiv QRUVW, \\
S & \equiv PQRST & \equiv TUVW & \equiv PQRSUVW, \\
U & \equiv PQRSTU & \equiv STVW & \equiv PQRVW, \\
PQ & \equiv RST & \equiv PQSTUVW & \equiv RUVW, \\
PS & \equiv QRT & \equiv PTUVW & \equiv QRSUVW, \\
PU & \equiv QRSTU & \equiv PSTVW & \equiv QRVW, \\
ST & \equiv PQR & \equiv UVW & \equiv PQRSTUVW.
\end{array}
$$

All other main effects and two-factor interactions show the same patterns as one of these. These patterns reflect the structure of the Defining Contrasts which split the factors into three groups

P, Q, R; S, T; U, V, W,

such that factors in a group of the same size have the same aliasing structure. That is the aliases for the main effect of Q, like those for P, include a four-factor, a five-factor and a six-factor interaction; the aliases for the two-factor interaction QT (or SW) will include a three-factor, a five-factor and a six-factor interaction.

The confounded effects can be three-factor interactions or higher-order interactions. We therefore consider the alias structure for three-factor interactions keeping the group structure for the factors in mind.

$$
\begin{array}{lll}
P\ \ SU \equiv QR & TU \equiv P & T\ VW \equiv QRS\ VW, \\
PQ\ \ U \equiv\ RSTU \equiv PQ\ ST & VW \equiv\ R\ VW.
\end{array}
$$

The first form of three-factor interaction includes one factor from each group. There are $3 \times 2 \times 3 = 18$ such three-factor interactions and they therefore account for 18 of the remaining 27 df. The second form of three-factor interaction has two factors from the first group with one from the third, and includes as an alias a similar three-factor interaction with the groups reversed. There are 3×3 such interactions and they account for the remaining

Table 15.21. *Analysis of variance structure for quarter replicate of* 2^8

df	
8	Main effects
28	Two-factor interactions
18	Three-factor interactions in the pattern PSU
9	Three-factor interactions in the pattern PQU
63	Total

		Block	
I	II	III	IV
(1)	pq	pr	qr
rsu	pqrsu	psu	qsu
qtw	ptw	pqrtw	rtw
qrstuw	prstuw	pqstuw	stuw
pqst	st	qrst	prst
psw	qsw	rsw	pqrsw
pruw	qruw	uw	pquw
pquv	uv	qruv	pruv
pqrsv	rsv	qsv	psv
ptuvw	qtuvw	rtuvw	pqrtuvw
prstvw	qrstvw	stvw	pqstvw
stuv	pqstuv	prstuv	qrstuv
rtv	pqrtv	ptv	qtv
qsuvw	psuvw	pqrsuvw	rsuvw
qrvw	prvw	pqvw	vw

Figure 15.26 A quarter-replicate of a 2^8 structure in four blocks of 16 units.

9 df. All the df in the analysis of variance are now identified (see Table 15.21). The con-founding problem requires the choice of two of these 27 three-factor interactions such that the generalised interaction of the two is also an alias of a member of the 27. We observe that one of the aliases for the second group is of the form PQ ST VW, i.e. two factors from each group, and deduce that two three-factor interactions from the first group with no letter in common will satisfy all the requirements.

The design is therefore defined as follows:

Defining Contrasts	PQRST, STUVW (implies PQRUVW)
confounded effects	PSU $\equiv$ QRTU $\equiv$ PTVW $\equiv$ QRSVW
	QTV $\equiv$ PRSV $\equiv$ QSUW $\equiv$ PRTUW
implying	PQSTUV $\equiv$ RUV $\equiv$ PQW $\equiv$ RSTW.

The principal block must be even with respect to all Defining Contrasts and confounded effects, and the resulting design is given in Figure 15.26.

Block

I	II	III
$p_0q_0r_0s_0$	$p_0q_0r_2s_2$	$p_0q_0r_1s_1$
$p_0q_1r_2s_1$	$p_0q_1r_1s_0$	$p_0q_1r_0s_2$
$p_0q_2r_1s_2$	$p_0q_2r_0s_1$	$p_0q_2r_2s_0$
$p_1q_0r_2s_1$	$p_1q_0r_1s_0$	$p_1q_0r_0s_2$
$p_1q_1r_1s_2$	$p_1q_1r_0s_1$	$p_1q_1r_2s_0$
$p_1q_2r_0s_0$	$p_1q_2r_2s_2$	$p_1q_2r_1s_1$
$p_2q_0r_1s_2$	$p_2q_0r_0s_1$	$p_2q_0r_2s_0$
$p_2q_1r_0s_0$	$p_2q_1r_2s_2$	$p_2q_1r_1s_1$
$p_2q_2r_2s_1$	$p_2q_2r_1s_0$	$p_2q_2r_0s_2$

Figure 15.27 A one-third-replicate of a 3^4 structure in three blocks of nine units.

Example 15.21 Consider again the one-third-replicate of the 3^4 structure, constructed in Section 14.3. Obviously blocks of 27 are too large and three blocks of nine units seems the simplest blocking structure. The df structure for the analysis of variance gives

	df
P	2
Q	2
R	2
S	2
Remainder	18

The aliasing patterns are listed in the earlier example and the nine effects, $PQ(\equiv R^2S)$, PQ^2, PR, PR^2 $(\equiv QS)$, $PS(\equiv QR^2)$, PS^2, QR, QS^2 and RS account for the remaining 18 df. The confounding effect must be chosen from one of these. If RS is chosen, then the 27 treatment combinations will be arranged in three blocks defined by

$$
\begin{array}{ll}
\text{I} & (r_0s_0,\ r_1s_2,\ r_2s_1), \\
\text{II} & (r_0s_1,\ r_1s_0,\ r_2s_2), \\
\text{III} & (r_0s_2,\ r_1s_1,\ r_2s_0),
\end{array}
$$

giving the design shown in Figure 15.27. Direct construction of the design by first allocating levels for P and Q to the three blocks of nine and then allocating the levels of R, followed by levels of S, would inevitably lead to the same design or to the design with R and S levels interchanged.

Example 15.22 Finally, consider again the design problem of Example 14.2, the half-replicate of the $2^3 \times 3^2$ structure for which three blocks of 12 units might be an appropriate blocking structure. This revised design problem reemphasises a result encountered previously in both approaches to confounding. When the restrictions are changed it is usually appropriate to consider the new problem directly rather than modify the original design. If we attempt to allocate the 36 combinations in the optimal half-replicate of Example 15.6 to three blocks of 12 by allocating all QRS combinations to each block, adding the T levels so as to avoid confounding QRT and minimising the confounding of ST, then the distribution of levels of P between blocks is extremely uneven (ten : two split in one block). The half-replicate of the $2^3 \times 3^2$ structure in three blocks of 12 must therefore be constructed directly.

	Block	
I	II	III
$p_0q_0r_0s_0t_0$	$p_1q_0r_0s_0t_1$	$p_0q_0r_0s_0t_2$
$p_1q_0r_0s_1t_1$	$p_0q_0r_0s_1t_2$	$p_1q_0r_0s_1t_0$
$p_0q_0r_0s_2t_2$	$p_1q_0r_0s_2t_0$	$p_0q_0r_0s_2t_1$
$p_1q_0r_1s_0t_0$	$p_0q_0r_1s_0t_1$	$p_1q_0r_1s_0t_2$
$p_0q_0r_1s_1t_1$	$p_1q_0r_1s_1t_2$	$p_0q_0r_1s_1t_0$
$p_1q_0r_1s_2t_2$	$p_0q_0r_1s_2t_0$	$p_1q_0r_1s_2t_1$
$p_1q_1r_0s_0t_0$	$p_0q_1r_0s_0t_1$	$p_1q_1r_0s_0t_2$
$p_0q_1r_0s_1t_1$	$p_1q_1r_0s_1t_2$	$p_0q_1r_0s_1t_0$
$p_1q_1r_0s_2t_2$	$p_0q_1r_0s_2t_0$	$p_1q_1r_0s_2t_1$
$p_0q_1r_1s_0t_0$	$p_1q_1r_1s_0t_1$	$p_0q_1r_1s_0t_2$
$p_1q_1r_1s_1t_1$	$p_0q_1r_1s_1t_2$	$p_1q_1r_1s_1t_0$
$p_0q_1r_1s_2t_2$	$p_1q_1r_1s_2t_0$	$p_0q_1r_1s_2t_1$

	Block	
I	II	III
$p_0q_0r_0s_0t_0$	$p_0q_0r_0s_0t_1$	$p_0q_0r_0s_0t_2$
$p_1q_0r_0s_1t_1$	$p_1q_0r_0s_1t_2$	$p_1q_0r_0s_1t_0$
$p_0q_0r_0s_2t_2$	$p_0q_0r_0s_2t_0$	$p_0q_0r_0s_2t_1$
$p_1q_0r_1s_0t_1$	$p_1q_0r_1s_0t_2$	$p_1q_0r_1s_0t_0$
$p_0q_0r_1s_1t_2$	$p_0q_0r_1s_1t_0$	$p_0q_0r_1s_1t_1$
$p_1q_0r_1s_2t_0$	$p_1q_0r_1s_2t_1$	$p_1q_0r_1s_2t_2$
$p_1q_1r_0s_0t_2$	$p_1q_1r_0s_0t_0$	$p_1q_1r_0s_0t_1$
$p_0q_1r_0s_1t_0$	$p_0q_1r_0s_1t_1$	$p_0q_1r_0s_1t_2$
$p_1q_1r_0s_2t_1$	$p_1q_1r_0s_2t_2$	$p_1q_1r_0s_2t_0$
$p_0q_1r_1s_0t_0$	$p_0q_1r_1s_0t_1$	$p_0q_1r_1s_0t_2$
$p_1q_1r_1s_1t_1$	$p_1q_1r_1s_1t_2$	$p_1q_1r_1s_1t_0$
$p_0q_1r_1s_2t_2$	$p_0q_1r_1s_2t_0$	$p_0q_1r_1s_2t_1$

Figure 15.28 Two alternative designs for a half-replicate of a $2^3 \times 3^2$ structure in three blocks of 12 units.

Plainly all main effects can be allocated to be unconfounded, and it would seem possible to try to keep the two-factor interactions PQ, PR, PS, PT, QR, QS, QT, RS, RT unconfounded. In constructing the design there is no point in avoiding confounding three-factor interactions between the three blocks because, as was pointed out in the earlier example, there are not enough df to estimate three-factor interactions. However, all combinations of levels for each three-factor interaction should be included as nearly equally in the total 36 combinations as is possible to avoid introducing aliases of important effects for other important effects.

Two alternative designs are shown in Figure 15.28. The choice between them depends on the relative importance of obtaining some information on ST, and of correlation between the estimates of P and QR. The first design shown in Figure 15.28, is a different half-replicate from that used earlier, but it still includes all combinations of PQST, of PRST and of QRST and the partial defining contrasts are still PQR and PQRST. The ST interaction has 2 df completely confounded between blocks and the partial aliasing of P with QR (and, of course, Q with PR and R with PQ) occurs to the same extent as previously.

The second alternative shown in Figure 15.28 allows some information on the ST interaction which is partially confounded with blocks at the cost of increasing the correlation between $\hat{P}$ and $\widehat{QR}$.

I	II	III	IV	V	VI	VII	VIII
(1)	p	q	r	s	t	qr	qt
prt	rt	pqrt	pt	prst	pr	pqt	pqr
qrst	pqrst	rst	qst	qrt	qrs	st	rs
pqs	qs	ps	pqrs	pq	pqst	prs	pst

Figure 15.29 Division of 2^5 combinations into eight columns confounding PQR, PRS and PST.

15.9 Confounding in row-and-column designs

Confounding has been discussed so far only in the context of a single blocking classification. However, the principles of confounding can also be applied when there are two orthogonal systems of blocks, which, as in Chapter 8, will be referred to as rows and columns. Both the mathematical approach of Sections 15.6 and 15.7 and the intuitive approach of Sections 15.1 and 15.2 can be employed to produce confounding systems for double blocking systems. We shall illustrate the uses of the two philosophies through examples.

The general theory of confounding for two blocking classifications is less developed than for a single blocking structure. The standard mathematical theory of confounding in a single blocking classification is sufficient for single replicates of factorial structures which have equal numbers of levels for all factors, using a rectangular array both of whose dimensions are powers of the number of levels per factor. This is, of course, a very restricted set of situations, and the results may be of limited practical importance, but they do illuminate the general pattern of such problems. Two confounding systems must be defined, one for the rows and one for the columns, such that the two sets of confounded effects are entirely distinct. The allocation of treatment combinations to rows and to columns in accordance with the double confounding requirements provides a uniquely defined design. It should be noted that one result of combining two confounding systems which involve higher-order interactions for rows and for columns is to produce particular patterns of occurrence of levels of individual factor levels in the two-way array. The model for analysis is not affected by these patterns, but they should be looked at to check that there is no reason for concern about the properties of the combined design.

Example 15.23 If a 2^5 structure is to be investigated using a unit structure in four rows × eight columns, then a column confounding system of seven effects and a separate row confounding system of three effects are required. Suitable sets of effects could be

$$\text{PQR, PRS, PST, QS, QRST, RT, PQT,}$$

for columns, and

$$\text{PQRS, PRT and QST}$$

for rows. The treatment combinations in the eight columns are quickly derived as shown in Figure 15.29.

Because each of the row confounded effects is orthogonal to all of the column confounded effects, the combinations in each column must be rearrangeable into four rows such that combinations are:

Replicate 1

Column

	1	2	3	4	5	6	7	8
Row 1	(1)	qs	rst	pqrs	qrt	pr	pqt	pst
2	prt	pqrst	ps	qst	pq	t	qr	rs
3	qrst	rt	q	pt	s	pqst	prs	pqr
4	pqs	p	pqrt	r	prst	qrs	st	qt

Replicate 2

Column

	9	10	11	12	13	14	15	16
Row 5	(1)	ps	qt	rst	pqst	prt	qrs	pqr
6	prs	r	pqrst	pt	qrt	st	pq	qs
7	qst	pqt	s	qr	p	pqrs	rt	prst
8	pqrt	qrst	pr	pqs	rs	q	pst	t

Figure 15.30 Two replicates of a 2^5 structure arranged in four rows × eight columns with different confounding systems.

(1) even with respect to PQRS and PRT,
(2) even with respect to PQRS and odd with respect to PRT,
(3) odd with respect to PQRS and even with respect to PRT,
(4) odd with respect to PQRS and PRT.

The resulting unrandomised design is shown as the first replicate in Figure 15.30. The randomisation procedure would use permutations of rows and columns as in Chapter 9. A consequence of the method of construction with the combination (1) in row 1 column 1 is that the element in row i and column j is the 'product' of the two combinations in row 1 and column j, and row i and column 1, where the 'product' is generalised in the same sense as the generalised interaction of multiple confounding theory. Note that this pattern inevitably leads to patterns for the levels of each factor so that in each column the higher level of P occurs either in rows 2 and 4, or in rows 1 and 3. If the randomisation were such that the rows were in the order 1, 3, 2, 4 and the columns were in the order 1, 2, 3, 5, 4, 6, 7, 8, then the upper level of P would occur in the top right quarter or the bottom left quarter only, which might sometimes be regarded with concern.

In the analysis of variance, effects confounded with rows or columns cannot be estimated. The analysis of variance structure for the design is therefore as shown in Table 15.22. Designs with more than one replicate can be constructed by combining several single replicate designs with partial confounding. However, if the column (or row) effects might be expected to be consistent across rectangles, then it would be preferable not to employ partial confounding for *both* rows and columns and this creates additional problems.

Example 15.24 Next consider the same treatment structure but with two replicates with each replicate having a different confounding system. In the design shown in Figure 15.30 the first replicate uses the same confounding systems as in the previous example,

columns	PQR	PRS	PST	QS	RT	QRST	PQT
rows		PQRS	PRT	QST.			

Table 15.22. *Outline analysis of variance for* 2^5
treatments in rows and columns

Source	df
Rows	3
Columns	7
Main effects	5
Two-factor interactions (not QS and RT)	8
Error	8
Total	31

The second replicate has confounding systems:

columns	PQS	QRS	PST		PR	QT	PQRT	RST
rows		PRS	QRT	PQST.				

Because the design involves partial confounding, some of the SS in the analysis of variance are calculated from only one of the two replicates. The use of two replicates also allows information on some of the three-factor interactions to be calculated if that is considered useful.

Example 15.25 Now suppose that the two replicates are to be arranged within a unit structure of 8 rows × 8 columns. It would be possible to use the same column confounding system (PQR, PRS, PST, QS, QRST, RT, PQT) with one row confounding system (PQRS, PRT, QST) for the first replicate in the first four rows and a different row confounding system (PRST, PQS, QRT) in the last four rows. However, this still leaves the two-factor interactions QS and RT totally confounded with columns, even though with eight units in each column it would seem reasonable to hope that it would not be necessary to confound any two-factor interactions.

To confound groups of only three effects between columns and between rows, assuming that this is possible because each column or row contains eight combinations, we may proceed as follows. Suppose PQR, PST and QRST are to be confounded between columns, and PRT, QST and PQRS with rows, then the design is not fully defined since each system has only four classes. The allocation defined by the confounding systems is shown in Figure 15.31. Each 'row' and 'column' combination contains two treatment combinations. If the full 8 × 8 unit structure is divided into four quarters, each 4 × 4, then the allocation in Figure 15.31 can be split into two 4 × 4 structures. The obvious way of splitting the pairs is on the even/odd value for the five-factor interaction PQRST. Thus the 'even' treatment combination of each pair is allocated to the quarters in the north-west and south-east corners. After randomising the orders of rows and of columns the resulting design is shown in Figure 15.32.

The confounding method of choosing three effects to confound with rows, three to confound with columns and one to confound with quarters (before randomisation) implies that there is a further confounding of effects with half-rows in quarters, or half-columns in quarters. The effects confounded in this manner include main effects P(= PQRST × QRST) and T(= PQRST × PQRS) and the results can still be seen after randomisation. In each column

		PQR even PST even	PQR even PST odd	PQR odd PST even	PQR odd PST odd
PRT	QST				
even	even	(1)	pr	pqrs	qs
		pqt	qrt	rst	pst
even	odd	qrst	pqst	pt	rt
		prs	s	q	pqr
odd	even	st	prst	pqrt	qt
		pqs	qrs	r	p
odd	odd	qr	pq	ps	rs
		prt	t	qst	pqrst

Figure 15.31 Basic confounding system for two replicates of 2^5 structure in eight rows × eight columns.

	Column 1	2	3	4	5	6	7	8
Row 1	prt	t	ps	qr	qst	pq	pqrst	rs
2	pqt	qrt	pqrs	(1)	rst	pr	pst	qs
3	qrst	pqst	q	prs	pt	s	rt	pqr
4	pqs	qrs	pqrt	st	r	prst	p	qt
5	prs	s	pt	qrst	q	pqst	pqr	rt
6	qr	pq	qst	prt	ps	t	rs	pqrst
7	(1)	pr	rst	pqt	pqrs	qrt	qs	pst
8	st	prst	r	pqs	pqrt	qrs	qt	p

Figure 15.32 Final design for 2^5 in eight rows × eight columns.

t always occurs in the first three rows and the last row (columns 1, 2, 5 and 7), or in rows 4–7 (columns 3, 4, 6, 8). However, if the additive model for the row-and-column design correctly represents the variation within this blocking structure, then this 'confounding' is not practically relevant because the 'confounding blocks' are not represented by terms in the assumed model. (This is one design where randomisation theory would diverge from the infinite population model theory.)

Because the randomisation procedure confounds effects with semi-columns × rows or with semi-rows × columns, the effects confounded with these 'blocks' should be assessed relative to the variation between semi-rows within columns or semi-columns within rows, requiring two further levels of variation within the analysis of variance, similar to the criss-cross design (see Chapter 18). If this is felt to be important, then a better choice than PQRST for confounding between quarters would be PQT. The analysis of variance structure for the design of Figure 15.32 is straightforward allowing all main effects and two-factor interactions plus six non-confounded three-factor interactions to be estimated.

Example 15.26 The final example, quoted at the beginning of the chapter, is a design problem of a more general form, encountered during consultancy, and is best solved through the intuitive approach, even though it involves confounding both between rows and between columns. The experimental problem is to investigate absorption by rabbits of sugar occurring in various forms and amounts in the rabbits' diets. The experimental treatments selected have a 2^3 structure, being two concentrations (C) × two forms (F) × presence or absence of

			Rabbit		
	1	2	3	4	5
Time period 1	(1)	acf	a	f	c
2	ac	f	cf	c	af
3	cf	a	ac	af	(1)
4	af	ac	(1)	acf	cf
5	acf	c	f	a	ac

Figure 15.33 A 2^3 design in 5 rows × 5 columns.

an additive (A). The experimental units are short periods of three weeks for each rabbit. Five rabbits are available for the experiment and each may be treated and observed for the same five time periods. There will definitely be consistent differences between rabbits, and there may be consistent differences between the five time periods. The design problem is to arrange a 2^3 treatment structure within a 5 × 5 row-and-column structure. Main effects should be most efficiently estimated (least affected by row or column effects) but two-factor interactions are also of interest.

From the general results for row-and-column designs in Chapter 8, we know that precision of main effect comparisons depends mainly on the frequency of occurrence, together within a row or column, of the two levels to be compared. Therefore we seek to construct a design such that, for each main effect, the two levels will occur as nearly equally as possible in each row or column. In other words each row and each column should include two or three replicates of each level for each of the three factors. A design which satisfies these requirements is shown in Figure 15.33. It is not known whether this design is, in any sense, optimal, but it appears to be sensible, and illustrates the possibility of constructing particular designs to satisfy particular requirements.

By maximising the 'equality' of occurrence for main effect levels, interaction contents may be less even. For example the CF interaction comparison $(cf + (1)) - (c + f)$ has $+$ and $-$ terms allocated in rows in the proportions ($\frac{3}{2}, \frac{1}{4}, \frac{3}{2}, \frac{3}{2}$ and $\frac{2}{3}$), while the proportions in the columns are ($\frac{3}{2}, \frac{3}{2}, \frac{3}{2}, \frac{2}{3}$ and $\frac{2}{3}$). These proportions are obviously not very disparate so that the design will provide considerable information about two-factor interactions.

Since all treatment effects are not orthogonal to rows or columns the order of fitting is important. The analysis of variance structure is exactly as would be expected and is shown in Table 15.23.

Exercises

15.1 Construct sensible designs for a 2 × 3 × 3 factorial structure, assuming main effects are most important, followed by two-factor interactions, and explaining the advantages of your designs, for:
    (i) six blocks of six units,
    (ii) four blocks of nine units,
    (iii) three blocks of 12 units,
    (iv) nine blocks of four units,
    (v) seven blocks of five units.
You may wish to assess any of your designs by using a statistical package or matrix program to calculate variances of effects using a dummy analysis of variance.

Table 15.23. *Analysis of variance structure for rabbits experiment*

Source	df
Rows	4
Columns	4
C	1
F	1
A	1
CF	1
CA	1
FA	1
CFA	1
Error	9
Total	24

Table 15.24. *Data for Exercise 15.2*

Replicate 1	Block I	a_0b_0 41	a_1b_1 21	a_2b_2 13
	Block II	a_0b_1 54	a_1b_2 25	a_2b_0 20
	Block III	a_0b_2 44	a_1b_0 61	a_2b_1 24
Replicate 2	Block IV	a_0b_0 52	a_1b_2 30	a_2b_1 24
	Block V	a_0b_1 44	a_1b_0 39	a_2b_2 26
	Block VI	a_0b_2 47	a_1b_1 32	a_2b_0 38

15.2 The data in Table 15.24 are taken from a 3^2 experiment on lettuce emergence, confounded in blocks of three. Use a standard computer package to calculate the analysis of variance and obtain standard errors for comparisons between pairs of the nine treatments.

15.3 Construct a design to compare six treatments in nine blocks of four units each, with no treatment occurring twice in a block when the treatment set is a 2×3 factorial structure, avoiding any confounding of the main effect for the two-level factor and with minimal confounding for the main effect of the three-level factor. Use a standard computer package to obtain variances of least squares estimates of main effect differences, and comment on the efficiency of estimation of each main effect difference.

15.4 An experimenter is working on the effects of various chemicals on fungal growths on leaves. The natural experimental unit is a half-leaf, and he can either use a leaf (= two half-leaves) as a block, or use a pair of leaves (= four half-leaves) as a block. Clearly the random variance will be larger in the latter case. The experimenter proposes to examine the effects of five chemicals using a 2^5 factorial structure, with the two levels for each chemical being presence or absence. He is interested in the main effect of each chemical and also in whether there are any two-factor interactions.

Devise suitable confounding schemes for a single replicate of a 2^5 factorial

(i) in 8 blocks of four units,

Table 15.25. *Two-way treatment totals*

	n_0	n_1	n_2	p_0	p_1	p_2	Total
k_0	139	107	69	134	99	82	315
k_1	121	77	121	106	105	108	319
k_2	92	98	95	100	87	98	285
p_0	134	120	86	340	291	288	
p_1	103	106	82				
p_2	115	56	117				
Total	352	282	285				

$$\sum y^2 = 37425$$

Block

I		II		III	
$n_0p_0k_0$	41	$n_0p_0k_2$	40	$n_0p_0k_1$	53
$n_0p_1k_2$	11	$n_0p_1k_1$	38	$n_0p_1k_0$	54
$n_0p_2k_1$	30	$n_0p_2k_0$	44	$n_0p_2k_2$	41
$n_1p_0k_1$	11	$n_1p_0k_0$	61	$n_1p_0k_2$	48
$n_1p_1k_0$	21	$n_1p_1k_2$	39	$n_1p_1k_1$	46
$n_1p_2k_2$	11	$n_1p_2k_1$	20	$n_1p_2k_0$	25
$n_2p_0k_2$	12	$n_2p_0k_1$	42	$n_2p_0k_0$	35
$n_2p_1k_1$	21	$n_2p_1k_0$	24	$n_2p_1k_2$	37
$n_2p_2k_0$	13	$n_2p_2k_2$	46	$n_2p_2k_1$	58
	171		354		394

Figure 15.34 Design and data for a single replicate of a 3^3 experiment.

(ii) in 16 blocks of two units, avoiding confounding any main effect,

(iii) in 16 blocks of two units confounding one main effect, and as few two-factor interactions as possible.

If sufficient material is available for five replicates, explain why (ii) is undesirable. Also discuss possible partial confounding schemes in five replicates for (i) and (iii), and explain how you would choose between schemes based on (i) and (iii) in terms of the random variances σ_1^2 for (i) and σ_2^2 for (iii).

15.5 A factorial experiment with five two-level factors A, B, C, D and E is arranged in blocks of eight plots per replication, certain interactions being confounded with block differences. The same interactions are confounded in each replication. The treatment combinations tested in one of the blocks are

a, b, cd, ce, ade, bde, abcd, abce,

where, for example, the symbol ade denotes the combination of factors A, D and E at their upper levels, and B and C at their lower levels. Which interactions are confounded?

15.6 A single replicate of a 3^3 experiment on the effects of three levels of nitrogen, three levels of phosphorus and three levels of potash on germination of lettuce seedlings gave the results as in Figure 15.34 and Table 15.25.

Use a standard computer package to calculate the analysis of variance and report your conclusions.

15.7 In a 2^6 experiment with factors A, B, C, D, E and F all three-factor and higher interactions may be assumed negligible as may the two-factor interactions involving either A or C. Construct a quarter-replicate which allows main effects and other two-factor interactions to be separately estimated. Suggest a suitable blocking scheme for a quarter-replicate in two blocks of eight units.

15.8 An experiment is to be designed to compare different recipes for a cake mixture. Three treatment factors are proposed:

$$A = \text{level of a milk product} \quad \text{(two levels)},$$
$$B = \text{level of a starch product} \quad \text{(two levels)},$$
$$C = \text{level of an emulsifier} \quad \text{(three levels)}.$$

Each of the 12 cake mix combinations is to be replicated four times. On each of two days cakes could be baked at each of 12 different times, and at each time two cakes could be baked, one in the left-hand side of the oven, the other in the right-hand side. There may be consistent differences between the two sides, and also between the 12 times within a day. The unit structure is therefore 12 rows (times) × two columns (sides) repeated on two days. How do you allocate the 12 treatment combinations so that the three main effects and the AB interaction are all estimated as efficiently as possible?

16

Quantitative factors and response functions

16.0 Preliminary examples

(*a*) A research student was proposing an extensive investigation of the decay of seed germination over time. The change of germination rate over two years was to be examined. Seeds would be treated in various different ways, and seeds from many different sources would be used. The major design questions were how often to test the germination rate during the two years, and how many seeds to use on each test occasion. The student, when questioned about the pattern of decay with time, was adamant that all previous data showed clearly that the relationship of germination rate with time was linear. For each combination of seed source and treatment, about 2000 seeds would be available. How should the 2000 seeds be sampled over the two-year period?

(*b*) In a rather similar investigation into the decline of strength of weldings used for oil rigs, again looking at the pattern over a long period of stress, the proposed design on which comments were requested was that which was regarded by the engineers concerned as too obvious to require statistical advice. As in the seed germination investigation, it was believed absolutely that the relationship of strength with level of stress was linear. The only analysis which was contemplated was to fit a straight line regression of the variable measuring strength on the length of time for which the stress was applied. The 'obvious' design proposed by the engineers was to use equal replication for eight, equally spaced, durations of stress. Can the despised statistician offer any improvement?

16.1 The use of response functions in the analysis of data

When an experiment involves treatments which are levels of a quantitative factor, it is essential to examine the pattern of response of the observed variable to the changing stimulus of the factor. It is almost always appropriate, as a first step, to plot mean yields against the factor level. This will often be only a first step, and frequently we should seek to draw inferences about the pattern of response by fitting an appropriate response function, and interpreting the response in terms of the values of the parameters of the response function. The analysis may be taken further, with the comparison of fitted response functions for several levels of other factors included in the treatment structure.

First we consider the set of available and appropriate response functions. There are different levels of appropriateness. There is always a level in the modelling of a process at which any simple algebraic model is inadequate to represent the true mechanisms governing the relationship between the response and the stimulus. At the other extreme, there are models which essentially attempt no more than a crude smoothing of the data to provide a very simple summary of the results. Between these two extremes there is the main area of the development of algebraic models which are judged by whether their behaviour is compatible with general or specific patterns of response believed to hold for the system being investigated, and by whether the models fit the data, using statistical assessment of the fit.

The simplest response model used is the straight line

$$E(y) = \alpha + \beta x.$$

This is hardly ever considered as a practically realistic model in spite of the two examples at the beginning of this chapter, but is sometimes useful for calculating an average rate of increase of y with x over a limited range of x values for which the linear relationship is approximately true. Replacing x by x^λ greatly increases the possible shapes of the model, but with λ unknown the model becomes non-linear and hence more difficult to work with. We will discuss non-linear models shortly. Sometimes the possible values of λ are restricted to a small set of rational numbers, leading to the increasingly popular *fractional polynomial* models. These are somewhat simpler to deal with, but still non-linear.

The simplest, and historically the most used, curved relationship is the quadratic

$$E(y) = \alpha + \beta x + \gamma x^2.$$

The disadvantages of this model are that it is too restrictive to be realistic. Specifically, if extrapolated beyond the range of the data, it rapidly gives impossible values. For the most common response pattern with β positive and γ negative, and a maximum value of y at $x = -\beta/2\gamma$, negative values of y are implied when x differs from this value by more than $(\beta^2 - 4\alpha\gamma)^{\frac{1}{2}}/2\gamma$. A second disadvantage is the symmetry of the quadratic about the optimum. It will rarely be credible in a practical situation that the mechanisms governing the increase of y to its maximum value will be mirrored by the mechanisms controlling the decrease beyond the maximum. The third disadvantage of the quadratic curve is that it does not include an asymptotic form of response, i.e. a response in which y increases (or decreases) monotonically but at a gradually reducing rate so that, as x increases, y tends to a limiting value.

The symmetry disadvantages can be overcome by considering powers of x, perhaps using fractional polynomials. To obtain curves which also satisfy the other objections, we have to use models which are more complex in a fundamental way than those considered so far. The straight line, and the quadratic, with or without prespecified powers, are all linear response functions in the sense that the parameters occur in a linear form, though if a power is treated as a parameter to be estimated the model is no longer linear. This linearity of parameters is important because it leads to a very simple form of estimation using the general linear model ideas of Chapter 4. This ease of estimation was extremely important before computers were available, and, although non-linear models were used, it was necessary to devise clever and complex methods of fitting the models to data. Consequently, the quadratic response was used frequently when not appropriate because of the difficulty of fitting a more appropriate model.

There are now fewer computational difficulties in the use of non-linear models, and the quadratic model should retreat to a less prominent position, being mainly used as a smoothing curve of rather greater flexibility than the straight line. The two major families of non-linear curves which provide greater flexibility in modelling response patterns are based on exponential and on reciprocal relationships, respectively.

Exponential relationships are used extensively for asymptotic responses. The simplest form is

$$E(y) = A(1 - e^{-kx}),$$

the Mitscherlich response used extensively for the response of crop yield to varying fertiliser level. A corresponding form for decay curves in chemical and biological processes, or in medical survival studies, is

$$E(y) = Ae^{-kx}.$$

This model also arises in kinetic studies of first-order chemical reactions, in which the treatment variable x is time and the response y is the amount of unconverted substrate. In these situations A, the initial amount of substrate, is usually known and indeed fixed by the experimenter, so that k is the only unknown parameter. Whether time can really be considered to be a treatment factor depends on how the experiments are actually conducted. If a single reaction is run and observed at several different times, then we have essentially a single experimental unit with repeated measurements of the response. On the other hand, if measuring the substrate at a particular time involves stopping the experiment, then each observation will be from a different experimental unit and time is just like any other treatment factor.

This example allows us to illustrate another complication with non-linear models. In practice, chemists estimate k by least squares by taking logarithms of both sides of the equation. This is equivalent to assuming that the model is

$$y = Ae^{-kx}e^{\epsilon'},$$

i.e. it has multiplicative errors with a log-normal distribution. Then, after transformation, we obtain

$$\log y = \log A - kx + \epsilon'$$

and we once again have a simple linear regression model. However, there is no reason why this should be the correct error structure. Perhaps a more obvious assumption is

$$y = Ae^{-kx} + \epsilon,$$

i.e. additive normal errors. In this case the model parameters can only be sensibly estimated by non-linear least squares or closely related methods. Of course, there is no reason why this assumption must be correct either, although it might often be made thoughtlessly. At the stage of planning the experiment some tentative assumption about the error structure will usually be made based on past experience in similar experiments. It is important that this is considered carefully, rather than just assuming the first error structure that comes to mind.

Numerous extensions of these models have been used for particular situations. A general form including both the previous particular cases is

$$E(y) = A + be^{-kx}.$$

By letting $z = e^{-x}$, this can be written as

$$E(y) = A + bz^k,$$

so we can see that first-order fractional polynomial models also fit into this class. More exotic forms of exponential relationships include

$$E(y) = xAe^{kx}$$

used by Carmer and Jackobs (1965), the Gompertz curve

$$E(y) = y_0 A^{Bx}$$

used extensively in growth studies and, perhaps the most exotic of all,

$$E(y) = A - B\exp(1 - kx^\theta)$$

used by Reid (1972) to describe the response of ryegrass swards to a very wide range of nitrogen levels.

Reciprocal relationships, of which the simplest form is

$$E(y) = \beta_0 + \beta_1 (x - x_0)^{-1},$$

have been used in many areas of research, and have been justified on theoretical grounds in some cases of biological research and in crop density–yield relations in agronomy (the 'law of constant final yield'; Hozumi, Asahira and Kira, 1972). In enzyme kinetic studies, biochemical theory implies that many enzymes have a steady-state reaction rate which is related to the substrate concentration by the Michaelis–Menten equation, usually written in the form

$$E(y) = \frac{V_{\max}x}{K_{\mathrm{m}} + x},$$

where $V_{\max}$ represents the maximum reaction rate when the system is saturated with substrate ($x \to \infty$) and K_{m} represents the substrate concentration at which the reaction rate reaches half of its maximum value. This application has provoked considerable discussion about the appropriate form of errors to assume (Ruppert, Cressie and Carroll, 1989; Nelder, 1991). The Michaelis–Menten curve is also much used in yield–density relations (after Hozumi *et al.*, 1972).

Reciprocal relationships have been generalised to form the class of inverse polynomials

$$x/E(y) = \text{polynomial in } x$$

by Nelder (1966). Particular cases which have been found useful are the inverse linear

$$1/E(y) = \beta_0/x + \beta_1,$$

which is a reparameterisation of the Michaelis-Menten equation, and the inverse quadratic

$$1/E(y) = \beta_0/x + \beta_1 + \beta_2 x,$$

which gives a non-symmetric curve with an optimum (at $x = (\beta_0/\beta_2)^{\frac{1}{2}}$). An alternative form of non-symmetric optimum curve is

$$1/E(y)^\theta = \beta_0/x + \beta_1,$$

which has been justified, on theoretical biological grounds, for yield–density relations for agronomic studies.

One other group of relationships which continues to receive attention (in spite of determined efforts by *some* statisticians) is the 'broken stick' model in which the response is modelled by two straight lines

$$E(y) = \alpha_0 + \beta_0 x \qquad\qquad x \le x_1,$$
$$E(y) = \alpha_0 + \beta_0 x_1 + \beta_1(x - x_1) \qquad x \ge x_1,$$

where x_1 is a fourth parameter.

16.2 Design objectives

Fitting a response model to experimental data may be appropriate for various different reasons. Correspondingly, we can define several different objectives for model fitting. As the objective of the fitting varies, so will the choice of treatment levels (the design) which best satisfies the objective. Various authors have produced lists of possible objectives, notably Box and Draper (1975). To identify which objectives are most important, we must re-examine our motives in fitting response models. These may include:

(*a*) supplying a summary of data for the purpose of presentation,
(*b*) examining the appropriateness of a postulated model,
(*c*) estimating the parameters or functions of parameters of a well-established response function to provide biological or chemical information,
(*d*) providing a smoothed version of the data to give predictions of response, or a description of the pattern of response, over a range of levels of the stimulus factor,
(*e*) as a step in identifying optimal levels of a set of stimulus factors.

Of these, (*a*) has no implications for design beyond a general requirement for the data to give a good representation of the system under study. For the other four, we can separate (*b*) and (*c*), which are strongly related to a particular model or models, from (*d*) and (*e*), which are primarily related to prediction. The models in (*d*) and (*e*) are used as a means of retaining most of the data information in a concise form, but with no implications of interest in the parameters individually. We consider (*d*) and (*e*) in the next chapter, which is concerned with the exploration of response surfaces representing the response to several variables.

We can further subdivide (*c*) and can relate (*b*) to (*c*). Most of the models contemplated for fitting to data will not have more than four parameters. While it is possible to construct models with more parameters, it is extremely difficult to obtain a sufficiently informative set of data to be able to make inferences about more than four parameters. For many data sets an attempt to fit a model with four parameters leads to the conclusion that the estimated correlation between the fitted values of two of the parameters is so high that the data do not contain independent information about both parameters, but only about some combination of them.

The major division in experimental design for parameter estimation is between choosing a design to provide maximum information about a single parameter or a single combined function of parameters, or a design to provide maximum information about the set of parameters as a whole. Algebraic results are more easily available (in general) for the latter, but the danger of treating all parameters as being of equal importance is that information on the more important parameters may be markedly reduced.

The examination of the appropriateness of the fitted model implies that there is some idea of how the model might be inappropriate. This in turn implies an alternative model and the need to discriminate between the original and alternative models. If the alternative model is essentially the original model with an additional parameter, then the problem of optimal discrimination between the two models is equivalent to optimising the estimation of the additional parameter. If the alternative model does not include the original model as a special case, then discrimination cannot be reduced to a particular example of estimation. A formal approach to discrimination is often unfruitful because, if two alternative curves are both believed to provide a realistic description of the pattern of response which is expected, then the differences between the curves are likely to be relatively slight. This is illustrated in Figure 16.1(a), in which two similar asymptotic curves, one exponential and the other a linear inverse polynomial, are superimposed. It is clear that incredibly precise data will be needed to discriminate between the two models.

A further illustration of the similarity of alternative models for response functions is provided in Figure 16.1(b). Three models which have been used for describing fertiliser response curves, in which the data show a maximum, are displayed. The three models are:

(1) $y = (\beta_0 + \beta_1 x)/(1 + \beta_2 x + \beta_3 x^2)$,
(2) $y = (\beta_0 + \beta_1 x + \beta_2 x^2)/(1 + \beta_3 x)$,
(3) a pair of straight lines (the broken stick model of the previous section).

These three curves were included in a wide range of models which were compared by Sparrow (1979) for fitting a set of 83 experimental sets of data on the response of spring barley to nitrogen fertiliser with nine levels of nitrogen in each data set. A typical set of data is shown in Figure 16.2. The three curves considered here gave very similar patterns of fit, the average residual mean squares being 30, 32 and 34 $((kg/ha)^2 \times 10)$ for the three models. Sparrow concludes that all three curves 'represent the response of spring barley to nitrogen with some degree of success' and 'the three models differed little in their average predicted optimal yields', and even after such a very extensive set of experiments it was not possible to distinguish any clear preference between the curves. Such an experience is extremely common from comparative studies of response curves.

The next three sections of this chapter examine the choice of levels for specific parameter estimation, for general parameter estimation and for discrimination. The penultimate section is concerned with the practically important question of considering several criteria and examining the robustness of different optimal designs to alternative criteria. Finally, a different experimental approach to investigating response functions in specific circumstances is considered.

16.3 Specific parameter estimation

The simplest response function is a straight line relationship

$$E(y) = \alpha + \beta x,$$

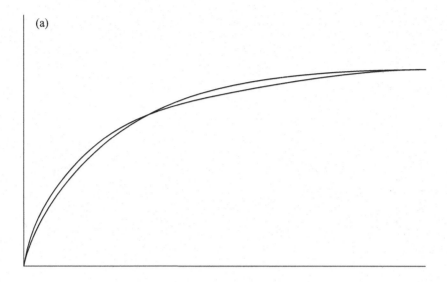

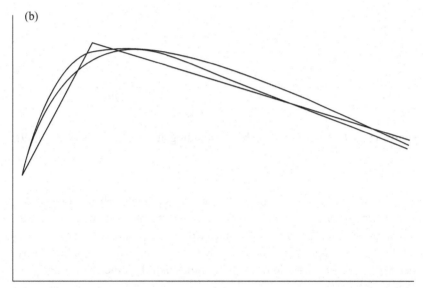

Figure 16.1 Illustration of the similarity of response curves for different models:
(a) two asymplotic responses from exponential and linear inverse polynomial
models; (b) three response curves showing a maximum, from models (1), (2) and (3)
used in Sparrow (1979).

and the simplest design question concerns the choice of values of x to optimise the estimation
of β, the slope of the relationship. We must minimise the variance of $\hat{\beta}$, which is

$$\sigma^2 \Big/ \sum_i (x_i - \bar{x})^2,$$

which, since σ^2 is assumed not to depend on the choice of treatments, requires choosing the
xs to maximise

$$S = \sum_i (x_i - \bar{x})^2.$$

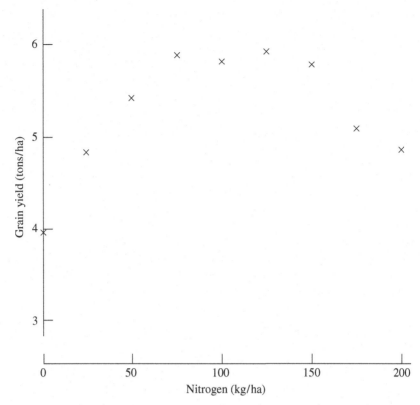

Figure 16.2 Response of spring barley to nitrogen: data from Sparrow (1979).

In discussing maximisation, it is assumed that the total number of observations is fixed, since, without this restriction, precision can always be improved by using more observations. It is immediately apparent first that S may be increased by increasing the range of x values, and second that, for a fixed range, S is maximised by taking half the permitted total observations at the upper end of the permissible range of x values and the other half at the lower end of the permitted range.

This result illustrates immediately the practical difficulties implicit in considering criteria for optimal design. An experimenter may be initially confident that the expected response in his experiment is a straight line relationship. Often when presented with the resulting optimal design requiring observations at only two levels of x, the experimenter's confidence in the straightness of the relationship will waver, and a compromise design will be selected. Nonetheless, the study of optimal designs for specific criteria is valuable because it indicates the ideal design under extreme assumptions, and provides a well-defined basis from which a compromise solution may be determined.

The optimal design for the simple criterion of optimising the estimation of the slope of a straight line relationship is easily obtained. There are a number of other interesting criteria for parameters or functions of parameters of the simple response functions discussed in the first section of this chapter. These include:

(a) for the quadratic model, $E(y) = \alpha + \beta x + \gamma x^2$, the estimation of either the linear or quadratic parameter, β or γ;

(b) for the same quadratic model, the estimation of the position of the maximum yield, $x_m = -\beta/2\gamma$, the criterion being either the minimisation of the variance of the estimate of x_m or the maximisation of the expected response at the predicted optimum;

(c) for the exponential model, $E(y) = A(1 - e^{-kx})$, the estimation of either the asymptote parameter A or the curvature parameter k;

(d) for the inverse linear model, $E(y) = x(\beta_0 + \beta_1 x)^{-1}$, the estimation of either of the two parameters β_0 or β_1, or the ratio β_0/β_1 which is equivalent to K_m in the Michaelis–Menten parameterisation;

(e) for the inverse quadratic model, $E(y) = x(\beta_0 + \beta_1 x + \beta_2 x^2)^{-1}$, the estimate of the position of the optimum, $x = (\beta_0/\beta_1)^{\frac{1}{2}}$;

(f) for any of the models (c), (d) or (e), the estimation of the x value for which the slope of the relationship is equal to some value, C, corresponding to the cost per unit of the input variable x.

Apart from (a), for which the algebraic argument follows exactly the same pattern as that for the slope of a straight line relationship, no exact algebraic results are possible for any of these criteria, since all involve either non-linear parameters or non-linear combinations of parameters. Such results as are available are based on approximate or asymptotic variances of parameters or parameter combinations, and depend on the experimenter having some prior knowledge about the value of the quantity to be estimated. The possible approaches are illustrated through two examples.

Example 16.1 Suppose, for the quadratic model,

$$E(y) = \alpha + \beta x + \gamma x^2$$

with $\gamma < 0$, the interest is in estimating the value of x for which y is a maximum,

$$x_m = -\beta/2\gamma.$$

The obvious estimate is obtained by plugging in the the least squares estimates of β and γ:

$$\hat{x}_m = -\hat{\beta}/2\hat{\gamma},$$

and the variance–covariance matrix of $(\hat{\alpha}, \hat{\beta}, \hat{\gamma})$ can be used to obtain an approximate, asymptotic, variance for $\hat{x}_m$.

Ideally, we would consider a completely general set of n observations, $x_1, x_2, \ldots x_n$, but the problem is then too general for a simple analytic solution. The set of possible designs is therefore restricted to those with $(k + 1)$ equally spaced levels of x, spread between zero and an upper limit d, each level being replicated r times. The values of k, d and r may be chosen subject to an overall limit on the total number of observations,

$$(k + 1)r = n.$$

The approximate variance of $\hat{x}_m$ is

$$V(\hat{x}_m) \simeq (1/4\gamma^4)\{\gamma^2 \operatorname{Var}(\hat{\beta}) + \beta^2 \operatorname{Var}(\hat{\gamma}) - 2\beta\gamma \operatorname{Cov}(\hat{\beta}, \hat{\gamma})\},$$

where the variance–covariance matrix of $(\hat{\alpha}, \hat{\beta}, \hat{\gamma})$ is

$$
\sigma^2 \begin{bmatrix} n & \sum x & \sum x^2 \\ \sum x & \sum x^2 & \sum x^3 \\ \sum x^2 & \sum x^3 & \sum x^4 \end{bmatrix}^{-1}.
$$

Inserting the relevant algebraic sums for $\sum x^i$ and simplifying we obtain

$$
V(\hat{x}_m) \simeq [3k\sigma^2/\{n(k+2)\gamma^2 d^2\}]
$$
$$
\times [1 + \{15k^2/(k-1)(k+3)\}(1 - 2x_m/d)^2].
$$

Considering this variance as a function of d and k, we see first that the terms in k occur as $k/(k+2)$ or $k^2/(k-1)(k+3)$, both of which are increasing functions of k, provided k is at least 2 (and of course $k = 1$ is of no interest, since we must estimate three parameters, α, β and γ). Hence, the approximate variance is minimised by taking $k = 2$, i.e. by using just three observations, each replicated $n/3$ times. With $k = 2$, the approximate variance reduces to

$$
V(\hat{x}_m) \simeq 3\sigma^2/(2n\gamma^2 d^2)\{1 + 12(1 - 2x_m/d)^2\}.
$$

The absolute minimum is obtained by taking d infinitely large. However, investigation shows that there is a local minimum when

$$
d = \{24/(9 + \sqrt{3})\}x_m \simeq 2.24x_m,
$$

and, in fact, the approximate variance can be reduced below the variance for this value of d only by taking d larger than $4.6x_m$. The conclusions from this example are:

(i) Only three different x values should be used when the objective is to fit a three-parameter quadratic response function. To use more reduces the efficiency of estimation of parameters.

(ii) The range of values over which observations should be taken will depend on a rough guess of the position of the optimum. Without such a guess, no proper choice can be made.

(iii) The optimal pattern for the three observations using equally spaced observations is to use $(0, x, 2x)$, where x is taken to be about the guessed optimal position, with a tendency to choose x higher than the guessed positions, rather than lower.

Note that, if we wish to consider a range of values from c to d, where c is a non-zero lower limit, then, because a change of origin leaves a quadratic function as a quadratic function, the value of d should be chosen so that the midpoint of the (c, d) interval is slightly above the guessed optimum. However, in the original formulation of the approximate variance, d^2 is replaced by $(d - c)^2$, showing that, if c can be chosen freely, the variance is minimised by choosing c as small as possible, which, in many practical circumstances, is zero. To state these results another way, which might correspond more to the advice often given to industrial experimenters:

(i) Three levels, centred at the expected optimum and covering as large a range as possible, should be used.

(ii) If the possible range of values is restricted on one side of the optimum (usually the lower end at zero), then the three levels should be centred about 10% further from this limit than the expected optimum.

In practice, experimenters are extremely reluctant to use values as high as twice their guessed optimum, and the statistician will have to argue strongly to make the range of observations sufficiently wide to provide efficient estimates. The experimenters' reluctance is, in part, correct, since it stems from an unexpressed doubt about the validity of the quadratic response function (or, for that matter, any specified response function) over a wide range. Therefore, the choice of design must reflect a compromise between pressures. A wide range of x values is required to reduce the variance of parameter estimates. But the range of x values should be restricted so that the assumed response model is a good approximation. This is the basis of the response surface methodology described in the next chapter.

Example 16.2 The second example shows a method of solution through computational rather than analytic methods. For most practically relevant questions of design for response functions, the criterion of primary interest is a variance of either a non-linear parameter or a non-linear combination of linear parameters. Therefore, the variance to be minimised is an approximate, asymptotic variance, and a crucial component in the choice of the factor levels is the assumption that the approximation is good, and that it is equally good for all sets of x observations. The adequacy of the approximation can be assessed through simulation, and unfortunately it appears that in some particular cases the usual first-order approximation may become inadequate for precisely those designs (choices of x values) which minimise the approximate variance.

The use of simulation to check results derived from an approximate variance function suggests the direct use of simulation. If the efficiency of any design is to be assessed by the simulated variance of the estimate of the parameter criterion, then the optimality criterion reduces to a search procedure. We require to find the design for which the simulation variance of the estimate of the parametric combination is minimised. Since the simulation variance is not exact, we are searching for the optimum of a function which we cannot observe exactly, but only with error. The size of the simulation sample may be varied to reflect the precision required. Thus, when the design is some way from the optimal design, only a small sample is required, but, near the optimal design, larger samples are needed to obtain more precise information about the true variance function. In an asymptotic model such as the exponential

$$E(y) = A(1 - e^{-kx})$$

the estimation of the optimum position is not necessary, since y increases monotonically with x. However, if the input to the system, x, has a non-zero cost, then the interest is in the value of x for which dy/dx is equal to the cost of x. The economic yield to be maximised is

$$E(y) = A(1 - e^{-kx}) - cx,$$

and the optimum value of x is given by

$$dy/dx = Ake^{-kx} - c = 0,$$

i.e.

$$x_m = -(1/k) \log(kA/c).$$

Table 16.1. *Asymptotic and simulation variances for location of maximum: (a) approximate asymptotic variance for x_m; (b) simulation of variances for x_m*

(a)

	x_1/x_2			
	0.2	0.4	0.6	0.8
$x_2 = 12.5$	0.0307	0.0231	0.0382	0.1367
$x_2 = 25$	0.0048	0.0055	0.0125	0.0590
$x_2 = 37.5$	0.0026	0.0044	0.0140	0.0890
$x_2 = 50$	0.0022	0.0056	0.0258	0.2278

(b)

	x_1/x_2			
	0.2	0.4	0.6	0.8
$x_2 = 12.5$	0.0412	0.0288	0.0330	0.1780
$x_2 = 25$	0.0073	0.0092	0.0168	0.0886
$x_2 = 37.5$	0.0050	0.0048	0.0141	0.0957
$x_2 = 50$	0.0030	0.0052	0.0466	0.2480

In a study of optimal designs for the exponential model, Francis (1978) considered various criteria, and in particular examined designs for estimating the value x_m. Francis compared the precision of estimates of x_m using an approximate formula for the asymptotic variance of x_m and also performing a simulation study. The study included various sets of parameter values, and the results considered here are for $k = 0.1, A = 200$ and $c = 0.9$. Twenty simulations were performed for each of 16 design pairs of (x_1, x_2) values using ten observations at x_1 and ten at x_2 with $\sigma^2 = \frac{1}{6}$. The results for the approximate asymptotic variance and the observed simulation variance are shown in Table 16.1(a) and (b). Various patterns emerge from these results. It is clear that the approximate asymptotic variance gives an underestimate of the sampling variance. However, the general pattern of optimality is similar for the two criteria with the optimal x_t/x_2 ratio decreasing as x_2 increases. The results suggest that the optimal x_1/x_2 ratio is higher for the simulated variances than for the asymptotic variances. To establish whether the asymptotic variance formula gives an accurate estimate of the optimal x_1/x_2 values for any x_2 level, a more extensive simulation study in the neighbourhood of the optimum would be needed. However, a reasonable estimate of the optimal ratio x_1/x_2 can be obtained from the investigation of either the asymptotic variance or the simulation variance. When x_2 is small, about 12.5, the optimal ratio is about 0.4, and when x_2 is large, about 50, the optimal ratio diminishes to a bit less than 0.2.

This and the other examples considered here show that the design which minimises the required variance depends on a prior estimate of (at least some of) the model parameters. In many cases it is reasonable to assume values which are accurate enough for the purposes of designing a reasonably good experiment. However, sometimes it is clear that a design which is best for some plausible prior estimate is substantially inferior for other plausible prior estimates. In such cases, it can be beneficial to specify a joint prior distribution of values of the parameters and seek to minimise the expectation of the relevant variance over this prior distribution. Since the computation of the expectation usually requires sampling from the prior distribution, this idea fits in naturally with the simulation approach illustrated in the previous example. The implementation requires only the modification that the values of A and k are sampled from the prior for each simulation run, along with errors being sampled as above. However, the extra variance introduced means that considerably more simulation runs will be needed. However, widely available computers are now powerful enough to run these simulations, at least for fairly small experiments where the asymptotic approximation is poorest, so we expect to see more of this approach in the future.

16.4 Optimal design theory

In linear models, the information about the precision of parameters is contained in the variance–covariance matrix of the parameter estimates. The variance of a single parameter or the variance of a linear combination of parameters, such as $\alpha + 3\beta - 2\gamma$, may be derived by evaluating that variance directly from the variance–covariance matrix. However, interest is not always confined to a single criterion, and there will often be interest in several aspects of the parameters of the model.

One approach to this diversity of interest is to assume that all parameters of the model are of equal interest, and to consider how to estimate the parameters as a whole as efficiently as possible. The variance–covariance matrix of the set of parameters provides a possible measure. In the same way that a confidence interval for a single parameter is calculated from the variance for the parameter estimate, a confidence region for a set of parameters may be derived from the variance–covariance matrix of the parameters, and for a fixed number of treatments the volume of the region is proportional to the determinant of the variance–covariance matrix. The smaller the volume of the region, the smaller the 'uncertainty' about the parameters and, correspondingly, the greater the precision of estimation of the parameters as a whole.

The minimisation of the confidence region can be achieved by minimising the determinant of the variance–covariance matrix, referred to as the generalised variance, or equivalently by maximising the determinant of the least squares matrix, $|A'A| = |C|$. The two simplest, and only important, one-dimensional response functions with linear parameters, for which the generalised variance may be examined, are the linear and quadratic polynomials.

For $y = \alpha + \beta x$, the least squares matrix is

$$C = \begin{bmatrix} n & \sum x \\ \sum x & \sum x^2 \end{bmatrix}$$

and the generalised variance is

$$G = \sigma^2 \Big/ \left\{ n \sum x^2 - \left(\sum x \right)^2 \right\}.$$

It may be noted that minimising G is equivalent to minimising the variance of $\hat{\beta}$, which we considered in the previous section. This equivalence of the generalised variance and a specific criterion variance is exceptional.

For the quadratic model

$$E(y) = \alpha + \beta x + \gamma x^2,$$

the least squares matrix is

$$C = \begin{bmatrix} n & \sum x & \sum x^2 \\ \sum x & \sum x^2 & \sum x^3 \\ \sum x^2 & \sum x^3 & \sum x^4 \end{bmatrix}$$

and the generalised variance is

$$G = \sigma^2 / \left(S_0 S_2 S_4 + 2 S_1 S_2 S_3 - S_0 S_3^2 - S_1^2 S_4 - S_2^3 \right),$$

where $S_i = \sum x^i$. The variances of $\hat{\beta}$ and $\hat{\gamma}$ are

$$\text{Var}(\hat{\beta}) = G\left(S_0 S_4 - S_2^2\right)$$

and

$$\text{Var}(\hat{\gamma}) = G\left(S_0 S_2 - S_1^2\right),$$

and clearly both provide criteria different from that of the generalised variance.

There is a large amount of general theory deriving from studies of the generalised variance criterion, also known as D-optimality. There are other forms of optimality which may be considered as alternatives to D. Those most used in practice, A-, G- and I-optimality, also consider the parameters symmetrically but express the precisions of parameters in different combined forms. Perhaps the most natural is A-optimality, which minimises the average of the variances of the parameter estimates. This criterion is, however, dependent on the specific parameterisation used, as well as on the particular coding used for the treatment factor, so these have to be chosen very carefully for each specific experiment.

Although the concept of generalised variance is defined exactly only for linear models, there is an equivalent concept for non-linear models. This is defined as the determinant of the matrix of approximate, or asymptotic, variances and covariances at prior estimates of the parameters. These are derivable from the second differentials of the sum of squared deviations of yield from its expected value, $\sum \{y - E(y)\}^2$, or equivalently from the likelihood, when normal distribution assumptions are made. The adequacy of the approximation may be assessed by simulation in exactly the same way that is necessary for a specific parameter criterion and averaging over prior distributions might be beneficial and can be implemented by simulation.

A further general comment about the generalised variance criterion is that its use is not restricted to quantitative factors. The concept of the overall precision of a set of parameters is applicable to any design problem since the determinant of the variance–covariance matrix is interpretable as the volume of the multi-dimensional confidence region of the parameters.

There are, for the response function models which are the subject matter of this chapter, two important conclusions which emerge from consideration of the generalised variance criterion. First, and most important, for any linear or non-linear model, the optimal design in the generalised variance sense rarely requires more distinct observations than there are parameters in the assumed model, although when prior distributions are used the optimal design sometimes has one or two additional points. This supports the patterns observed in the investigations of specific parameter criteria. For the estimation of the slope of a straight line response, only two observation values were required; for the estimation of the optimum position in the quadratic, only three observation values were required. In general, for a p-parameter response function, and the associated p-dimensional generalised variance criterion, the optimal design usually concentrates information on p observation values, the use of more than p observation points dissipating the available information. The loss of information through the use of more than p observations is considered further in the penultimate section of this chapter.

The second aspect of the generalised variance criterion that is important is that it is a general criterion, and consequently must provide less information than is possible about specific parameters, or combinations of parameters. There can be no absolute rule for the choice between general design optimality and specific design optimality. In general, the more specific the interest in a particular parameter, the less appropriate will be the use of a general criterion. When several parameters are of interest, a general criterion will be appropriate, but experimenters should select a criterion carefully. It will usually be possible to define a more relevant criterion than the generalised variance.

16.5 Discrimination

The main concern of an experimenter presented with the optimal design for estimating the slope of a linear regression, i.e. half the observations at each end of the permissible range, is the absence of observations at intermediate points, and in particular at the centre of the range. This concern arises quite correctly from a need to check the assumed model. More generally, there is a class of design criteria deriving from the desire to discriminate between models. Often in biological research there are alternative models used to represent the response to a range of stimuli. The desire to compare alternative models, and to design an experiment to determine which model is appropriate, or, more reasonably, which is not appropriate, is a proper stimulus for an experiment.

There are, therefore, two reasons for considering discrimination as a criterion for choosing a design: either as a subsidiary criterion, necessary before the estimation which is the primary objective can be confidently applied, or as the primary criterion in establishing appropriate classes of models. The former is relatively simple, and is equivalent to a different estimation criterion. The latter is unrealistic, in that it is inconceivable that after identifying the appropriate model, experimenters would not also want to estimate its parameters. The difference, therefore, at least when models are nested is only in the relative importance of discrimination and estimation. Here we discuss the case where discrimination is a secondary criterion.

The testing of an assumed model is most simply represented as a comparison between the assumed model and a more complicated, but related, model. In this situation, the more complicated model will include the assumed model as a special case. Often there will be a single additional parameter in the more complicated model. For example, the natural expression of the doubt about the relevance of a linear regression model, given that there is good reason for believing that the linear regression model might be relevant, is to include a quadratic term, γx^2, in the model. The check of the adequacy of the linear regression may be achieved by considering the SS for fitting the quadratic term, in addition to the linear regression. In general, the adequacy of an assumed model may be checked by assessing the improvement in the fit due to an additional parameter and, when the additional parameter is an extra, linear, term in the model, the appropriate criterion is the extra SS for fitting that term.

However, the criterion of maximising the SS due to fitting an additional parameter is equivalent to the criterion of minimising the variance of the estimate of the extra parameter. For linear models, this can be seen immediately from the results of Chapter 4. The SS for fitting a single additional parameter θ_2, is

$$\hat{\theta}_2'(C^{22})^{-1}\hat{\theta}_2$$

and the variance of the estimate of θ_2 is $\sigma^2 C^{22}$, where C^{22} is the diagonal element corresponding to the parameter θ_2, in the variance matrix for the fitted parameters. The logic clearly applies to non-linear models also, though no corresponding exact result exists.

Example 16.3 Consider the discrimination against a quadratic model as an alternative to a linear regression model. The terminology is the same as that used in previous sections. The expected value of the SS for fitting γ, allowing for the fitting of α and β, is

$$E(\hat{\gamma}^2)\{(S_0 S_2 S_4 - S_1^2 S_4 - S_0 S_3^2 + 2S_1 S_2 S_3 - S_2^3)/(S_0 S_2 - S_1^2)\}.$$

Equivalently, the variance of $\hat{\gamma}$ is

$$\text{Var}(\hat{\gamma}) = \sigma^2(S_0 S_2 - S_1^2)/(S_0 S_2 S_4 - S_1^2 S_4 - S_2^3 + 2S_1 S_2 S_3 - S_0 S_3^2).$$

Suppose observations are taken at three equally spaced values $(0, x, 2x)$, with n_1, n_2 and n_3 replications, respectively, then

$$\text{Var}(\hat{\gamma}) = \sigma^2(n_1 n_2 + 4n_1 n_3 + n_2 n_3)/(4n_1 n_2 n_3 x^4).$$

It is rapidly verified that, given a fixed total, $n = n_1 + n_2 + n_3$, the variance is minimised by taking

$$n_1 = n_3 = n/4, \quad n_2 = n/2.$$

This may be recognised as sensible by considering that the quadratic parameter measures the difference between the response at the middle of the range and the average response at the two ends of the range. The resources are equally split between the middle and the ends.

16.6 Designs for competing criteria

If an experiment has a single objective, then it is comparatively easy for the statistician to determine the best design in the sense of satisfying that objective. The objective may be specific or general. The principle of how to find an optimal design for any single criterion is hopefully clear from the examples in the previous sections, though considerable computation may be necessary to achieve the design.

Table 16.2. *Values of four criteria for several designs*

			Criteria			
Observations			Var($\hat{\alpha}$)	Var($\hat{\beta}$)	Var($\hat{\alpha} + 5\hat{\beta}$)	Var($\hat{\alpha} + 8\hat{\beta}$)
0	0	10	0.500	0.015	0.375	0.660
0	10	10	1.000	0.015	0.375	0.360
0	5	10	0.833	0.020	0.333	0.385
2	5	8	1.722	0.056	0.333	0.833
6	8	10	8.333	0.125	1.458	0.333
5	9	10	4.905	0.071	0.976	0.333
0	0	2	0.500	0.375	7.375	20.500
0	1	5	0.619	0.071	0.976	3.286

However, experimenters rarely restrict themselves to a single criterion but seek to answer several questions. Statisticians may regret this tendency but should also recognise that it is not only natural but also scientifically proper. The practical design problem therefore requires consideration of more than one objective, and it is necessary to examine the suboptimality of designs across a range of objectives.

Example 16.4 The principles of suboptimality are illustrated for a trivial example in which the response model is a straight line. The philosophy is the same for other response models but the details are less clear. The generalised variance optimal design is identical with that for optimising the variance of the slope estimate. The variances for four criteria for each of eight three-point designs with all points lying within the range (0, 10) are shown in Table 16.2. The results for individual criteria are fairly obvious. For the slope (or equivalently the general) criterion, the optimal design requires observations only at the two ends of the range. For the intercept criterion two observations must be at zero, and, for each criterion concerned with prediction at a particular point, the optimal design requires the mean of the three design points to be at the point for which prediction is required. If we try to select a single design from those considered in Table 16.2 which minimises the maximum ratio of variance/minimum variance for the set of four criteria considered in this example, then the design with equally spaced values spanning the complete range should be chosen. The optimal design for the general criterion is not much worse, but the crucial distinction is between designs spanning the possible range and those with a reduced range. Each of the latter is very poor for at least one criterion.

However, the more important aspect of this simple example is the degree of suboptimality for one criterion when a design chosen for a different criterion is used. There is no design which is optimal for any of the four criteria, which does not give an inefficiency of at least a factor of 2 for at least one other criterion. Therefore, before accepting a design which is optimal for one criterion, it is necessary to consider the efficiency of the design for other criteria of interest. It will rarely be appropriate to use a design which does not cover the possible range of values.

The specific conclusions from the above example should not be used to say that a particular design is good 'in general'. We concluded that the design with equally spaced points was

Table 16.3. *Efficiency of estimation of the linear slope and of the discrimination power against a quadratic effect for various designs*

Design observations					Variance efficiency $S_0 S_2 / 2000^2$	Power efficiency $(S_0 S_4 - S_2^2)/(500 S_0)$
-1	-0.5	0	0.5	1.0		
1000	0	0	0	1000	1.00	0.00
900	0	200	0	900	0.90	0.36
800	0	400	0	800	0.80	0.64
700	0	600	0	700	0.70	0.84
600	0	800	0	600	0.60	0.96
500	0	1000	0	500	0.50	1.00
400	400	400	400	400	0.50	0.70
600	300	200	300	600	0.675	0.65
200	400	800	400	200	0.30	0.54
800	100	200	100	800	0.825	0.50

the best compromise for the competing objectives of interest in this experiment. We cannot extrapolate from this to conclude that this design is good in any more general sense. In fact, studies of designs under many different criteria which attempt to conclude that, for example, a *D*-optimal design is a generally good choice, because it is not too inefficient under many criteria are misguided. The only conclusion that can be drawn from such a study is that the criteria which have been included in the study are not too different from the *D* criterion. In order to find an appropriate design for an experiment with complex objectives, there is really no alternative to carrying out the kind of study illustrated here.

Example 16.5 Consider again the problem of the student investigating the decay rate of seed germination with time. The optimal design, assuming a linear regression model and the criterion of minimising the variance of the slope, is to test 1000 seeds at the beginning of the two-year period and 1000 at the end of the two years. This proposed design was greeted with some dismay by the student, and even after some discussion about why the design was optimal he was not prepared to accept the design although he understood the reasonableness of the design concept. Essentially he needed protection against the failure of the linearity assumption. From consideration of the optimal design to discriminate between the straight line and the simplest alternative, the quadratic, we know that this would require 1000 seeds tested at the middle of the two-year period with 500 at each end. For estimating the slope, this design, which is optimal for discrimination, is only 50% efficient. Essentially, the two criteria are completely incompatible. The efficiencies (i) for estimating the slope compared with the optimal design (1000, 0, 1000), and (ii) for the power of discrimination compared with the optimal design (500, 1000, 500), are shown in Table 16.3.

The design finally chosen involved a compromise, as all practical designs must. Seeds were tested at six-monthly intervals, with 800 at the beginning and end, 200 at the middle (one year) and 100 each after six and 18 months. This design is 82.5% efficient for estimating the slope, it gives a reasonable power for detection of the non-linearity (which turned out not to be detectable), and it allowed the student to be 'doing something' throughout his research period!

Historically, little work has yet been done on compromise designs for particular problems, although this is an area of much current research interest. We believe that there is sufficient evidence to identify some general principles for the practical selection of levels.

(1) It is essential to know what kind of model is intended for the analysis of results, and to be able to define approximately the range of levels of x for which the model should be appropriate.
(2) The levels chosen should cover the maximal range of levels of x, for which the assumed model is believed to be appropriate, except in cases when the true response at one or both ends of the range is known, in which case they should be moved some way towards the middle.
(3) The number of levels should be the same as the number of parameters in the intended model, or one or two more. For any single criterion most optimal designs appear to need no more levels than the number of parameters in the model, and this is of course the minimum number of levels necessary to fit the model, although for Bayesian optimality one or two additional points are sometimes needed. The extra levels provide both for a conflict of interests between the competing criteria, and some protection against major failure of the model. The argument for dispersing the fixed total resources among many excess levels to detect any of a large number of possible discontinuities of the model is faulty because there will not be adequate precision at any point to detect a discontinuity, while the dispersal of resources reduces precision for the major criteria, sometimes to the extent of being unable to get useful information.

In practice, as mentioned earlier, models with more than four parameters for a single stimulus variable are not informative, and the number of levels for a quantitative factor should almost invariably be between three and six. In general these levels should be replicated equally, though this recommendation will not be appropriate for compromise designs when the relative importance of the two criteria is clearly unequal, as in Example 16.5.

This advice is, of course, for physical experiments. In so-called computer experiments, in which the response is the output of a deterministic, but highly complicated, calculation, replication is meaningless and there should always be as many different levels of each factor as there are runs. This is an area of increasing importance in applications and much methodological research, and will no doubt continue to be for some years to come. However, it seems to us to be misleading to call such studies experiments, given that there are no treatments and that randomisation, replication and blocking or control are all meaningless, and we will not discuss them further in this book.

16.7 Systematic designs

There is one form of experimentation where the advice to use a small number of levels may be ignored. In some agricultural crop experiments the requirement that the conditions of growth for the harvested crop shall be representative of the conditions for large-scale production of the crop for a particular treatment may lead to harvesting only a small part at the centre of

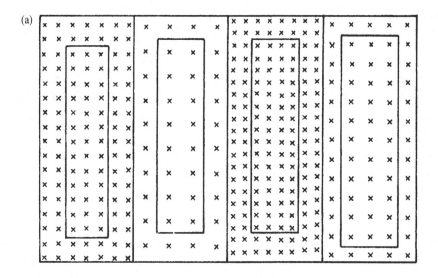

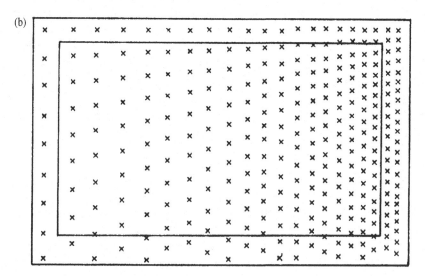

Figure 16.3 Efficient use of area with a systematic design: harvested area is shown within rectangles for (a) four spacings in randomised design, (b) a systematic design.

each experimental plot, the surrounding area being a guard area. In the particular case of experiments to compare different crop spacings, the harvested proportion of an experiment to compare four densities in a randomised block design may be as low as 40% as shown in Figure 16.3(a). Systematic designs have been used for such experiments, following the ideas of Nelder (1962) and have also been used for several other types of treatment, including fertilisers and crop protection chemicals.

The concept of a systematic design is that over a series of subplots the level of the quantitative factor varies systematically but slowly (for example 10% change) through the sequence of subplots. Each particular treatment level is then surrounded by treatment levels differing only slightly, and therefore guard areas are unnecessary. Examples of a Nelder fan

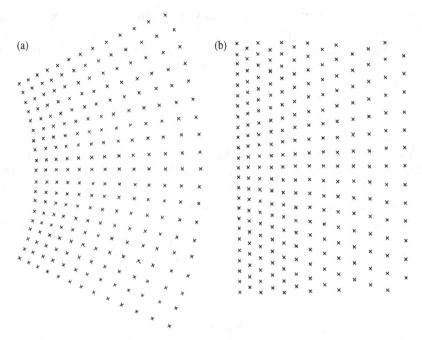

Figure 16.4 Systematic designs for changing plant density (a) fan design;
(b) parallel rows.

design and a row design for systematic density variation are shown in Figure 16.4. When
a crop would normally be planted on a rectangular lattice, the planting arrangement about
individual plants is distorted by the systematic arrangement, as can be seen in Figure 16.4(b).
The fan design retains a constant planting arrangement while the density is varied. Fans
are rather awkward to arrange in the field but they have successfully been used in many
experiments particularly for tree crops. For comparison with the randomised block design
the harvested area for a systematic design is shown in Figure 16.3(b), and may be as much
as 80%. The inefficiency arising from using many levels must be weighed against the greater
proportion of the experimental resources used in the harvested data.

The analysis for systematic designs will be discussed in Section 18.5. Because the treatment levels are not randomised there may be doubts about the validity of an analysis of
variance. However, for most systematic designs the analysis of variance would not be the
most appropriate initial step in the analysis. Instead the fitting of a response function to
represent the effect of the systematic factor for each set of data should usually be used. The
subsequent analysis is based on comparisons of parameters of the fitted response functions.
Such comparisons may be between replicates of the systematic plot or between different
treatments applied to the different whole plots within which the systematic treatment level
variation occurs.

In general, the use of systematic treatment plots will occur within an experiment in which
there are other treatment factors which are applied to whole plots. The systematic treatment
then resembles a split plot factor, except that rather than comparing individual split plot
treatments the fitted response function represents the information from each main (or whole)

plot. The allocation of the other treatments to whole plots, the replication, blocking and randomisation of whole treatment plots are exactly the same as for ordinary, non-systematic experiments. In addition, the direction of the systematic change of the treatment level within each whole plot can, and should, be randomly selected.

Systematic designs are an extremely useful addition to the practical statistician's library of designs. They are not the same as the systematic designs which were used in the early twentieth century and which were, quite properly, replaced by the randomised designs of Fisher. Their use is limited, chiefly in the early stages of research programmes, but it would be foolish to discard them simply because they do not satisfy either of the general principles that treatment allocation should be randomised, or that the numbers of treatment levels should not exceed four. Many statisticians, aware of the importance of randomisation in the analysis of experimental data, have been slow to accept the advantages of the use of systematically arranged factor levels. The opposition to the use of systematic components of experimental designs is, we believe, based on a narrow view of experimentation, and of the methods of analysis which are appropriate for drawing conclusions from experimental data.

Exercises

16.1   How inefficient was the proposed design in Section 16.0(b) using eight equally spaced levels to estimate the slope of a straight line relationship?

16.2   Assuming a quadratic response function, $E(y) = \alpha + \beta x + \gamma x^2$, for an experiment in which y is observed once at each of three levels, $0, k$ and 1, where $0 < k < 1$, find the variance–covariance matrix of the least squares estimates of the parameters α, β and γ. Find:
(a) the algebraic form of the generalised variance;
and the variances of the estimates of:
(b) γ,
(c) the slope of the response at $x = 0.5$,
(d) the change of y between $x = 0$ and $x = 0.5$,
(e) the value of $E(y)$ at $x = 1$.
Discuss and explain the results, obtaining optimum values of k where these exist.

16.3   Assuming the model $E(y_i) = \alpha + \beta_1 x_{1i} + \beta_2 x_{2i} + \varepsilon_i$, n_1 observations are taken at $x_1 = x_2 = 0$, n_2 at $(1, 1)$ and n_3 at $(2, 0)$. Obtain expressions for the variances and covariances of $\hat{\beta}_1$ and $\hat{\beta}_2$ and hence show that for $n_1 + n_2 + n_3 = N$ (fixed):
(a) β_1 is estimated most precisely if $n_2 = 0$, $n_1 = n_3 = N/2$;
(b) β_2 is estimated most precisely if $n_2 = N/2$, $n_1 = n_3 = N/4$;
(c) the sum of variances is least when $n_1 = n_3 = N/(\sqrt{2} + 2)$ and $n_2 = \sqrt{2}N/(\sqrt{2} + 2)$;
(d) $\hat{\beta}_1$ and $\hat{\beta}_2$ are independent only if $n_1 = n_3$.

16.4   In an experiment to investigate the effect of temperature on the period of pupation of a certain species of butterfly, a total of $4n$ pupae are to be studied at three different temperatures $t_0, t_0 + 1$ and $t_0 + 2$. Equal numbers of pupae, n, are incubated at temperatures t_0 and $t_0 + 1$, whilst the remaining $2n$ are incubated at temperature $t_0 + 2$. Assuming that the relationship between time of pupation y and temperature t is of the form

$$E(y) = \alpha + \beta(t - t_0) + \gamma(t - t_0)^2 + \varepsilon,$$

where the ε are independent normally distributed errors with zero mean and common (unknown) variance σ^2, show that the least squares estimates of β and γ are given by

$$\hat{\beta} = 2\bar{y}_1 - \bar{y}_0 - \bar{y}_2,$$
$$\hat{\gamma} = \frac{1}{2}(\bar{y}_2 - \bar{y}_1),$$

where $\bar{y}_0, \bar{y}_1$ and $\bar{y}_2$ are the averages of the observations at temperatures $t_0, t_0 + 1$ and $t_0 + 2$, respectively.

If $\gamma > 0$, show that the minimum expected value of y is at a temperature $t_m = t_0 - \beta/2\gamma$, and that the variance of the estimator $\hat{t}_m = t_0 - \hat{\beta}/2\hat{\gamma}$ is approximately

$$\sigma^2(\beta^2 + 6\gamma\beta + 11\gamma)^2/(8n\gamma^4).$$

17

Multifactorial designs for quantitative factors

17.0 Preliminary examples

(*a*) An experiment was to be conducted to optimise the pretreatment and drying conditions for the production of high-quality potato cubes by a process of high-temperature puffing (Varnalis *et al.*, 2004). Four treatment factors, namely the blanching time, the sulfiting time, the initial drying time and the puffing time, were to be studied. No mechanistic model for the effects of the factors was available, so as an approximation it was decided to fit a second-order polynomial response surface model. It was expected that there could be day to day differences in response, but only six runs could be made per day. Six blocks (days), each containing six experimental units (runs) was considered to be a reasonable size of experiment. Hence, a suitable set of treatments and replications has to be chosen for the 36 runs and arranged in blocks of size 6.

(*b*) An experiment is to be conducted to investigate the effect of the levels of two additives on the quality of a cake production process (Bailey, 1982). The resources available are 25 experimental units using the same five ovens on each of five days. The blocking structure must therefore be a 5×5 row-and-column design. The design choices concern the treatment levels and their replication, and particularly the pattern of allocation of treatment combinations to experimental units. There is a particular interest in the linear $\times$ linear interaction term and the linear and quadratic main effects of each factor must also be regarded as important.

17.1 Experimental objectives

In the previous chapter, we considered parametric response functions for individual treatment factors. In this chapter, we always consider several factors and, in contrast to the philosophy of the previous chapter, the interest is in the overall response surface, rather than the response to individual factors. With several factors, there is rarely a theoretical model which can be assumed to describe the system, so purely empirical models are used, in which the individual parameters usually have little scientific meaning. Some specific characteristics of the surface are of special interest, but the general pattern of response is also important.

A specific characteristic which is often important is the optimum position, i.e. the combination of factor levels for which the expected response is a maximum (or minimum). The response to be maximised may be a simple yield or output, or it may be modified to allow for costs, but the overall principle is unchanged. Essentially, we are searching for the highest point on an unknown surface without assuming an established parametric model. In the immediate region of the optimum, the surface may be approximated by a simple model, the simplest sensible model being a second-degree polynomial, and such a model is often assumed, implicitly or explicitly. However, until the position of the optimum has been clearly identified it may not be known whether the planned experiment is in the neighborhood of the optimum.

A second characteristic of response surface exploration is that we are interested in predicting the pattern of yields, usually over quite a wide range of factor levels, and usually in a region about the optimum. Since prior information about the form of the surface is not available, interest may lie in any direction from the point at which experimentation is commenced. Thus, the axes of the factors, whose values are being varied to provide the exploration of the surface, do not necessarily define more interesting directions than other directions. Generally, the design of an experiment to explore the surface will cover a region of the surface and will have a more-or-less clearly defined centre which should correspond to the position which is believed, *before the experiment*, to be of primary interest. This may be an estimated position of the optimum, or economic optimum from some previous experiment, or simply a combination of 'standard' levels.

In constructing designs to explore response surfaces, we should not forget other ideas in experimental design, and one particularly important idea which is extremely relevant is that of blocking. In experiments to investigate response surfaces, as in other forms of experimentation, it is usually necessary to have replication of observations, and this should lead to consideration of blocking the experimental units. In addition, the set of combinations of factor levels for which the response is to be measured may be quite extensive and, for a sensible grouping of units into blocks, it will frequently not be possible to include all the different experimental treatments in a single block or even in two blocks. The choice of allocation of treatments to blocks has to be made efficient in terms of the particular form of analysis to be employed.

Another related aspect of design concerns the possibility of sequential experimentation. For investigations of industrial chemical processes, it may be sensible to plan each observation sequentially using the information from the results from all preceding observations. Alternatively, and more commonly, it may be appropriate to take small groups of observations sequentially, with each group being designed to take account of the results of previous groups. Again, when groups of observations are planned, the form of analysis to be applied must be considered to ensure that the group is of sufficient size to enable the analysis to be performed. Sequential ideas of experimentation can be important, even when observations must be acquired in large groups separated by long periods, as in agricultural crop trials.

Before these aspects of the design of response surface experiments are considered in detail, we must consider the form of analysis. The interest in the general pattern of the surface, with the intent of predicting yield over the region explored by the design, and probably with particular interest in the position of the optimum, implies the fitting of a multivariable response equation. In addition to the estimates of the parameters of the response surface, a measure of the lack of fit of the surface should be obtainable. This will require both an estimate of pure error and the use of more factor level combinations than there are parameters to be estimated for the response surface equation. The choice of experimental treatment combinations should provide lack-of-fit information sensitive to those departures from the form of the proposed response equation which might be thought more likely, so that the lack-of-fit test should be reasonably powerful.

The most frequently used response surface equation is the general second-degree polynomial:

$$y = \beta_0 + \sum_i \beta_i x_i + \sum_i \beta_{ii} x_i^2 + \sum_i \sum_{j>i} \beta_{ij} x_i x_j. \tag{17.1}$$

The advantages of this model are that it is extremely simple to fit, it is familiar to most experimenters in a variety of contexts and it has a simply defined optimum, given by the set of equations derived from the derivatives of y with respect to each x_i:

$$2\beta_{ii} x_i + \sum_{j \neq i} \beta_{ij} x_j = -\beta_i,$$

for each value of i, where $\beta_{ji} = \beta_{ij}$.

The practical use of the fitted second-degree polynomial and the associated equations for estimating the values of x_i required for the maximum yield is more difficult than it might appear. Clearly if there is a simple well-defined optimum and the observations are exactly those given by the quadratic model, then the fitted surface will be identical with the true model and the optimum position will be estimated exactly. However, the second-degree polynomial is only an approximation to the true model and observations always include a random component due to the particular experimental units. The set of random unit effects can very easily lead to a set of observations showing a different pattern from the assumed model. Not only will there be shifts of position, but frequently the fitted surface which summarises the observations will not have a simple optimum. When this happens the solution to the above equations will not define the position of the maximum but some other point where all the derivatives are zero.

In the one-dimensional second-degree response function the only practically interesting alternative to a maximum is a minimum. However, when there are two or more factors the random variation involved in the observations frequently leads to other apparent patterns. The simplest is the saddle point in two dimensions for which the surface should be envisaged from two perpendicular directions. In one, each vertical slice of the surface will show a maximum, in the other a minimum, hence the 'saddle'. A contour diagram of a saddle point surface is shown in Figure 17.1. In more than two dimensions, there are modifications of the simple saddle point, and the larger the number of dimensions, the greater is the probability of some form of saddle point. A detailed example of fitting a second-degree response surface with saddle point problems is considered in Chapter 20.

Although the second-degree polynomial is the most frequently used model, it is not the only one, nor is it always the best. As computational facilities improve alternative families of response surfaces, which do not have the restrictive characteristics of second-degree polynomials discussed in Section 16.1, should become more frequently used. Other functions which have been used are the generalised exponential

$$y = A\left(1 - e^{-\sum_i k_i x_i}\right)$$

and the general inverse second-degree polynomial. This latter is written in the form

$$y^{-1} = P(x_1, x_2, \ldots, x_k) \bigg/ \prod_{i=1}^{k} x_i,$$

Figure 17.1 Contours for a saddle point surface.

where $P(x_1, \ldots, x_k)$ is the general second-degree polynomial, or, alternatively,

$$y^{-1} = \beta_0 + \sum_i \beta_i x_i + \sum_i \gamma_i x_i^{-1} + \text{cross product terms.}$$

17.2 Response surface designs based on factorial treatment structures

Before considering specific requirements of response surface designs, we review some of the available treatment structures, with particular emphasis on the use of factorial structures. The first two designs which might be considered for response surface investigation are the 2^n and 3^n factorial structures.

The 2^n structure does not allow estimation of quadratic terms in the second-degree polynomial because, with only two levels of a factor, there can be no information on the change of slope. Therefore, the 2^n structure will provide information only on the gradient of the surface in all possible directions. For more than four variables, the 2^n structure uses many more combinations than are necessary for the estimation of the linear surface equation. If quadratic terms are omitted, then the polynomial response equation takes the form

$$y = \beta_0 + \sum_i \beta_i x_i + \sum_i \sum_{j>i} \beta_{ij} x_i x_j.$$

The pattern of increase of the number of treatment combinations (2^n), and the number of parameters to be fitted $\{1 + (n^2 + n)/2\}$, with the number of factors, n, is shown in Table 17.1, which demonstrates clearly the wastefulness of the 2^n series in terms of the information per treatment combination. In spite of this inefficiency, and the restriction to information about gradients in all possible directions, the 2^n experiment does offer some considerable benefits in the initial stages of an investigation of a response surface. If many factors are being

Table 17.1. *Numbers of treatments and parameters for two-level factorial designs*

n	2	3	4	5	6
Combinations	4	8	16	32	64
Parameters	4	7	11	16	22

Table 17.2. *Numbers of treatments and parameters for two-level factorial and fractional factorial designs*

n	2	3	4	5	6	7	8
Replicate	1	1	1	$\frac{1}{2}$	$\frac{1}{2}$	$\frac{1}{2}$	$\frac{1}{4}$
Combinations	4	8	16	16	32	64	64
Parameters	4	7	11	16	22	29	37

Table 17.3. *Numbers of treatments and parameters for three-level factorial designs*

n	2	3	4	5
Combinations	9	27	81	243
Parameters	6	10	15	21

used, fractional replicates may be used since the model to be fitted includes only main effect and two-factor interaction terms. Thus, any fractional replicate in which all main effects and two-factor interactions are not aliased with each other may be used. A half-replicate becomes appropriate when $n = 5$, and a quarter-replicate when $n = 8$. The minimum number of combinations required for estimation of all two-factor interactions, using the smallest feasible fractional replicate, may be compared with the number of parameters (see Table 17.2).

Although 2^n structures, using fractional replicates for large n, are suitable for initial investigations, the lack of information about quadratic effects makes the structures impractical for later experiments closer to the optimum, where the curvature becomes more important than linear trends. The quadratic parameters can be estimated from the 3^n structure and the equation to be fitted will be the full quadratic model, (17.1). However, the 3^n experimental structures suffer even more severely from the increasing size of the number of combinations compared with the number of parameters. The combinations and parameter numbers are shown in Table 17.3. Clearly, therefore, for $n = 3$ or more, the simple 3^n structure is extremely inefficient.

Table 17.4. *Numbers of treatments and parameters for central composite designs, with those using fractional replicates in brackets*

n	2	3	4	5	6
Combinations	9	15	25	43(27)	77(45)
Parameters	6	10	15	21	28

The first designs that were developed to replace the 3^n structures were called central composite designs because they were made up from several components. The initial component was a 2^n structure (or fractional replicate of a 2^n). The two levels were interpreted as being on either side of a central value for each factor, the combination of central values being thought of as the origin, or centre, of the design. The use of observations at the centre, although it increases the number of levels for each factor to three, is not sufficient to estimate all the quadratic coefficients. Indeed, when only a centre point is added to the 2^n structure, all quadratic coefficients will be aliased since they are estimated as the difference between the yield at the centre and the average of the yields at the two extreme levels. Therefore, additional points are added to the design on each axis (or equivalently for each factor). The three components of the design, levels being defined relative to a zero level at the origin, are thus:

(i) all 2^n combinations of ± 1 for each factor;
(ii) origin observations $(0, 0, 0, \ldots, 0)$;
(iii) axial observations $(\pm \alpha, 0, 0, \ldots, 0), (0, \pm \alpha, 0, \ldots, 0), \cdots, (0, 0, 0, \ldots, \pm \alpha)$.

The relationship between the number of observations for a central composite design, and the number of parameters, for various numbers of factors, n, is shown in Table 17.4, the numbers of combinations in brackets being for half-replicates of the basic 2^n component. The choice of the value of α is available to the experimenter. Usually α will be chosen to be between 1 and 1.5, with $\alpha = 1$ often having practical advantages, as it means that only three levels of each factor are needed. In calculating the numbers of observations we have included only a single observation at the origin. There are sometimes advantages in using more observations at the origin than at other combinations.

The central composite designs satisfy the general requirements of response surface designs which are that:

(i) the parameters of the model to be fitted can be estimated,
(ii) the number of treatment combinations is not allowed to become too large, and
(iii) the observations are spread fairly evenly over the region within which information about the surface is required.

The last characteristic is very important, but is difficult to appreciate because it involves an assessment of the spread of points in n-dimensional space. The situation for two factors provides only a partial insight into the problem and the three-factor case is already difficult to represent.

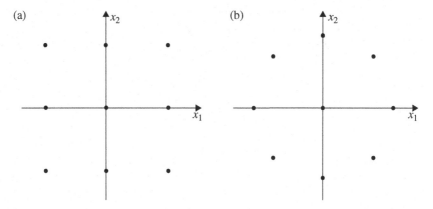

Figure 17.2 Two-factor designs: (a) factorial; (b) central composite.

In the case for two factors, the three-level factorial and standard composite designs are shown in Figure 17.2(a) and (b), the value of α for the composite design being chosen to be $\sqrt{2}$ (for reasons which will be discussed in the next section). It can be seen that the difference between the designs is that the 'corner' points in Figure 17.2(a) are relatively far away from the centre of the design, in contrast with the more even spread of distances from the centre in Figure 17.2(b). The choice between the designs depends on the relative interest in the different regions or directions.

For three factors the 3^3 factorial and the classical central composite design are represented in Figure 17.3(a) and (b). The major differences which should be perceived in the two diagrams are, first, the non-directional effect of the composite design and, second, the absence in the second design of observations with exactly two out of the three factors differing from the origin values.

The selection of components for inclusion in a composite design is somewhat arbitrary. This can be seen by considering the third design in Figure 17.3(c), which is composed solely of the points in Figure 17.3(a), which were omitted when forming Figure 17.3(b), together with the origin point. The cuboctahedron design of Figure 17.3(c) is an entirely sensible alternative design to consider for investigating a response surface. If there are three centre points, this is an example of a class of three-level response surface designs known as Box–Behnken designs.

In fact, especially with more factors, many different designs can be obtained by combining subsets with different numbers of factors at non-zero levels. The particular combinations of subsets used will depend on the experimental objectives and the different weights given to estimating the second-degree model, check for lack of fit and estimating error variance. These subset designs are discussed in more detail in Gilmour (2006). Computer search algorithms can also be used to find response surface designs which are optimal according to the D or some other criterion. Although easily obtained, these designs suffer from an inability to estimate error or check for lack of fit, so usually have to be modified for practical experimentation. Much of the discussion in the later sections of this chapter will centre on the classical central composite designs. They have been widely used in practical experimentation and are known

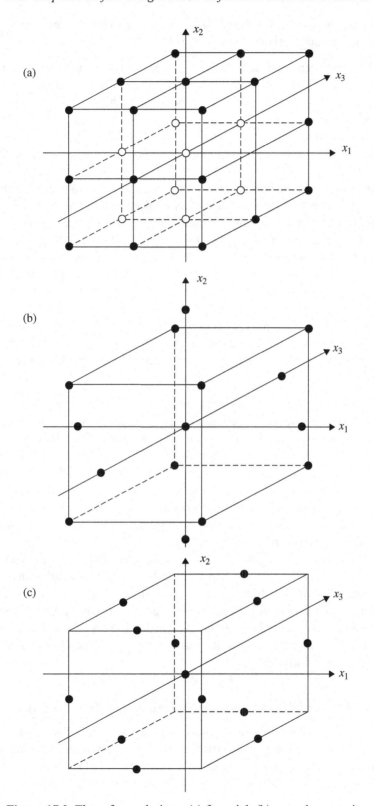

Figure 17.3 Three-factor designs: (a) factorial; (b) central composite;
(c) cuboctahedron.

to provide practically useful information. However, it is important to recognise that they are certainly not the only appropriate form of design.

Example 17.1 High-temperature puffing is a process that leads to better quality dehydrated potatoes. Varnalis *et al.* (2004) reported the results of an experiment to optimise the pretreatment and drying conditions for the production of high-quality potato cubes. Four factors were to be studied, namely the blanching time, sulfiting time, initial drying time and puffing time, each at three levels. Only six runs could be made per day and day to day differences in the response were considered likely. Therefore, it was decided to use six blocks, each of six runs.

The design made up of the points of the type $(\pm 1, \pm 1, \pm 1, 0)$ plus four centre points is very efficient, gives relatively good estimation for all effects and is, unusually, orthogonal for estimating all effects including the quadratic effects. This design was recommended by Edmondson (1994) and studied further by Davis and Draper (1995). It does, however, have only 3 df for estimating pure error and so for this experiment a design made up of the central composite design points, with two replicates of the axial points and four centre runs, which is almost as good in terms of efficiencies and allows 11 df for pure error, was preferred.

A search algorithm was used to arrange this design into six blocks of size 6 and the design in Table 17.5 was obtained. This design has efficiency factors of 96.2–96.4% for linear parameters, 99.7% for quadratic parameters and 74.4–100% for cross-product terms. The pattern of efficiency factors was considered acceptable and this design was used in the experiment.

Various modifications of the basic central composite designs have been suggested and used, particularly in fertiliser response investigation. These usually reflect particular knowledge about the pattern of response. A very simple example is the situation where it is known that the requirements for two fertilisers are closely and positively related. Then it will be sensible to define two new treatment factors, the first being the sum of the two original factors (possibly an unequally weighted sum), and the second the ratio of the two original factors. If it is believed that the critical ratio of the two fertilisers is 3 : 2 for $X_1 : X_2$, then an appropriate design for a 3^2 factorial structure is that shown in Figure 17.4(a), the corresponding central composite design being shown in Figure 17.4(b).

An alternative family of designs to the central composite designs developed for fertiliser trials by Rojas (1963, 1972) are the San Cristobal designs. These all include the set of 2^n factorial combinations. In addition the type 1 San Cristobal designs have axial points asymmetrically placed at -1 and α, where α is larger than 1 and generally larger than the α values used in the corresponding central composite. The type 2 designs are even more asymmetrical with axial points only in the positive direction. The two types of design are illustrated in Figure 17.5(a) and (b).

17.3 Prediction properties of response surface designs

In the previous section alternative designs have been discussed in a general fashion, emphasising the need for a subjective approach to choosing a design. It would be wrong, however, to imply that there are no technical considerations which might influence choice. Two criteria which can provide further information are concerned with predictions of yield, and with

Table 17.5. *Design used for potato puffing experiment*

Block	Factors			
	X1	X2	X3	X4
1	−1	−1	−1	−1
	1	−1	1	−1
	−1	0	0	0
	0	1	0	0
	0	0	−1	0
	0	0	0	1
2	−1	−1	−1	1
	1	−1	1	1
	1	0	0	0
	0	1	0	0
	0	0	1	0
	0	0	0	−1
3	−1	−1	1	−1
	−1	1	−1	1
	1	−1	−1	1
	1	1	1	−1
	0	0	0	0
	0	0	0	0
4	−1	−1	1	1
	−1	1	1	−1
	1	−1	−1	−1
	1	1	−1	1
	0	0	0	0
	0	0	0	0
5	−1	1	−1	−1
	−1	1	1	1
	1	0	0	0
	0	−1	0	0
	0	0	1	0
	0	0	0	−1
6	1	1	−1	−1
	1	1	1	1
	−1	0	0	0
	0	−1	0	0
	0	0	−1	0
	0	0	0	1

testing the fit of the assumed model. For any particular design, the variance of the predicted yield at a point $(x_1, x_2, \ldots)$ on the surface may be calculated. This prediction variance can be plotted to display the variation of precision for prediction over the region of interest. The plots in Figure 17.6(a) and (b) are for the 3^2 and central composite designs shown in

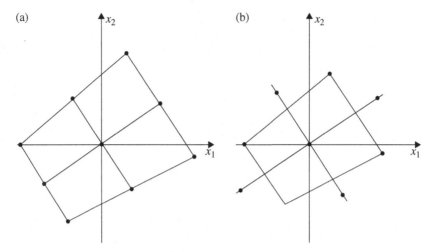

Figure 17.4 Two-factor designs with $(X_1 \times X_2)$ and (X_1/X_2) as factors:
(a) factorial; (b) central composite.

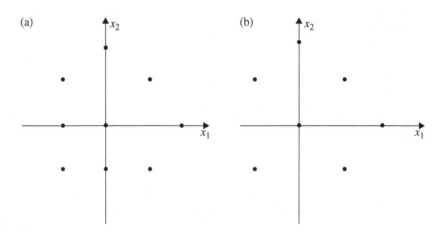

Figure 17.5 San Cristobal designs: (a) type 1; (b) type 2.

Figure 17.2(a) and (b). The plots show the contours of prediction variance, with the minimum variance represented by the lowest contour value. Only one-quarter of each design is shown since the other quarters are exact repetitions. For the 3^2 factorial design the prediction is most precise in areas towards the centre of each quarter, though the variance is not much larger at the centre of the complete design. For the central composite design the contours are circular about the centre of the whole design, and the most precise prediction is obtained at about three-quarters of the distance from the centre to the outer ring of observations. Further control over the prediction variance can be achieved through variation of the number of observations at the origin, and, in Figure 17.6(c) and (d), the effects on prediction variance of increasing the replication at the origin from one in Figure 17.6(b) to three in Figure 17.6(c) and five in Figure 17.6(d) are shown. To make the prediction variances comparable in all four diagrams the variances in Figure 17.6(c) and (d) have been scaled to allow for the greater number of observations in the designs with additional centre replication. The effect of extra

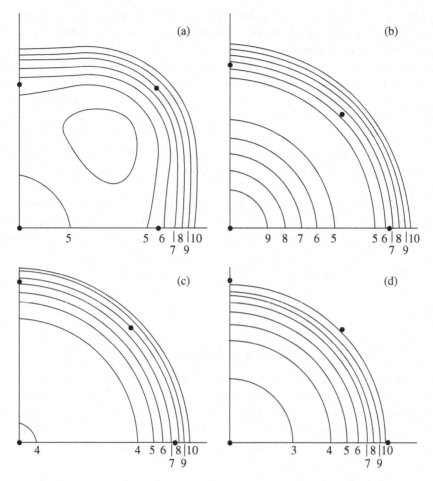

Figure 17.6 Contours for variance ($\hat{y}$): (a) 3^2; (b) central composite; (c) central composite with three origin points; (d) central composite with five origin points.

centre replication is inevitably to improve precision near the centre with some corresponding loss of precision towards the outside of the design.

In Figure 17.6(b), the contours of prediction variance are circular, and this arises from the choice of α in the composite design of Figure 17.2(b). The characteristic of circular contours for prediction variance is termed 'rotatability'. For the general central composite design for n factors, using a full replicate of the basic 2^n, it can be shown that the value of α, the distance of the axial point from the origin, should be chosen to be $2^{n/4}$ for rotatability. For $n = 2$ and 4, the distance of the axial point from the origin is identical to that for the 2^n treatment combinations. For $n = 3$ and 5, the two types of distance are very similar, but as n increases further the axial points are placed relatively further away. For most practical purposes, however, the observations can be thought of as placed on an approximately spherical shell about the origin.

In considering the variation of prediction variances, it should be noted that there are three important components of the design. The relative distances of various points of the design from the origin determine the shape of the contours. The relative replication of the origin points determines how the prediction variance varies with distance from the origin. And the range of the design (the distance of points from the origin) determines the region over which an acceptably low level of prediction variance can be maintained. Just as in the estimation designs of Chapter 16, where for a particular assumed response model more efficient estimates are obtained by using only a minimum number of different observations, so in the composite design there is no real benefit of introducing both inner and outer rings of observations. In any particular direction through the origin, the information about the quadratic response is essentially provided by three values, two at the circumference of the shell and one at the origin, and all necessary changes in patterns of prediction variances can be achieved through controlling the origin replication and the distance of the observations from the origin.

It can be argued that too much emphasis on prediction of the expected response across the region of experimentation misses the more important predictions of differences in response between different points on the response surface. If, as is often the case in industrial experiments, the centre of the design space $(0, 0, \ldots)$ represents standard operating conditions, then the expected response at the origin is known before the experiment. Interest at $(x_1, x_2, \ldots)$ should focus on the prediction of the difference in expected response between this point and the origin. Therefore we should concentrate on $V[\hat{y}(x_1, x_2, \ldots) - \hat{y}(0, 0, \ldots)]$.

17.4 Lack of fit and confirmatory runs

Another important general characteristic of response surface designs is the capacity to test the fit of the model. In central composite designs, this is achieved by comparing the lack of fit SS (with df equal to the difference between the number of distinct observation points and the number of fitted parameters) with the pure error SS provided either by overall replication of the design observations, or by replication of origin observations. When $n = 2$ or 3, there will often be several replications of the complete design, but for $n = 5$ or 6, the design may be restricted to a single replicate. However, for $n = 5$ or 6, the origin replication will usually be quite substantial to achieve reasonable prediction variances in the neighbourhood of the origin, and this replication also permits a test of fit.

The test of fit using a variance ratio statistic from the lack-of-fit and pure error MS is of a very general nature. If any particular alternative form of model is suspected, then it should be possible to choose a design to optimise discrimination against the particular alternatives. In practice, this is rarely considered, and the general test is used.

If a central composite design with $\alpha = 1$, a Box–Behnken design or some other three-level subset (or other) design is used, then the general lack-of-fit test uses only higher-order interaction contrasts, since the pure cubic terms cannot be estimated with only three levels. Even if a few higher-order interactions are included in the model to improve the fit, predictions and the estimation of the location of the maximum point on the response surface rely on an assumption of symmetry in the shape of the response surface. This is a strong and, often, an unrealistic assumption. Especially if the optimum is predicted to be some distance from any of the design points, it is good practice, where possible, to make some confirmatory runs at and near the predicted optimum, usually along with some at standard operating conditions. It is common to find that the predicted maximum is not achievable and sometimes, when the fitted response surface falls steeply away from the maximum, the difference can be very large indeed. Nevertheless, even if the true maximum is not estimated very precisely, the results obtained at the predicted maximum can often represent a large improvement over standard operating conditions.

17.5 Blocking response surface designs

As in all experiments, the design of a response surface experiment must involve consideration of the natural variation amongst the experimental units. The composite designs most commonly used for response surface exploration involve quite large numbers of observations, and the fitted response equation will include a large number of parameters estimated from a complex set of equations. The need for efficient blocking in response surface designs has sometimes been given little importance, with the result that large blocks are often used in practice. When blocking has been considered, the requirement of orthogonal estimation of parameters has been given heavy weight because of the need to be able to fit the response equation with relatively simple calculations. It is clear from general consideration of blocking principles and the interpretation of mildly non-orthogonal effects that more efficient designs can be achieved through the use of non-orthogonal blocking.

The principles behind the choice of treatment allocation to blocks are essentially similar to those of Chapters 7 and 15. Ideally the comparisons of interest should be orthogonal to block differences. Where this ideal cannot be achieved, the level of non-orthogonality should be reduced to the minimum possible. To illustrate this general philosophy, we consider the two-factor composite design in detail, and then examine briefly a range of combinations of treatment set size and block size.

The two-factor composite design contains nine observations. Suppose that it is decided to duplicate the origin observation, giving ten observations per replicate. Suppose further that it is felt that blocks of ten observations will give an unnecessarily large value of σ^2, and it is decided to use blocks of five observations. Consider the model to be fitted:

$$y = \beta_0 + \beta_1 x_1 + \beta_2 x_2 + \beta_{11} x_1^2 + \beta_{22} x_2^2 + \beta_{12} x_1 x_2$$

and the ten observations:

$$y_1 \text{ at } (0, 0), y_2 \text{ at } (0, 0),$$
$$y_3 \text{ at } (1, 1), y_4 \text{ at } (1, -1), y_5 \text{ at } (-1, 1), y_6 \text{ at } (-1, -1),$$
$$y_7 \text{ at } (\sqrt{2}, 0), y_8 \text{ at } (-\sqrt{2}, 0),$$
$$y_9 \text{ at } (0, \sqrt{2}), y_{10} \text{ at } (0, -\sqrt{2}).$$

The least squares estimates of the parameters, ignoring block effects, are:

$$\hat{\beta}_0 = \frac{1}{2}(y_1 + y_2),$$

$$\hat{\beta}_1 = \frac{1}{8}\{y_3 + y_4 - y_5 - y_6 + \sqrt{2}(y_7 - y_8)\},$$

$$\hat{\beta}_2 = \frac{1}{8}\{y_3 - y_4 + y_5 - y_6 + \sqrt{2}(y_9 - y_{10})\},$$

$$\hat{\beta}_{11} = \frac{1}{32}\{-8(y_1 + y_2) + 2(y_3 + y_4 + y_5 + y_6) + 6(y_7 + y_8) - 2(y_9 + y_{10})\},$$

$$\hat{\beta}_{22} = \frac{1}{32}\{-8(y_1 + y_2) + 2(y_3 + y_4 + y_5 + y_6) - 2(y_7 + y_8) + 6(y_9 + y_{10})\},$$

$$\hat{\beta}_{12} = \frac{1}{4}(y_3 - y_4 - y_5 + y_6).$$

It is now apparent that splitting the treatments into two blocks:

$$I = (y_1, y_3, y_4, y_5, y_6)$$

and

$$II = (y_2, y_7, y_8, y_9, y_{10})$$

has no effect on the parameter estimates, because if block effects b_1, b_2 ($= -b_1$) are included in the model the expected values of the parameter estimates do not involve b_1, and b_2. In this example, then, the treatments can be allocated to blocks so that all parameters are estimated orthogonally to the blocks. In passing, it is worth noting that the parameters β_0, β_{11}, β_{22} are not mutually orthogonal, the correlations of the estimates being $(4/7)^{1/2}$ for β_0 and β_{11} and $(3/7)^{1/2}$ for β_{11} and β_{22}. Note also that, for any other required pattern of block sizes, it is not possible to achieve treatment allocation so that estimation of parameters is orthogonal to blocks. However, if blocks of four were required so that a double replicate would occupy five blocks, then there are various choices of block allocations such that block effects have little influence on parameter estimates. For example a basic set of five blocks

$$(y_1, y_2, y_7, y_8),$$
$$(y_1, y_2, y_9, y_{10}),$$
$$(y_3, y_4, y_5, y_6) \text{ twice,}$$
$$(y_7, y_8, y_9, y_{10})$$

provides orthogonal estimation for β_1, β_2 and β_{12}. An alternative set of five blocks

$$(y_1, y_3, y_7, y_8),$$
$$(y_1, y_4, y_6, y_7),$$
$$(y_2, y_3, y_6, y_9),$$
$$(y_2, y_5, y_9, y_{10}),$$
$$(y_5, y_4, y_8, y_{10})$$

leaves all estimates non-orthogonal to blocks but at mild levels of non-orthogonality.

In general, when a second-degree polynomial equation is fitted to data from a composite design, some consistent patterns in the estimation of parameters can be identified:

(i) the $\hat{\beta}_0$ estimate involves origin observations;
(ii) the $\hat{\beta}_i$ estimates involve two sets of comparisons, one for the 2^n component of the design, and one for the axial points;
(iii) the $\hat{\beta}_{ii}$ estimates involve all combinations, with the dominant comparisons being those between the axial points and the origin points;
(iv) the $\hat{\beta}_{ij}$ estimates involve comparisons only within the 2^n component of the design.

Hence, any orthogonal blocking system should keep the 2^n component of the design in blocks separate from the axial component. By examining the numbers of treatment combinations in the two components, sensible structures for blocking composite designs of various sizes can be developed. The sizes of the different components of the design are shown in Table 17.6. The numbers of origin observations suggested is quite arbitrary and will be determined by

Table 17.6. *Sizes of different components of central composite designs*

n	2	3	4	5	6
2^n component	4	8	16	32(16)	64(32)
Axial	4	6	8	10	12
Origin points	2	3	4	5	6

the requirements for the prediction variance. Usually, there should be an origin observation in each block.

To attempt to tabulate all possible reasonable designs is outside the philosophy of this book. Instead we make some suggestions for $n = 3$ and $n = 4$ and hope that the general principles will be clear.

$n = 3$ Two blocks of between seven and nine observations keeping the 2^3 and axial group in separate blocks would be very suitable. If smaller blocks are required, then the 2^3 can be split into two groups of four, confounding the three-factor interaction, which is not used in estimating the second-degree polynomial. The axial points can be split into pairs.

$n = 4$ Three blocks of eight combinations each with an added origin point is ideal. The four-factor interaction in the 2^4 component may be confounded. Blocks of four involve confounding one two-factor interaction (though the fractional partial confounding ideas of Section 15.4 may be appropriate here). The possibilities of (a) confounding the 2^4, (b) splitting pairs of axial points off the axial group, and (c) origin points, allow almost any block size to be catered for. Thus,

Block size = 4: Confound the 2^n in four blocks of four. Four further blocks each containing two axial points + two origin points.

Block size = 5: 2^4 confounded in four blocks of four with an origin point in each block. Two further blocks each with two pairs of axial points plus an origin point.

Block size = 6: 2^4 confounded in four blocks of four, two blocks also including an axial pair, and two including two origin points. A fifth block with two axial pairs plus two origin points.

Block size = 7: 2^4 confounded in four blocks of four with each block also containing an axial point pair and an origin point.

It can be seen that, as the number of factors increases and the sizes of the blocks decrease, it can become difficult to construct good block designs. There are simply too many parameters to keep track of simultaneously. Instead, we can use computer algorithms for arranging a given treatment design in blocks. That of Trinca and Gilmour (2000) is the closest in spirit to the methods described here, as it directly tries to minimise the average variance inflation caused by nonorthogonality. However, other algorithms based on the D- or A-optimality criterion can also be used. Although the designs obtained differ in their detailed properties, they have all been shown to have advantages over designs obtained by using techniques such as fractional partial confounding.

17.6 Experiments with mixtures

In some response surface experiments the factors which may be varied in the experiment are inevitably not independent, because an increase in the level of one factor necessitates an overall reduction in the total of the levels of the other factors. Many chemical processes, including production of edible materials, are based on mixing together several components. Experimentation may be required to investigate the effects of changing the proportions of different components. But since the total amount of material used in a process is fixed, the levels of the various factors to be investigated must sum to a fixed total.

The general principles developed in this chapter and the previous one apply also to experiments on the response functions and surfaces when the factors are components in a mixture. However, the technical details do require new formulation. The subject of mixture experiments has an extensive literature and there is a review of the subject in the book by Cornell (2002).

The fundamental difference between mixture experiments and experiments in which all factors may be varied independently is most clear when only two or three factors are considered. Suppose there are two factors in an investigation into the nutritional benefit of different types of bread. Families may be classified by the proportion of bread they consume which is made from white flour (factor A) or the proportion made from wholemeal flour (factor B). If different diets based on 25%, 50%, 75% and 100% white bread are to be compared, then it follows that in these diets there will be 75%, 50%, 25% and 0% wholemeal bread. The two factors are exactly related, and varying the levels of one inevitably produces a corresponding variation in the levels of the other. Effectively there is only one factor for variation.

Suppose, however, that bread is classified into three classes: white flour (W), partially refined flour (PR), wholemeal flour (WM). We might now consider ten treatments:

	W:	PR:	WM:
(1)	25%;	0%;	75%;
(2)	50%;	0%;	50%;
(3)	75%;	0%;	25%;
(4)	100%;	0%;	0%;
(5)	25%;	25%;	50%;
(6)	25%;	50%;	25%;
(7)	25%;	75%;	0%;
(8)	50%;	25%;	25%;
(9)	50%;	50%;	0%;
(10)	75%;	25%;	0%.

Now changing the level of one factor does not inevitably change the level of another (for example (1) → (2) does not change PR, (2) → (5) does not change WM) but the levels of the three factors are interrelated. The advantage of the choice of the ten treatments listed may be seen by examining Figure 17.7(a). The triangle represents the set of all possible diets in terms of the subdivision into W, PR or WM. With the restriction of including at least 25% W, the set of observations is spread evenly over the remaining area, in agreement with the general philosophy of factorial experimentation. If we attempt to impose a complete factorial structure for two factors, say W and WM, then the maximum levels of the two factors cannot

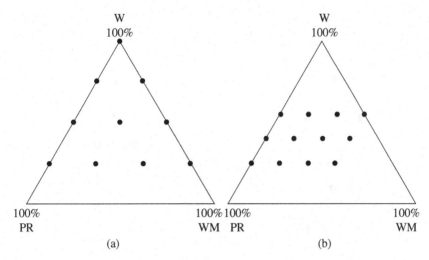

Figure 17.7 Mixture designs: (a) simplex design with at least 25% W; (b) factorial for W and WM factors. W = white flour, WM = wholemeal flour, PR = partially refined flour.

total more than 100%. The form of experiment that would result from a complete factorial design of, for example, 25%, 40%, 55% W, with 0%, 15%, 30%, 45% WM is represented in Figure 17.7(b) and plainly covers a much more restricted set of mixtures than the design of Figure 17.7(a).

The design in Figure 17.7(a) is described as a simplex design and is the form of design widely used for experiments on mixtures. It is not necessary for the levels of factors to vary widely, or even similarly. Some of the possible variations of form of simplex are shown in Figure 17.8. The concept of simplex designs generalises immediately to four or more factors. A very simple simplex in five factors with three levels each, and a more extensive design in four factors with four levels each, are tabulated in Table 17.7. In general, a simplex design with n factors and r levels of each factor results in

$$\frac{(n + r - 2)!}{(r - 1)!(n - 1)!}$$

combinations.

As in response surface designs without restrictions on the sets of levels of factors it will often be found useful to add centre points to simplex designs. Of the three examples previously considered only the bread example already includes a centre point observation (50% W, 25% PR, 25% WM). Unless the number of factors is one less than the number of levels per factor a centre point will not occur. For the designs of Table 17.7 we might add centre points:

$$(24, 19, 19, 19, 19) \text{ for Table } 17.7(a),$$

and

$$(25, 25, 25, 25) \text{ for Table } 17.7(b).$$

Such centre points may be replicated more frequently than other combinations to concentrate the predictive precision towards the centre of the design.

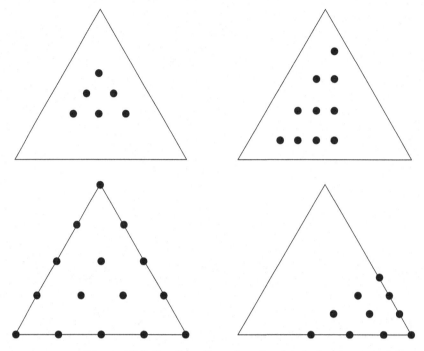

Figure 17.8 Possible alternative simplex designs.

As with factorial structures it is important to emphasise that it is not necessary to use a complete simplex design. The example of nitrogen applications for winter wheat can be viewed as containing a set of mixture designs at different levels of total nitrogen, with incomplete simplices at each level. This representation of the winter wheat experiment is displayed in Figure 17.9 with the levels of total nitrogen represented by different digits ($1 = 80, 2 = 160, 3 = 240, 4 = 320$).

In the same way that the interdependence of factor levels leads to the replacement of the complete factorial structure by a simplex design symmetric in the factors, the response function models used for mixture designs must also differ from those for factorial structured treatments. The simple linear response model in three variables is

$$y = \beta_0 + \beta_1 x_1 + \beta_2 x_2 + \beta_3 x_3.$$

Because the factor levels are interrelated,

$$x_1 + x_2 + x_3 = 1.$$

One variable could be expressed in terms of the other two and the model written in terms of only two variables. However, this destroys the symmetry of the model, and a simple form retaining symmetry is obtained by rewriting the model in the form

$$y = \beta_0(x_1 + x_2 + x_3) + \beta_1 x_1 + \beta_2 x_2 + \beta_3 x_3$$
$$= \gamma_1 x_1 + \gamma_2 x_2 + \gamma_3 x_3,$$

Table 17.7. *Two simplex designs for experiments with mixtures: (a) simplex design for five factors with three levels each; (b) simplex design for four factors with four levels each*

(a)

A	B	C	D	E		A	B	C	D	E
20	15	15	15	35		30	15	15	15	25
20	15	15	25	25		30	15	15	25	15
20	15	15	35	15		30	15	25	15	15
20	15	35	15	15		30	25	15	15	15
20	15	25	15	25		40	15	15	15	15
20	15	25	25	15						
20	35	15	15	15						
20	25	25	15	15						
20	25	15	25	15						
20	25	15	15	25						

(b)

A	B	C	D		A	B	C	D
10	10	10	70		30	10	10	50
10	10	30	50		30	10	30	30
10	10	50	30		30	10	50	10
10	10	70	10		30	30	10	30
10	30	10	50		30	30	30	10
10	30	30	30		30	50	10	10
10	30	50	10		50	10	10	30
10	50	10	30		50	10	30	10
10	50	30	10		50	30	10	10
10	70	10	10		70	10	10	10

where $\gamma_i = \beta_0 + \beta_i$. A similar reformulation for the second-degree model leads to

$$y = \gamma_1 x_1 + \gamma_2 x_2 + \gamma_3 x_3 + \gamma_{12} x_1 x_2 + \gamma_{13} x_1 x_3 + \gamma_{23} x_2 x_3,$$

and extensions to third-degree polynomials or the use of reciprocal terms can also be helpful.

By their nature, mixture designs are mainly concerned with the more general exploration of response surfaces discussed in this chapter rather than the particular parameter estimation of the previous chapter. The other important ideas of this chapter, concerning blocking of observations, controlling the prediction precision and testing the fit of the model are all relevant to mixture designs. For details the reader is referred again to Cornell (2002), though careful thought about the principles considered in this chapter should make the modifications required for application to mixture designs clear.

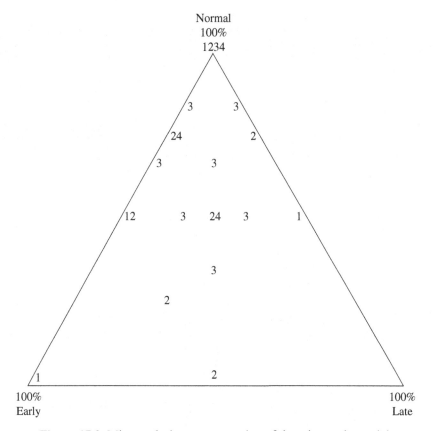

Figure 17.9 Mixture design representation of the winter wheat trial.

17.7 Non-linear response surfaces

Example 17.2 In the winter wheat experiment already mentioned twice in this chapter, there were 24 treatments in two blocks of 24 plots each. The experiment was repeated for three years at each of three farm sites: Home Farm, Upton Grey and Easton Manor. The mean yields, in tonnes per hectare from the three sites for one year, are given in Table 17.8. After examination of the data for the pairs of replicates and of the general pattern of response it was decided to fit models of the form

$$1/y = \beta_0 + \beta_1(z+c)^{-1} + \beta_2(z+c)^{-2} + \beta_3(x_E + d)^{-1} + \beta_4(x_L + f)^{-1},$$

where z, x_E and x_L are the levels of total nitrogen, early nitrogen and late nitrogen, respectively. The model includes eight parameters, three of which (c, d, f) represent base levels of nitrogen available in the soil, and are inherently non-linear parameters. The pattern of replicate variation suggested strongly that variation of $1/y$ was homogeneous. The models wrere therefore fitted by multiple regression with five linear and three non-linear parameters using the mean of the $1/y$ values from the two replicates as the dependent variable.

Table 17.8. *Design and data for winter wheat experiment*

	Nitrogen (kg/ha)			Mean yields (tonnes/ha)		
				Eastern	Home	Upton
Total	Early	Normal	Late	Manor	Farm	Grey
0	0	0	0	5.20	3.86	5.06
80	0	80	0	8.20	6.60	7.40
80	40	40	0	8.10	7.07	6.84
80	80	0	0	8.34	7.56	7.11
80	0	40	40	8.44	6.00	6.95
160	0	160	0	8.25	8.44	8.12
160	0	120	40	8.38	7.61	7.68
160	40	120	0	9.12	9.07	8.91
160	40	80	40	9.94	8.92	8.38
160	80	80	0	9.58	8.95	9.00
160	80	40	40	8.78	9.08	8.88
160	80	0	80	9.32	8.74	8.74
240	0	240	0	8.00	8.26	8.30
240	0	200	40	9.33	8.36	8.28
240	40	200	0	9.23	9.36	9.01
240	40	160	40	9.86	9.22	8.02
240	40	120	80	10.00	9.08	8.85
240	80	160	0	9.82	9.52	8.04
240	80	120	40	9.30	9.48	9.30
240	80	80	80	10.01	9.42	9.18
320	0	320	0	9.60	8.68	7.52
320	40	240	40	9.94	9.36	9.07
320	80	240	0	9.66	9.70	9.06
320	80	160	80	9.95	9.52	9.25

The estimation of the non-linear parameters was pursued by grid search methods to determine a minimum residual SS. In fact, it was found that the complete model contained more parameters than could be estimated properly and models including both β_1 and β_2 were not fitted. Some of the grid search results for the residual SS for four models,

(1) model $1/y = \beta_0 + \beta_1(z + c)^{-1}$,
(2) model $1/y = \beta_0 + \beta_1(z + c)^{-1} + \beta_3(x_E + d)^{-1}$,
(3) model $1/y = \beta_0 + \beta_1(z + c)^{-1} + \beta_3(x_E + d)^{-1} + \beta_4(x_L + f)^{-1}$,
(4) model $1/y = \beta_0 + \beta_2(z + c)^{-2} + \beta_3(x_E + d)^{-1} + \beta_4(x_L + f)^{-1}$,

are tabulated in Table 17.9 for the Home Farm data set. It is clear that the parameters d and f are very poorly determined. Similar results are obtained for the other two data sets and it was decided to set $d = f = 20$. The residual SS were then examined for a range of values of c for revised models

(2a) model $1/y = \beta_0 + \beta_1(z + c)^{-1} + \beta_3(x_E + 20)^{-1}$,
(3a) model $1/y = \beta_0 + \beta_1(z + c)^{-1} + \beta_3(x_E + 20)^{-1} + \beta_4(x_L + 20)^{-1}$,
(4a) model $1/y = \beta_0 + \beta_2(z + c)^{-2} + \beta_3(x_E + 20)^{-1}$,

Table 17.9. *Residual SS for four models for the winter wheat experiment: (a) model (1); (b) model (2); (c) model (3); (d) model (4)*

(a)

c	10	20	30	40	50	60
RSS	364	238	180	163	172	194

(b)

			c		
d	20	30	40	50	60
5	111	65	55	64	85
10	110	65	54	64	85
20	109	64	54	64	85
30	109	65	54	64	85
40	110	65	55	65	86

(c)

	$c =$	30			40			50		
f	$d =$	10	20	30	10	20	30	10	20	30
10		59	58	59	46	46	46	54	54	55
30		59	58	59	46	45	46	54	53	54
40		59	58	59	46	45	46	54	53	54

(d)

				c			
d	60	70	80	90	100	110	160
10	88	64	51	46	48	56	67
20	88	64	51	46	48	56	67
30	88	64	51	46	49	56	68

with the results for the three sites as shown in Table 17.10. Testing the hypothesis of invariance of c-values over the three sites gives the analysis of residual SS for model (4a), see Table 17.11, where the df are based on the numbers of linear and non-linear parameters fitted and are only approximate. The variation of c is clearly non-significant, assuming that the approximate F-test for the ratio of the MS is at all appropriate. Similar results are obtained for the other two models.

Comparison of the different models for the three sites shows that the inclusion of β_4 is never significant. The model with the quadratic term for total nitrogen is slightly superior to

Table 17.10. *SS for revised models: (a) model (2a); (b) model (2b); (c) model (2c)*

(a)

c	Home Farm	Easton Manor	Upton Grey
20	110	67	178
30	64	65	154
40	54	71	143
50	64	81	140
60	85	93	141

(b)

30	58	58	145
40	46	65	135
50	54	75	133

(c)

60	88	67	166
70	64	66	154
80	51	68	145
90	46	71	139
100	48	77	135
110	56	83	134
120	67	90	134

Table 17.11. *Analysis of residual SS for winter wheat experiment*

	SS	df
Separate c-values	246	59
Single c-value	256	61
Difference	10	2

that with a linear term and the fitted models selected are

$$\text{Home Farm: } 1/y = 0.094 + 1267(z + 90)^{-2} + 0.183(x_E + 10)^{-1},$$
$$\text{Easton Manor: } 1/y = 0.094 + 682(z + 90)^{-2} + 0.099(x_E + 10)^{-1},$$
$$\text{Upton Grey: } 1/y = 0.105 + 653(z + 90)^{-2} + 0.111(x_E + 10)^{-1}.$$

The residual MS from these fits may be compared with the error MS from the analysis of variance for the two blocks × 24 treatment designs giving the results in Table 17.12. Since the dependent variable used for the model fitting is the mean of two replicates the residual

Table 17.12. *Summary of results for winter wheat experiment*

	Model fitting			Analysis of variance
	RSS	df	RMS	EMS
Home Farm	46	20.3	2.3	4.3
Raston Manor	71	20.3	3.5	6.3
Upton Grey	139	20.3	6.8	9.8

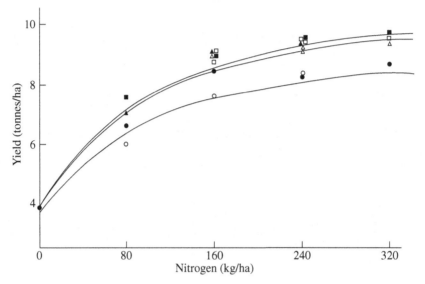

Figure 17.10 Fitted models and mean yields for winter wheat data for various
levels of early and late nitrogen: •(0, 0), ○(0, 40), △(40, 0), ▲(40, 40, or 80),
■(80, 0), □(80, 40 or 80).

MS should be approximately half of the corresponding error MS, and with the possible
exception of the Upton Grey data the models clearly provide an adequate fit. The general fit
for the Home Farm data is confirmed in Figure 17.10, though there is a suggestion that the
fitted inverse polynomials bend too gently.

As the number of factors increases, it becomes more difficult to fit non-linear response
surface models without running into numerical convergence problems. However, one class of
very flexible, but easily fitted, models is fractional polynomials. Originally suggested by Box
and Tidwell (1962) as transformations of the predictor variables, these consider replacing x_i
by $x_i^{\alpha_i}$, (or $\log x_i$ for $\alpha_i = 0$), where α_i is usually chosen from a limited set of values, such
as $\{-1, -0.5, 0, 0.5, 1, 2\}$, which allow simple interpretation. Royston and Altman (1994)
suggested a simple method of fitting such models using iterated least squares and Gilmour
and Trinca (2005) suggested initially trying non-linear least squares and only using Altman
and Royston's method if it fails. This allows inferences and model selection to be done in
an asymptotically valid way before simplifying for the purposes of interpretation. Fractional

polynomial response surfaces have not yet achieved much popularity, but we believe that this will change as more complex systems are studied in industrial and biological experiments.

Exercises

17.1 To investigate the effect of two fertilisers, nitrogen (N) and phosphate (P), on the yield of maize, an experiment consisting of 13 observations was performed using the following coded pairs of levels (x_1 and x_2) of N and P:

$$
\begin{array}{lccccccccccccc}
x_1 & -1 & -1 & 1 & 1 & -\sqrt{2} & \sqrt{2} & 0 & 0 & 0 & 0 & 0 & 0 & 0 \\
x_2 & -1 & 1 & -1 & 1 & 0 & 0 & -\sqrt{2} & \sqrt{2} & 0 & 0 & 0 & 0 & 0.
\end{array}
$$

The experimenter assumed that the yield of the ith observation may be represented by

$$
y_i = \beta_0 + \beta_1 x_{1i} + \beta_2 x_{2i} + \beta_{11} x_{1i}^2 + \beta_{12} x_{1i} x_{2i} + \beta_{22} x_{2i}^2 + \varepsilon_i,
$$

where the ε_i are independent and identically distributed with zero mean and variance σ^2.

Find the least squares estimators of the coefficients of the model, and show that the estimated yield for any combination of levels (x_1, x_2) has variance

$$
\sigma^2 \left(\frac{1}{5} - \frac{3\rho^2}{40} + \frac{23\rho^4}{160} \right),
$$

where $\rho^2 = x_1^2 + x_2^2$. Comment on the implications of this result.

17.2 A food technology researcher is to conduct an experiment in the laboratory of a food company to try to learn how to extend the lifetime of baked pastry products. The laboratory is available to her for five days, which would be used as blocks, and she can make at most five runs each day. The responses would not become available until after the experiment, so that it all has to be designed in advance. She decides to vary four factors, namely the amount of additive A, the amount of additive B, the mix speed and the baking time. For convenience, these are to be used at three levels each.

Ignoring blocking, consider various subset designs and recommend one to the experimenter. Explain your choices.

Arrange your chosen design in blocks by hand. If you have access to a suitable program, see if it can do better.

On arrival at the food company, the experimenter finds that they have given her a mixer which does not have variable speed. She asks you whether she can go ahead and use the design in the other three factors, or whether it is better to spend a day coming up with a better design to be run in the remaining four days. How would you respond?

17.3 In the investigation of a three-factor response surface, a cuboctahedron design is used, consisting of the 12 points $(0, \pm1, \pm1)$, $(\pm1, 0, \pm1)$, $(\pm1, \pm1, 0)$, plus four observations at the centre point $(0, 0, 0)$. Show how to fit the second-order surface:

$$
Y = \beta_0 + \sum_{i=1}^{3} \beta_i x_i + \sum_{i=1}^{3} \sum_{j=1}^{3} \beta_{ij} x_i x_j.
$$

Obtain variances of the estimates of βs and indicate which estimators are correlated. What test of the adequacy of the second-order fit is available?

17.4 The expected value of a response depends linearly on factor values $x_1, \ldots, x_k$, the errors of the responses being independently distributed with constant variance, σ^2. Prove that, whatever

Table 17.13. *Two response surface designs*

Design	Design points	No. of observations
A	$(0, -1)$	$n/8$
	$(1, 1)$	$3n/8$
	$(-1, 1)$	$3n/8$
B	$(1, -1)$	$n/8$
	$(-1, -1)$	$n/8$
	$(1, 1)$	$3n/8$
	$(-1, 1)$	$3n/8$

design is used with n observations, the variance of the estimated response at the centroid of the design points is σ^2/n.

With $k = 2$, the designs A and B shown in Table 17.13 both have their centroids at $(0, \frac{1}{2})$. Verify directly that the estimated response at this point has variance σ^2/n. Make a thorough comparison of the designs with respect to other properties that would be relevant.

17.5 A design is required for an experiment on three quantitative treatment factors, T_1, at five equally spaced levels, and T_2 and T_3 each at two levels, in two blocks of ten plots each. The equation

$$E(y) = \beta_0 + \beta_1 x_1 + \beta_{11} x_1^2 + \beta_{111} x_1^3 + \beta_{1111} x_1^4 + \beta_2 x_2 + \beta_3 x_3$$
$$+ \beta_{12} x_1 x_2 + \beta_{13} x_1 x_3 + \beta_{23} x_2 x_3 + \beta_{123} x_1 x_2 x_3$$

is to be fitted to the results; x_1, x_2, x_3 are the levels of the three factors. Construct a design confounding β_{23} as little as possible and leaving all other effects in the equation unconfounded.

Calculate the loss of information on β_{23}.

Determine which effects, if any, will be biased by variation in the quadratic effect of T_1 between different levels of T_2 and T_3. (The orthogonal polynomial quadratic coefficients for five equally spaced levels are 2, -1, -2, -1, 2.)

18

Split-unit designs

18.0 Preliminary examples

Both of the following examples come from situations where the experimenter designed an experiment without consulting a statistician. When the experimenter came to consult the statistician, with the experimental data, the first problem was to identify the structure of the design and then to provide a suitable analysis.

(*a*) The first experiment was concerned with the influences on the production of glasshouse tomatoes of differing air and soil temperatures. Eight glasshouse compartments were available, and these were paired in four 'blocks'. Each compartment contained two large troughs in which the tomatoes were grown and in each half of each trough the soil temperature could be heated to a required level or left unheated. In one compartment of each block the minimum air temperature was kept at 55 °F and in the other the minimum air temperature was 60 °F. In each trough, one half was maintained at the control soil temperature while the other half was at an increased temperature, the increased temperature being 65 °F for one trough, and 75 °F for the other trough. The design layout for one pair of compartments is shown in Figure 18.1. Yields of tomatoes from each half-trough were recorded.

(*b*) The second experiment was also concerned with heating, this time the heating characteristics of different forms of plastic pot situated in cold frames and of different forms of covers for the frames. Six wooden frames were divided into four quarters and four different forms of pot were allocated to the four quarters randomly in each frame. The frames were covered either by glass or by polythene, three frames having glass covers and three having polythene covers. All the heat was derived from solar energy. After one week of observation, the covers were exchanged so that each frame that had previously been covered by glass was now covered by polythene and vice versa. The pot positions were unaltered. The design is represented in Figure 18.2. For each pot, in each week, the soil temperature was recorded.

18.1 The practical need for split units

In many investigations in which two or more treatment factors are being investigated, the size of experimental unit which is appropriate for one treatment factor is different from that appropriate for another treatment factor. In the chemical industry, a process may have several stages, starting with a large mixing stage from which the material is subdivided into smaller quantities for the later chemical interactions; if the constituents of the original mixture are varied, the experimental unit must be the large container used for the initial stage, but for treatment factors involving changing the conditions for the subsequent stages of the process, the unit may be a subdivision of the large initial experimental unit.

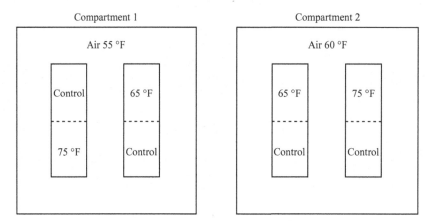

Figure 18.1 Design layout for one pair of compartments for the tomato experiment on the effects of air and soil heating.

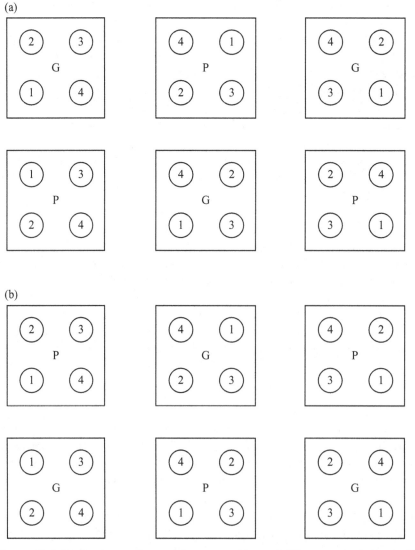

Figure 18.2 Design layout for experiment on effects of different materials on soil heating in pots: (a) week 1; (b) week 2. P = polythene, G = glass.

| | | Block I | | | | | Block II | | |
| | | Large unit | | | | | Large unit | | |
1	2	3	4	5	1	2	3	4	5
p_2q_1	p_5q_1	p_3q_2	p_1q_3	p_4q_1	p_3q_2	p_4q_3	p_2q_2	p_5q_1	p_1q_2
p_2q_3	p_5q_3	p_3q_1	p_1q_1	p_4q_3	p_3q_3	p_4q_2	p_2q_1	p_5q_3	p_1q_1
p_2q_2	p_5q_2	p_3q_3	p_1q_2	p_4q_2	p_3q_1	p_4q_1	p_2q_3	p_5q_2	p_1q_3

| | | Block III | | | | | Block IV | | |
| | | Large unit | | | | | Large unit | | |
1	2	3	4	5	1	2	3	4	5
p_5q_1	p_1q_3	p_3q_3	p_2q_2	p_4q_3	p_2q_3	p_4q_1	p_1q_2	p_3q_1	p_5q_2
p_5q_2	p_1q_2	p_3q_1	p_2q_3	p_4q_2	p_2q_1	p_4q_3	p_1q_1	p_3q_2	p_5q_3
p_5q_3	p_1q_1	p_3q_2	p_2q_1	p_4q_1	p_2q_2	p_4q_2	p_1q_3	p_3q_3	p_5q_1

Figure 18.3 Experimental plan for a split plot design with four blocks each divided into five large plots, each further divided into three split plots.

In experiments on dairy cattle, the experimental treatments may include different treatments of the pasture and variation of milking conditions. Pasture treatments must be applied to groups of animals if they are to be representative of farming practice, but milking method treatments may be applied to individual animals. This division of treatments into those necessarily applied to groups of subjects and those applied to individuals also occurs in psychological experiments. In field crop experiments, or glasshouse experiments, some treatments such as cultivation or irrigation methods or electrical heating conditions must be applied to large areas for the results to have practical relevance, but treatments such as varying the variety of the crop or involving chemical treatment of individual plants or parts of plants may be advantageously applied to smaller units, such as individual plants or even individual leaves. In many industrial experiments some factors have levels which are hard to set and it is natural to apply them to long runs of the process, whereas levels of other factors are easier to set and so they can be applied to much shorter runs.

When the levels of one treatment factor must be applied to large units and the levels of a second treatment factor may be applied to small units, though they could also be applied to large units (which are simply amalgamations of small units), then there are two possibilities open to the experimenter. The first is to use only large units. This will almost inevitably place a severe restriction on the total number of units that can be available for the experiment, and will lead to few replicate units for each treatment combination. The second possibility is to use different sizes of units for different treatments. The latter approach may be thought of in two ways: based on using small units, but with restrictions on the allocation of treatment combinations so that combinations involving the same level of a 'large-unit' factor are grouped together on small units forming a large unit; alternatively the large units used for one treatment factor can be envisaged as being split into smaller units for the levels of other factors. Diagrammatically the design for four blocks each containing 15 small units grouped into five large units, with each small unit split into three subunits, is shown in Figure 18.3. The name 'split plot' design was coined for agricultural field crop experiments taking the two-stage procedure for design as the central concept. The more general term 'split-unit' design retains that concept. Other names for such designs, especially those which are non-orthogonal, are 'bi-randomisation designs' and, for the broader class of such structures, 'multistratum designs'.

18.1.1 Models for split-unit designs

As with the incomplete block designs discussed in Chapter 7 information about treatments occurs in two different strata of the experiment. Unlike the incomplete block design, where the information in both strata was about the same treatment comparisons, the different strata in split-unit designs contain information about different sets of treatment comparisons. The design of Figure 18.3 may be thought of as consisting of two randomised block designs. In the first, involving variation between the large units, the subdivision of the large units into small units is ignored, and the yields are assumed to be influenced by the block differences and by the differences between the treatment levels applied to the large units. The model for a yield from block i and the large unit receiving treatment level p_j in block i is

$$z_{ij} = \mu + b_i + p_j + \eta_{ij}. \tag{18.1}$$

The second level of variation is that between split units within each large, or main, unit. At this level the large units are viewed as blocks, and all the influences on the large units are subsumed into 'block' effects, with treatment effects due to the different treatments applied to the different split units within each large unit. The large unit effects are, of course, themselves attributable partly to blocks and partly to main unit treatments as described in model (18.1), but for consideration of the variation between split units, this is not relevant. However, the treatment effects on split-unit yields are not simply the main effects of the split-unit treatments but are modified by the interaction between the split-unit factor levels and the particular main unit treatment. The model for a yield from the split unit receiving treatment level q_k, within the main unit m_{ij}, is therefore

$$y_{ijk} = \mu + m_{ij} + q_k + (pq)_{jk} + \varepsilon_{ijk}. \tag{18.2}$$

The two models (18.1) and (18.2) can be placed in context by considering an overall model which can be constructed by describing the logical stages of the design of the experiment:

$y_{ijk}\ldots$	the yield in block i for main unit treatment p_j and split-unit treatment q_k
$=$	depends on
μ	the general level of average yield
$+b_i$	and the effects of block i
$+p_j$	and the effects of treatment p_j,
$+\varepsilon'_{ij}$	these treatments being allocated randomly to main units
$+q_k$	and the effects of treatment q_k
$+(pq)_{jk}$	modified by the treatment p_j
$+\varepsilon_{ijk}$	the q_k treatment levels being allocated randomly to split units within each main unit.

The resulting model

$$y_{ijk} = \mu + b_i + p_j + \varepsilon'_{ij} + q_k + (pq)_{jk} + \varepsilon_{ijk} \tag{18.3}$$

leads directly to the submodels (18.1) and (18.2) if we write

$$z_{ij} = y_{ij.} \text{ with } \eta_{ij} = \varepsilon'_{ij} + \varepsilon_{ij.}$$

Table 18.1. *Analysis of variance structure for a split-unit design*

Source	SS	df	MS
Blocks	$\sum_i \left(Y_{i..}^2/pq\right) - \left(Y_{...}^2/bpq\right)$	$b-1$	
P main effect	$\sum_j \left(Y_{.j.}^2/bq\right) - \left(Y_{...}^2/bpq\right)$	$p-1$	
Main unit residual	By subtraction	$(b-1)(p-1)$	E_a
Main unit total variation	$\sum_{ij} \left(Y_{ij.}^2/q\right) - \left(Y_{...}^2/bpq\right)$	$bp-1$	
Q main effect	$\sum_k \left(Y_{..k}^2/bp\right) - \left(Y_{...}^2/bpq\right)$	$q-1$	
PQ interaction	$\sum \left(Y_{.jk}^2/b\right) - \sum_j \left(Y_{.j.}^2/bq\right)$ $- \sum_k \left(Y_{..k}^2/bp\right) + \left(Y_{...}^2/bpq\right)$	$(p-1)(q-1)$	
Split unit residual	By subtraction	$(b-1)p(q-1)$	E_b
Total	$\sum_{ijk} Y_{ijk}^2 - \left(Y_{...}^2/bpq\right)$	$bpq-1$	

and

$$\mathrm{m}_{ij} = \mathrm{b}_i + \mathrm{p}_j + \varepsilon'_{ij.}$$

The least squares estimates and the corresponding analysis of variance may be derived by minimising the SSs

$$S_1 = \sum_{ijk} \{y_{ijk} - \mu - \mathrm{m}_{ij} - \mathrm{q}_k - (\mathrm{pq})_{jk}\}^2 \tag{18.4}$$

and

$$S_2 = \sum_{ij} (y_{ij.} - \mu - b_i - p_j)^2.$$

However, the interrelationship of the two analyses of variance is simpler if instead of S_2 we minimise

$$S_2' = \sum_{ijk} (y_{ij.} - \mu - b_i - p_j)^2. \tag{18.5}$$

This change has no effect on the parameter estimates because k does not occur within the summation, but it has the advantage that the total SS about the mean derived from (18.5) is the same as the SS for fitting m_{ij} in (18.4) so that the two analyses of variance are linked. The form of the analysis of variance structure is as in Table 18.1.

The two residual MS E_a and E_b allow estimation of the variances of the two random components in the model (18.3),

$$\sigma^2 = \mathrm{E}\left(\varepsilon_{ijk}^2\right) \text{ and } \sigma'^2 = \mathrm{E}\left(\varepsilon_{ij.}'^2\right).$$

The expected values of E_b and E_a are

$$\mathrm{E}(E_b) = \mathrm{E}\left(\varepsilon_{ijk}^2\right) = \sigma^2$$

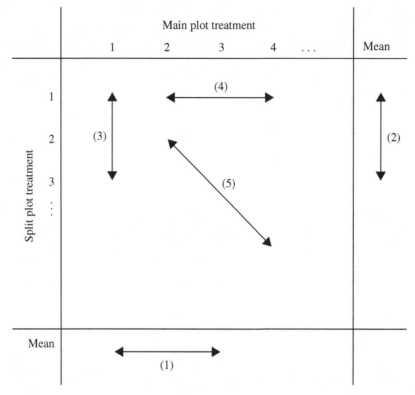

Figure 18.4 Pattern of SEs for a two-way table of means for a split-plot design
(definitions of SE (1), (2), (3), (4), (5) in text).

and

$$E(E_a) = qE\left(\eta_{ij}^2\right) = q\left(\sigma'^2 + \sigma^2/q\right)$$
$$= \sigma^2 + q\sigma'^2.$$

Hence σ^2 is estimated by E_b, and σ'^2 is estimated by $(E_a - E_b)/q$.

18.1.2 SEs for comparing treatment means

The fact that information about treatment differences exists at two different levels leads to a rather more complex structure of SEs than the usual fully randomised factorial experiment. The basis for calculating all SEs is the complete model (18.3). Consider the table of two-way treatment means shown in Figure 18.4. The arrows show the different comparisons of interest.

(1) $y_{.1.} - y_{.2.}$ is a main effect comparison of main unit treatment levels. From (18.3) the comparison is:

$$(y_{.1.} - y_{.2.}) = (p_1 - p_2) + (\varepsilon'_{.1} - \varepsilon'_{.2}) + (\varepsilon_{.1.} - \varepsilon_{.2.}),$$

all other terms either cancelling or being eliminated by the usual constraints. Hence

$$\mathrm{Var}(y_{.1.} - y_{.2.}) = 2\sigma'^2/b + 2\sigma^2/bq = 2(\sigma^2 + q\sigma'^2)/bq,$$

which is estimated by $2E_a/bq$ which is the usual form of variance for a randomised block design, with the extra divisor q.

(2) $y_{..1} - y_{..2}$ is a main effect comparison of split-unit treatment levels. From (18.3):

$$y_{..1} - y_{..2} = (q_1 - q_2) + (\varepsilon_{..1} - \varepsilon_{..2})$$

and the variance is:

$$\mathrm{Var}(y_{..1} - y_{..2}) = 2\sigma^2/bp,$$

which is estimated by $2E_b/bp$, again the usual randomised block form of variance.

(3) $y_{.11} - y_{.12}$ is a comparison of split-unit treatment levels for level 1 of the main-unit treatments:

$$y_{.11} - y_{.12} = (q_1 - q_2) + \{(pq)_{11} - (pq)_{12}\} + (\varepsilon_{.11} - \varepsilon_{.12}),$$

for which the variance is

$$\mathrm{Var}(y_{.11} - y_{.12}) = 2\sigma^2/b$$

estimated by $2E_b/b$.

(4) $y_{.11} - y_{.21}$ is a comparison of main-unit treatment levels for level 1 of the split-unit treatments:

$$y_{.11} - y_{.12} = (p_1 - p_2) + (\varepsilon'_{.1} - \varepsilon'_{.2}) + \{(pq)_{11} - (pq)_{21}\} + (\varepsilon_{.11} - \varepsilon_{.12})$$

for which the variance is

$$\mathrm{Var}(y_{.11} - y_{.21}) = 2\sigma'^2/b + 2\sigma^2/b.$$

This is the first case where the estimate of the variance is not simple, being

$$2\{E_a + (q-1)E_b\}/(bq).$$

(5) $y_{.11} - y_{.22}$ involves different main-unit treatment levels and different split-unit treatment levels:

$$y_{.11} - y_{.22} = (p_1 - p_2) + (\varepsilon'_{.1} - \varepsilon'_{.2}) + (q_1 - q_2)$$
$$+ \{(pq)_{11} - (pq)_{22}\} + (\varepsilon_{.11} - \varepsilon_{.22})$$

for which the variance is

$$\mathrm{Var}(y_{.11} - y_{.22}) = 2\sigma'^2/b + 2\sigma^2/b,$$

exactly the same as for the previous comparison (4).

The form of the variance estimate for these last two comparisons causes major problems of statistical inference, because the estimate involves two different error MS with different df and different expectations. The linear combination $E_a + (q-1)E_b$ does not have a χ^2 distribution and therefore, strictly, we are not able to calculate confidence intervals or significance levels for comparisons (4) or (5). In practice if we use the smaller of the two error df which will always be that for E_a, $(b-1)(p-1)$, then we will be acting in a cautious fashion, obtaining confidence intervals that are unnecessarily large, and requiring unnecessarily extreme values in determining significance. Such a procedure is termed 'conservative' and is usually regarded as desirable in statistics, though whether it is sensible to emphasise the importance of avoiding type I errors rather than type II errors is often debatable in research work. In any case the complexity of the SEs and the consequent difficulty with df provide a major drawback to the use of the split-unit design. REML analysis automatically produces estimated SEs, but these are dependent on strong distributional assumptions.

18.2 Advantages and disadvantages of split-unit designs

Although the main reason for using split-unit designs, and indeed the whole reason for their existence, is practical necessity, there are some experimenters and statisticians who maintain that there are situations where split-unit designs are, in theory, to be preferred to fully randomised factorials. The argument is simple. In a split unit design, information about treatment differences occurs at two levels. The form of the model (18.3) implies that the expected error MS of main-unit comparisons will be larger than that for split-unit comparisons. Since the total set of units is the same as would have been used for a fully randomised factorial treatment set, the single error MS for that experiment would have been the weighted average of the two error MS for the split-unit experiment. Consequently, in the split-unit experiment, comparisons based on the split-unit error MS will be more precise than the same comparisons would have been in the fully randomised factorial experiment, while comparisons based on the main-unit error MS will be less precise than in the fully randomised factorial experiment.

This form of argument is formally correct, but can be misleading, and it is sometimes simplified further to a form in which it contains only the most miniscule grain of truth. Split-unit designs, it is said, are appropriate when one factor main effect is of little interest, and the interest is in the other factor main effect and the interaction effect, which are estimated in terms of split-unit comparisons.

Now, the grain of truth is that, indeed, the precision of information on split-unit treatment main effects and interactions in the form of F statistics within the analysis of variance should be expected to be better than that of the main-unit treatment main effects. However, the disadvantages of the split-unit design are many and four disadvantages are discussed here in detail.

1. The information occurs at two levels and the estimation of both error variances is on fewer df than would be available to estimate the single error variance for the fully randomised factorial.

2. If the main-unit error MS, E_a, the split-unit error MS, E_b, and the fully randomised factorial error MS, s^2, are compared, then, since s^2 is a weighted average of E_a and E_b, and the weights are the df, $(b-1)(p-1)$ for E_a, $(b-1)p(q-1)$ for E_b, the difference $(E_a - s^2)$ is usually much greater than the difference $(s^2 - E_b)$. In other words the gain in precision for some effects is much less than the loss for others.

   To illustrate the scales of these two disadvantages, the values of the two sets of error df, and the single-error degrees of freedom are shown in Table 18.2, together with the ratio

Table 18.2. *Error df*

b	p	q	Main unit	Split unit	Randomised factorial	$\dfrac{E_a - s^2}{s^2 - E_b}$
2	2	6	1	10	11	10.0
2	6	2	5	6	11	1.2
2	3	4	2	9	11	4.5
2	4	3	3	8	11	2.7
3	2	4	2	12	14	6.0
3	4	2	6	8	14	1.3
4	2	3	3	12	15	4.0
4	3	2	6	9	15	1.5
2	2	9	1	16	17	16.0
2	9	2	8	9	17	1.1
2	3	6	2	15	17	7.5
2	6	3	5	12	17	2.4
3	2	6	2	20	22	10.0
3	6	2	10	12	22	1.2
3	3	4	4	18	22	4.5
3	4	3	6	16	22	2.7
4	3	3	6	18	24	3.0
2	2	12	1	22	23	22.0
2	12	2	11	12	23	1.1
2	3	8	2	21	23	10.5
2	8	3	7	16	23	2.3
3	2	8	2	28	30	14.0
3	8	2	14	16	30	1.1
2	4	6	3	20	23	6.7
2	6	4	5	18	23	3.6
4	2	6	3	30	33	10.0
4	6	2	15	18	33	1.2
3	4	4	6	24	30	4.0
4	3	4	6	27	33	4.5
4	4	3	9	24	33	2.7

of

$$(E_a - s^2)/(s^2 - E_b)$$

for all possible split-unit designs with 24, 36 or 48 plots, and with an explicit replication level of four or less.

It is plain from Table 18.2 that the df for the main-unit error are too small for effective estimation of variance except when q = 2, i.e. when there are two split-unit treatment levels. In the few cases where there are at least eight df for the main-unit error, the ratio of the loss/gain for the main-unit and split-unit comparisons is usually very low, and the gain and loss of precision are then similar. However, this almost invariably occurs with only two split plots in each main plot when it might be expected that the gain in precision would be relatively small. The last line of Table 18.2 provides a case when the main-plot error df rise to nine, with three split-unit treatments. However, the ratio of loss/gain is 2.7, emphasising that the precision of main unit comparisons is very much reduced.

3. If instead of considering F ratios in the analysis of variance we consider the tables of treatment mean values, which will usually be the main focus of interest for the experimenter, then the advantages of split-unit designs for investigating interaction become less clear. Particularly when there are substantial interaction effects the experimenter will interpret the results in terms of comparisons between pairs or groups of treatment combinations. Consider again Figure 18.4 in which the five types of comparison for which SEs were derived in the previous section are illustrated. In terms of precision, compared with the fully randomised factorial allocation within each block, the main effect comparison (2) and the interaction comparison (3) may be expected to be more precise using a split-unit design. Main effect comparisons (1) will be much less precise. Interaction comparisons (4) and (5) will be less precise than in the fully randomised factorial allocation because of the reduction in df for part of the combined SE. Essentially when considering the comparison of treatment mean values, the practical assessment of interaction effects necessarily involves main effect differences, and hence the benefits of split-unit designs for estimating interaction effects largely disappear.

4. The split-unit design, as has been made clear in the previous section, requires four different SEs in the presentation of results, which causes particular problems in the interpretation of graphical presentation of means. And, of course, the necessity of using SEs for which exact t-tests and confidence intervals are not possible, and which are based on two estimated variances of which one may have few df, also makes interpretation and presentation more complex.

In conclusion, the argument for using split-unit designs, because of their greater precision for interactions and main effects of the split-unit factor, is an over-simplification. In many cases using split-unit designs will lead to no useful estimates of the main effects of the main-unit factor, or of comparisons of levels of the main-unit factor for particular levels of the split-unit factor. In addition, as we have already seen in Chapter 15, there are much better alternative designs using confounding, if it is thought desirable to design the experiment to have different comparisons at different levels of precision. The only sound advice we can offer is to avoid split-unit designs except when they are essential for practical reasons, as described at the beginning of this chapter.

18.3 Extensions of the split-unit idea

The split-unit concept offers a wide range of extensions, all deriving from practical requirements of particular experiments. The warning against using split-unit designs, for reasons other than that of practical necessity, applies even more strongly to the extensions discussed in this section. Nevertheless, designs using many different forms of splitting units are used, some by necessity, others by accident.

Split-unit designs have been discussed in the context of an initial randomised block design, but split-unit versions of any of the designs discussed in Chapters 7 and 8 can be easily constructed. The main-unit part of the analysis of variance is that for the parent design. The split-unit analysis involves no block or row-and-column component, since it is entirely

Block I
Main plot

1		2		3		4	
Split plot		Split plot		Split plot		Split plot	
1	2	1	2	1	2	1	2
$p_3q_1r_2$	$p_3q_2r_1$	$p_4q_2r_1$	$p_4q_1r_3$	$p_2q_1r_2$	$p_2q_2r_1$	$p_1q_1r_2$	$p_1q_2r_2$
$p_3q_1r_1$	$p_3q_2r_3$	$p_4q_2r_2$	$p_4q_1r_1$	$p_2q_1r_1$	$p_2q_2r_3$	$p_1q_1r_1$	$p_1q_2r_3$
$p_3q_1r_3$	$p_3q_2r_2$	$p_4q_2r_3$	$p_4q_1r_2$	$p_2q_1r_3$	$p_2q_2r_2$	$p_1q_1r_3$	$p_1q_2r_1$

Block II

$p_4q_1r_2$	$p_4q_2r_3$	$p_1q_1r_1$	$p_1q_2r_2$	$p_3q_2r_2$	$p_3q_1r_3$	$p_2q_2r_1$	$p_2q_1r_3$
$p_4q_1r_3$	$p_4q_2r_2$	$p_1q_1r_3$	$p_1q_2r_1$	$p_3q_2r_3$	$p_3q_1r_1$	$p_2q_2r_2$	$p_2q_1r_2$
$p_4q_1r_1$	$p_4q_2r_1$	$p_1q_1r_2$	$p_1q_2r_3$	$p_3q_2r_1$	$p_3q_1r_2$	$p_2q_2r_3$	$p_2q_1r_1$

Block III

$p_1q_2r_3$	$p_1q_1r_2$	$p_3q_1r_3$	$p_3q_2r_3$	$p_4q_2r_1$	$p_4q_1r_3$	$p_2q_1r_2$	$p_2q_2r_1$
$p_1q_2r_2$	$p_1q_1r_1$	$p_3q_1r_1$	$p_3q_2r_2$	$p_4q_2r_3$	$p_4q_1r_2$	$p_2q_1r_1$	$p_2q_2r_3$
$p_1q_2r_1$	$p_1q_1r_3$	$p_3q_1r_2$	$p_3q_2r_1$	$p_4q_2r_2$	$p_4q_1r_1$	$p_2q_1r_3$	$p_2q_2r_2$

Figure 18.5 Experimental plan for a split-split plot design.

composed of comparisons within main units. Variances of comparisons between split-unit factor levels, whether main effect comparisons or at a particular level of the main unit, all involve only the split-unit error MS, and are in the simple form of the variance of a difference between two sample means considered in the previous section. Variances of comparisons between levels of the main-unit factor have the same efficiency as those for the parent design, except that, when the comparisons are made at a particular split-unit treatment level, there is an additional error component from the split-unit error MS.

The first major extension of the split-unit concept involves splitting the split units again for a third treatment factor. The design now consists of three levels. Blocks are divided into main units to which the levels of the first treatment factor are applied; the main units are split into split units to which the levels of the second treatment factor are applied; and the split units are split into split-split units to which the levels of the third treatment factor are applied. The design is illustrated in Figure 18.5.

The model is a direct extension of the simple split-unit model:

$$y_{ijkl} = \mu + b_i + p_j + \varepsilon''_{ij} + q_k + (pq)_{jk} + \varepsilon'_{ijk} + r_l + (pr)_{jl}$$
$$+ (qr)_{kl} + (pqr)_{jkl} + \varepsilon_{ijkl}.$$

The analysis of variance structure is in three strata, and is summarised in Table 18.3.

The expectations of E_a, E_b and E_c are $\sigma^2 + r\sigma'^2 + qr\sigma''^2$, $\sigma^2 + r\sigma'^2$ and σ^2, respectively. Hence, $\sigma^2, \sigma'^2, \sigma''^2$ can be estimated and SEs calculated for all comparisons of treatment means.

18.3.1 Criss-cross designs

There is a second type of extension of the split-unit idea which is often found, and this is the criss-cross, or strip-plot or strip-block, design. A two-factor treatment structure is used and, as in the simple split-unit design, the levels of factor P require large units from practical necessity. In addition, the levels of factor Q also require large units. Again, it would be possible to use a large basic unit, but this will usually lead to a very low level of replication

Table 18.3. *Outline analysis of variance for split-split unit design*

Source	df	MS
Blocks	$b - 1$	
P main effect	$p - 1$	
Main-unit residual	$(p - 1)(b - 1)$	E_a
Main-unit total	$bp - 1$	
Q main effect	$q - 1$	
PQ interaction	$(p - 1)(q - 1)$	
Split-unit residual	$(b - 1)p(q - 1)$	E_b
Split-unit total	$bpq - 1$	
R main effect	$r - 1$	
PR interaction	$(p - 1)(r - 1)$	
QR interaction	$(q - 1)(r - 1)$	
PQR interaction	$(p - 1)(q - 1)(r - 1)$	
Split-split-unit residual	$(b - 1)pq(r - 1)$	E_c
Split-split-unit total	$bpqr - 1$	

Block I

p_4q_3	p_4q_4	p_4q_2	p_4q_1
p_1q_3	p_1q_4	p_1q_2	p_1q_1
p_2q_3	p_2q_4	p_2q_2	p_2q_1
p_5q_3	p_5q_4	p_5q_2	p_5q_1
p_3q_3	p_3q_4	p_3q_2	p_3q_1
p_6q_3	p_6q_4	p_6q_2	p_6q_1

Block II

p_2q_1	p_2q_3	p_2q_2	p_2q_4
p_6q_1	p_6q_3	p_6q_2	p_6q_4
p_4q_1	p_4q_3	p_4q_2	p_4q_4
p_3q_1	p_3q_3	p_3q_2	p_3q_4
p_1q_1	p_1q_3	p_1q_2	p_1q_4
p_5q_1	p_5q_3	p_5q_2	p_5q_4

Block III

p_1q_2	p_1q_4	p_1q_3	p_1q_1
p_2q_2	p_2q_4	p_2q_3	p_2q_1
p_4q_2	p_4q_4	p_4q_3	p_4q_1
p_5q_2	p_5q_4	p_5q_3	p_5q_1
p_6q_2	p_6q_4	p_6q_3	p_6q_1
p_3q_2	p_3q_4	p_3q_3	p_3q_1

Figure 18.6 Experimental plan for a criss-cross design.

for all treatment combinations, because of the limitation on total resources. The alternative approach is to consider a set of small units, arranged in a rectangular array in each block, and to apply the levels of P to whole rows of units, and the levels of Q to whole columns. The resulting design is shown in Figure 18.6; the origin of the name should be obvious.

Table 18.4. *Outline analysis of variance for criss-cross design*

Source	df	MS
Blocks	$b - 1$	
P main effect	$p - 1$	
Row main-unit residual	$(p - 1)(b - 1)$	E_a
Row main-unit total	$bp - 1$	
Q main effect	$q - 1$	
Column main-unit residual	$(b - 1)(q - 1)$	E_b
Column main-unit total	$bq - 1$	
PQ interaction	$(p - 1)(q - 1)$	
Split-unit residual	$(p - 1)(b - 1)(q - 1)$	E_c
Total	$bpq - 1$	

The criss-cross design requires a complex analysis of variance, and uses information in a very inefficient manner. Except when practically essential, it should not be used, but it does occur sufficiently frequently, sometimes without deliberate intent, for the complexity of analysis to be discussed in detail. Some of the ways in which criss-cross and related designs occur by accident in highly structured experiments are discussed in the next section.

The information from a criss-cross design is in three strata. Levels of P are applied to whole rows and therefore main effects of P are estimated from comparisons between rows within blocks. Similarly, main effects of Q are estimated from comparisons between columns within blocks. However, neither comparisons between rows, nor comparisons between columns, provide any information about the interaction effects. To obtain interaction information, we must consider the variation between the split units, which are splits of both row units and column units. The full model is

$$y_{ijk} = \mu + b_i + p_j + \varepsilon'_{ij} + q_k + \varepsilon''_{ik} + (pq)_{jk} + \varepsilon_{ijk}.$$

One conceptual difficulty with this model is that there is no independent randomisation at the level of the smallest unit; the randomisations of rows and columns completely determine the pattern of treatment occurrence. Nevertheless, it is clear that there are three levels of information, and consequently three sections to the analysis of variance. The analysis of variance structure is shown in Table 18.4. Note that the block SS is a component of both the row main-unit variation and the column main-unit variation (and must be included in the calculation of the column main-unit residual), but can only occur once in the total analysis of variance.

The variances, σ^2, σ'^2 and σ''^2 associated with the three error terms $\varepsilon_{ijk}, \varepsilon'_{ij}, \varepsilon''_{ik}$ can be estimated from the three residual MS. The expected values of E_a, E_b and E_c are $q\sigma'^2 + \sigma^2$, $pq''^2 + \sigma^2$ and σ^2, so that estimates of σ^2, σ'^2 and σ''^2 are

$$E_c,$$
$$(E_a - E_c)/q$$

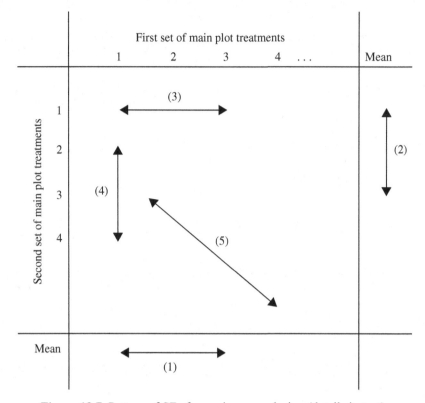

Figure 18.7 Pattern of SEs for a criss-cross design (details in text).

and

$$(E_b - E_c)/r.$$

The spread of information over the three levels leads to a profusion of standard errors for comparisons within the two-way table of treatment means. Consider the two-way structure of treatment means displayed in Figure 18.7.

The SE of the main effect comparisons $(p_1 - p_4, q_2 - q_3)$ are as usual derived from the corresponding residual MS, E_a and E_b. The other three comparisons in Figure 18.7 require different combinations of estimated error variances. Comparison (3) is estimated by

$$y_{.11} - y_{.21},$$

which can be written, from the general model:

$$y_{.11} - y_{.21} = (p_1 - p_2) + \{(pq)_{11} - (pq)_{21}\} + (\varepsilon'_{.1} - \varepsilon'_{.2}) + (\varepsilon_{.11} - \varepsilon_{.21}),$$

from which

$$\text{Var}(y_{.11} - y_{.21}) = 2\sigma'^2/b + 2\sigma^2/b,$$

which is estimated by

$$2\{E_a + (q - 1)E_c\}/bq.$$

Similarly, the variance for comparison (4) is estimated by

$$2\{E_b + (p - 1)E_c\}/bp.$$

Comparison (5) is

$$y_{.23} - y_{.45},$$

for which the error terms involved are

$$(\varepsilon'_{.2} - \varepsilon'_{.4}) + (\varepsilon''_{.3} - \varepsilon''_{.5}) + (\varepsilon_{.23} - \varepsilon_{.45}),$$

whence

$$\mathrm{Var}(y_{.23} - y_{.45}) = 2(\sigma'^2 + \sigma''^2 + \sigma^2)/b,$$

which is estimated by

$$2\{pE_a + qE_b + (pq - p - q)E_c\}/bpq.$$

Clearly, assigning exact df to these estimates of SE is impossible, and indeed having adequate (≥ 10) df for each of E_a, E_b and E_c is rarely possible. However, if the need arises because of practical reasons, then the criss-cross design with all its attendant difficulties of interpretation may be useful. Further ramifications of split-unit ideas may be produced at will, and may sometimes be necessary practically, but there are rarely sufficient df for the main-unit components of analysis to be of much value.

18.3.2 Analysis for a complex split-unit design

Example 18.1 To demonstrate the practical analysis of split-unit and criss-cross designs we consider some data from an experiment investigating the long-term effects on pasture composition of different patterns of grazing. Three treatment factors are included:

Period (P): the length of the period when plots were grazed, with three levels (3, 9, 18 days).
Spring grazing cycles (SP): the number of cycles. Either two (two periods grazing with two (long) gaps), or four (four periods grazing with four (shorter) gaps).
Summer grazing cycles (S): two or four cycles, as for spring grazing cycles.

The experimental design is a 3×3 Latin square for the three periods, with each of the nine main plots split twice in a criss-cross design for the two grazing cycle factors. Please note that the authors were not consulted about this design: nor was any other statistician! The design is shown in Figure 18.8. The data values, which are percentage areas covered by the principal grass, are also included in Figure 18.8.

The calculation of the analysis of variance requires, first, the set of main plot totals with the associated totals for rows, columns and periods, which are shown in Table 18.5(a). The other totals required are for the $3 \times 2 \times 2$ treatment combinations, with all subtotals, given in Table 18.5(b). The model on which the analysis is based includes four levels of variation: for

Table 18.5. *Main plot totals and treatment totals: (a) main plot totals; (b) treatment totals*

(a)

	Main plot totals (period length)			Total
	116.3(18)	172.5(9)	136.5(3)	425.3
	161.1(9)	181.4(3)	106.8(18)	449.3
	163.1(3)	85.5(18)	150.0(9)	398.6
Total	440.5	439.4	393.3	1273.2

(b)

Grazing cycle		Period			
Spring	Summer	3	9	18	Total
2	2	160.8	146.4	91.7	398.9
2	4	186.7	177.3	100.2	464.2
4	2	59.2	56.3	43.0	158.5
4	4	74.3	103.6	73.7	251.6
2	Total	347.5	323.7	191.9	863.1
4	Total	133.5	159.9	116.7	410.1
Total	2	220.0	202.7	134.7	557.4
Total	4	261.0	280.9	173.9	715.8
Total	Total	481.0	483.6	308.6	1273.2

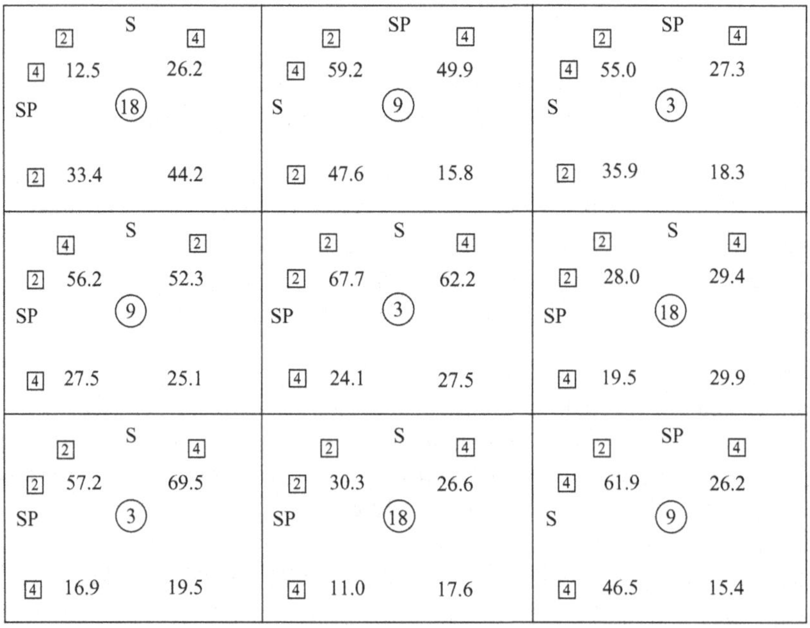

Figure 18.8 Design and data for 3 × 3 Latin square with each main plot split in a criss-cross design.

Table 18.6. *Analysis of variance for grazing experiment*

	SS	df	MS	E(MS)
Rows	107.20	2	53.60	
Columns	120.95	2	60.48	
Periods	1676.49	2	838.24	
Error	214.38	2	$107.19 = E_a$	$\sigma^2 + 2\sigma'^2 + 2\sigma''^2 + 4\sigma''''^2$
Total	2119.02	8		
Spring grazing cycle	5700.25	1	5700.25	
PSP	823.20	2	411.60	
Error	478.33	6	$79.72 = E_b$	$\sigma^2 + 2\sigma''^2$
Spring half plots total	9120.80	17		
Summer grazing cycle	696.96	1	696.96	
PS	80.78	2	40.39	
Error	366.80	6	$61.33 = E_c$	$\sigma^2 + 2\sigma'^2$
Summer half-plots total	3263.56	17		
SPS	21.47	1	21.47	
PSPS	51.74	2	51.75	
Error	177.35	6	$29.56 = E_d$	σ^2
Total	10515.90	35		

main plots, for two sets of half-plots for main effect comparisons of the two grazing cycle factors, and for the quarter-plots for grazing cycle combinations

$$y_{ijlm} = \mu + r_i + c_j + p_{k(ij)} + \varepsilon'''_{ij} + sp_l + (p, sp)_{kl} + \varepsilon''_{ijl}$$
$$+ s_m + (p, s)_{km} + \varepsilon'_{ijm} + (sp, s)_{lm} + (p, sp, s)_{klm} + \varepsilon_{ijlm}.$$

The analysis of variance provides a summary of the results as shown in Table 18.6.
    The initial conclusions which should be noted from the analysis of variance are:

(i) The major treatment effects are those of
        spring grazing cycle,
        grazing period length,
        summer grazing cycle, and
        period × spring grazing cycle.
    All other interaction effects may be ignored.
(ii) The pattern of error MS follows the usual expectation with remarkable consistency. The whole plot error MS (E_a) is the largest, the lowest level error MS (E_d) is the smallest and the two half-plot error MS are intermediate.

To present the results we require SEs for comparing whole plot treatment means, both forms of half-plot treatment means and the two-way table of means for period × spring grazing cycle. The comparisons and their variance may be derived from the full model for y_{ijlm}.

Periods

$$y_1 - y_2 = (p_1 - p_2) + (\varepsilon_1''' - \varepsilon_2''') + (\varepsilon_{1.}'' - \varepsilon_{2.}'') + (\varepsilon_{1.}' - \varepsilon_{2.}') + (\varepsilon_{2..} - \varepsilon_{2..}),$$

$$\mathrm{Var}(y_1 - y_2) = 2\sigma'''^2/3 + 2\sigma''^2/6 + 2\sigma'^2/6 + 2\sigma^2/12,$$

which is estimated by

$$2E_a/12 = 2 \times 107.19/12 = 17.86.$$

Spring grazing cycles

$$y_{..1.} - y_{..2.} = (sp_1 - sp_2) + (\varepsilon_{..1}'' - \varepsilon_{..2}'') + (\varepsilon_{..1.} - \varepsilon_{..2.}),$$

$$\mathrm{Var}(y_{..1.} - y_{..2.}) = 2\sigma''^2/9 + 2\sigma^2/18,$$

which is estimated by

$$2E_b/18 = 2 \times 79.72/18 = 8.86.$$

Summer grazing cycles The pattern is exactly the same as for spring grazing cycles and the variance is estimated by

$$2E_c/18 = 2 \times 61.33/18 = 6.81.$$

Periods $\times$ spring grazing

$$y_{11} - y_{12} = (sp_1 - sp_2) + \{(psp)_{11} - (psp)_{12}\} + (\varepsilon_{11.}'' - \varepsilon_{12.}'') + (\varepsilon_{11.} - \varepsilon_{12.})$$

$$\mathrm{Var}(y_{11} - y_{12}) = 2\sigma''^2/3 + 2\sigma^2/6),$$

which is estimated by

$$2E_b/6 = 2 \times 79.72/6 = 26.57$$

and

$$y_{11} - y_{21} = (p_1 - p_2) + \{(psp)_{11} - (psp)_{21}\} + (\varepsilon_1''' - \varepsilon_2''')$$
$$+ (\varepsilon_{11}'' - \varepsilon_{21}'') + \varepsilon_{11.} - \varepsilon_{21.}),$$

$$\mathrm{Var}(y_{11} - y_{21}) = 2\sigma'''^2/3 + 2\sigma''^2/3 + 2\sigma^2/6,$$

which is estimated by

$$2(E_a + E_b - E_c + E_d)/12 = 2 \times 155.14/12 = 25.86.$$

The presentation of results is therefore as in Table 18.7.

The major effect on the percentage of the principal grass is caused by the number of cycles during the spring grazing period. The use of four cycles reduces the percentage to less than half that observed for two cycles of grazing. This effect is most pronounced for the shorter grazing periods, being largest for three-day grazing periods and much smaller for 18-day grazing periods. The percentages of the principal grass are generally lower for the 18-day grazing periods. Finally it should be noted that the number of cycles during the summer grazing period also has a significant effect but in the opposite direction to that of the spring grazing cycle effect. Using four grazing cycles for the summer grazing tends to increase the percentage of the principal grass, but the increase of 9% is less than half the decrease of 25% apparently due to the four grazing cycles in the spring grazing period. Clearly the system being explored is complex.

Table 18.7. *Summary of results for grazing experiment:*
(a) spring grazing; (b) summer grazing

(a)

	Period			Mean
	3	9	18	
Two cycles	57.9	54.0	32.0	48.0
Four cycles	22.2	26.6	19.4	22.8
Mean	40.1	40.3	25.7	
SE for the difference between two period means				= 4.2 (2 df)
SE for the difference between two cycle means				= 3.0 (6 df)
SE for within table vertical comparisons				= 5.2 (6 df)
SE for within table other comparisons				= 5.1 (? df)

(b)

Two cycles 31.0
Four cycles 39.8
SE for difference between two cycle means = 2.6 (6 df)

18.4 Identification of multiple strata designs

Many experiments whose design is built up gradually, rather than being totally planned at the outset, involve multiple strata comparisons without deliberate intent on the part of the experimenter, and it is often necessary to sort out the actual structure of the experiment, and hence of the analysis of variance, from a rather confused description. In Chapter 6 we considered an example where small discs were cut from leaves and treated in one of eight different ways. The experimenter, consulting a statistician about the analysis, stated simply that each of the eight treatments was replicated ten times, and offered an analysis of variance in the form

treatments   7 df $F = 500$
error   72 df

After much questioning, it was revealed that the design of the experiment involved eight leaves, one from each of eight plants, the ten discs for a single treatment being taken from a single leaf, each treatment using a different leaf. Two levels of variation, between discs within a leaf, and between leaves, are involved, and the design and analysis of variance can be regarded as a split-unit design of a non-standard (and non-useful) form, giving the two-stage analysis of variance as shown in Table 18.8. The analysis of variance structure identifies clearly the absence of any information about the precision of treatment comparisons. The treatments must be compared in terms of the main-unit error variable, about which there is no information; there is abundant information about the split-unit error variance but there are no treatment comparisons for which that variance is relevant.

Table 18.8. *Structure of analysis of variance for experiment on leaves*

	df
Treatments	7
Main-unit (leaf) error	0
Main-unit (leaf) total	7
Split-unit (disc) error	72
Total	79

Table 18.9. *Analysis of variance structure for leaves experiment*

	df
Plants (blocks)	9
New and old (age)	1
Leaf pairs error	9
Leaf pairs total	19
Upper × lower (side)	1
Age × side	1
Leaves error	18
Leaves total	39
Chemicals	1
Age × chemicals	1
Side × chemicals	1
ASC	1
Discs error	36
Total	79

The experimenter's original description could have implied many different designs. For example, the design could have been defined in three stages:

(i) Ten plants, each providing two pairs of leaves, one pair of old leaves and one pair of new leaves (old/new is one factor of the 2^3 treatment set).

(ii) Each leaf provides two discs, one inocculated with chemical C_1 and one with chemical C_2.

(iii) The two leaves of a pair, each providing two discs, are allocated one for inocculation into the upper side of the leaf and the other for inocculation in the lower side of the leaf.

The analysis of variance for such a design would have had four strata: plants, leaf pairs, leaves, and discs, only three of which provide information about the treatments. This should be recognisable as a split split-unit design, with plants as blocks and leaf pairs as main units, leading to the analysis of variance structure as shown in Table 18.9.

Before the analysis of variance and subsequent interpretation of treatment means can be calculated the statistician will often have to diagnose the design actually used. As we have already emphasised in Chapters 5 and 6, it is essential to identify the different levels of replication and thus of variation in an experiment, and to use the appropriate levels of variation for each treatment comparison. This recognition will frequently lead to analyses of the split-unit family.

18.4.1 The consequence of extreme balancing

In many experiments, experimenters go to considerable lengths to balance all possible patterns, and this often leads to criss-cross structures with multiple strata in two-way blocking structures. The second example at the beginning of this chapter is such a design, with frames as blocks, positions within each frame as one form of main plot, and times within each frame as the other form of main plot. Another example from a psychology experiment illustrates the process of recognising this type of structure in more detail.

Example 18.2 The experiment involved 16 groups with four individuals per group. The 16 groups were blocked into four sets of four groups each. The principal treatments were four group stress environments (S); each group was subjected to each stress treatment in turn, the order of the four treatments being arranged according to a 4×4 Latin square within each set of four groups. The individuals within each group were two males and two females, and two drug levels (high and low) were allocated so that the four sex and drug combinations all occur in each group. Each individual remained on the same drug level throughout the four stress treatments and, of course, remained the same sex! To assess the performance of individuals under stress and drug treatment, four test papers were used, all the individuals in a set of four groups having the same test at a particular time, the test papers being changed between sets and times according to a Latin square design.

This is a very complex design, built up gradually by introducing ideas of control and of treatment comparisons sequentially. The three major stages of the design are represented in Figure 18.9.

To unravel the design and the corresponding analysis structure, we consider the different treatment comparisons and the stratum from which information is available.

(i) Test papers – this may not be of interest, but if it is, then information comes from the 16 combinations of four sets × four times.

(ii) Stress environments – these are applicable to groups at different times and therefore information comes from the 64 combinations of 16 groups × four times.

(iii) Test × stress – from the same stratum as stress because stress is applied to groups within sets.

(iv) Sex, drug, sex × drug – these are applied to individuals consistently over time so that time has no relevance to these comparisons, and the relevant stratum is individuals within groups.

(v) All other interaction effects can only be assessed by considering the stratum of individual × time combinations which consist of the smallest possible units.

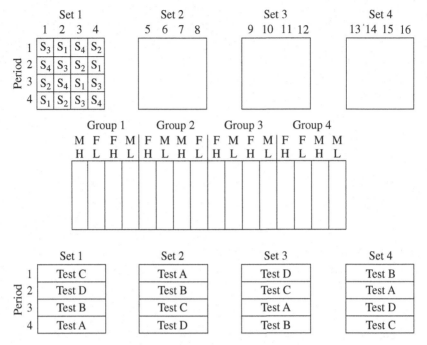

Figure 18.9 Diagrammatic representation of the design structure for the experiment of Example 18.2.

The structure of the resulting analysis of variance is set out in Table 18.10. Essentially, the individuals and times classifications provide a criss-cross structure, while individuals within groups within sets provide a split split-unit structure.

This example is an unusually complex one, but hopefully it demonstrates the Latin square and criss-cross structures which arise naturally when an experimenter attempts to control systematically the patterns of treatment allocation.

18.5 Systematic treatment variation within main units

We have discussed the use of repeated measurements from an experimental unit. When these measurements are based on separate samples from within an experimental field plot we have to consider how to select the samples from within the plot. The systematic selection of samples for different harvests within a plot will tend to give more precise comparisons than random sampling because the samples can be taken in identical fashion. Starting from the same end of each plot will provide a similar environment for the sequence of samples for different treatments. Systematic selection of samples is usually recommended for multiple harvests but randomised selection should be used if a split-plot analysis is to be applied to the data.

An analogous idea to systematic sampling from a plot is the application of systematically varying treatments to successive sections of each plot. The motivation for the use of a systematic arrangement of treatment levels stems from the desire to use resources efficiently.

Table 18.10. *Outline analysis of variance for stress testing experiment*

Stratum	Sources	df
Sets × times	Sets	3
	Times	3
	Tests	3
	Sets × times error	6
	Sets × times total	15
Groups × times	Groups within sets	12
	Stresses	3
	Tests × stresses	9
	Groups × times error	24
	Group × times total	63
Individuals	Sets	3
	Groups within sets	12
	Sex	1
	Drug	1
	Sex × drug	1
	Individuals error	45
	Individuals total	63
Individual × times	Stress × sex	7
	Stress × drug	3
	Stress × sex × drug	3
	Test × sex	3
	Test × drug	3
	Test × sex × drug	3
	Stress × test × sex	9
	Stress × test × drug	9
	Stress × test × sex × drug	9
	Error	99
	Total	255

In a conventional randomised design to compare three or four different spacings, or nutrient treatments, it is necessary to have substantial 'guard' or 'discard' areas round each differently treated plot, so that the harvested area in the middle of the plot is representative of a whole field receiving the particular treatment of that plot. The more different the conditions produced by the different treatments, the greater the need for large guard areas. In many experiments, the harvested areas from each plot may be no more than 30 or 40% of the total plot area.

Systematic split-plot treatments have been suggested and used to increase the harvested area proportion of the experiment for quantitative treatment factors of various types. Nelder (1962) and Bleasdale (1967) suggested various designs for spacing experiments, and many experiments have subsequently used systematic spacing factors. Fertiliser factor levels have been applied systematically by Cleaver, Greenwood and Wood (1970). Trials of insecticides and herbicides have used systematic placement of successive concentrations. Similar ideas have been used in some medical and animal trials, where treatments are applied in order of increasing concentration, because the effect of a subsequent treatment at a lower concentration may be masked by the residual effects of the previous treatment. In field crop experiments, systematic arrangements may be used for two treatment factors, the systematic change of level occurring in perpendicular directions for the two factors. In all cases the use of systematic treatment allocation must be considered cautiously, but particularly in the early stages of investigations. Experiments using systematic treatment arrangements can be extremely informative, provided some proper replication is also included.

The principle behind the use of systematic designs for field crop experiments is that if treatment levels are placed in increasing order, then the two treatment levels on either side of a particular level will be similar to that level, and guard areas can be eliminated, provided that the adjacent treatment levels are sufficiently similar, so that the environment of the particular level may be regarded as representative of application of that level in a large area. Traditionally, it has been suggested that changes of crop density or of fertiliser level of 10% should satisfy the requirement of similarity of adjacent levels, but this is a 'rule of thumb' which may be too lenient or too stringent in particular circumstances. Particular forms of systematic design for spacing arrangement experiments were discussed in Chapter 16.

As for repeated measurement data, it will usually not be appropriate to analyse data from experiments with systematic arrangements of split-plot factor levels by calculating the analysis of variance for a split-plot design as if the systematic treatment levels had been randomised within each main plot. Although such an analysis might be argued to be valid because the dominant contribution to σ^2 is believed to be the plant variation, rather than soil variation, doubts about the validity of assumptions would be difficult to satisfy. In any case the objectives of the experiment will usually be more appropriately satisfied by first fitting a yield response function over the levels of the systematically applied treatment factor for each main plot, and then analysing the fitted response functions.

The implications of this philosophy of analysis, which is essentially the same as that outlined in Chapter 4, for the analysis of repeated measurements, are that, when levels of a quantitative factor are applied systematically, there should be other treatment factors applied to main plots, and also some explicit replication of main plots, unless the number of combinations of main-plot factors is very large. Essentially, the analysis of variation of the fitted response functions uses the structure of the main plots only, so that the experimental design of the main plots must constitute a valid design, in particular providing adequate estimation of the main-plot error variance. The layout for a typical 'systematic' design is shown in Figure 18.10. The design has ten levels of the systematic split-plot factor level within main plots for a 2×3 factorial structure in three randomised blocks of six main plots each. Not only is the allocation of the six main plot treatments ($p_1q_1, p_1q_2, p_1q_3, p_2q_1, p_2q_2, p_2q_3$) to main plots randomised in each block, but the direction in which the split-plot treatments change systematically from low to high is also randomised for each main plot.

18.6 The split-unit design as an example of confounding

The structure of a split-unit design is essentially the same as that of a confounded experiment, although the terminologies that have evolved for the two types of design are different. In a confounded design, there are one or more replicates, each replicate including a single

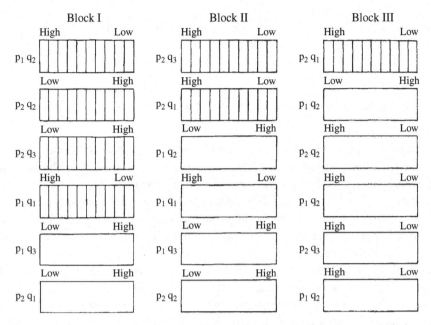

Figure 18.10 Experimental design layout using systematic treatment variation within a 2 × 3 main plot treatment structure.

observation for each treament combination. Each replicate is divided into blocks, each block containing the same number of units. The choice of treatment combinations to be allocated to each block of units determines which treatment effects are estimated in terms of differences between units within blocks, and which will be estimated in terms of differences between blocks, described as confounded with block differences.

Split-unit designs include several blocks, each block including one observation for each treatment combination. Each block is divided into main units and levels of one or more factors are applied to main units. Each main unit contains the same number of split units and the levels or level combinations for the remaining factors are allocated to the split units within each main unit. The choice of factors to be applied to main units and to split units determines which treatment effects will be estimated in terms of differences between split units within main units and which will be estimated in terms of differences between main units.

The equivalence of the two design structures is summarised as follows:

split unit designs		*confounded designs*
blocks	≡	replicates
main units	≡	blocks
split units	≡	units

Recognising split-unit designs as a special case of confounding can improve understanding of both types of design. The inefficiency of split-unit designs derives from the fact that comparisons of main-unit treatments are generally much less precise than comparisons of split-unit treatments. However, consideration of confounded designs shows that, if some

Block

	I	II	III	IV	V	VI
	$p_1q_0r_0$	$p_0q_1r_1$	$p_1q_1r_1$	$p_1q_0r_1$	$p_0q_0r_0$	$p_1q_0r_0$
	$p_0q_1r_0$	$p_0q_0r_0$	$p_0q_0r_1$	$p_1q_1r_0$	$p_0q_1r_1$	$p_0q_0r_1$
	$p_1q_1r_1$	$p_1q_1r_0$	$p_1q_0r_0$	$p_0q_1r_1$	$p_1q_0r_1$	$p_0q_1r_0$
	$p_0q_0r_1$	$p_1q_0r_1$	$p_0q_1r_0$	$p_0q_0r_0$	$p_1q_1r_0$	$p_1q_1r_1$

Figure 18.11 Three replicates of a $2 \times 2 \times 2$ structure in six blocks of four units.

treatment comparisons must be estimated more precisely than others, there is no necessity for the less precise comparisons to be main effects, as in the split-unit designs. In the development of confounded designs the less precise comparisons are chosen to be interactions of several factors. Thus, except when split-unit designs are used because they are required for practical reasons, we should not only prefer the fully randomised factorial to the split-unit design, but should consider whether the confounded design, losing information on the multifactor interactions, should not be preferred to both.

For example consider the choice of design for the three two-level factors, P, Q and R. We might decide that R is of less importance and consider a split-unit design with R as a main unit factor and P and Q as split-unit factors. Alternatively we could use a confounded design with PQR confounded between main units (or blocks in confounding terminology). To prefer the split-unit design it is necessary to argue that the three-factor interaction PQR is a more important effect than the main effect of R. In practice this argument would be difficult to believe, particularly since the interpretation of interaction effects through tables of means requires the main effect estimates.

One aspect of confounded designs which we have so far neglected and which the contrast with split-unit designs emphasises is the information about treatment effects which is available from comparisons between blocks. In the split-unit design, although the precision of main-unit treatment effects is reduced, the information about the effects of main-unit factors is important. In the confounded design, differences between blocks can provide estimates of those interaction effects which are confounded between blocks. To see how much information may be recovered from interblock information, we reconsider Example 15.4 with three two-level factors in blocks of four units.

Typically three or four replicates using 24 or 32 total observations would be used. Suppose 24 observations are to be used, with six blocks of four, as shown in Figure 18.11. Then the estimate of the three-factor interaction, PQR, will be given by the contrast of block totals

$$(B_1 - B_2 + B_3 - B_4 + B_6 - B_5),$$

with a multiplying factor of $1/12$ to convert the estimate to a per unit basis. This estimate is simply the average of three differences between block means and the precision of this estimate will depend on the variance of block differences. The situation is identical with that for three pairs of observations, each pair being for two treatments $T_1 (= +PQR)$ and $T_2 (= -PQR)$. The analysis of variance of block totals will be as shown in Table 18.11.

If the pairing of blocks into replicates is not done in a systematic way, then the between-replicate variation is pooled with the block error to give a block error SS, based on four df.

Table 18.11. *Outline analysis of variance for block totals for confounded design*

Source	df
PQR	1
Replicates (pairs)	2
Block error	2
Total block variation	5

The model implicit in this analysis may be derived from the model for the unconfounded design in three blocks of eight units:

$$y_{ijkl} = \mu + b_i + P_j + Q_k + R_l + (PQ)_{jk} + (PR)_{jl} + (QR)_{kl} + (PQR)_{jkl} + \varepsilon_{ijkl}.$$

When using six blocks of four units, the algebraic appearance of the model remains the same, but only four of the possible (j, k, l) combinations occur for each i value. The sets of four are chosen so that all possible combinations of any two of (j, k, l) do occur with each i, but the $(PQR)_{jkl}$ terms occurring with i are such that, in each block, the four combinations are either

$$\{(0, 0, 0), \ (0, 1, 1), \ (1, 0, 1) \text{ and } (1, 1, 0)\}$$

or

$$\{(1, 0, 0), \ (0, 1, 0), \ (0, 0, 1) \text{ and } (1, 1, 1)\}.$$

However, the definition of $(PQR)_{jkl}$ implies that

$$\sum (PQR)_{jkl} = 0,$$

where the summation is over any one suffix, so that all the combinations in the first group are equal, and so are those of the second group, the two common values being equal and opposite. Hence, all those combinations in the first group may be written as $-PQR$, and all those in the second as $+PQR$. The block totals may be expressed in the form

$$Y_{i...} = 4\mu + 4b_i \pm 4PQR + \sum_{jkl} \varepsilon_{ijkl},$$

where the sign of PQR depends on the value of i, being $+$ for $i = 1, 3, 6$ and $-$ for $i = 2, 4, 5$. For this model, estimates of μ and PQR may be obtained by rewriting the model

$$Y_{i...} = 4\mu \pm 4PQR + \eta_i$$

to obtain not only the obvious estimate

$$PQR = (Y_{1...} - Y_{2...} + Y_{3...} - Y_{4...} + Y_{6...} - Y_{5...})/24,$$

but also the analysis of variance structure as follows:

	SS	df
PQR	$24(PQR)^2$	1
Residual	S_r	4
Total	$\sum Y_{i..}^2 - Y_{...}^2/6$	5

A variance for the estimate of PQR is obtained from the residual MS ($S_r/4$) which provides an estimate of Var(η).

The parallel between split-unit designs in which main effects are estimated from comparisons between main units, and other confounded designs, in which high-order interactions are estimated from comparisons between blocks, emphasises two aspects of design. First that in both designs information about treatment effects occurs at two levels of variation, and the information at both, or in more complex designs all, levels of variation should be considered. Second that whenever information is split between different levels of variation, there is a choice of how the information is distributed between different treatment effects.

18.7 Non-orthogonal split-unit designs

Although the properties of the split-unit designs discussed thus far in this chapter have been limited to split-unit treatment effects being orthogonal to block effects and to main-unit effects, this is a reflection of the historical development of these designs and is substantially the result of the facilities for analysis being similarly restricted at the time when the designs and the analysis of results were being developed. In fact there is no need for the restriction that all sets of effects be orthogonal. At the lowest level this means that there are no problems when observations are missing, but there are also various modifications to the basic, orthogonal split-unit designs which can be used. Not only is there no requirement for orthogonality between main-unit treatments and split-unit treatments, but there is no necessity for the number of split units to be the same for each main unit.

First we should be clear that with modern analysis facilities there are no difficulties in analysing multilevel designs in which treatment effects are not orthogonal to block or main-unit effects. The general principles and methods of analysis have been fully explained in Chapter 9. Therefore when there are sensible reasons for wanting to use split-unit designs with non-orthogonal treatment effects we should expect to be able to construct efficient designs, and we will consider the construction of various forms of non-orthogonal designs.

The effects of non-orthogonality are different in different situations. In the simplest situation where the record for one split-unit in one main unit in one block is missing, then all effects will have a very small level of non-orthogonality, but this will have virtually no impact on the estimation of effects. To explain this we need to recognise that although we have lost exact orthogonality for the combinations of main-unit treatments and split-unit treatments the departure from orthogonality is tiny, and the consequent correlations between the various different sets of estimates are for all practical purposes irrelevant. If we have a situation where one combination of the two-way table of main-unit treatments and split-unit treatments is missing because it is impossible or of no interest, then we could have a more substantial correlation between the two sets of main effects, and also between main effects and interaction effects. The impact of these correlations is to produce a mild level of non-orthogonality, which would mean that SS for different sets of effects would depend on the order in which the effects were fitted. However there is only one sensible order of fitting effects: blocks – main effects for the main unit treatment factor – main-unit effects – main effects for the split-unit treatment factor – interaction effects. Hence although the effects are mildly correlated, the analysis is otherwise unaffected; as the number of treatment combinations not included increases so do the correlations of effect estimates, but there are no changes to the pattern of analysis.

If we have a situation where the split-unit treatments are a hierachical set, with no necessary connection between the split-unit treatment levels for one main-unit treatment level and those for another main-unit treatment level, then the analysis will fit the split treatment effects within each main treatment, rather than split treatment main effects plus the interaction with main treatment effects. If the split-unit treatments are wholly different for each main-unit treatment, then there is actually no non-orthogonality; sometimes there will be some overlaps between the sets of split-unit treatment levels, and then there could be a non-orthogonal element to the analysis. However, if the analysis is completed ignoring any repetition of split unit treatments for different main-unit treatments (essentially a pure hierarchical analysis model), then the impact is simply to ignore any main effects of the split-unit treatment levels which occur for more than one main-unit treatment.

If the number of split-unit treatments is greater than the number of split units per main unit, then there will inevitably be non-orthogonality between main and split treatments, and this will be controlled in the design.

18.7.1 Particular forms of non-orthogonality

Often when planning to use combinations of two treatment factors there will be some treatment combinations which are impractical or undesirable. The design will then simply omit those treatment combinations. There are no other design considerations since, as has already been noted, the analysis is straightforward and the interpretation will inevitably be concerned only with those treatments which are thought appropriate.

An example of this situation would occur if the experiment was comparing different densities of two crops in a two-crop-mixtures experiment (also referred to as intercropping). If the different densities of each crop were achieved by having a constant within-row density and changing the gaps between the rows, then the number of rows of the second crop which could be fitted in the gaps between rows of the first crop would be limited by the size of the gap between the rows of the first crop. It would be quite natural to use the first crop density as the main plot treatments. The combinations of main-plot and split-plot treatments which could be included in the experiment would be less than the full factorial set.

Situations where the split-unit treatments are specific to each main-unit treatment are quite common. A fairly extreme example would be if the treatment structure discussed in Section 12.6, displayed in Table12.12, were to be arranged in a split-unit structure, with the different chemicals as main-unit treatments. This would be sensible if it was thought desirable to keep the different chemicals separated as much as possible. The two chemicals, O with two different surfactants and E, plus the two controls would be the five main-unit treatments. Within the main units for chemicals O and E, there could be split units for two sprayers (Ed and Ul for chemical O, and Ul and Con for chemical E, and then split-split units for the three concentrations. There are, of course, various other alternative ways of arranging the 20 treatments in multiple level designs, and as in the original discussion the replication of different treatments may vary.

Sometimes the number of possible split-unit treatments can be quite large and can be too large for the sensible size of a main unit. This is quite likely to occur whenever the number of split-unit treatments is greater than three or four and is ubiquitous in industrial experments with some factors whose levels are harder to reset than others. An example of this situation was an experiment on a protein extraction process to study the effects of five factors, namely the inflow feed position, the feed flow rate, the gas flow rate and the concentrations of two proteins, each at three levels. The levels of the last four factors could be set by simple adjustments, but setting the feed flow position involved taking the equipment apart and reassembling it and so could only be done once per day. It was decided to use 21 days for the experiment. Two combinations of levels of the other factors could be used each day, so that we have a split-unit structure with 21 main units and two subunits within each main unit. It does not take long to realise that it is not possible to find an orthogonal split unit design in this structure. Finding good non-orthogonal designs by hand is quite a challenge, but computer-based algorithms are becoming increasingly available – see Trinca and Gilmour (2001) for a description of the design that was used for this experiment.

18.7.2 Row–column designs with split units

The discussion of incomplete main units suggests that we might reconsider the ideas of row-and-column designs with treatments applied to split units within the row × column main units (the idea of the Trojan square considered in Chapters 8 and 9). There may not always be actual main-unit treatments applied to the row × column main units, but the concepts are clearly closely related to those of split-unit designs with incomplete main units just considered.

We will consider horticultural experiments within a glasshouse to illustrate the ideas. Typically in a glasshouse we might expect two forms of environmental variation which we should expect to allow for by blocking in two directions. In one direction, from one side of the glasshouse to the other there will be clear differences between the areas at the two sides and between the sides and the middle under the highest part of the roof. It would probably be reasonable to expect five (row) blocks (side – intermediate – middle – intermediate – side) to explain most of the variation in this direction. However, in the other direction, along the glasshouse, it is not clear how many blocks would be needed. The two ends will be different, but between the two end sections we could have one other block or two more blocks or three or four more blocks (between three and six column blocks). Within each row × column unit there could be two or more split units, with the obvious proviso that with more column blocks there will be fewer split units per row × column main unit.

Suppose that the total number of split plots which can be included in each row along the glasshouse is 12. We could arrange these in four column blocks with three split plots in each main (row × column) plot, or we could have six column blocks with two split plots per main plot. If we were very fortunate we might have twelve treatments to compare in this structure. The choice between the two alternative arrangements of split plots and main plots would depend on the expectations we have about the effects of the two alternative column blocking

Column	1	2	3	4
Row 1	A B C	F H J	D I L	E G K
Row 2	G I J	A D E	B H K	C F L
Row 3	E H L	C I K	A F G	B D J
Row 4	D F K	B G L	C E J	A H I
Row 5	A J K	H I L	B E F	C D G

Figure 18.12 Glasshouse example design.

systems, and the relative sizes of the within-main-plot error variances. However, the loss of efficiency from using just two split plots in each main plot is considerable (as it would be in incomplete block designs with two plots per block) and it would usually be thought desirable to use at least three split plots within each main plot.

The construction of designs with three or more split-plot treatments per row × column plot, with any numbers of rows and columns is not in general very difficult, though there are, as yet, no formal methods for construction and the designs constructed in an ad hoc manner are inevitably not guaranteed to have the maximum possible efficiency, though there are not yet methods for calculating that maximum efficiency. There are optimal designs for some situations with numbers of rows and of columns equal, and only two split units per main unit, and these have been mentioned in Chapter 8. An example design for the glasshouse situation with five rows (across the glasshouse) and four columns (along the glasshouse) with three split plots per main unit is shown in Figure 18.12. We do not claim that this design is the most efficient possible (and readers are invited to see whether they can improve on the design shown), but the design in Figure 18.12 is clearly at least nearly optimal with all twelve treatments in each row, all twelve treatments at least once in each column and with nine treatments occurring within a row × column main plot with ten of the other other eleven treatments, and only three pairs of treatments repeated in a main plot.

The structure of the design in Figure 18.12 could be modified or extended in different ways to fit other design problems. One very simple modification would be to replace one of the blocking systems by a set of main-plot treatment levels. Thus if we have four main-plot treatments in five blocks, the columns become the main-plot treatments and the rows become the blocks, and we have an orthogonal design for main plot treatments and blocks with all 12 split-plot treatments in each block and at least once with each main-plot treatment. This design is an example of the incomplete split-unit design discussed in the previous subsection.

An even more complex design is achieved if we apply main-plot treatments to the different row × column plots. If there are four main-plot treatments, it is obviously easy to allocate the four main-plot treatments to main plots so that each main-plot treatment occurs in each row and at least once in each column. Having each of the split-plot treatments at least once with each main plot is impossible; this is because the need to have each split-plot treatment in each row and in each column makes similar patterns to those required for the main-plot treatment allocation. The reader is invited to find a design with as many combinations of main- and split-plot treatments occurring as possible. If this design problem occurred in practice there would be a choice between using both rows and columns as blocking systems and having a design with considerable non-orthogonality between main- and split-plot treatment sets, or using only one of the blocking systems and having a design with a low level of non-orthogonality.

18.8 Linked experiments

One final form of experiment which could be considered as an example of multiple-level units is where the material resulting from one experiment is used as the resource for a second experiment. An obvious example, which has been quite extensively discussed in the statistical literature is concerned with wine production. The first level of experiment will be concerned with different ways of growing the grapes, including possibly different varieties of grape from which the wine will be made, and the second level of experiment will be concerned with different methods of producing the wine from the grapes produced in the first experiment. Other examples would be in growing apples for cider production, or almost any raw food material which will then be processed for a particular form of food production.

The fundamental question in all such linked experiments is to what extent the experiments should be linked. Because the material for the second experiment is derived from the output from the first experiment there will be properly defined information from the analysis of the results from the first experiment about the variability of the resource material for the second experiment.

There are two alternative ways in which the material from the first experiment can be used as input to the second experiment. In one pattern the material from each separate plot can be kept separately and used separately in the second experiment. Each plot from the first experiment is then a form of main unit in the second experiment. The treatments for the second experiment can then be applied as split-unit treatments in the second experiment. Alternatively the second-experiment treatments can be applied to different main-unit material from the first experiment using a form of confounded design. The split-unit design is quite simple to construct, but has the disadvantage of requiring rather large numbers of individual split units, which may be both too small and too numerous for easy handling in the second-experiment process. The confounded experiment design may be quite difficult to construct and arranging the second-experiment treatments to be orthogonal to the first-experiment treatments, to the first-experiment blocks, and to any second-experiment blocking structure will usually be impossible. Hence there will be a loss of information with the confounded design approach.

The alternative approach is to combine the output from the first experiment into one set of material from each treatment of the first experiment, and to use these combined sets of material as the input for the second experiment. This approach gives a simpler practical procedure for the second experiment, reduces the complexity of design for the second experiment and does not make the second experiment any less precise. Hence there seems no good design reason to use any other approach. It is possibly a bit more difficult to calculate the precision of the full results from the second experiment because the variability of the first experiment is not naturally retained as a component of the variation of the second experiment, and has to be added on to the analysis of results from the second experiment.

Table 18.12. *Yields and sums of squares for Excercise 18.1*

	Maize variety					
Cowpea variety	1	2	3	4	5	Total
A	80.1	50.6	41.5	46.5	57.8	276.4
B	59.1	3 4.2	41.2	45.3	53.7	233.5
C	100.5	85.7	71.7	90.9	68.9	417.7
D	36.6	49.6	74.8	54.3	56.9	272.2
Total	276.3	220.1	229.2	236.9	237.3	1199.8

$\sum y^2$ $= 27\,122$ $\sum (\text{cowpea} - \text{maize total})^2 = 78\,665$

$\sum (\text{block total})^2 = 487\,581$ $\sum (\text{block} - \text{cowpea total})^2 = 129\,092$

$\sum (\text{cowpea total})^2 = 379\,485$ $\sum (\text{block} - \text{maize total})^2 = 99\,199$

$\sum (\text{maize total})^2 = 289\,751$ $\sum (\text{grand total})^2 = 1\,439\,520$

It seems clear from the discussion in the two previous paragraphs that there are very few advantages in making the designs of the two experiments more interdependent than is necessary. Each experiment should be designed to be as efficient as possible, and any interdependence of the results should be allowed for in the analysis and interpretation of results, but without any attempt to change the two separate designs.

Exercises

18.1 In an experiment to investigate the benefits of growing maize and cowpeas together, a criss-cross design was used. Four cowpea varieties were allocated to the four rows of each block and five maize varieties to the columns of each block. Three blocks of this design were used, the orders of the cowpea varieties and of the maize varieties being randomised in each block.

Calculate the analysis of variance for the sets of total yields and SS in Table 18.12, and briefly summarise the conclusions to be drawn from the results.

18.2 An experiment on competition between grass and legumes was designed as a 2^4 structure in two blocks with eight main plots per block and two split plots per main plot. The main-plot treatments were two cutting regimes (C), two grass species (G) and two legume species (L); the split-plot treatments were two proportions, grass:legume (P).

The treatment totals for yield per plot (in a simplified form) are given in Table 18.13. Complete the analysis of variance, started in Table 18.14, separating treatment SS into single df components. Summarise the conclusions to be drawn from the experiment, presenting appropriate tables of means and all necessary SED between means.

If it is assumed that all three-factor and higher-order interactions are zero, obtain an estimate of the difference $(c_0 g_1 l_0 p_0 - c_1 g_1 l_0 p_0)$ and the SE of this estimate.

18.3 An experiment on air and soil heating effects on glasshouse tomatoes was designed in a split split-plot form. The main plots were whole glasshouse compartments in which the air was maintained at either 60°F or 55°F. Each compartment contained two large troughs to be regarded as the split plots and each trough was divided into two halves to make two split-split plots.

In each trough, one-half was heated and the other was unheated (control). In each compartment, the soil in the heated half of one trough was kept at a minimum temperature of 65°F, and

Table 18.13. *Treatment totals for Exercise 18.2*

	p_0	p_1
$c_0 g_0 l_0$	20	34
$c_0 g_0 l_1$	14	29
$c_0 g_1 l_0$	24	37
$c_0 g_1 l_1$	18	32
$c_1 g_0 l_0$	27	31
$c_1 g_0 l_1$	24	25
$c_1 g_1 l_0$	33	37
$c_1 g_1 l_1$	26	29

Table 18.14. *Partial analysis of variance for Exercise 18.2*

	SS
Blocks	4.5
Main-plot treatments	119.0
Main-plot total	132.5
P	144.5
P × main-plot treatments	62.5
Total	342.0

Table 18.15. *Mean yields for Exercise 18.3*

	Air 55 °F	Air 60 °F	Mean
Control	$y_{.1.1}$	$y_{.2.1}$	$y_{...1}$
Soil 65 °F	$y_{.112}$	$y_{.212}$	$y_{..12}$
75 °F	$y_{.122}$	$y_{.222}$	$y_{..22}$
Mean	$y_{.1..}$	$y_{.2..}$	

the soil in the heated half of the other trough had a minimum temperature of 75 °F. There were four blocks (pairs of compartments) and a diagrammatic representation of the design is shown in Figure 18.13.

The model for the yield variation can be written:

$$y_{ijk} = \mu + b_i + \varepsilon''_{ij} + \varepsilon'_{ijk} + \varepsilon_{ijkl} + \text{treatment effect},$$

where the suffices (i, j, k, l) correspond to blocks, compartments within blocks, troughs within compartments and half-troughs within troughs. There are only six distinguishable treatments for which the mean yields may be written as in Table 18.15.

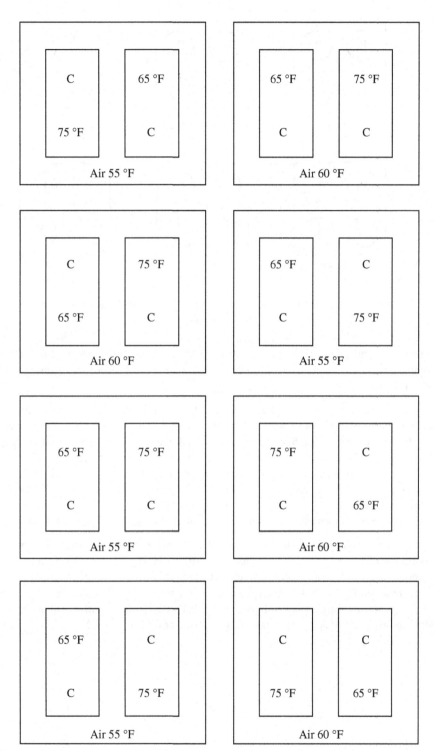

Figure 18.13 Full design layout for tomato air and soil heating experiment (Exercise 18.3).

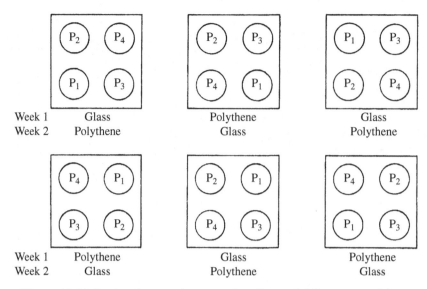

Figure 18.14 Design for experiment on the effects of different materials on soil heating.

For each (practically) interesting comparison (we suggest there could be eight) derive the estimated variance, using the error MS of 3.51 for main plots, 0.67 for split plots, and 0.62 for split split plots.

18.4 An experiment on the heating characteristics of four different types of plastic pot and two different forms of cover was conducted in six cold frames. In the first stage of the experiment, each frame was divided into four quarters, and each different type of pot allocated to one quarter in each frame. Three of the frames were covered with glass, and three with polythene. The design is represented diagrammatically in Figure 18.14. After a week, the covers on each frame are changed (G → P, P → G), and a further set of 24 observations obtained, the pots remaining in the same positions.

What is the correct structure for the analysis of variance of the complete set of 48 observations?

18.5 In a randomised block design levels of factor A are applied to plots running across the block in a north–south direction and levels of factor B to strips running across the block in an east–west direction. Each level of B is split for levels of C, each level of C running across all A levels, the randomisation of levels of C being independent for each level of B. Write down the model for this design and outline the analysis of variance. What is the estimated variance of the comparison between two levels of A for given level of C, averaged over levels of B?

Part IV

Coda

19

Multiple experiments and new variation

19.1 The need for additional variation

Thus far in this book we have considered the design of individual experiments and have been concerned to ensure that each experiment should provide answers to the questions which motivated the experiment as efficiently as possible. In general this has required that the variation in the experimental units be controlled so that the answers provided from each experiment should be as precise as possible. This will frequently require that there should be relatively little variation between the experimental units, i.e. the population from which the units are drawn will be narrowly defined.

However, if the population from which the units are taken is narrowly defined, then it follows logically that the results from the experiment would apply only to that narrowly defined population. This would usually be quite unacceptable to an experimenter who would hope to convince the wider world that the results from the experiment would apply for a much wider population. For example determining which variety of rice gives the best results in a highly controlled experiment on the paddy fields of a research institute is only going to be useful if that variety of rice is going to be the best for a large region within which the institute is located, and if farmers in that region believe in the results. Similarly if a new drug is shown to be an improvement on current practice, through a rigorously controlled clinical trial, the pharmaceutical company which has produced the drug will wish to promote the use of the drug across the whole population of the country, or even of several countries. A new washing powder is not going to sell well, if the benefits have been demonstrated for only one type of washing machine with one type of fabric.

The problem therefore is that an experiment will only give sufficiently precise answers if variation in that experiment is tightly controlled through the use of experimental units drawn from a narrowly defined population; while the results of the experiment are only going to be useful if they are applicable to a wide population. One solution is to introduce more variation into the larger experimental programme. For agricultural variety testing programmes this is often done by using multiple replicate experiments with as many as ten experimental sites spread around a country, and the same experiment at each site.

There is another difficulty with variety testing, or indeed with other agricultural experimental comparisons of management practices, and this is the variation from year to year. Year-to-year variation may be greater than place-to-place variation, particularly for variety comparisons. Since the objective for variety comparison trials is to predict the differences between varieties for future years, it is obviously necessary that the consistency of variety differences over years should be examined by repeating experiments over years, possibly in addition to replication across sites. Such repeated experiments may not be exact repeats since there will often be new candidate varieties for inclusion in subsequent years. Broadly, however, there will be a strong similarity between the varieties included in successive years' experiments.

An alternative to using many locations could be to have different conditions for different replicates in a single location. Instead of having different locations which provide variation in unspecified ways, an additional factor providing changing conditions is added to the experiment so that the interaction of the variety differences with this identified factor can be assessed. Introducing additional factors into experiments is also very useful in factorial industrial experiments. Because the initial experiment is factorial, adding additional 'noise factors' to represent known variation over which the factors of the initial experiment must be assessed is both simple and very important.

Another reason for designing sets of experiments is to overcome the suspicion of experimental results when very controlled trials are conducted at research institutions with very high levels of equipment and resources, or in high-resource-level laboratories. Because the potential users do not have access to such controlled facilities the results may be thought to be much more convincing when the experiments are removed from the controlled research situation. The scope for involving potential users of the results in the experimental process opens up further opportunities for the design of experiments.

Even if experiments have been planned independently to answer the same or similar questions, the experiments being performed in many different locations for different populations, there will be a need to develop methods of analysis for the combined experiments.

In the subsequent sections of this chapter we shall consider various different aspects of sets of experiments to answer similar questions. Some of the discussion will be concerned with the analysis of multiexperiment data, assuming that each experiment has been efficiently designed, but there are also some design issues which will be discussed.

19.2 Planned replication of experiments

The purpose of repeating experiments at different locations or in different years is to broaden the applicability of the conclusions. The use of different locations is to avoid the danger that the results are determined by the very narrow population of experimental conditions inherent in a very precise experiment at a single location. Repeating experiments over years attempts to avoid the danger that results are determined by the, possibly unusual, conditions in a single year. Repetitions over locations and over years may be combined. In general we shall discuss experiments repeated across different environments, using the variation of 'environments' to include both location differences and year differences, since most of the important aspects of repeated experiments apply to either locational or annual differences. However, when we come to consider the interpretation of the results, and in particular the estimation and prediction of variety differences for the future, then the differences between years cease to be relevant, and we consider only the variation between different sites.

The use of sets of experiments for a range of enironments occurs most frequently in the comparison of large numbers of varieties of agricultural or horticultural crops. Many such sets of experiments are organised every year by national or international research centres. The purpose of these regular sets of experiments is to provide a testing situation for new varieties of crops developed by breeding programmes. Examples of such sets of experiments are: (i) annual trials in England for varieties of horticultural crops, e.g. cauliflowers, organised by the National Institute for Agricultural Botany; and (ii) annual international trials for wheat varieties organised by CIMMYT, based in Mexico.

Questions which require to be answered in the process of designing such trials include:

(i) How many sites should be used?
(ii) How should the sites be chosen?
(iii) How many varieties should be tested for each environment?
(iv) Should the set of varieties be the same for each environment?
(v) How should each set of varieties be chosen?
(vi) Should there be 'control varieties' included in each experiment, and how should these be used?
(vii) What experimental design should be used for each environment?
(viii) How should the results of the set of experiments be analysed?

Before answering each of these questions, of course, we have to understand the purposes of the sets of experiments, and what the questions being asked of each set of experiments are. Many such sets of experiments have become established annual events, and it is quite easy to lose sight of the original purposes.

The trials exist to provide a testing ground for mainly new varieties developed through national or independent breeding programmes. All varieties submitted for these national or international trials will have been tested within the organisations which submit them to the trials, passing through initial selection trials.

(i) The number of sites is not an aspect for which statistical design principles offer any clear advice. The number will usually be determined by the range of regions for which the researcher would like results, and for which local sites and experiment managers can be found. However, it is worth pausing to reflect whether the number of sites used is sufficient to estimate the variance component for sites and whether, if not, any important conclusions will depend on this.

(ii) Ideally each site should be chosen randomly from the set of potential sites within a region. Random selection is preferable whenever there are several possible sites within a region. However, setting up experimental sites is a complex activity, and often the researcher will simply choose to use the sites for which the most efficient management of the experiment can be achieved. This is sometimes referred to as purposive sampling, but really it is a question of management efficiency.

(iii) The number of varieties to be tested in each environment must primarily reflect the objectives of the overall set of experiments. If there is a balance to be achieved between the number of varieties to be tested and the amount of replication at the site, then it will usually be sensible to use more varieties and less replication. But there is no clear statistical design principle to guide the choice of the number of varieties for each environment.

(iv) It is not necessary that the set of varieties should be exactly the same for each environment, though it is desirable that most, i.e. at least 90%, of the varieties should be repeated at each environment. There will always be arguments for including a few specific varieties for each local site, these being varieties which are known to be effective at that site, or varieties which have been developed specifically for the site. There may also be varieties in the set to be tested at every site which are known to be unsuitable for a particular site, and such varieties may be omitted for one site. The reason for requiring that a large majority of varieties are repeated at all sites is simply that the use of multiple sites to compare varieties makes no sense unless most varieties are tested at all sites.

(v) The selection of those varieties for inclusion in all environments is the most important part of the process of selecting varieties. The questions which the set of experiments is designed to answer, and the varieties which are available, will mainly determine how the main varieties are to be chosen. Varieties to be included only at a single site, or those to be omitted at a single site will be chosen as indicated in the previous paragraph.

(vi) It is likely that the set of varieties to be included in all environments will include a number of well-established varieties which will essentially act as controls for the combined experiment. Comparisons of varieties will use such control varieties as a standard. It is not therefore likely that any further control varieties outside the main set of varieties to be compared will be needed. It is possible that the design of experiments in some enviroments could be thought to be best achieved by the use of control varieties, but this will be discussed in the next paragraph.

(vii) This is the most important design question. Traditionally the design for each environment has been the same, either a randomised block design if the number of varieties is not too large, or, more recently, an alpha design using incomplete blocks in each of several replicates. It is not necessary that the design should have the same structure for each environment, though having the same design makes the analysis simpler. However, since the analysis will be done by computer, the simplicity of analysis is not a critical issue, and the design for each environment should be chosen to produce the most precise results that can be achieved in that environment. This will probably be an alpha design for each environment, but, particularly if the set of varieties is not the same at each environment, then different alpha designs will be appropriate at each environment.

It would even be possible to use complete block designs for some environments and alpha designs at others, or to use completely different forms of control of variation by using a set of control plots spread throughout the experimental area in such a way as to provide a precise estimate of the map of variation across the whole experimental area, and hence obtain estimates of the variety yields relative to the control variety. This latter approach to the control of variation has been used together with an effective analysis program for estimating spatial variation.

19.2.1 Analysis of sets of experimental results

The analysis of the results from a set of replicate experiments distributed over multiple environments is staightforward. First the results for each environment should be analysed. The analysis for each environment clearly reflects the blocking system for the environment, and will usually simply include an SS for blocks and then the SS for varieties adjusting for blocks. The analysis should include checking whether there are any aberrant values in the data set.

After the analyses for each environment have been calculated, it is usually felt useful to calculate an overall analysis for the set of sites in a single year, which would normally include block variation for each each site, the SS for the main effect variety variation, and then the SS for the interaction between varieties and sites. The interaction SS tells us how much deviation from the main variety effects is to be found across the sites as a set.

Similarly it is possible to calculate a combined analysis for several years for a single site. Also a combined analysis for all site/year environment combinations can be calculated with the environmental effects split into year effects and location effects if that is likely to provide more information.

In calculating the combined analysis we assume that the combined error SS is a meaningful concept. That is the individual error MS from the different environments are assumed to be sufficiently similar that the average error MS from the combined analysis is a reasonable estimate of overall error variation. Formally we could consider the combined error MS as an estimate of the common random variance, σ^2. Before this estimate of variance can be used, for example to 'test' the significance of the interaction mean square, we should decide formally how to test whether it is reasonable to assume a common error variance for all environments. In practice, such formal testing is of little practicality, since it is almost certain that the underlying random variances for the individual experiments are not equal.

However, believing that the different error variances are not estimates of a single common random variance does not mean that the average error variance is of no use. Provided that the error variation in the different experiments is broadly similar (i.e. mostly not different by a factor of more than 2), then the average error variance is both a meaningful quantity and a sensible basis for assessing the size of the environment × variety interaction MS, and also of the variety main effect MS. Formal testing of the hypothesis of zero effects may be of no value, but examining the relative sizes of the three MS (variety main effect, environment × variety interaction, and average error) is crucial to the interpretation of variety effects across the set of environments.

19.2.2 Interpretation of results

The purpose of repeating the comparison of varieties across a range of environments is to assess the extent to which the differences between varieties are consistent over the different environments, and to determine whether the recommended best varieties should be the same for the different environments. Again this is not primarily a question of formal testing of hypotheses of zero effects, but rather of producing the most sensible estimates of differences for the range of environments. It is extremely unlikely that the 'true' differences between varieties are identical at the different sites. It is also unlikely that the changes to the varietal differences between locations are larger than the average differences between the varieties. In more formal terms, there will usually be some evidence of interaction effects, but the interaction effects will tend to be rather smaller than the main effects of varieties.

How then do we produce the best estimates of variety differences for each site, and, in particular, produce recommendations of the best varieties for each site? This is the one stage when we shall consider variation between sites differently from variation between years. When we are estimating variety differences this will be of value for predicting future performance. There is an interest in estimating the varietal differences for each site, but there is no equivalent interest in estimating the varietal differences for each year (past year predictions are not interesting, but future ones are). Hence we are only considering different sites as different environments. At this stage we eliminate years by taking averages over years for each site. When the varieties at a particular site have varied through different years, we work with the variety main effects for that site, averaged over years.

We return to the question of producing the best estimates of varietal differences for each site using both the main effect diffferences averaged over sites and the individual site estimates. If we had done only one experiment at one site, then we would be content to accept the ordering of varieties produced by that single experiment, though we might also consider whether previous estimates of differences between the varieties from earlier trials should be used to modify the results. When the experiment at one site is part of a formal set of replicate experiments, it is more clearly apparent that we should use the varietal differences from other sites to modify the results at any individual site.

How should the variety differences at one particular site, and the variety differences averaged over all sites be combined to produce best estimates of differences for the one site? The obvious information to use is the relative sizes of the variety main effect MS and the variety × environment interaction MS. To assess the relative importance of the variety main effect and the variety × environment interaction we use the average error mean square.

Almost invariably it is observed that the interaction MS is less than the main effect MS, but usually larger than the average error MS. It is helpful to think how we would choose estimates of variety differences for a single environment by considering two extreme situations. First suppose that the interaction MS is no larger than the average error MS; then we should use the main effect differences to estimate differences for any particular environment, because we have no evidence to suggest that the variety effect is different in any particular environment and the main variety effect estimates provide the best estimate of variety differences in each and every environment. At the opposite extreme suppose the interaction MS is about the same size as the main effect MS or, most improbably, larger than the main effect MS; then there is no reason to involve the main effect differences to modify the differences in any particular environment and we should use the observed variety differences in each environment as the best estimate of the variety differences in that environment. When the interaction MS is somewhere between the average error MS and the main effect MS we should use an average of the main effect differences and the particular environment differences, giving more weight to the main effect differences as the interaction MS is closer to the average error MS.

If we define a measure, λ, of the size of the interaction MS relative to the main effect and error MS, in the form

$$\lambda = (\text{interaction MS} - \text{error MS})/(\text{main effect MS} - \text{error MS}),$$

then a reasonable rule for estimating a variety difference for a particular environment would be to use

$$\lambda \times \text{observed difference in environment} + (1 - \lambda) \times \text{main effect difference}.$$

This averaging process assumes that the interaction MS lies between the error MS and the main effect MS. If the interaction MS is less than the error MS, then λ is set to zero, and if the interaction MS is greater than the main effect MS, then λ is set to one.

The SEs for the combined estimates of variety differences can be calculated from the variances of the estimates of the particular environment difference and of the main effect. These should be provided by a combined analysis. Strictly the value of λ is also an estimate and the precision of that estimate should also be allowed for in calculating the SE for the

Table 19.1. *Variety means for each site*

Variety	Site 1	Site 2	Site 3	Site 4	Variety mean
A	22	29	21	28	25
B	15	32	21	20	22
C	17	24	17	26	21
D	14	24	17	21	19
E	17	18	18	15	17
F	11	17	14	22	16
Site mean	16	24	18	22	

Table 19.2. *Combined analysis of variance for multisite variety trial*

	SS	df	MS
Sites	60	3	20.00
Varieties	56	5	11.20
Variety × site	40	15	2.67
Pooled error	60	60	1.00
Blocks	12		

combined estimate; however, the uncertainty in the estimate of λ will usually be very small given that all the MS are based on fairly large numbers of df. Also where the varieties are not identical in all environments, there will be a tiny covariance arising from the non-orthogonal element in the estimation of the main effects and environment effects. Hence the variances, and hence the SE, for the combined estimates are only approximate, though the approximation will usually be close to the correct value.

The approximate variance for the combined estimate is calculated as

$$\lambda^2 \times \text{variance (particular estimate)} + (1 - \lambda)^2 \times \text{variance (main effect)}.$$

19.2.3 An example of multisite analysis

We shall illustrate this method for calculating estimated variety differences by using some artificial data. Suppose we have data for comparing six varieties at four sites, with four replicate complete blocks at each site. The set of variety means at each site is shown in Table 19.1. The relevant parts of the combined analysis of variance are shown in Table 19.2.

The analysis of variance shows that there are large differences between the varieties, that these differences are fairly consistent but that there is clear evidence of some interaction between varieties and sites. A particular illustration of the interaction effect is the different pattern of mean yields for varieties A, B and C at sites 2 and 4.

Table 19.3. *Combined estimates of variety means*

	Site			
Variety	1	2	3	4
A	21.2	29.0	22.7	27.2
B	17.5	27.0	20.2	23.3
C	17.0	24.8	18.7	23.5
D	14.8	23.2	17.0	21.0
E	13.7	20.5	15.5	18.3
F	11.8	19.5	14.0	18.7

Table 19.4. *Adjusted means based on main effects*

	Site			
Variety	1	2	3	4
A	21	29	23	27
B	18	26	20	24
C	17	25	19	23
D	15	23	17	21
E	13	21	15	19
F	12	20	14	18

The value of λ is $1.67/10.20 = 0.164$, and combined estimates of variety means at each site, using this value of λ, and adjusting to the site means are shown in Table 19.3.

We now have three sets of possible estimates for the variety differences for each site:

(1) the site means for each site separately (Table 19.1), ignoring information from the other three sites;
(2) the combined estimates in Tables 19.3; and
(3) the means using the variety main effects only, adjusted to each site mean (shown in Table 19.4).

The SED between two means for comparing two variety means at the same site for these three sets of estimates are:

(1) 0.71,
(2) 0.32,
(3) 0.35.

In this example the variety main effects are large, the interaction effects are relatively small, and the error variance is unusually small. Hence many comparisons of variety means in a site are significantly different in all three forms of estimate. Nevertheless the consequences of using the combined estimate do illustrate both the sensible nature of including interaction effects and also the smoothing of the more extreme interaction effects (such as variety B in site 2 or variety C in site 4).

19.3 Introducing additional factors in experiments

In the previous section we considered the design of multiple experiments at many sites, when the only requirement for the sites was that they should represent a range of different environments in which the results of the trials should be relevant. No more detailed knowledge of the different environments was assumed. In other situations where it is desirable that the results of a single experiment should be broadened to be relevant to a wider range of situations, there may be very clear ideas about the causes of variation between environments. For the purposes of variety testing it might be thought desirable to test varieties at two different levels of fertilisation, or using two different forms of crop management. For the evaluation of a substantial factorial structure of treatment factors in the development of a new washing powder, it might be thought desirable to test the combinations on several different makes of washing machine, or in regions with different natural water hardnesses.

There are essentially two different approaches to incorporating additional environmental factors into an existing experiment in order: (i) to broaden the environments for which the results of the experiment should be applicable, and (ii) to examine the extent to which the comparisons between treatments vary over different levels of external factors. These are to expand the experiment to include whole replicates at different levels of additional factors, or to directly include the additional factor(s) as treatment factors within the experiment. The latter approach is only practicable when the basic experiment has a factorial structure; the former is appropriate when the initial experiment has clearly defined replicates with one or more blocks per replicate.

When the basic experiment has several replicates, possibly at a number of sites, it is extremely simple to introduce one or more additional factors by applying different levels of the factor to different replicates. For example if the basic experiment has four replicates and the additional factor has two levels, then two replicates will have one level of the additional factor, and the other two replicates will have the other level. This will provide full information about the interaction of the additional factor with the treatments in the experiment, as well as giving information about the residual variation. If there are only two replicates, then allocating the two levels of the additional factor to the two replicates means that the interaction effect cannot be separated from the residual effect. However, if the experiment is repeated at several sites, then an estimate of residual variation can be obtained by assuming that the interaction effects are consistent over sites (in other words, the three-factor interaction for varieties × additional factor × sites is not important).

It is not necessary to have equal numbers of replicates with the different levels of the additional factor. If the basic experiment has three replicates, then one level of the additional factor can be allocated to two replicates and the other level can be allocated to the third replicate. This provides information about the interaction and also allows the residual variation to be estimated from the two replicates with the same level of the additional factor. Similarly if the additional factor has three levels and the basic experiment has four replicates, then allocating one of the three levels to two replicates and the other two levels to one replicate each will again provide information about the interaction and about the residual variation.

We illustrate the use of noise factors in an experiment.

19.3.1 The use of noise factors to represent additional variation

When the basic experiment has a factorial treatment structure it is possible to add additional factors with levels representing important potential variation in the population of environments in which the results of the basic experiment are to be relevant. In engineering applications, such factors are traditionally referred to as 'noise' factors. The noise factors may be treated in exactly the same way as other factors in the experiment, though the main effects of noise factors are usually of no direct interest. Thus adding a single two-level noise factor to a basic experiment which has six two-level factors gives an experiment with seven two-level factors. If the original experiment was a half-replicate, then the experiment including the noise factor will be a quarter-replicate, assuming that the total number of experimental units is to be unchanged. This does not essentially alter the problem of designing the experiment, though the design problem is clearly different.

Noise factors are particularly useful in robust product or process development, i.e. in learning how make a process or product in such a way that it will give consistent results across a range of environments. We therefore want to *exploit* interactions to choose levels of the factors which we can control in the field to compensate for the main effects of factors which we cannot control in the field (though we can and do control them in the experiment). This is sometimes represented as choosing levels of the treatment factors which minimise the variance across levels of the noise factors. Although attractive in the sense that it relates the experimental results to what is expected in the field, this approach has led to some mistaken decisions by missing specific interactions and can easily lead to confusion. As G. E. P. Box put it, 'interactions are interactions and variance is variance'. There is also the theoretical problem that the levels of the noise factors used in the experiment are not a random sample from a population of noise levels, so that any sample variance calculated is not estimating any sensible population variance.

The fact that the main effects of the noise factors are not of interest in themselves can sometimes be exploited by using a split-unit design, with the noise factors allocated to main units and the other factors of interest (sometimes called control factors) allocated to the subunits. Then the interactions between control factors and noise factors are estimated in the subunit level and so considerable information can be gained about them. If there are strong practical reasons for runnning the experiment in this way, then it should be done. Otherwise, it is probably best to stick to a completely randomised design or a block design with high-order interactions confounded. This is because the distinction between control factors and noise factors is not always clear cut. Many experimenters start to become interested in the main effects of noise factors if they are different from what is expected.

We illustrate the issues in choosing a design with an experiment on factors which affect the performance of washing powders. Chemists wanted to compare the effects of two types of bleach (A) and two types of surfactant (B) and decided to run 16 washes with different combinations of two levels. However, they wanted to test the effects of these two factors not only in ideal, laboratory, conditions, but also across a range of environments. They decided to include four noise factors in the experiment, namely the wash temperature (C), the water hardness (D), the size of the load (E) and whether or not a prewash was used (F), each noise factor being included at two levels.

Testing six two-level factors in 16 runs implies the use of a quarter-replicate and a regular fraction can be used based on its defining contrast (or more correctly, based on what this contrast implies about which effects are estimable). The usual advice would be to use the fraction with defining contrast $I \equiv ABCD \equiv ABEF \equiv CDEF$, since this avoids aliasing any main effects with two-factor interactions. However, here the main effects of the noise factors are not of much interest, while interactions between control factors and noise factors are. These are aliased in pairs, AC with BD, AD with BC, AE with BF and AF with BE. Another option might be to use the fraction defined by $I \equiv CD \equiv CEF \equiv DEF$, which leaves AC and AD aliased, BC and BD aliased and all other control × noise interactions clear of low-order effects. Another option would be the fraction defined by $I \equiv CD \equiv ABCEF \equiv ABDEF$. Either of these is equivalent to combining the wash temperature and water hardness into a single noise factor whose levels represent good (high temperature, soft water) and bad (low temperature, hard water) washing conditions.

19.4 Practical context experiments

There are a number of situations where experiments are designed within the normal processes of agriculture, medicine or industrial activity. One reason is that if the experiments are too divorced from the actual processes, then those working in the normal activity of agriculture, medicine or industry may be disinclined to be persuaded by the experimental results. In clinical trials another reason is that it is practically effective to recruit subjects for a trial through the normal process of treating patients. In industrial experiments a further reason is that often the only available machinery for conducting the trial is that used for normal production.

In agriculture, and in particular in developing countries, there has been a substantial gap between the research at research stations, usually located close to major cities, and the work of agricultural extension officers who have been trying to persuade farmers to adopt improved practices or new varieties, based on results from the research institutes. Because the research institutes are seen as remote from the experience of farmers, and the research environments are not familiar to farmers, the research institute results are not believed to be relevant to the farmers' situation.

19.4.1 On-farm experiments

Some on-farm experiments are essentially small-scale experiments of the same general form as the larger research station experiments. The number of treatments will be smaller, between two and six, the total number of experimental units at a single site will be small, probably between ten and twenty, and the experiment will be clearly managed by the research scientist, with the farmers following a pre-set procedure, and with the research scientist vsiting the farms on a regular basis. Blocking should involve small blocks since the variation on each farm can be much larger than on a research station and therefore the potential gain from blocking will be considerable. However, it is not always easy to identify how to design the blocks at each site, and for simplicity a randomised complete block design will often be used, which may not be a major disadvantage given that the number of treatments is small.

However, the more interesting on-farm experiments are those where the research scientist allows, and indeed encourages, the participating farmers to make decisions about all aspects of the experiment. Essentially, the whole process becomes experimental. To illustrate the concept, we shall describe in some detail an on-farm experiment on chickens and egg production in Kenya, the experiment being designed and managed by Dr Joseph Ndegwa. The purpose of the experiment was to evaluate the effectiveness of a range of management practices on chicken and egg production on individual farms. The farms are typical household areas within villages, and the owning and management of hens in each household is a form of subsistence farming.

Five regions of Kenya were included in the experiment. In each region, three villages were chosen and in each village ten farmers were chosen. The choice of the regions was determined by the potential for working with farmers in the region to improve results; the choice of villages in each region was partly to provide a representative sample, with a small element of random choice. The selection of farmers was largely on the basis of those farmers who were prepared to be involved. Both elements of selection could have benefitted from a greater degree of random choice, with clearly defined lists of the possible villages and of the possible farms within each village; however, the potential for bias due to the selection method is quite clear and should be taken into account when assessing the results of the experiment.

The farmers were all given training about the advantages of different forms of treatment for their hens and then made their own choices of treatments for their own farm. The treatment choices related to: (i) the forms of housing to prepare for their hens, (ii) the forms of supplementary feed, (iii) the use or not of vaccination against prevalent forms of disease, and in addition they decided on their strategies for the breeding of additional chickens and the selling or retention of hens and chickens, and the use of hens for household food. Although farmers made all their own choices about the treatments, the diversity of choices and of combinations of choices of treatments within the large number of farms allowed the effects of different treatment components to be assessed through multiple regression analysis of the various outputs. Essentially any management practice that can take different forms can be varied and recorded for use in the interpretation of the experimental results.

The outputs were mainly egg production, but also the flock sizes through the three years of the experiment and the sales and eating of chickens. These were all recorded by the farmers, with regular inspection of the system and recording methods by the experiment organiser, Dr Ndegwa.

The advantages of this form of experimentation are that the process and the results can all be observed not only by the participating farmers but also by their neighbours. The environments within which the experimental results were obtained are plainly those for which the results are to be used in the future. Because the experimental treatments were chosen by the farmers, they were applied in what the farmers believed were the most appropriate ways, and therefore had direct relevance to the participating farmers and their neighbours. And sufficient control could be exercised by the experiment organiser that the experimental results are valid predictors of the conclusions to be expected if the treatments are applied in the future, and the population of farmers to whom the preferred methods of management can be recommended should have no difficulty in believing that the results would apply in their farms.

19.4.2 Final stage clinical trials

Many final stage clinical trials, in which a new drug, or new form of management of drugs is compared with a standard form of treatment, will use a design where subjects are recruited through general practitioner surgeries. A particular example is a trial for the treatment of asthma, in which the standard treatment involves two drug inhalers, one prophylactic drug used at the beginning and end of each day, and a second drug for dealing with any asthmatic difficulties during the day (or night); the new treatment is a combined drug inhaler to be used at the beginning and end of each day, and as necessary during the day. Obviously the use of a single inhaler provides a much more convenient form of drug administration. The final stage clinical trial was required to demonstrate that the control of asthma was at least as effective with the combined drug inhaler as with the two separate inhalers.

In this particular trial subjects were recruited at a large number of general practitioner surgeries, with target numbers per surgery of 15 or fewer, and a total number of subjects of several hundred, randomly divided between the standard treatment and the new treatment. Randomisation was controlled, without the knowledge of the doctors participating in the trial, by the trial organiser and the random allocation of subjects to the combined or separate inhaler systems was independent for each location. One important aspect of trials where subjects are recruited at many different locations, beyond the direct control of the trial organiser, is that the information provided to each potential subject about the purposes of the trial, the procedure for each subject, the dangers inherent in the trial treatments, and the possibilities for withdrawing from the trial, must be very clearly described in the patient information sheets. This will be rigorously examined in the ethical assessment of trial, which will also be concerned with the design efficiency, and the scientific justification for the trial.

19.5 Combined experimental analysis

Previously in this chapter we have discussed the design and analysis of experiments which are conducted at many different locations and at different times. In these multienvironment experiments the design of the complete set of specific location experiments will normally be planned as a whole, though some local details may be determined later in the process. The analysis is relatively straightforward because the experiments are designed as a set.

However, there are many other situations where it is necessary to consider the joint analysis of results from various experiments where the experiments have been planned and designed individually. Typically the questions posed for each experiment will be similar, but the population of experimental units will differ between experiments, there will be differences in the management of each experiment and the formal design of each experiment will be different for each experiment.

An example which illustrates the problems and the solutions concerns a horticultural situation which involved several statisticans having to argue the appropriateness of their analysis in a legal court. Strong claims had been made about the benefits of a new fertiliser product for growing horticultural crops. Scientists in the horticultural research institute became convinced that the fertiliser provided no benefit, having conducted a number of trials to compare the new fertiliser with an inert control. The belief that the feriliser was of no value was made in a very public forum, and a legal case was brought by the manufacturers of the fertiliser, alleging improper representation on the part of the research scientist.

There was a large number of sets of experimental results from experiments conducted by many different scientists in different countries over a number of years to assess the benefit of the new fertiliser. In all cases the basic design consisted of a direct comparison of the new fertiliser with an inert control, but the interpretation of the different experiments by different people appeared to lead to contradictory conclusions. Much of the argument was based on the use of significance statements derived from the analysis of individual experiments.

The apparent differences between the conclusions from different experiments could be the result of real changes in the effect of the fertiliser in different experiments, but it could also be the consequence of the use of different levels of resource in different experiments. It became apparent that significance tests provided no help in assessing the beneficial effect, if any, of the fertiliser. The crucial analysis of the results from all the experiments was based on the complete set of estimates of the fertiliser effect from all the experiments and an examination of the distribution of the effects. This demonstrated clearly to the statisticians involved that the distribution of effects was centred at zero, with a spread respresenting both the possible variation between experiments and the random variation of individual experiments. By the end of the trial this view was accepted by the court.

19.5.1 *Meta analysis for multiple clinical trials*

The most important area where sets of similar experiments are analysed together is the final stages of clinical trials in which normally two treatments are compared, one being a proposed new treatment and the other being the present standard treatment. It is commonly observed that at this stage of the clinical trial process several different trials will be organised, usually in different countries or areas of large countries, that the trials will be similar in organisation with differences representing variation in general treatment practice in the different locations, and it is desirable to learn about the benefits of the new treatment from consideration of the results from all the trials. The same need for combining results from many different trials arises with epidemiological trials investigating the differences attributable to different environmental or behavioural conditions, a well-known example being causes of cancer, such as smoking or asbestos.

Meta analysis, as such combined analysis of multiple trials has traditionally been labelled, has been very substantially concerned with testing the null hypothesis of a zero difference between the two treatments. However, there are alternatives and we shall concentrate on analyses which are primarily concerned with estimation rather than hypothesis testing. To understand why we prefer the estimation approach it is necessary to examine carefully the situations where meta analysis will be applied.

Consider first any single experiment at this final stage of the clinical trial or epidemiological investigation process. By this final stage of the clinical trial process, there will usually be a strong a priori belief that the new treatment does have a benefit, though it will not usually be absolutely clear that the benefit is large enough to justify changing from the standard to the new treatment. If there were not a belief that the new treatment is beneficial the resources to conduct such a trial would probably not be made available. But without strong evidence that the benefits of the new treatment justify the probable increased cost of the new treatment it is unlikely that the new treatment would be adopted for regular use.

Hence the objective of any trial at this stage is not to question the hypothesis that there is no difference between the treatments but to estimate precisely the amount of benefit. Indeed the use of significance testing in a single trial at this stage is essentially testing whether the trial is of sufficient size to detect the effect, rather than testing whether the effect is zero.

With an estimation approach the methodology is simple, and is essentially the same as that which has already been explained in Section 19.2.3. Each individual trial provides an estimate of the effect of the new treatment. The set of effect estimates from the various trials provides both an estimate of the average, or main, effect for the comparison between the two treatments, and the variation of the treatment effect across trials, or the environment × treatment effect. An analysis of variance for the set of trials divides the total variation into treatment main effect, environment × treatment interaction effect, and residual SS, and MS.

As with the variety comparison multiple trials discussed in Section 19.2, the relative sizes of the treatment main effect MS, the environment × treatment interaction MS, and the residual MS provide information about the extent of the variation of the treatment effect across environments. We would usually expect that the interaction MS would be larger than the residual MS, and smaller than the main effect MS. There are several reasons for expecting interactions between environments and treatments: the treatment effect may vary between populations in the different environments; the actual methodology for applying the new and standard treatments may vary between trials; the definition of the acceptable subjects for the trial may vary between trials; and the overall organisation and timing of the trials may vary. All of these variations are properly represented in the causes of the interaction variation because they can also be expected to apply when and if the new treatment replaces the standard treatment.

From the analysis of the set of trials we can obtain an estimate of the main effect of the difference between the treatments, using all the information from all trials, and usually achieving a very precise main effect estimate. We can also obtain estimates of the effect for each environment using both the estimate of the effect from the trial in that environment, and the main effect from the complete set of trials. The weighting of the two estimates is again exactly in the same form as for the variety trials, using MS for main effect, interaction effect and residual to define λ in the form

$$\lambda = (\text{interaction MS} - \text{residual MS})/(\text{main effect MS} - \text{residual MS})$$

and then calculating the estimate for a particular environment as

$$\text{local estimate} \times \lambda + \text{main effect estimate} \times (1 - \lambda).$$

Variances, and hence SEs, for combined estimates can be calculated from the variances for main effect and local effects, in the same way as in Section 19.2.3.

20

Sequential aspects of experiments and experimental programmes

20.1 Experimentation is sequential

No experiment is an island, complete of itself, as John Donne might have said if he had been interested in formal scientific experimentation. There must, presumably, have been first experiments in all scientific fields, but in modern experimentation the design of each new experiment depends in varying degrees on the information gained from previous experiments and/or other, non-experimental, studies. This information comes in many different forms which are considered in the next section.

Sometimes experiments are designed quite deliberately in sequences to achieve a final result. This happens most obviously in selection programmes where it is known that the programme will start with a very large number of possible alternatives, which will be whittled down during a sequence of experiments to one, or a small number of, best choices. Screening programmes operate in a similar fashion.

In the development of new pharmaceutical drugs there is a very formal sequence of experiments which must be completed before the new drug can be accepted as available to medical science. This process moves from an essentially screening initial stage through to a very precisely controlled final clinical trial where the ideal drug dosage will be examined for a large number of randomly allocated subjects under controlled circumstances for a broad population.

Another area where experimentation is sequential is in the search for optimal conditions when several factors may be varied. The number of factors can become very large and the process may be partly a screening process to identify which factor levels have an effect followed by a search for the optimal conditions.

There are also some situations where the conduct of experiments is spread over a substantial time period, which produces two sequential aspects. First, because the experiment extends over a long time there is the possibility of examining the results at intermediate stages; this can produce some particular problems to be solved. Also the randomisation procedure can be modified to be influenced by the randomisations earlier in the process.

The design aspects of all these types of sequential experimentation are examined in the subsequent sections of this chapter. There are no new concepts of design for individual experiments, but rather new ways of using designs already discussed in earlier chapters, with some additional insights into the properties of patterns of the designs which have been introduced.

20.2 Using prior information in designing experiments

Information from previous experiments will be used in various forms in the design of any new experiment. Information on the results for different treatments in possibly several previous experiments can inform the choice of treatments in the new experiment. Information on variability in previous experiments is crucial to the design of any new experiment. And information on the effectiveness of factors which must be considered in blocking or recording as covariates will again be available from many previous experiments.

Consider first the way in which the results from previous experiments will affect the choice of treatments. Suppose that in a previous experiment the results for three equally spaced levels of a quantitative factor have shown that the performances of the second and third levels are almost equal while the performance for the first level is markedly lower. Then it would be natural to again examine three levels in the new experiment, using the second and third levels from the previous experiment with an additional level midway between the two; this should enable the optimal level to be deduced with maximum precision.

If the experimentation is progressively selecting better varieties of a crop, then obviously the varieties performing best in the most recent experiment are going to be the most obvious choices for the new experiment.

If previous industrial experiments have examined the effects of several factors, the information about the relative sizes of the main effects of different factors, and which interactions appear to be important will help to determine which factors should be used in the next experiment. The previous experimental information could also indicate the relative sizes of interaction effects and should help to determine the ideal levels of any quantitative factors.

There is usually a great deal of information about the variability of results to be expected for well-experimented populations. For example the variability of cauliflower yields was examined by one of us for some 50 experiments, for various forms of treatment, in the UK during the early 1960s and the variability of mean yield per plot was discovered to be almost exactly inversely proportional to the number of plants per plot; this enabled the variability of future experiments to be very precisely predicted. Similarly a research scientist working on the effects of various treatments on egg production of hens was able to predict the variability of yields from a group of hens on the basis of the experimental history of the error MS in previous experiments. In the same way the variability to be expected in clinical trials for almost any specific form of measurement can be very accurately forecast. In industrial, and many laboratory based, experiments, the same equipment is used for many experimental studies and again considerable information about the usual variation between runs might be available.

The importance of having a reliable forecast for the error variance for a future experiment based on information from previous experiments is, of course, that it enables the experimenter to choose the appropriate level of replication to achieve any required precision of results.

The ability of statisticians to forecast variability for future experiments is often a surprise to experimenters, who assume that the experiment they are going to do is new and that nothing much is known about the likely results. The experimenter, not surprisingly, thinks about the treatments to be compared and therefore perceives the experiment as quite different from anything that has been done before. However, the forecasting of variability depends on the use of experimental subjects who are similar to those previously used, which is extremely likely, and the measurements being of the same form as previously, which again is highly likely. The concepts of newness of an experiment are quite different for experimenter and statistician, and this can generate some misunderstandings when considering the variability to be expected.

The third aspect where we can learn from previous experiments about the design for a new experiment is the whole area of blocking control and choice of covariates. Results from previous experiments show us how effective particular systems of blocking are at reducing or controlling the error variance, and this information should enable us to choose blocking systems for future experiments more successfully. Equivalently, the effectiveness of different covariates in reducing error variance in past experiments gives a good indication of the probable success using those same covariates in future experiments.

20.3   Sequences of experiments in selection programmes

There is a large section of research in agriculture and horticulture which is concerned with the selection of improved varieties of crops. The input to selection trials comes from breeding programmes, from both commercial and research organisations. The numbers of potential new varieties developed through breeding programmes are enormous and there is a need for a systematically designed set of trials to produce one or two new improved varieties from often many thousands of candidates.

It is obvious that a selection process will be more efficient if it is designed to have several stages, with the resources per candidate being small in the initial stages when large numbers of candidates are being compared, and the resource per candidate being much greater when the relatively few best candidates are being compared in the later stages of the process. This is because we do not want to spend much resource on candidates which are not good, so that more resource can be available for making the more precise comparisons of the candidates which are more likely to be the successful choice.

In fact, some theoretical research has shown that in a multistage selection process in which the objective is to select a very small number of best varieties from a very large population of candidates, the total resources should be divided equally between the stages, with the proportion selected from one stage to the next stage being the same for each pair of successive stages; this implies of course that the amount of resource per candidate will increase proportionally by the same factor for each pair of adjacent stages.

This essentially simple and intuitively reasonable rule was developed by Curnow (1961) and Finney (1958). They used the most sensible criterion for the design of the selection system to optimise, which is the expected average yield of the varieties ultimately selected. It would, of course, be possible to have other criteria for success, such as maximising the probability of selecting the best variety, but since the purpose was that farmers should grow the selected varieties, and should achieve the greatest possible yields, the criterion used by Curnow and Finney is the most practically sensible.

The construction and design of a multistage selection programme should proceed as follows. The first information needed is how many initial candidates there will be, and how many varieties are to be selected at the end of the process. The major decision in the overall design is how many stages of selection are to be included. If the number of stages (trials) is t, then the design structure will require that t successive reductions by a proportion, p, will reduce the candidates from the initial number, N, to the final number, n of selected varieties. The formal calculation of p is $\log p = (\log n - \log N)/t$.

For example if there are $N = 10\,000$ initial candidates, to be reduced to a final set of $n = 5$ varieties, a four-stage procedure would reduce the numbers for the three trials following the initial trial to 1495, 223 and 33, the proportion, p, being 0.1494.

Provided these numbers are acceptable for the planning of the sequence of trials, then we can proceed to the planning of individual trials. If the numbers do not seem sensible it would be necessary to change the number of stages, either increasing the number so that the proportion proceeding to the next stage is larger or reducing the number so that the proportion proceeding is smaller. Alternatively it would, of course, be possible to decide that the initial number of candidates should be reduced or the final target number of varieties be changed to help give an acceptable set of intermediate numbers.

When the number of candidate varieties for each stage has been determined, the individual trials can be designed, remembering that the total resources used should be the same for each stage. Thus, suppose it was decided that for the final stage of selection there should be four replicate plots for each of the 33 remaining varieties, with 100 plants per plot. Then the total number of plants required for the final trial would be 13 200. this would imply that for the initial trial with 10 000 candidates, there should be just one plant per candidate; for the second stage with 1495 remaining varieties there would be about nine plants per variety, and for the third stage with 223 candidates remaining there would be about 59 plants per variety.

This is a second point in the planning at which the experimenter needs to decide that these figures for numbers of plants per variety at each stage are sensible. If they are thought to be reasonable, and the numbers in the previous paragraph are fairy typical for this kind of design problem, then the next stage is to design the individual trials. This is where some of the design concepts for variety trials discussed in Chapter 7 are useful.

For the first stage trial with 10 000 candidates each with one plant, the appropriate design would be a completely random allocation of candidates to field positions with plants of a control variety spread at regular intervals across the experimental area. This gives some check on any large-scale variation across the trial area, which may be adjusted for by fitting a response function for the control variety yields.

For the second stage trial with 1495 varieties each with nine plants, it is probably appropriate to form one plot of nine plants per variety, and to again use an unreplicated trial with control variety plots spread across the trial area.

For the third stage trial with 223 varieties with 59 plants per variety, it would clearly be desirable to have two or three plots for each variety (with either 30 plants per plot with two replicates or 20 plants per plot with three replicates). If it is decided to have three replicates, the most suitable design would be an alpha design possibly with about 15 blocks per replicate.

For the final stage trial with 33 varieties to be compared using four plots of 100 plants per plot for each variety, an appropriate design would obviously be an alpha design with five or six blocks per replicate.

The actual numbers in a real multistage trial would almost always vary from the planned numbers, and of course there are many points in the planning process where it could be decided that the implications of the rules for constant proportional reduction of candidates between stages, and equal resources for each stage might not be acceptable. However, the general pattern should not deviate too far from the theoretically most efficient system.

20.4 Sequences of experiments in screening programmes

Screening is different from selection. In selection the principal measurement is always quantitative, usually yield, and the problem in selecting the varieties with the highest average yields is caused by the variation of individual plant yields within each variety population. In screening the problem is to detect those candidates, usually a small proportion, which have any effect. If the effect is essentially a qualitative one, the problem is simply the large number of candidates which have to be examined to detect the very few which have an effect. When the effect is quantitative, there may also be another problem because the method of making a measurement is inherently variable so that it is not clear when an effect has been observed.

For screening when the effect to be detected is qualitative, it is not immediately apparent how statistical design can be helpful. However, there is a subset of situations where candidates can be tested in groups with the result of the test being positive if one or more of the candidates in the group is positive, and the result being negative if none of the candidates is positive. When a group gives a positive result then, of course, all of the candidates in the group must be tested again to discover which are positive. However, when the group test is negative all candidates in the group can be eliminated, with a large saving in the number of tests required.

The methods of group testing, which have been discussed briefly at the end of Chapter 12, are inherently sequential, and testing candidates in groups of sizes eight, reduced to four, then two and finally one is obviously a sequential activity where the pattern of future testing depends only on the previous results. It is important that whenever sets of candidates emerge, having been included in a group test, but about which nothing is known (as a consequence of the results of other group tests), then such candidates reenter the testing process within a group of eight. For example a group test of eight candidates produces a positive result; the first subgroup of four also produces a positive result; then the properties of the candidates in the second subgroup of four are unknown in exactly the same way that candidates not yet tested are unknown; such candidates are returned to the untested population.

For screening when the effect to be tested is quantitative, the design of the sequence of trials uses two concepts already considered in this section and the previous one on selection. Where the variation in the measurement process is large, then the screening problem becomes very similar to a selection process, and the concepts of reducing the candidates by a consistent proportion between successive stages, and of using similar levels of resource at each stage will be appropriate. However, the concept of group testing is also relevant, and applying several treatments simultaneously to the same experimental unit can provide information on whether any of the treatments are effective; the use of group testing methods tends to be more effective when the variation in the measurement process is smaller. Combining the two concepts of consistent reduction and group testing can be effective, though practically more complex to implement in the intermediate situation where the variation of measurement is neither large nor small. For example a large number of candidate treatments could be applied in groups to experimental units in sets of three, and the treatments in groups which appeared to have substantial effects could be included as individual treatments in the next stage trial.

20.5 Sequences of experiments in pharmaceutical trials

In the development of new drugs in the pharmaceutical industry there is a fairly standard and consistent sequence of experimental stages. These experiments are conducted extremely formally because when a drug is finally ready to be presented for licensing it is necessary to produce evidence that all stages of the development process have been correctly completed. Some of the stages use laboratory experiments, in some cases using small rodents, others are applied to human subjects, using clinical trials for which ethical approval of all aspects of the trial is required.

The first stage will almost always be some form of screening trial in which large numbers of potential drugs are screened to identify any which appear to have evidence of positive effect. The screening can use live subjects or in vitro methods using chemical reactions or may be based on computer models predicting performance characteristics of the drugs based on physical or chemical properties. Screening trials, as has already been mentioned, inevitably involve very large numbers of candidates from which only a very small number of potentially successful drugs will emerge. Consequently it is necessary that the information on each drug in a screening trial will be very small and therefore the screening trials will produce numbers of false positives, while hopefully avoiding rejecting too many positive candidates. This may lead to the use of multiple stages in the screening process, since, as has been noted in the previous sections, the objective is to eliminate non-useful drugs with the minimum of effort.

After the screening process one or more drugs should emerge which merit further detailed examination. This stage will usually use small rodents or other live animal subjects. The object is to verify that there is a detectable change produced as a result of the application of the drug. Sufficient replication must be used to ensure that only drugs which have a real benefit proceed to the next stage. Usually experiments at this stage compare a single new drug with a non-active placebo to assess whether the drug is sufficiently promising to merit further examination of its properties.

One of the more tricky aspects of developing a new drug is the identification of the level of concentration (or dose) of the drug which is required. This is particularly difficult because the process of moving from treating small mammals to large humans involves a scaling up of the amount of the drug applied, and the scaling factor involves various possible measures of increased size. Hence it may be necessary to have experiments to determine the most appropriate dose level both with small mammals, and, later in the overall experimental process, with human subjects. These trials usually involve a gradual increase of the dose level in preset steps to identify an optimum level in terms of the beneficial effects, which is well within the limits beyond which harmful side-effects occur. Dose level trials are very tricky to design and require great care, particularly when applied to human subjects after a scaling-up of the dose level.

Finally there are the clinical trials, using human subjects, where the precise benefit of the new drug must be determined, and the possibility of adverse side effects tested. Ethical requirements for such clinical trials are very tightly controlled and a major requirement will always be that the statistical design must be highly efficient so that the number of human subjects exposed to a drug whose effects are still not absolutely established should be no larger than is absolutely necessary. This may in some situations lead to the use of within-subject comparisons using cross-over trials as discussed in Chapters 5 and 8.

There may be more than one clinical trial stage, with smaller trials used first to confirm that the drug is producing a genuine effect, using very tightly controlled sets of subjects. Then larger trials with a much wider range of subjects representing a wider population to establish the size of the beneficial effect more precisely. Possibly then there will also be a multisite, multinational trial used as discussed in the previous chapter.

The design of most of the clinical trials will usually be simple, and non-contentious, involving decisions about stratification. The only case where this is unlikely to be true is in the cross-over form of trials, or more generally any trial where subjects receive different treatments at different times. The subjects essentially act as blocks, and it is very easy for medical researchers to produce trial designs which are very inefficient in the use of within-subject information, which inevitably will be the primary source of information.

One of the really important aspects of the sequence of experiments which are involved in the development of an effective drug is that the objectives of the different experiments at different stages of the sequence change according to the stage at which the experiment is required. It is often assumed that clinical trials have the objective of testing the hypothesis that there is no difference between the effects of the two treatments. This is true for possibly two stages, one very early with small mammals, and probably the first stage after the dose fixing trials, but definitely not for most. Obviously the screening trials, and also the dose fixing trials, are not concerned with hypothesis testing. However, once it is believed that the drug has a beneficial effect, and this will apply after the first formal clinical trials with human subjects, the purpose of the later clinical trials will be to determine the size of the beneficial effect, and not to repeatedly test a null hypothesis.

The ultimate objective of the whole research process is to establish the amount of benefit which the drug can be expected to produce for an individual patient. The criterion which the licensing authority (NICE in the UK) will use when the final judgement on the licensing of a drug is being considered is whether the benefit is economically sufficient to justify allowing the sale of the drug.

20.6 Sequential nature within clinical trials

In some experiments the recruitment of subjects is sequential, often with a period of many months, and this has encouraged other ways of looking at some clinical trials sequentially. If subject recruitment occurs sequentially, then it follows that the results of the trial also become available in a sequential fashion. It also follows that it is possible to examine the results from the trial at an intermediate stage, whereas the normal procedure would be to look at the results only when the trial is complete. The temptation to look at the results before they are complete is obviously that information about the success of the trial would become available sooner.

The reason why intermediate examination of results is usually regarded as undesirable is that multiple examination of the results at several, or even many, intermediate time points gives a greater probability of observing a significantly extreme result on at least one of the intermediate examination points, when the hypothesis of a zero true difference between the two treatments is true. This means that when multiple intermediate examinations of the results are allowed the probability of observing a significantly extreme result when the null hypothesis is true is not that which was assumed in the design of the experiment.

Therefore the calculation of significance, when multiple assessment points are used requires a substantial new theory. Experiments in which multiple examination of results, which may be taken to the logical extreme of examining the results after each new result becomes available are often referred to as sequential experimentation, and there is a very substantial literature on this subject. Essentially if a graph of the cumulative proportion of positive results, against the number of subjects for which results are thus far available is plotted, then two boundaries are constructed the upper one representing a positive significant result, and the lower one a null result. The path of the cumulative proportion will eventually cross one of the boundaries, or will come to the end of the maximum number of subjects permitted in the experiment.

One advantage of using a sequential experiment design is that the result will usually become clear (a boundary is crossed) when rather fewer subjects have been recruited than would be the case in a normal fixed size trial. This is clearly beneficial, and there has been a lot of research on sequential trials in this form; much more information on such sequential experimentation can be found in the book by Whitehead (1997).

The benefits of sequential clinical trials apply only to those situations where the objective of the experiment is the testing of a null hypothesis. There is no equivalent benefit when the objective is the precise estimation of the size of the beneficial effect, because although intermediate examination of results is possible, such examinations cannot stop the trial earlier than was planned in the trial design. As we have already explained in Section 20.5, hypothesis testing is appropriate in the first clinical trial with human subjects, possibly also in a second trial, but the later, usually larger, trials should properly be concerned with estimation rather than with testing a hypothesis which is no longer believed to be true.

One other aspect of clinical trial design which can be modified to take advantage of the sequential nature of recruitment of subjects is the randomisation procedure. Because subjects are recruited sequentially, it is possible to adjust the randomisation rules to make it more likely that the numbers of subjects are more nearly equal for the two treatments. Modifications to the randomisation procedure to allow sequential changes to the probabilities for treatment allocation are discussed in Section 11.7.

20.7 Sequences of experiments in response optimisation

Another situation in which sequences of experiments are necessary is in the determination of optimal conditions, most commonly for an industrial process. The number of factors for which the optimal combination is required may be quite substantial. The variable to be optimised is almost always a simple quantitative measurement of output or yield, though sometimes some combination of measurements may be optimised if there is a need to represent a situation when characteristics other than the simple yield must be allowed for in defining the best performance. We shall assume that a single measurement variable to represent the yield to be maximised has been defined before the sequence of experiments is begun.

An experimental programme to determine the optimal conditions, when many factors can be changed, splits naturally into two parts. The first part, which may require several experiments in sequence, is concerned with moving as rapidly as possible towards the region where the highest yields are to be found. The classical methodology has been well established for many years, and this first stage is concerned with experimental designs to find the direction of steepest ascent. The second stage, which may again require several experiments in sequence, is concerned with estimating the second-degree polynomial surface which best fits the experimental data, and with estimating the point of maximum yield for the quadratic polynomial. The most difficult decision in the optimisation process is the determination of when to change from steepest ascent experiments to experiments to estimate the polynomial surface.

Experiments to estimate the steepest ascent direction will always include two factor levels for each factor. The most obvious design to use is a 2^n single replicate for n two-level factors. From this design we can fit linear main effects and linear × linear interaction terms to estimate the first-order surface, and thence the direction of steepest ascent. However, when the number of factors is large we do not need a full replicate and can estimate all the necessary effects from a fractional replicate as described in Chapter 14. It is often useful to include a third level of each factor, with a few runs at the centre point of the factor space to detect lack of fit of the linear response surface fitted.

One decision that is needed when designing successive two-level factorial designs is how to change the levels of the factors from one experiment to the next. Ideally we would like to make as big a change as possible to move as quickly along the line of steepest ascent, while avoiding the two dangers of overshooting the optimum and missing a real large change in the direction of steepest ascent. These dangers would tend to suggest that the centre for each new design should not be more than two steps along the direction of steepest ascent (the step being the difference between the two levels of a factor). It is also necessary to decide the initial steplength, and the change of steplength from one experiment to the next. Initial steplengths are simply best guesses on the part of the experimenter. It may be sensible to double steplengths in the initial stages of the experimental process, with less substantial changes as the process develops.

Experiments to estimate the second-degree polynomial surface will require at least three levels of each factor. As discussed in Chapter 17, the 3^n factorial designs are not usually the most efficient for estimating a second-degree polynomial surface in the region of the optimum, and the central composite designs described in that chapter are the most appropriate designs for estimating the quadratic polynomial. The estimation of the optimum combination of factor levels is then straightforward.

The decision of when to change from steepest ascent experiments to second-degree polynomial estimation experiments is, as has already been noted, difficult and there are no rules to guide the experimenter. The best advice is to observe the gradients of the steepest ascent direction estimated in successive two-level experiments and to use the rate at which the gradient declines as an indicator of the approach to the region around the optimum yield. Of course, if the estimated gradients are increasing, then there is no question of changing from the two-level designs to the central composite designs. Conversely, if the gradient declines extremely, then the change to the central composite designs has probably been delayed too long, though since the two-level design can easily be augmented to produce a composite design, and the two-level design uses fewer observations, the delay is not critical. It may indeed be a good plan to err in the direction of continuing with the two-level design longer rather than switching to the central composite.

Finally it is necessary to decide to end the search. The best indicator of when to stop is when the estimated optimum is well inside the experimental region of the latest central composite design after ignoring factors which do not seem to have any important effect.

20.8 Continuous on-line experimentation

In some situations, particularly occurring in industrial processes, it is natural to make observations on different treatment factor combinations one at a time, in a sequential manner, determining the treatment for the next observation on the basis of all the results so far. The design methodology then becomes very similar to that used for optimising a complex mathematical function, and is essentially a form of organised trial and error investigation. Of course, unlike a mathematical function, individual observations will include an element of random error, and therefore the search procedure cannot be exactly defined. Nevertheless for many situations the level of error may be relatively small, so that the methods for function optimisation will operate efficiently, and all that is needed to deal with error problems is to restart the process occasionally to avoid an observation with an unusual sized error from dominating the process for a long period.

There are many optimisation methods available, but probably the one most familiar to statisticians is that of Nelder and Mead (1965), which involves a simplex exploring the response surface variation of the yield output. In this methodology an initial starting combination of factor levels must be defined, and also the steplengths for all factors. The initial set of observations, assuming that there are n treatment factors, includes $(n + 1)$ initial observations, comprising the initial starting combination and n further observations changing the value for each factor in turn by the defined steplengths. Thereafter the factor combination giving the worst result is replaced by a new combination defined in terms of the other n observed combinations until the simplex homes in on the optimum combination.

0

Designing useful experiments

0.0 Some more real problems

Many of the problems considered in previous chapters have occurred during statistical consultancy sessions and are appropriate to motivate this chapter. Typical examples are the peppers experiment in Chapter 7 and the rabbits experiment in Chapter 15. Three more problems occurred in the work of one of us (RM) in quick succession within three weeks in early 1984 and illustrate a wide range of practical design problems.

(a) Problem 1: the moving cups. In assaying the chemical concentration of liquid samples, a set of about 24 samples can be automatically assayed in a single batch. The samples are placed in cups which are held in position on a circular disc, and the disc rotates so that each cup in turn appears beneath the assay machinery. It is required to investigate changes in the concentration of chemicals over time, and therefore the 24 samples are assayed over a time period of between one and two hours. Two shapes of cup are available, and two sizes of sample are possible within each cup. Three chemicals, sodium (Na), chlorine (Cl) and potassium (K) are of interest, and it is proposed to have two levels (zero and some) for each chemical. The cups may be covered during the run of 24 samples until each is sampled, but the covers, if used, must be used for all 24 samples in the run. Given that the change of chemical concentration with time is the primary interest, and that the experimenter wishes to examine the effects on this change of varying cup shape, sample size, cover, Na, Cl and K levels, how should an experiment consisting of 8–10 complete runs, of 24 samples each, be planned?

(b) Problem 2: mowing and species. The experiment is to examine the effect on germination of weed seeds of different mowing treatments for grass, within which the weed seeds are spread. A set of 15 mowing treatments is proposed, including various combinations of mower type, height of blades, mowing frequency and collection of clippings (a subset of a $3 \times 2 \times 2 \times 2$ factorial structure). There are ten weed species whose germination rates are to be examined. The minimum practical plot size for a mowing treatment is 1.5 m × 1 m and, within that plot, there is space for two areas within which seeds of two different species can be broadcast. Given that the primary objective is the comparison of seed germination rates for different mowing treatments, both overall and also for each weed species, with comparison of germination rates for the different weed species a secondary objective, how should the experiment be designed? The total resources available allow five complete replicates of the 15×10 combinations (750 miniplots). What size plot for each mowing treatment should be used, how many weed species per plot, and how should the weed species be chosen for each plot?

538

(*c*) Problem 3: recognising daddies. This is a design problem only in retrospect, but it is a design problem nonetheless, since before the analysis can be performed the design must be identified. Each of 32 young children were asked to draw two pictures of a person. Each picture was then scored for various aspects of anatomical accuracy. Each child was asked to draw (i) a man, and (ii) their daddy; the drawing was to use either (i) coloured crayons, or (ii) black pencil on white paper; and the order of the two requests could be (i) daddy first, and man second or (ii) man first and daddy second. Each of the four possible patterns:

(i) 1st, man, colour;	2nd, daddy, black and white
(ii) 1st, man, black and white;	2nd, daddy, colour
(iii) 1st, daddy, colour;	2nd, man, black and white
(iv) 1st, daddy, black and white;	2nd, man, colour

was used with each pattern given to eight children. Given the set of two scores for each of 32 children, how should the effects of man/daddy, colour/black and white, and order be estimated, together with their interactions? And how does the analysis alter when, as actually happened for one set of scores, three of the 32 children failed to produce drawings?

0.1 Design principles or practical design

Any book on experimental design develops the concepts on which the subject is based in a logical, systematic manner so that the principles can be clearly understood. In doing so, the authors inevitably produce a sequence of examples, which are essential to the orderly progress of the subject, but which, inevitably, offer themselves as a set of recipes of available designs. We have tried in this book to put a greater emphasis on principles and on the intuitive application of those principles, with rather fewer standard design examples, but the book still contains a large number of potential recipes. The problem with this seemingly inevitable format is that it is not appropriate to the statistician, or experimenter, when they are considering how to design an experiment to satisfy a particular set of practical conditions.

The designer of an actual experiment is required to produce a design appropriate to a very particular set of circumstances. But, except where the designer is very experienced, he or she will not be able to assess the entire spectrum of design ideas, and make decisions about the design, from the basis of a comprehensive knowledge of how all the principles of design might relate to this particular problem. Instead 'designing an experiment' has very frequently been interpreted as picking a design recipe which hopefully fits the particular circumstances and, if necessary, making compromises in the objectives of the experiment and the structure of the experimental material in order to fit the problem into the recipe's requirements. As evidence of this tendency to Procustean design, it is only necessary to observe the very large proportion of experiments, perhaps as high as 85%, which take the form of randomised complete block designs. If the principles of blocking are being correctly applied, then we have to accept that in every case (*a*) the natural block structure of the experimental units consists of blocks of equal size, and (*b*) the natural block size which gives the most efficient comparison of treatments happens to be exactly the same as the number of treatments required for the objectives of the experiment. This is a level of coincidence that any statistician (and most other scientists) should find hard to accept.

Ideally, a book about designing experiments should start from the problems which the experimenter has to solve, and should develop the subject from the viewpoint of the needs of the experimenter. Now this pattern patently cannot be used since, without an understanding of the basic principles, the solution of problems would proceed in a disorganised and unintelligible fashion. The standard format of books on design is inevitable, and therefore correct. But it is not sufficient. A book must also present the other point of view, and it is the purpose of this chapter to try to describe how experiments should be designed. It is Chapter 0 because it should come first, but it is not placed first because it requires all the other chapters, or at least some parts of them, to precede it. The stimulus for this chapter comes from a dinner conversation with Richard Jarrett, to whom one of us (RM) is very much indebted.

The process of designing an experiment is described from the viewpoint of a statistician being fully involved with the research scientist in the design of an experiment. We believe that this is the only efficient approach to designing experiments. No non-statistician has enough knowledge, without expert statistical advice, to design the best possible experiment on all occasions. And, unless the statistician is prepared to be deeply involved in the work of the research scientists and to have extensive discussion with the experimenter he or she will not always be able to give fully relevant advice.

The need for an experiment arises from a question or set of questions to which the research scientist wants to find answers. The questions will imply the form of experimental unit which is to be used, though there may often be some flexibility in the precise definition of an experimental unit. Thus, questions about drugs purporting to cure cancer in humans have, at some stage, to be answered by trying the drugs on individual men and women: a person is a unit! Different pasture systems must be tried on substantial areas of land, and the systems assessed either by harvesting the grass or by grazing the pasture with animals. The experimental unit could be a defined area of pasture, or it could be the area of pasture plus a group of animals grazing the pasture. To answer questions about optimal conditions for a chemical process, the experimental unit will require the operation of the process on a particular machine for a specified length of time, the length of time being chosen by the experimenter.

Having recognised the questions to which the experiment must give answers and the form of experimental unit to be used, then the next two stages are taken in parallel. First, the available resources of experimental units must be assessed. Second, the formulation of experimental treatments must be made to fit the questions. As has been emphasised previously, the separation of these two aspects of design is essential. The third, and final, stage consists of combining the two components, units and treatments, to produce the experimental design.

0.2 Resources and experimental units

The experimenter, having considered the general objective of the experiment, has identified the natural experimental unit. There are then two aspects of the set of units which must be considered. The first is the total amount of resources that can be made available for the experiment. Sometimes there is a clear and simple limit. The number of animals available for a nutrition experiment may be restricted to a set of, say, 25, in the required condition at the time of experiment; there may be a further limitation that the animals are available for 16 weeks only. Sometimes the limit is not so clearly defined, but it may be assumed by those responsible for allocating resources that the number of units for a particular type of experiment could not be greater than, say, 100. For example in a chemical industrial experiment in which the unit is a day's production, it might be assumed that no experiment could exceed two months, because of the restrictions on normal practice imposed by experimental conditions. A further type of limitation is that imposed by the back-up facilities available for chemical analysis of samples from the experimental units, and by storage space for material awaiting analysis. Whatever the cause of the restrictions on resources, it is important to recognise the limits from the outset of the planning of the experiment, so that the choices available to experimenter and statistician are clearly defined.

The second consideration is the structure of the experimental units. Structure can be identified in two forms: the inevitable structure and the suspected structure. Inevitable structure may arise from the choice of units. If it is decided that the unit should be a period of time for a particular animal, and the same pattern of periods is to be used for each of the available animals, then there will be a two-dimensional structure of animals × periods. Alternatively if each patient tests a sequence of different drugs, without control of the times at which the drugs are taken, then there is a one-dimensional structure of units within patients. It is important in recognising structure to distinguish between the double classification crossed structure, in which the pattern of units for an animal or patient is identical for each subject, and the simple single classification structure, in which the unit classification is not consistent across subjects.

Suspected structure is concerned with using the practical experience of the experimenter and statistician to try to anticipate patterns of variation in the output or analysis of the units. Thus, in animal or medical trials, there are many characteristics of the animal or patient which could be recorded before the experiment, and which might be thought to be relevant to the future performance of the animal or patient; age, sex, size, condition and a wide range of variables measuring previous performance or environment might all be candidates for defining suspected structure relevant to the design of the experiment. In experiments with field crops, the basis for suspected structure becomes less well defined, and plots may be expected to produce more or less similar yields according to geographical closeness, in addition to similarities arising from physically defined variables such as moisture and water levels, height, slope, and so on. Potential characteristics defining suspected structure for other forms of experimentation have been discussed in Chapters 2 and 7.

In preparing the definition of the set of units to be available for the experiment, it is essential that the initial identification of structure be made *without consideration of the detailed choice of treatments to be used*. The clear identification of the structure of the available units is probably the major factor in determining the precision of the final experiment, and unnecessary compromises introduced into the structure of the experiment by anticipating the treatment allocation stage may make the experiment less precise.

The considerations of structure may result in various degrees of definition of the unit structure to carry forward to the treatment allocation stage. Thus possible definitions of structure could include:

(i) Forty-eight units available in a 16 × 3 cross-classification structure, the 16 super-units being themselves grouped into four groups of three and one of four.

(ii) Between 40 and 60 units which will be grouped into three major groups of between 10 and 25 units, with a further desirable restriction that at most five units can be managed in a similar way at the same time, so that groups of up to five units should be defined within the major groups.

(iii) Seventy-five units available in 14 groups containing, respectively, eight, eight, seven, six, six, six, six, five, five, four, four, four, three, three units.

(iv) Up to 60 units available with a desirable group structure in which each group should contain at most eight units, though nine or even ten units might just be acceptable.

(v) About 80 units likely to be available (spread over a considerable time period) with classification into groups defined by four major factors giving 16 groups, the size of which might be expected, in advance, to vary between 0 and 15.

Any less specific description of the information about resources and structure of the available units will usually indicate that the experimenter and statistician have not thought sufficiently about the units to be used in the experiment.

This discussion of suspected structure has been mainly in terms of only a single grouping of units, but there will often be more than one possible basis of grouping, or blocking. The experimental units may therefore be identified within a two-dimensional structure, with each unit being classified into groups in each of two separate classifications, each of which is comprehensive in that all units are classified in each classification. In many situations, it will not be clear how many blocking classifications are sensible. What can always be achieved, however, is a list of possible classification systems, and a rough ordering of likely importance, ranging from essential to almost certainly negligible. In most practical situations one or two blocking classifications will appear to be clearly important, and sometimes there will be perhaps one or two further classifications that might be important.

At the conclusion of the consideration of the structure of the experimental units, the units will usually have been classified in a single set of blocks or a double-blocking classification (as discussed in Chapter 8). A blocking classification system may be based on a combination of blocking factors if there is no reason to believe that differences between levels of one blocking factor will be consistent over different levels of another blocking factor. The use of a double-blocking classification implies such consistency. It will rarely be appropriate to use no blocking system or to use a treble or higher-dimensional blocking system. Information on classifications which are not used to provide blocking structure should always be recorded for possible later use as covariates. The information about multiple blocking patterns should be identifiable in the same way which we have illustrated for a single blocking system.

0.3 Treatments and detailed objectives

In Chapter 12, we discussed several types of treatment structure and the objectives implied by each type. We now approach the problem from the more valid viewpoint of considering the objectives first, and then deducing the treatments. Scientifically, this latter approach is obviously correct, and it must be said that it is astonishing how many scientists arrive at a statistician's office for discussions of experimental design or, more frequently, for analysis of experimental data, with well-defined treatments, but with no clear idea of the questions for which the treatments should provide answers. A reasonable test of whether the experimental treatments have been chosen appropriately to answer well-defined questions is whether the detailed form of analysis of treatment variation is clearly defined. If the experimenter does not know how to analyse and interpret the treatment variation, then he or she has not properly considered the objectives of the experiment.

So how should the objectives of an experiment be identified? First, it is important to realise that experiments are rarely isolated events, but are usually part of an overall research structure. Individual experiments must be planned in the context of an overall structure, and this means not only that each experiment must respond to the results of previous experiments, but also that the prospect of future experiments should not be ignored in the planning of the current experiment. Overall, research structures are broadly classifiable into two classes; in the first, the objective is to seek an optimum form of operation of a system; the second is concerned with developing an understanding of the mechanics of a system. In addition there is a third stream of research experiments arising from requirements to satisfy regulations concerning testing of materials.

The search for an optimal mode of operation is more technological than scientific, and can be characterised by a reasonably well-defined objective, and by definition of the ranges of the components of the system, or factors, which can be used. Thus, in considering the growth of a mixture of two crops, maize and beans, simultaneously on the same piece of land, the research project should allow for variation of the spatial pattern of arrangement of the two crops, nutrient provision, management of the crops, particularly timing and control of pests, and varieties or genotypes of the crops. A breeding programme might be attached to the research project or it may be that only varieties previously available are to be considered. There may be limits on the costs of material inputs or management materials to make the investigation relevant to the farmers for whom the research findings are intended.

A quite different example would be in the examination of the effects of drugs, for example alcohol, on the performance of human beings. Here, it is not so much an optimum but a reasonable limit that is the primary interest. The research will be limited to some particular aspects of performance, and this limitation will have implications for the set of conditions under which performance is observed. For example we may be interested in the effect of alcohol on the performance of drivers, and this would eliminate consideration of persons below the minimum age for driving, and would also eliminate from consideration individuals who were unable to drive or who were restricted in possible alcohol intake. We might wish to consider the variation of forms and amounts of alcohol, of the condition of the persons receiving the alcohol and the environment within which the driving was to be performed.

In any experiment within an optimum seeking research programme, the objectives for each particular experiment must be considered. A major question will usually be how many aspects of variation, or factors, to consider simultaneously in an experiment. This will depend on the stage of the experiment within the research programme. At the beginning of the programme, it is important to include many factors, because failure to investigate the interaction of factors at an early stage can easily lead to wasting resources through pursuing too narrow a line of research before eventually checking on the effects of other factors. It is difficult to conceive of an investigation with too many factors at an early stage. The number of levels per factor will normally be small at early stages, because only broad patterns of response can be examined.

In development stages of research programmes, experiments will predominantly be concerned with between two and four factors, with other factors included only to provide checks against changes of interaction patterns. The number of levels may increase but, unless it is proposed to consider very complex models in the analysis of the data, it is probable that three or four levels chosen to optimise the information required about the effect of each factor will be adequate for most purposes. The systematic designs of Chapters 17 and 18 depart from this pattern in order to use the resources more efficiently. The majority of experiments involving quantitative factors will be concerned directly with the estimation of the optimal levels of the factors and three levels, sufficiently widely spaced to allow precise estimation of the optimum, will usually be sufficient.

Late in the research programme, the major concern will be to provide overall information on the optimal conditions, and on the deviations from the optimum, and here the multifactor designs for estimating response surfaces will often be appropriate, with many factors included, partly to provide simultaneous information about the effects of all important factors, but also to continue to provide some checks on interaction information.

In summary then, in technological experiments the important decisions about treatments concern the number of factors and numbers of levels. The factorial structure is the natural one from which to commence consideration of experimental treatments, but increasingly, as the knowledge of the research programme expands, complete factorial structures will be found to include combinations known to be ineffective, and these should be omitted. In general, experimenters have not sufficiently accepted the requirement for using many factors, particularly in the early stages of investigations, and this leads to much inefficient use of resources. The numbers of levels should usually be two (earlier) or three (later) except for response surface designs, which, because they relate to the whole surface and not to individual factors, may have up to five levels.

In research programmes whose primary purpose is to develop understanding of underlying mechanisms, the objectives are different, and the experiments will be correspondingly different. One major difference will be in the pattern of observations made for each unit. For the technological experiment, there will usually be a single measure of 'yield' or 'output', possibly with some subsidiary measures. In the scientific experiment, there will often be a series of measurements, often spread over time, and this will tend to reduce the total number of units that can be managed. It may be necessary to increase the number of levels of a quantitative factor to allow for uncertainty about the appropriate form of response model. Nonetheless, it is difficult to believe that large numbers of levels are particularly desirable, since this leads to a diffusion of information. It is of little use having information about very many levels, to guard against the occurrence of a sudden change which may occur at any level, because the precision of information at each level is likely to be so poor that detection of any particular sudden change is improbable.

This discussion may seem too much concerned with general good sense that the principles of use to an experimenter may be very opaque. More specifically then the questions an experimenter should ask are:

(a) Where does the present experiment occur in the overall research programme? Are the implications of previous experiments well understood? What are the questions arising from previous research that should be answered in this experiment? And how much future experimentation should be expected?

(b) What are the precise questions that the experiment should answer? There will usually be several questions. Various possible forms of question are:
   (i) Does the difference between A and B cause a change in yield? How big is this change?
   (ii) Is this change affected by levels of other factors?
   (iii) What is the yield response to change in the stimulus level of a factor?
   (iv) Which of a range of alternative procedures produces the highest yield?

In most experiments, the total number of questions asked should correspond to between 10 and 50 df. The idea of relating df to questions arises from the analysis to be performed on the experimental results, but it is essential that this is considered at the design stage. A question of form (i) is a single df question. Those of type (ii) may be single or multiple df, depending on the degree of complexity of interactions that is to be allowed. A single treatment factor may produce a single df question of type (i), or more than one. For example if an additional constituent is considered for a recipe, it may be added at either of two times. There are then two questions: does the constituent affect the finished product; and does the timing affect the product? In terms of treatment contrasts, these may be represented as $(-2, +1, +1)$, $(0, -1, +1)$.

For yield response to a quantitative factor, the number of df is equal to the number of parameters in the proposed response function. For multi-factor response functions, the number of df is equal to the number of parameters. If a large number (n) of alternative procedures (synthesised drugs, breeding lines) are being compared, then $(n - 1)$ df are implied.

0.4 The resource equation and the estimation of the error variance

The ideas of the resource equation have already been presented briefly in Chapter 1. However, it seems right to reiterate the concept at this point as we consider the general principles for designing practical experiments to fit particular practical situations and requirements. In particular, it is important to consider how much information is needed about the error variance so that we can determine how much information about treatment comparisons can be obtained.

The total information in an experiment involving N experimental units may be represented by the total variation based on $(N - 1)$ df. In the general experimental situation, this total variation is divided into three components, each serving a different function:

(a) The treatment component, T, being the sum of df, corresponding to the questions to be asked.
(b) The blocking component, B, representing all the environment effects to be included in the analysis model. Usually this will be $(b - 1)$ df for the b blocks, but other alternatives are $(r - 1) + (c - 1)$ for row-and-column designs, or more generally, all effects in multiple control or multiple covariate designs. It may also be desirable to allow a few df for unforeseen contingencies such as missing observations.
(c) The error component, E, being used to estimate s^2 for the purpose of calculating SEs for treatment comparison, and, more generally, for inference about treatments.

We can express this division of information as *the resource equation* in terms of the df used in each component T, B and E:

$$T + B + E = N - 1.$$

Now, to obtain a good estimate of error, it is necessary to have at least 10 df, and many statisticians would take 12 or 15 df as their preferred lower limit, although there are situations where statisticians have yielded to pressure from experimenters and agreed to a design with only 8 df. The basis for the decision is most simply seen by examining the 5% point of the *t*-distribution, and using sufficient df so that increasing the error df makes very little difference to the significance point, and hence to the interpretation.

The implications for design are obvious. It is necessary to have at least 10 df for error, to estimate σ^2 and, in choosing the number of treatment question df, we should not normally allow E to fall below 10. But equally, if E is allowed to be large, say greater than 20, then the experimenter is wasting resources by not asking enough questions. If the experiment has too many df for error, then ways should be found of asking more questions about the treatments.

The failure of an experimental design through having too many df for error (E is too large) occurs much more frequently than allowing E to fall too low, and it is as much a failure of design. To achieve an efficient design, with E smaller and T larger, we have only to consider the factorial concept and look to see how we can examine interactions between the major treatment effect(s) and other treatment or environmental effects. The basic economic advantage of factorial experiments arises from asking many questions simultaneously. To insist, as many scientists do, on asking only one or two questions in each experiment, is to waste resources on a grand scale.

One final comment on this question of efficient use of resources. It is, of course, possible to keep the value of E in the region of 10–20 by doing several small experiments, rather than one rather larger experiment in which the total number of experimental units is much less than the sum of the units in the separate small experiments. The use of multiple small experiments instead of one large experiment is inefficient because of the need to estimate σ^2 for each experiment. In designing the large experiment, the experimenter must ensure that B is large enough to allow all the important environmental effects to be included in the analysis model, providing precise estimation of all treatments effects; the experimental design must allow E to be sufficiently large to estimate σ^2 well, but not so large that the design becomes inefficient; and T must use all the remaining df by including in the analysis model as many effects as can be imagined to be usefully estimated.

0.5 The marriage of resources and treatments

One of the recurring themes of this book has been that the available computing facilities have very substantially reduced the restrictions of possible designs, and the main focus of this modern freedom of design is at the stage of allocating treatments to units within a natural blocking structure. The reader should, by this stage, be expecting to be told that, having decided the natural blocking structure and the relevant treatment structure embodying all the desired questions, the allocation problem is simple. And he or she would be correct, because it is. Nevertheless it is necessary to consider how to combine the ideas of the previous chapters when solving a particular problem.

Table 0.1. *Outline analysis of variance for a 2^4 design*

Main plot		Split plot	
Source	df	Source	df
Blocks	2	Q	1
P	1	R	1
Error	2	S	1
		Two-factor interactions	6
		Three-factor interactions	4
		Four-factor interactions	1
		Error	28

Table 0.2. *Outline analysis of variance for an alternative design*

Main plot		Split plot	
Source	df	Source	df
Blocks	2	R	1
P	1	S	1
Q	1	RS	1
PQ	1	Other two-factor interactions	4
Error	6	Three- and four-factor interactions	5
		Error	24

First, we should check whether there are, within the set of experimental treatments, restrictions on the possible size of units to which some factor levels can be applied. Is it necessary to consider split-unit designs, and if so, what are the restrictions imposed by the practical nature of the experiment? If some treatments require larger units than the proposed experimental unit size, then information about treatment comparisons will be divided between (at least) two strata. Each stratum should have adequate error df, and this could lead to using more factors as main plot factors than is strictly necessary. For example suppose a 2^4 treatment structure is proposed, with three replicates, in three blocks of 16 units. If one factor, P, requires larger units than the basic unit (of which there are 48), then the natural design is to use two main plots in each of three blocks. Each main plot will be split into eight split plots for the other eight treatment combinations giving an analysis of variance structure as shown in Table 0.1.

This design provides virtually no information about the precision of estimation of the effect of P. Now, suppose that, although the levels of P require units bigger than the basic unit, a main plot of four basic units is adequate for a level of P. Then better information about P can be obtained by having four main plots within each block, and four split plots within each main plot. To use such a design, other treatments must be allocated to main plots, and the unthinking split-plot user might decide to allocate factor Q also to main plots, giving an analysis of variance structure as shown in Table 0.2.

Table 0.3. *Outline analysis of variance for the third design approach*

Main plot		Split plot	
Source	df	Source	df
Blocks	2	Q	1
P	1	R	1
QRS	1	S	1
PQRS	1	Two-factor interactions	6
Error	6	Other three-factor interactions	3
		Error	24

	Block I		
Main plot 1	Main plot 2	Main plot 3	Main plot 4
pqrs	p	(1)	qrs
pq	prs	qs	r
pr	pqs	rs	q
ps	pqr	qr	s
	Block II		
qs	ps	prs	s
(1)	pq	pqr	qrs
qr	pqrs	pqs	q
rs	pr	p	r
	Block III		
pqs	q	pqrs	rs
prs	r	ps	qr
p	qrs	pq	(1)
pqr	s	pr	qs

Figure 0.1 Split-plot design with P and QRS as main-plot effects.

However, the clear message from Chapter 15 is that the effects, in addition to P, allocated to main plots should be those of least interest, since main plots are likely to be less precise. Therefore, the other effects to be allocated to main plots should be the three-factor interaction QRS, and the interaction with P, PQRS. The design is then as shown in Figure 0.1, in which block I has not been randomised to display structure more clearly, but blocks II and III have been randomised. The analysis of variance structure is as shown in Table 0.3. Usually it would be decided, before the experiment, to 'pool' the SS for QRS and PQRS with the main-plot error to provide 8 df for error in the main-plot analysis.

The purpose of this example is to illustrate that the ideas of different chapters must be considered together. The necessity of a split-plot design does not mean that the ideas of confounding should be ignored.

Similarly, if an experiment has two levels of one factor, and a large number of levels of a second factor (testing breeding lines under two environments), and the two levels of the first factor must be applied to large units (for example of size = six basic units), the necessity of a split-plot design does not require that all levels of the second factor must appear as split plots within each main plot. Suppose there are 18 levels of the second factor, then the simple

Block I

Main plot

1	2	3	4	5	6
p_0q_1	p_0q_7	p_0q_{13}	p_1q_1	p_1q_3	p_1q_5
p_0q_2	p_0q_8	p_0q_{14}	p_1q_2	p_1q_4	p_1q_6
p_0q_3	p_0q_9	p_0q_{15}	p_1q_7	p_1q_9	p_1q_{11}
p_0q_4	p_0q_{10}	p_0q_{16}	p_1q_8	p_1q_{10}	p_1q_{12}
p_0q_5	p_0q_{11}	p_0q_{17}	p_1q_{13}	p_1q_{15}	p_1q_{17}
p_0q_6	p_0q_{12}	p_0q_{18}	p_1q_{14}	p_1q_{16}	p_1q_{18}

Block II

Main plot

1	2	3	4	5	6
p_0q_1	p_0q_2	p_0q_3	p_1q_1	p_1q_2	p_1q_3
p_0q_4	p_0q_5	p_0q_6	p_1q_6	p_1q_4	p_1q_5
p_0q_7	p_0q_9	p_0q_8	p_1q_{10}	p_1q_8	p_1q_7
p_0q_{10}	p_0q_{12}	p_0q_{11}	p_1q_{12}	p_1q_9	p_1q_{11}
p_0q_{13}	p_0q_{14}	p_0q_{15}	p_1q_{14}	p_1q_{13}	p_1q_{16}
p_0q_{16}	p_0q_{17}	p_0q_{18}	p_1q_{15}	p_1q_{18}	p_1q_{17}

Figure 0.2 Experimental plan for a 2×18 structure in two blocks each containing six main plots, each containing six split plots.

split-plot design has two blocks each containing two main plots, each of which contains 18 split plots. Each main plot is now very large, and variation between split plots within main plots will be unnecessarily large. If it is felt that main plots should include no more than six split plots for the within-main-plot variation, σ^2, to be acceptably small and properly estimable, then the proper design will consist of two blocks, each block containing six main plots, three for level 1 of the first factor, and three for level 2; the levels of the second factor will be allocated in groups of six to the total of 12 main plots, so that the concurrence of pairs of levels is as even as possible, which requires that most pairs of levels occur once together with a few zero and double joint occurrences.

A possible design is shown in Figure 0.2 (shown without randomisation). Note that testing the 18 levels as split-plot treatments implies that the design allocation of these 18 to main plots of six split plots each is resolvable within each set of three main plots with the same main plot factor level. The joint occurrences are therefore less even than would be achieved in a non-resolvable design.

Leaving the topic of split-plot designs, but remembering that the comments relevant to other designs apply in some measure to split-plot designs also, we need to bring together the various ideas relating to small block sizes for large numbers of experimental treatments. The two situations discussed at length in previous chapters concern the allocation of treatments to small blocks, when all treatment comparisons are of equal importance (Chapters 7 and 8) and when there is a clear priority ordering for factorial effects with main effects more important than two-factor interactions, which in turn are more important than three-factor interactions (Chapter 15 again).

In practice, it will often be true that neither of these situations exactly fits the experimental requirements. The general situation is that there are a number of treatment comparison questions, of which there will be a subset felt to be rather more important than the rest. If the

		p_0q_0	p_0q_1	p_1q_0	p_1q_1
	r_0	×	×	×	×
s_0	r_1	×	×	×	×
	r_2	×	×	×	×
	r_0	×	×		
s_1	r_1	×	×		
	r_2			×	×

Figure 0.3 Treatment set of 18 from 24 'possible' combinations from a $2^3 \times 3$ structure.

natural block size pattern does not allow all experimental treatments to occur in each block, then the following steps may be helpful.

(1) Decide if any subset of treatments is sufficiently important that replication of individual experimental treatments should vary.

The relative replication of the experimental treatments should be determined to reflect this relative importance. A simple example of this is the use of a positive control treatment when all treatments are to be compared directly with the control. Here, the relative replication of the control should be between 1 and $\sqrt{n}$, where there are n treatments for comparison with the control. Alternatively, the control may be required only for general background information, and then the relative replication of the control may be less than one (i.e. only one or two control plots in the whole experiment). More complex situations might involve, say, a $2 \times 2 \times 3$ factorial structure plus six additional treatments consisting of a subset of the $2 \times 2 \times 3$ structure for a changed level of some new factor. The structure of the treatments might be as in Figure 0.3.

The experimenter then has to decide in terms of the important comparisons whether any or all of the s_1 combinations require duplication. To leave the replication of the s_1 combinations at the same level as the s_0 combinations means, for example, that the $(s_0 - s_1)$ comparison is less precise that those of $(p_0 - p_1)$ or $(q_0 - q_1)$.

Although the question of relative replication should be considered, we believe that in only a few cases will it be important to use differential replication. There are two reasons for this. It is often difficult to determine the relative precision required for different comparisons at all precisely. The achieved variation of precision (as measured by variances of treatment differences) is always relatively less than the variation of replication. And one side effect of unequal treatment replication is an overall reduction in precision. To illustrate this for a very simple three-treatment set with a total of 24 observations, consider different replications of the treatments, and the standard errors of treatment differences as set out in Table 0.4.

It can be seen that using an unequal replication pattern gives small improvements in precision for one or two comparisons, at the cost of a larger loss of precision for other comparisons. The greater concentration of resources on some comparisons is largely ineffective partly because of the general loss of efficiency resulting from the unequal replication. It is therefore usually sensible to assume equal replication of treatments.

(2) Decide which of the two basic models is more nearly relevant either (*a*) all treatment comparisons are important, or (*b*) some treatment comparisons are of negligible importance.

Table 0.4. *Replications and SE of treatment differences*

Replication			SE			
t_1	t_2	t_3	$t_1 - t_2$	$t_1 - t_3$	$t_2 - t_3$	$t_1 - (t_2 + t_3)/2$
8	8	8	0.5	0.5	0.5	0.43
12	6	6	0.5	0.5	0.57	0.41
16	4	4	0.56	0.56	0.71	0.43
12	8	4	0.46	0.57	0.61	0.41

The second situation will almost always occur when the treatments have factorial structure, with the unimportant effects being higher-order interactions. For (*a*) the experimenter should consider whether there are comparisons which are more important and should make a list in order of decreasing importance. For (*b*), list the major comparisons, the negligible comparisons and an intermediate group which, while clearly not of negligible importance, are also less important than the major comparisons.

(3) For (*2a*), allocate treatments to units using the principles of making joint occurrence of treatments as nearly equal as possible but, where there must be some treatments with an extra replicate or some pairs with an extra joint occurrence, choosing from the list of the more important comparisons.

(4) For (*2b*), allocate treatments to units, confounding effects of negligible importance with blocks and, if further effects must be partially confounded in different replicates or fractions of replicates, choosing from the intermediate group of effect comparisons.

(5) After the treatment allocation, examine the actual precisions that are predicted for treatment comparisons, using a computer program to obtain an analysis of variance of dummy data, with σ^2 set equal to 1 to obtain the variance–covariance structure of treatment comparison estimates. Consider where the predicted variance pattern differs from that which might be desirable and try modifying the design by switching particular treatments or sets of treatments.

This last step is rarely attempted, but the facilities allowing the examination of the expected precision of comparisons over a range of alternative designs should now be possible with most computing facilities, and should become a regular part of the design process.

0.6 Three particular problems

We return now to the three problems posed at the beginning of this chapter.

0.6.1 Problem 1 The moving cups

The experiment is required to investigate the change of chemical concentration with time, within a run of a mechanical assay process, with additional factors of cup shape, sample size, use of covers and levels of Na, Cl and K.

Eight or ten runs of the assay process with 24 samples per run are available, and variation between runs is thought to be large compared with variation within runs. Therefore, the change with time must be estimated from within-run differences. Although the change with time is probably well approximated by a linear regression, it was decided to use three times (three samples) for each combination of factors tested within a run, so that any curvature in the trend could be assessed. It was therefore decided that for each run the set of 24 samples should consist of three samples for each of eight combinations from the six factors additional to the time factor. Two main questions remained: how to arrange the three times × eight combinations within the 24 samples for each run, and how to choose the eight combinations for each run.

The objectives for the allocation of the three times × eight combinations are to optimise the precision of the estimation and comparison of the time change effects. If the three times × eight combinations are allocated randomly to the 24 samples, then the time interval from which the time effect is estimated for each combination will vary in an uncontrolled fashion (and could even be small for *all* combinations). Even if the 24 samples are split into three groups of eight, and the eight combinations allocated randomly within each group (early, middle, late), the precision of the regression effect will vary over combinations. The estimation and comparison of time-trend regression effects is made most precise by allocating the eight combinations randomly to the first eight samples, and repeating the order of combinations in the second and third eights.

The choice of eight combinations for each run requires the application of ideas of split units and confounding. The use of covers must be for a complete run or not at all. Therefore, covers is a main unit (run) factor. Should we choose two other main unit factors, and use three main unit and three split unit factors, or should we use four, or even five, factors within each run? Although the time repetition within each run is a complicating aspect of the design, it is completely irrelevant to the choice of combinations. To obtain maximum precision of main effects, and two-factor interactions of the five factors, and of the interactions of the time change effect with these factors, all five factors should be varied within each run, confounding two three-factor and one four-factor interaction. The confounded interactions actually selected were Na × K × Cl, Na × shape × size, and the four-factor interaction, K × Cl × shape × size. The design pattern for one run is shown in Figure 0.4.

The allocation of combinations of levels of the six factors to the eight runs is shown in Figure 0.5, the labels used in the figure being N for Na, K for K, C for Cl, L for cover (lid), S for shape and V for volume (sample size).

The analysis of results is based on the linear and quadratic components of response to time (late − early; late + early − 2 (middle)). Each component is calculated for each of the eight combinations in each run. The analysis of variance structure is shown in Table 0.5.

0.6.2 Problem 2   Mowing and species

The primary interest was in the comparison of germination rates under the 15 mowing treatments, but the secondary interest, in the comparative germination of the ten weed species, was also a major objective of the experiment. No specific knowledge of the experimental area was available except that the variation between plots would be smaller for plots which were closer together.

Before proceeding further, it is necessary to ask why so many different treatments were needed in the experiment. When the proposed mowing treatments were examined, they were found to involve four factors, each with two or three levels. It was proposed to compare three

Table 0.5. *Outline analysis of variance for the moving cups experiment*

	df
Covers	1
Main-unit (run) error	6
Main effects (Na, K, Cl, S, V)	5
Two-factor interactions (including covers)	15
Three-factor interactions (excluding NaKCl, NaSV)	18
Split-unit error	18
Total	63

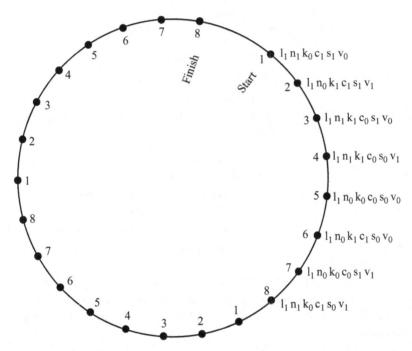

Figure 0.4 Design plan for one run of the moving cups problem.

mower types, two heights of cutting blades, two mowing frequencies, and the collection or non-collection of clippings. Of the potential 24 combinations, only three of the four combinations of height and frequency were feasible and one mower type did not permit collection of clippings; hence, the 15 combinations were the complete set of sensible factorial combinations and all were required to be able to assess main effects and two-factor interactions for each factor or pair of factors. The ten weed species resulted from an extensive selection process of the possible species and represent the major types of weeds occurring in grass. The statistician may often doubt whether as many species are needed as are often proposed in this kind of experiment, because of doubts whether the number of questions being asked really require as many as ten species, with 9 df for comparison. However, ultimately the statistician

Run 1	Run 2	Run 3
$1_1 n_1 k_0 c_1 s_1 v_0$	$1_0 n_1 k_1 c_0 s_0 v_1$	$1_0 n_0 k_1 c_0 s_0 v_0$
$1_1 n_0 k_1 c_1 s_1 v_1$	$1_0 n_0 k_1 c_1 s_0 v_0$	$1_0 n_0 k_0 c_1 s_1 v_1$
$1_1 n_1 k_1 c_0 s_1 v_0$	$1_0 n_1 k_1 c_0 s_1 v_0$	$1_0 n_1 k_1 c_1 s_0 v_1$
$1_1 n_1 k_1 c_0 s_0 v_1$	$1_0 n_1 k_0 c_1 s_0 v_1$	$1_0 n_1 k_0 c_0 s_1 v_0$
$1_1 n_0 k_0 c_0 s_0 v_0$	$1_0 n_0 k_0 c_0 s_0 v_0$	$1_0 n_1 k_0 c_0 s_0 v_1$
$1_1 n_0 k_1 c_1 s_0 v_0$	$1_0 n_1 k_0 c_1 s_1 v_0$	$1_0 n_0 k_0 c_1 s_0 v_0$
$1_1 n_0 k_0 c_0 s_1 v_1$	$1_0 n_0 k_0 c_0 s_1 v_1$	$1_0 n_0 k_1 c_0 s_1 v_1$
$1_1 n_1 k_0 c_1 s_0 v_1$	$1_0 n_0 k_1 c_1 s_1 v_1$	$1_0 n_1 k_1 c_1 s_1 v_0$

Run 4	Run 5	Run 6
$1_1 n_0 k_0 c_1 s_1 v_0$	$1_0 n_1 k_1 c_1 s_0 v_0$	$1_1 n_1 k_1 c_1 s_1 v_0$
$1_1 n_0 k_0 c_1 s_0 v_1$	$1_0 n_0 k_0 c_1 s_1 v_0$	$1_1 n_1 k_1 c_1 s_0 v_1$
$1_1 n_1 k_1 c_1 s_0 v_0$	$1_0 n_1 k_1 c_1 s_1 v_1$	$1_1 n_1 k_0 c_0 s_0 v_1$
$1_1 n_0 k_1 c_0 s_1 v_0$	$1_0 n_0 k_0 c_1 s_0 v_1$	$1_1 n_0 k_0 c_1 s_1 v_1$
$1_1 n_1 k_1 c_1 s_1 v_1$	$1_0 n_1 k_0 c_0 s_0 v_0$	$1_1 n_1 k_0 c_0 s_1 v_0$
$1_1 n_0 k_1 c_0 s_0 v_1$	$1_0 n_0 k_1 c_0 s_0 v_1$	$1_1 n_0 k_0 c_1 s_0 v_0$
$1_1 n_1 k_0 c_0 s_1 v_1$	$1_0 n_1 k_0 c_0 s_1 v_1$	$1_1 n_0 k_1 c_0 s_1 v_1$
$1_1 n_1 k_0 c_0 s_0 v_0$	$1_0 n_0 k_1 c_0 s_1 v_0$	$1_1 n_0 k_1 c_0 s_0 v_0$

Run 7	Run 8
$1_1 n_0 k_1 c_1 s_1 v_0$	$1_0 n_1 k_0 c_1 s_0 v_0$
$1_1 n_1 k_0 c_1 s_1 v_1$	$1_0 n_0 k_1 c_1 s_1 v_0$
$1_1 n_0 k_0 c_0 s_0 v_1$	$1_0 n_1 k_1 c_0 s_1 v_1$
$1_1 n_0 k_0 c_0 s_1 v_0$	$1_0 n_1 k_0 c_1 s_1 v_1$
$1_1 n_0 k_1 c_1 s_0 v_1$	$1_0 n_0 k_0 c_0 s_1 v_0$
$1_1 n_1 k_1 c_0 s_0 v_0$	$1_0 n_0 k_0 c_0 s_0 v_1$
$1_1 n_1 k_1 c_0 s_1 v_1$	$1_0 n_0 k_1 c_1 s_0 v_1$
$1_1 n_1 k_0 c_1 s_0 v_0$	$1_0 n_1 k_1 c_0 s_0 v_0$

Figure 0.5 Overall design for the moving cups problem.

must accept the experimenter's decision on the number of species, after the statistician has made plain the benefits of concentration of resources on a smaller number of species to answer more specific questions more precisely.

The treatment structure consisted of 15 mowing treatments × 10 weed species giving 150 combinations in a complete replicate. A mowing treatment required a minimum plot size of 1.5 m × 1 m. Each weed species required a plot of half that size. Three philosophies of design were discussed:

(a) Use large plots for each mowing treatment, sufficient to include all ten weed species. This is a split-plot design with mowing treatments as main plots, species as split plots. It is clearly inappropriate because, in addition to the general disadvantages of split-plot designs, the treatment comparisons of primary interest are applied to the main plots, and will therefore be the least precise.

(b) Divide the ten weed species into 5 pairs. Use 75 main plots for the 15 mowing treatments × 5 pairs, the two species in each pair being split-plot treatments. This gives comparisons between mowing treatments as precisely as is achievable. Of the 45 comparisons between weed species, 5 (between the two species of each pair) are liable to be considerably more precise than the other 40. After consideration of the particular species, the experimenter concluded that it was not reasonable to select 5 pairs such that those within-pair comparisons were of considerably more importance than the remaining 40 comparisons.

Set	1	AB	CD	EF	GH	IJ
	2	AC	BG	DI	EJ	FH
	3	AD	BH	CE	GI	FJ
	4	AE	BD	CF	GJ	HI
	5	AF	BJ	CH	DG	EI
	6	AG	BI	CJ	DF	EH
	7	AH	BC	DJ	EG	FI
	8	AI	BF	CG	DE	HJ
	9	AJ	BE	CI	DH	FG

Figure 0.6 Division of pairs of weed species into nine basic sets of five pairs each.

(c) Still using 75 main plots in each replicate, five for each mowing treatment, use different pairings of species for each mowing. Choose pairings of weed species so that over the whole experiment the pairings of species would occur as evenly as possible. Since there were 45 pairs of species, which could be grouped in nine sets of five pairs, such that each species occurred once in each set, sets of five pairs could be allocated to the different mowing treatments so that each set was used once or twice in each replicate. This design pattern made the main effect comparison of species as even as possible. The precision of mowing treatment comparisons would be as for (*b*), except that there would be a small level of non-orthogonality of species and mowing treatment effects in the main plot level of the analysis.

It was decided to use design philosophy (*c*). The experimental resources allowed five replicates. The identification of nine sets of five pairs, such that each set included all ten species and no pair was repeated in different sets, was a simple incomplete block design problem, and a solution is shown in Figure 0.6.

It was decided to allocate sets to mowing treatments, in such a manner that:

(i) for each mowing treatment, the sets were different in each of the five replicates;
(ii) in each replicate, the nine sets each occurred once or twice;
(iii) over the entire experiment, the nine sets each occurred eight or nine times. The allocation of sets to mowing treatments in the five replicates is shown in Figure 0.7.

0.6.3 Problem 3 Recognising daddies

The problem was to identify the design, and hence the analysis, of the experiment. The pattern of the experiment is displayed in Figure 0.8.

The set of eight possible observations consists of all possible combinations of three factors, each with two levels:

colour: C or B
daddy: d or m
order: 1 or 2.

The construction of the experiment ensures that each main effect is not influenced by differences between children. We may recognise children as blocks, in the design terminology of this book, and the two observations for each child as experimental units (within blocks). Examination of Figure 0.8 reveals not only that main effects are not confounded but that all

Mowing treatment	Replicate				
	I	II	III	IV	V
1	1	8	4	5	7
2	7	9	1	4	6
3	3	2	7	1	4
4	3	8	4	2	6
5	9	5	3	2	1
6	6	7	9	8	4
7	5	4	2	6	8
8	1	3	5	7	9
9	8	2	9	7	5
10	4	3	8	5	9
11	6	1	8	9	3
12	2	9	5	4	7
13	4	7	6	1	3
14	2	6	7	3	1
15	5	1	6	8	2

Figure 0.7 Allocation of sets of weed species pairs to mowing treatments in each replicate.

Children	Picture 1	Picture 2
1	Cd 1	Bm 2
2	Cd 1	Bm 2
⋮	⋮	⋮
8	Cd 1	Bm 2
9	Cm 1	Bd 2
10	Cm 1	Bd 2
⋮	⋮	⋮
16	Cm 1	Bd 2
17	Bd 1	Cm 2
18	Bd 1	Cm 2
⋮	⋮	⋮
24	Bd 1	Cm 2
25	Bm 1	Cd 2
26	Bm 1	Cd 2
⋮	⋮	⋮
32	Bm 1	Cd 2

Figure 0.8 Structure of data for the 'recognising daddies' problem.

three two-factor interactions are confounded between blocks (children), whereas the three-factor interaction is not confounded. The design is that which would have been constructed by methods of Chapter 15 if it had been determined that main effects must not be confounded. It seems likely that the design actually evolved along the philosophical lines of Chapter 15, ensuring that each child experienced each level of each factor (inevitable for order).

The analysis and interpretation of the results are of some interest, because they can be approached in various ways. The information, and hence the analysis, occurs at two levels:

Table 0.6. *Outline analysis of variance for recognising daddies*

Source	df
Between children	31
colour × daddy	1
colour × order	1
daddy × order	1
'main-unit' error	28
Within children	32
colour	1
daddy	1
order	1
colour × daddy × order	1
'split-unit' error	28

comparisons within children and between children. In the analysis of variance structure for a complete set of 32 children, eight for each possible pair of treatments, all treatment effects are clearly orthogonal and the structure is shown in Table 0.6.

The simplest method of calculating the analysis is to calculate the sum and difference of scores for each child (as in Section 13.3). There will be four types of sum and four types of difference corresponding to the four possible patterns of treatment combination pairs:

$$S_1 = (1, m, c) + (2, d, b),$$
$$S_2 = (1, m, b) + (2, d, c),$$
$$S_3 = (1, d, c) + (2, m, b),$$
$$S_4 = (1, d, b) + (2, m, c),$$

$$D_1 = (1, m, c) - (2, d, b),$$
$$D_2 = (1, m, b) - (2, d, c),$$
$$D_3 = (1, d, c) - (2, m, b),$$
$$D_4 = (1, d, b) - (2, m, c).$$

For each sum type and each difference type, the mean and variance are calculated. The effects are then estimated from the means, as follows:

$$\text{colour} = (c - b)(d + m)(1 + 2) = (\bar{D}_1 - \bar{D}_2 + \bar{D}_3 - \bar{D}_4)/4,$$
$$\text{daddy} = (c + b)(d - m)(1 + 2) = (-\bar{D}_1 - \bar{D}_2 + \bar{D}_3 + \bar{D}_4)/4,$$
$$\text{order} = (c + b)(d + m)(1 - 2) = (\bar{D}_1 + \bar{D}_2 + \bar{D}_3 + \bar{D}_4)/4,$$

$$CD = (c - b)(d - m)(1 + 2) = (-\bar{S}_1 + \bar{S}_2 + \bar{S}_3 - \bar{S}_4)/4,$$
$$CO = (c - b)(d + m)(1 - 2) = (\bar{S}_1 - \bar{S}_2 + \bar{S}_3 - \bar{S}_4)/4,$$
$$DO = (c + b)(d - m)(1 - 2) = (-\bar{S}_1 - \bar{S}_2 + \bar{S}_3 + \bar{S}_4)/4,$$

$$CDO = (c - b)(d - m)(1 - 2) = (-\bar{D}_1 + \bar{D}_2 + \bar{D}_3 - \bar{D}_4)/4.$$

The four sample variances for differences should be averaged to give s_d^2 and the three sample variances for sums should be averaged to give s_s^2. In each case, before pooling, the variances should be examined to check that pooling is not unreasonable. The variances of each effect can then be calculated as $s_d^2/32$ or $s_s^2/32$ for effects estimated in terms of differences, or sums, respectively.

If the numbers of children for each of the four patterns of treatment pairs are not equal, then the orthogonal analysis of variance cannot be calculated. However, the calculation of means and variances for sums and for differences for each of the four types is still the appropriate first step. Effects can be estimated from the four means as previously, and the pooled variance estimates can be calculated using weighted averages. The variance of each effect must reflect the numbers of children for each type, n_1, n_2, n_3, n_4, and will be

$$s_d^2(n_1^{-1} + n_2^{-1} + n_3^{-1} + n_4^{-1})/16.$$

Note the similarity of the analysis of this experiment to that for two-period, two-treatment, cross-over designs, discussed in Chapter 8.

Finally, to obtain a set of comparable estimates of mean performance for the eight experimental treatments, we calculate the mean effect

$$\text{mean} = (c + b)(d + m)(1 + 2) = (\bar{S}_1 + \bar{S}_2 + \bar{S}_3 + \bar{S}_4)/4$$

and use the inverse relation between treatment combinations and effects

$$
\begin{bmatrix} c & d & 1 \\ c & d & 2 \\ c & m & 1 \\ c & m & 2 \\ b & d & 1 \\ b & d & 2 \\ b & m & 1 \\ b & m & 2 \end{bmatrix} = (1/8)
\begin{bmatrix}
+1 & -1 & +1 & -1 & +1 & -1 & +1 & -1 \\
+1 & +1 & +1 & +1 & +1 & +1 & +1 & +1 \\
+1 & -1 & -1 & +1 & +1 & -1 & -1 & +1 \\
+1 & +1 & -1 & -1 & +1 & +1 & -1 & -1 \\
+1 & -1 & +1 & -1 & -1 & +1 & -1 & +1 \\
+1 & +1 & +1 & +1 & -1 & -1 & -1 & -1 \\
+1 & -1 & -1 & +1 & -1 & +1 & +1 & -1 \\
+1 & +1 & -1 & -1 & -1 & -1 & +1 & +1
\end{bmatrix}
\begin{bmatrix} \text{mean} \\ \text{order} \\ \text{daddy} \\ \text{DO} \\ \text{colour} \\ \text{CO} \\ \text{DC} \\ \text{CDO} \end{bmatrix}.
$$

Since some of the variances of the effects are calculated in terms of s_d^2 and some in terms of s_s^2, the variances of comparison of the treatment combination estimates will be combinations of these. The particular form of the variance will depend on the particular comparison and, of course, on the numbers of observations of each type.

0.7 Computer design packages and catalogues of designs

There are quite a number of computer packages, or components of packages, whose purpose is to produce experimental designs to fit specified requirements. There are also various catalogues of particular classes of designs, which are available to help the choice of an appropriate design. It is obviously desirable that we should consider the extent to which these computer packages and catalogues provide solutions for experimenters searching for appropriate designs for the experiments they wish to plan.

Our belief is that these packages and catalogues provide a relatively small contribution to the overall designing of practical experiments, but in some circumstances they can be very useful. In practice it is necessary to do most of the planning of a design before the design packages and catalogues are any help. The reason for this is that to use the design packages, it is necessary to know which design recipe (randomised block, Latin square, incomplete block (balanced or partially balanced or just efficient), split unit, complete factorial, fractional replicate, confounded factorial, response function design, response surface design, cross-over design, saturated factorial, Trojan square, Youden square, randomised or systematic, D-optimal, repeated measurement, alpha design, etc.) is to be used.

It should be clear to the reader of this book, and particularly of this chapter, that the major part of designing an experiment is working out the right structure for units, the right structure for treatments and how to combine the two structures in a sensible design structure. At the end of this process a design recipe, or combination of recipes, will have been identified. In some cases it would be appropriate for this recipe to be presented to a design package, with full details of the unit and treatment structures and the design package would produce the full practical design with all appropriate randomisations included. However, this is appropriate or necessary in only some cases. Similarly there will be some cases where the use of a catalogue of designs would be helpful in producing the final randomised design. To understand why computer packages and catalogues will be necessary and sufficient in only some cases we have to look again at the design process.

The really difficult part of designing an experiment is the determination of the correct structures for units and for treatments and the interrelationship between the two, leading to the design recipe. By this stage of the book we expect that the reader would appreciate that for many of the recipes, the construction of the best design, given the recipe, is quite simple and straightforward. Obviously any randomised complete block design can be constructed and randomised without any use of a computer package or catalogue. The same is true for complete split-unit designs. In general it is also true that for any small design, say those with no more than 36 experimental units, the construction of an experimental design for incomplete block, or fractional replicate designs, or for confounded designs, should all be simple enough for anyone who has read this book thoroughly.

However, when the designs are larger it may be quite difficult to identify the most efficient design within the class of designs for a particular recipe, and this is where the computer packages and catalogues may be useful. To use a computer program or catalogue, it is necessary that the design recipe, including the unit structure and treatment structure should have been completely defined. It is also necessary to define the criterion for determining opimality; often this is implied in the structure of the computer package, or in the classifications within a catalogue; it is essential that the user of any computer package or catalogue understands what criterion for optimality is being used.

For incomplete block designs there are packages and catalogues which may be useful, as has already been noted in Chapter 7. One situation is where it has been decided to use an alpha design for a variety trial using small blocks. The optimal alpha design for a given number of varieties, block size and number of replicates can be obtained using the alpha design approach of Patterson and Williams (1976) and their design package, Alphagen. The optimality criterion is based on all comparisons between varieties being equally important and is essentially the average variance of variety differences. This is the obvious criterion to use.

There are also catalogues of both balanced and partially balanced incomplete block designs. Most of these were developed when the method of analysis of the results was limited by the absence of general computer analysis programmes, and consequently the classifica-tions of designs are determined mainly by the patterns for the partial balance, rather than by a criterion based on variances of estimated differences. Nevertheless the best designs in most catalogues will usually be close to optimality based on average variance of treatment differences.

For two-way blocking designs, construction of designs is more difficult and computer packages should be able to provide help in finding good designs. For Latin squares there are lots of potential aids to good design construction. For two-way designs with incomplete rows and/or columns the design recipes are simple to define and computer packages should have no difficulty in finding good designs.

For designs with factorial structures, there is an additional problem and this concerns the optimality criterion. Unlike the simple incomplete block and two-way blocking designs, it is no longer sufficient to asssume that all treatment comparisons should be equally precisely determined. With factorials of any description, whether complete or incomplete, confounded or unconfounded, the analysis will almost always be in terms of main effects, two-factor interactions, and higher-order interactions, with usually a clear priority ordering of those effects as we have discussed frequently in Chapters 13, 14 and 15. Many computer packages searching for optimal designs have some form of optimality criterion, such as D-optimality, assumed, and this may not correspond to the differential prioritisation for different groups of effects in the experimenter's particular situation. Therefore it is important that the experi-menter has thought about the relative importance of the precision for different effects and has understood to what extent the computer package is able to reflect the chosen set of relative importances.

One area where optimality criteria have been extensively used is in the choice of response surface designs. Usually a second-order polynomial model will be assumed, and using a D-optimality criterion is assumed to provide a good approximation to most alternative criteria in the choice of the optimal design. This difference from the more general main effect and two-factor interaction model is because only the linear $\times$ linear interaction term will be included in the model, and this is always likely to be the most important component of the two-factor interaction. Thoughtless use of D-optimal designs should be avoided, however, since, for example, they often do not allow estimation of pure error variance, which is usually considered an important part of the analysis.

One area where catalogues of designs are almost essential is the saturated and, particularly, the supersaturated designs discussed in the final section of Chapter 14. These designs are so complicated that it is very difficult to develop a simple intuition for choosing the most efficient design, and quite heavy computing is necessary.

Another area where experimenters, not very confident in their understanding of design principles, may easily get help is in the randomisation of the design recipe to produce a final design. The ideas of randomisation have been thoroughly discussed in Chapters 10 and 11, and do not need repetition here.

It is also worth noting that computer packages and catalogues specify the design recipes for which optimal designs can be found often within quite tight restrictions. It will often happen that some of the design recipes which could emerge from following the approach of this book, which is to impose as few restrictions as possible, may not be covered by the computer packages. When this occurs, as it might, for example, with some of the irregular fractions of Chapter 14, then, as discussed in that chapter, the way to use a computer package to choose a design could be to try out the analysis for several alternative designs, setting the error variance to 1 and using dummy data, to compare the variances for different treatment comparisons.

One other problem with the limitations of design recipes within computer packages could be that the range of possibilities might be more narrow than the range of designs which could result from faithfully following the requirement to identify the right unit and treatment structures. A very simple example would be where (a) the unit structure is identified as 25 units in 5 rows × 5 columns, and (b) the treatment structure for a quantitative factor for which linear and quadratic effects are to be estimated consists of just three levels. Most computer packages (and quite a lot of statisticians!) would look at the unit structure and assume that a Latin square design is appropriate. The requirement for only three treatments would then be rejected as incompatible with the Latin square design. Now Latin square designs are excellent designs when all the circumstances are appropriate, but it would be wrong to require that there should be five treatments when the more efficient estimation of the linear and quadratic effects would be achieved with just three levels. Hence the correct design should have the three levels arranged within the 5 × 5 row-and-column array, despite the fact that some computer design packages (and quite a few statisticians!) would have difficulty in finding an efficient design.

Exercises

0.1 An agricultural research worker wishes to study the effect of an insecticide on the yield of a crop. He proposes to compare two methods of application (spray or powder – factor M), and two times of application (early or late – factor T). He can use about 50 plots in either of two ways:

   (i) Regard absence and presence of insecticide as a third factor, X, in a formal 2^3 design on factors X, M, T in six randomised blocks of eight, although all combinations of M, T in the absence of insecticide are necessarily identical.

   (ii) Use only one treatment without insecticide plus the four combinations for methods of using the insecticide, to give a set of five treatments, and arrange in two 5 × 5 Latin squares.

   Discuss critically the interesting and important contrasts between treatments for each design; using σ_1^2, σ_2^2, to represent variances per plot for (i), (ii), obtain the variance for each contrast. Hence indicate what guidance on the choice of design could be obtained from a fairly trustworthy guess at the ratio σ_1^2/σ_2^2. Identify clearly the circumstances in which one design appears to be a clear preference, and those in which the choice depends upon which aspect of the results is of chief concern to the investigator.

0.2 Eighteen experimental units, which are arranged in three rows of six, are available for an experiment on three treatments A, B, C. Four designs are under consideration:

   (i) completely randomised with six replications of each treatment;

   (ii) completely randomised with eight replications of A, five of B and five of C;

   (iii) six randomised blocks of three units each;

   (iv) two 3 × 3 Latin squares, both of which have the same rows.

   For each design show the form of the analysis of variance and indicate the calculations for the SS. Compare the precisions with which treatment differences are estimated from the four designs.

Animals

(a)

			1	2	3	4	5	6	7	8	9	10
Time	1		A	B	C	D	E	A	B	C	D	E
	2		B	C	D	E	A	B	C	D	E	A
	3		C	D	E	A	B	D	E	A	B	C

(b)

			1	2	3	4	5	6	7	8	9	10
Time	1		A	B	C	D	E	A	B	C	D	E
	2		B	C	D	E	A	C	D	E	A	B
	3		D	E	A	B	C	B	C	D	E	A
	4		E	A	B	C	D	E	A	B	C	D
	5		C	D	E	A	B	D	E	A	B	C

Figure 0.9 Two alternative designs to compare five diets for cattle using ten animals with either (a) three periods or (b) five periods.

If any of the four designs is wholly unsatisfactory, explain clearly why you would never use it. Indicate the merits of those not rejected and give an example of a situation in which each might be used.

0.3 Ten animals are available for a trial to compare five diets for cattle. Two designs are considered. The first consists of a double Youden square with three treatment periods for each animal; there will be recovery periods between consecutive treatment periods which will hopefully eliminate residual effects. The second design uses two 5×5 Latin squares balanced for residual effects which must be allowed for because of the closeness of consecutive treatment periods. Both designs are given in Figure 0.9. Obtain the SEs of estimated differences between direct treatment effects for each design, and discuss how an experimenter should choose between the two designs.

0.4 In a consumer trial ten men are available to compare the effectiveness of five brands of razor blades. Each individual is supplied with two blades of different brands and identical instructions for face preparation. Each morning for a week he uses these two brands of blades, one brand for each side of his face. The smoothness of each shave is measured and the average value for each blade is used as a measure of the quality of the blade. Suggest an appropriate design for this experiment if the trial is to be completed over the one-week period. Give details of the analysis and state clearly any models you use and assumptions you make.

0.5 It is proposed to investigate the effects of two chemicals (A, B), each at two concentrations $(A_1, A_2; B_1, B_2)$ on fungal growth on leaves. The basic experimental unit is a single leaf. It is believed that growth of the fungus is influenced by the age of leaf. The difference in growth for leaves of different ages is of some interest in itself, but it may also affect the differences between the chemical treatments. These effects are, however, of secondary interest compared with the comparisons of the chemicals. It is known that leaves from different plants may differ substantially; also leaves in a pair on a plant are likely to be particularly similar, and leaves of a similar age will behave more similarly than those whose ages differ more substantially.

Three designs have been suggested as follows:

(i) From each of eight plants use four leaves of differing ages (the same ages for each plant). Regarding the 32 leaves as an 8 plant $\times$ 4 age array, the four treatments $(A_1, A_2; B_1, B_2)$ are allocated so that each treatment appears twice for each leaf age and once for each plant.

(ii) From each of four plants use eight leaves (four old, four young) and apply the four treatments to leaves so that all eight combinations of chemical $\times$ concentration $\times$ age occur for each plant.

(iii) From each of four plants take four pairs of leaves (two old pairs, two young pairs). Allocate each chemical to one old pair and one young pair for each plant and allocate the two concentrations of each chemical to the two leaves of the pair.

Identify the appropriate form of analysis of variance for each design and discuss the relative merits of the three designs.

0.6   A parallel line bioassay is carried out to compare seven distinct test preparations (A, B, C, D, E, F, G) simultaneously with a standard, S, using two doses of each and 32 subjects per preparation. Outline the partitioning of df for the analysis of variance and mention any additional assumptions inherent in the usual analysis. Describe any changes in design and in the partitioning that would be appropriate to each of the following circumstances:
   (i) Only 128 subjects can be treated on one day, and responsiveness may change markedly from day to day.
   (ii) As (i), but also only eight combinations of preparation and dose level can be handled on one day.
   (iii) As (i), but preparation G can be omitted. The 256 subjects are to be allocated approximately optimally for estimating the potency of each of A, B, . . . , F, relative to S.
   (iv) Four doses of each of the eight preparations are to be used instead of two, only 64 subjects can be treated per day, and only eight combinations of preparation and dose can be handled on each of four days.
   (v) As (i), but time of treatment within a day may affect responsiveness. Only one person does the work and he requires about three minutes per subject.

0.7   In an experiment to investigate the effects of nutrition level (N) and housing temperature (T) on the rate of growth of piglets, a factorial experiment is to be used, with litters of eight or nine piglets as blocks. A primary object of the experiment is to estimate the linear effects of nutrition (the rate at which yield increases with increasing nutrition level) and also the linear $\times$ linear interaction of N and T. Three designs are considered with different numbers of levels of a factor, and with possibly different ranges of levels of a factor (range $=$ highest level $-$ lowest level). The designs considered are
   (i) A 2×2 factorial with factor level ranges N_2, T_2 in four blocks each containing two replicates of all four treatment combinations.
   (ii) A 3×3 factorial with equally spaced levels over ranges N_3, T_3, in four blocks each containing a single replicate.
   (iii) A 4×4 factorial with equally spaced levels over ranges N_4, T_4; two replicates each consisting of two blocks of eight units, each with the $N'N''T'T''$ interaction confounded (N', N'', T', T'' are dummy two-level factors for the four-level factors N, T).

For each design identify comparable estimates of the linear effect of nutrition and of the linear $\times$ linear interaction and find estimated variances of those estimates. For what values of N_2, T_2, N_3, T_3, N_4, T_4 are the estimates from the three designs equally precise?

Discuss briefly the experimenter's choice of design.

0.8   The following comes from a letter requesting help for the design of an experiment:
   In the trial we are to study: high and low oil silages fed at high and low levels, the silage to be treated with or without antioxidants, and diets are to be supplemented or not supplemented with limiting amino acids including tryptophan which is broken down in acid conditions.
   In addition to these 16 treatment combinations there are to be four control treatments, being high and low oil fish meals fed at high and low levels. All control feeds will include antioxidants and will be supplemented with amino acids.

We have two sets of cages available for this work: A 16-cage flat deck system to take up to ten chicks for three weeks and a 16-cage metabolism unit to take four chicks to three weeks. We plan to use both sets of cages and put four chicks per cage for two full experimental periods and one cage unit for a further experimental period.

I would appreciate any comments you have regarding this experimental design.

How would you reply?

References

Bailey, R. A. (1982) The decomposition of treatment degrees of freedom in quantitative factorial experiments. *Journal of the Royal Statistical Society, Series B*, **44**, 63–70.

Bartlett, M. S. (1978) Nearest neighbour models in the analysis of field experiments (with discussion). *Journal of the Royal Statistical Society, Series B*, **40**, 147–174.

Besag, J. E. (1974) Spatial interaction and the statistical analysis of lattice systems (with discussion). *Journal of the Royal Statistical Society, Series B*, **36**, 192–236.

Besag, J. E. and Kempton, R. A. (1986) Analyis of field experiments using spatial statistics. *Biometrics*, **42**, 231–251.

Bleasdale, J. K. A. (1967) Systematic designs for spacing experiments. *Experimental Agriculture*, **3**, 73–85.

Box, G. E. P. and Draper, N. R. (1975) Robust designs. *Biometrika*, **62**, 347–352.

Box, G. E. P. and Tidwell, P. W. (1962) Transformation of the independent variables. *Technometrics*, **4**, 531–550.

Butler, N. A., Mead, R., Eskridge, K. M. and Gilmour, S. G. (2001) A general method of constructing $E(s^2)$-optimal supersaturated designs. *Journal of the Royal Statistical Society, Series B*, **63**, 621–632.

Bryan-Jones, J. and Finney, D. J. (1983) On an error in 'Instruction to Authors'. *Horticultural Science*, **18**, 279–282.

Carmer, S. G. and Jackobs, J. A. (1965) An exponential model for predicting optimum plant density and maximum corn yield. *Agronomy Journal*, **57**, 241–244.

Cleaver, T. J., Greenwood, D. J. and Wood, J. T. (1970) Systematically arranged fertiliser experiments. *Journal of Horticultural Science*, **45**, 457–469.

Cochran, W. G. and Cox, G. M. (1957) *Experimental Designs*, Second edition. New York: Wiley.

Cornell, J. A. (2002) *Experiments with Mixtures*, Third edition. New York: Wiley.

Corsten, L. C. A. (1958) Vectors, a tool in statistical regression theory. *Mededelingen van de Land-bouwhogeschool te Wageningen*, **58**, 1–92.

Cox, D. R. (1958) *Planning of Experiments*. New York: Wiley.

Cullis, B. R. and Gleeson, A. C. (1989) Efficiency of neighbour analysis for replicated variety trials in Australia. *Journal of Agricultural Science, Cambridge*, **113**, 233–239.

Cullis, B. R. and Gleeson, A. C. (1991) Spatial analysis of field experiments – an extension to two dimensions. *Biometrics*, **47**, 1449–1460.

Curnow, R. N. (1961) Optimal programmes for varietal selection. *Journal of the Royal Statistical Society, Series B*, **23**, 282–318.

Darby, L. A. and Gilbert, N. (1958) The Trojan square. *Euphytica*, **7**, 183–188.

Davis, T. P. and Draper, N. R. (1995) A note on remnant three-level second order designs. Technical Report 954, Department of Statistics, University of Wisconsin-Madison.

Dorfman, R. (1943) The detection of defective members of large populations. *Annals of Mathematical Statistics*, **14**, 436–440.

Dyke, G. V. and Shelley, C. F. (1976) Serial designs balanced for effects of neighbours on both sides. *Journal of Agricultural Science, Cambridge*, **87**, 303–305.

Edmondson, R. N. (1994) Fractional factorial designs for factors with a prime number of quantitative levels. *Journal of the Royal Statistical Society, Series B*, **56**, 611–622.

Edmondson, R. N. (1998) Trojan square and incomplete Trojan square designs for crop research. *Journal of Agricultural Science, Cambridge*, **131**, 135–142.

Eskridge, K. M., Gilmour, S. G., Mead, R., Butler, N. A. and Travnicek, D. A. (2004) Large supersaturated designs. *Journal of Statistical Computation and Simulation*, **74**, 525–542.

Finney, D. J. (1958) Statistical problems of plant selection. *Bulletin of the International Statistical Institute*, **36**, 242–268.

Fisher, R. A. and Yates, F. (1963) *Statistical Tables for Biological Agriculture and Medical Research*. Edinburgh: Oliver & Boyd.

Francis, L. (1978) Experimental designs for the two-parameter exponential response curve. MSc dissertation, University of Reading.

Freeman, G. H. (1979) Some two-dimensional designs balanced for nearest neighbours. *Journal of the Royal Statistical Society, Series B*, **41**, 88–95.

Gilmour, A., Cullis, B. and Verbyla, A. (1997) Accounting for natural and extraneous variation in the analysis of field experiments. *Journal of Agricultural, Biological and Environmental Statistics*, **2**, 269–293.

Gilmour, S. G. (2006) Response surface designs for experiments in bioprocessing. *Biometrics*, **62**, 323–331.

Gilmour, S. G. (2006) Supersaturated designs in factor screening. In *Screening* (eds. S. M. Lewis and A. M. Dean), pp. 169–190. New York: Springer.

Gilmour, S. G. and Trinca, L. A. (2005) Fractional polynomial response surface models. *Journal of Agricultural, Biological and Environmental Statistics*, **10**, 50–60.

Gleeson, A. C. and Cullis, B. R. (1987) Residual maximum likelihood (REML) estimation of a neighbour model for field experiments. *Biometrics*, **43**, 277–287.

Gordon, T. and Foss, B. M. (1966) The role of stimulation in the delay of the onset of crying in the new-born infant. *Journal of Experimental Psychology*, **16**, 79–81.

Green, P. J., Jennison, C. and Seheult, A. H. (1985) Analysis of field experiments by least squares smoothing. *Journal of the Royal Statistical Society, Series B*, **47**, 299–315.

Hills, M. and Armitage, P. (1979) Two-period cross-over clinical trial. *British Journal of Clinical Pharmacology*, **8**, 7–20.

Hozumi, K., Asahira, T. and Kira, T. (1972) Intraspecific competition among higher plants: VI. Effect of some growth factors on the process of competition. *Journal of the Institute of Polytechnics of Osaka City University*, **D7**, 15–28.

John, J. A. and Williams, E. R. (1995) *Cyclic and Computer Generated Designs*, Second edition. London: Chapman & Hall.

Kempthorne, O. (1952) *The Design and Analysis of Experiments*. New York: Wiley.

Kempton, R. A. and Howes, C. W. (1981) The use of neighbouring plot values in the analysis of variety trials. *Applied Statistics*, **30**, 59–70.

Kerr, M. K. and Churchill, G. A. (2001) Statistical design and the analysis of gene expression microarray data. *Genetics Research*, **77**, 123–128.

Kuipers, N. H. (1952) Variantie-Analyse. *Statistica*, **6**, 149–194.

Lee, Y., Nelder, J. A. and Pawitan, Y. (2006) *Generalized Linear Models with Random Effects*. London: CRC Press.

McCullagh, P. and Nelder, J. A. (1989) *Generalized Linear Models*, Second edition. London: Chapman & Hall.

Maindonald, J. H. and Cox, N. R. (1984) Use of statistical evidence in some recent issues of DSIR agricultural journals. *New Zealand Journal of Agriculture*, **27**, 597–610.

Morse, P. M. and Thompson, B. K. (1981) Presentation of experimental results. *Canadian Journal of Plant Science*, **61**, 799–802.

Nelder, J. A. (1962) New kinds of systematic design for spacing experiments. *Biometrics*, **18**, 283–307.

Nelder, J. A. (1966) Inverse polynomials, a useful group of multi-factor response functions. *Biometrics*, **22**, 128–141.

Nelder, J. A. (1991) Generalized linear models for enzyme kinetic data. *Biometrics*, **47**, 1605–1615.

Nelder, J. A. and Mead, R. (1965) A simplex method for function minimization. *The Computer Journal*, **7**, 308–313.

O'Neill, R. and Wetheril, G. B. (1971) The present state of multiple comparison methods (with discussion). *Journal of the Royal Statistical Society, Series B*, **33**, 218–241.

Patterson, H. D. and Thompson, R. A. (1971) Recovery of inter-block information when block sizes are unequal. *Biometrika*, **5**, 545–554.

Patterson, H. D. and Williams, E. R. (1976) A new class of resolvable incomplete block designs. *Biometrika*, **63**, 83–92.

Pearce, S. C. (1963) The use and classification of non-orthogonal designs (with discussion). *Journal of the Royal Statistical Society, Series B*, **25**, 353–377.

Pearce, S. C. (1975) Row-and-column designs. *Applied Statistics*, **24**, 60–74.

Pocock, S. J. (1979) Allocation of patients to treatment in clinical trials. *Biometrics*, **35**, 183–197.

Rayner, A. A. (1969) *A First Course in Biometry for Agriculture Students*. Pietermaritzburg: University of Natal Press.

Reid, D. (1972) The effects of long-term application of a wide range of nitrogen rates on the yields from perennial ryegrass swards with and without white clover. *Journal of Agricultural Science, Cambridge*, **79**, 291–301.

Rojas, B. A. (1963) The San Cristobal design for fertiliser experiments. *Proceedings of the International Society of Sugar Came Technologists*, Mauritius.

Rojas, B. A. (1972) The orthogonalised San Cristobal design. *Proceedings of the International Society of Sugar Came Technologists*, Louisiana.

Royston, P. and Altman, D. G. (1994) Regression using fractional polynomials of continuous covariates: parsimonious parametric modelling (with discussion). *Applied Statistics*, **43**, 429–467.

Ruppert, D., Cressie, N. and Carroll, R. (1989) A transformation/weighting model for estimating Michaelis–Menten parameters. *Biometrics*, **45**, 637–656.

Sobel, M. and Groll, P. A. (1959) Group testing to eliminate effectively all defectives in a binomial sample. *Bell Systems Technology Journal*, **38**, 1179–1252.

Sparrow, P. E. (1979) Nitrogen response curves of spring barley. *Journal of Agricultural Science, Cambridge*, **92**, 307–317.

Trinca, L. A. and Gilmour, S. G. (2000) An algorithm for arranging response surface designs in small blocks. *Computational Statistics and Data Analysis*, **33**, 25–43. Erratum (2002), **40**, 475.

Trinca, L. A. and Gilmour, S. G. (2001) Multi-stratum response surface designs. *Technometrics*, **43**, 25–33.

Varnalis, A. I., Brennan, J. G., MacDougall, D. B. and Gilmour, S. G. (2004) Optimisation of high temperature puffing of potato cubes using response surface methodology. *Journal of Food Engineering*, **61**, 153–163.

Whitehead, J. R. (1997) *Sequential Clinical Trials*, Second edition. New York: Wiley.

Wilkinson, G. N. (1970) A general recursive procedure for analysis of variance. *Biometrika*, **57**, 19–46.

Wilkinson, G. N., Eckert, S. R., Hancock, T. W. and Mayo, O. (1983) Nearest neighbour (NN) analysis of field experiments (with discussion). *Journal of the Royal Statistical Society, Series B*, **45**, 151–211.

Williams, R. M. (1952) Experimental designs for serially correlated observations. *Biometrika*, **49**, 151–167.

Wit, E., Nobile, A. and Khanin, R. (2005) Near-optimal designs for dual channel microarray studies. *Applied Statistics*, **54**, 817–830.

Yates, F. (1935) Complex experiments (with discussion). *Supplement to the Journal of the Royal Statistical Society*, **2**, 181–247.

Yates, F. (1936) A new method of arranging variety trials involving a large number of varieties. *Journal of Agricultural Science, Cambridge*, **26**, 424–455.

Index

Items in the index are referenced to the section or chapter in which they occur.

Printed in the United States
by Baker & Taylor Publisher Services